Jorge Luis García-Alcaraz
Aidé Aracely Maldonado-Macías
Guillermo Cortes-Robles
Editors

Lean Manufacturing in the Developing World

Methodology, Case Studies and Trends from Latin America

Springer

Editors
Jorge Luis García-Alcaraz
Departamento de Ingeniería Industrial y Manufactura
Instituto de Ingeniería y Tecnología
Universidad Autónoma de Ciudad Juárez
Chihuahua
Mexico

Guillermo Cortes-Robles
Institute of Technology of Orizaba
Av. Instituto Tecnologico
Orizaba
Mexico

Aidé Aracely Maldonado-Macías
Autonomous University of Ciudad Juarez
Chihuahua
Mexico

ISBN 978-3-319-04950-2 ISBN 978-3-319-04951-9 (eBook)
DOI 10.1007/978-3-319-04951-9
Springer Cham Heidelberg New York Dordrecht London

Library of Congress Control Number: 2014934662

Printed on acid-free paper

Springer is part of Springer Science+Business Media (www.springer.com)

Preface

This book presents the state-of-the-art contributions of Lean Manufacturing (LM) applications in the developing world, including essentially input from Latin American authors. These contributions are not easily found in the literature nowadays and, because of its quality and importance, they need a forum to be published. Lean Manufacturing is a multidimensional approach that embraces a wide variety of management practices in a unified system. These practices comprise just-in-time, quality systems, cellular manufacturing, work teams, and supplier management, among others. Some other practices have been adopted more recently such as Human Factors and Ergonomics approaches. The essence of Lean is that these practices, in a synergistic way, can generate an efficient, high-quality system that produces finished products according to customers' demand with almost no waste.

This book was conceptualized for an audience of mainly graduate and undergraduate students; however, it can also be consulted by engineers and companies managers who search for the state-of-the-art applications of Lean Manufacturing within a wide diversity of sceneries and conditions. The book is intended to be an excellent source for divulgation of applied researches, lean concepts, and practices that have been successfully applied in the developing world domain. The book contains some trends in LM, concepts, and study cases organized into 27 chapters distributed conveniently for readers in four parts.

Part I: Introduction

Part I entitled Introduction is composed by only Chap. 1 from Salinas-Coronado et al. which first presents the main concepts, common techniques, tools, and methodologies related to Lean Manufacturing. Also, this introduction Part includes an application of Lean in a production process for the automotive industry in the Mexican maquiladora sector from the state of Baja California.

Part II: Main Lean Manufacturing Tools and Techniques

Part II is composed of 11 chapters dealing with LM management practices aiming to transform organizations into more quick to respond and competitive entities. This Part also contains a variety of applications of the most effective and popular Lean tools. Chapter 2 by Tapia-Esquivias et al. illustrates a troubleshooting proposal for solving problems in LM environment, as well as common tools of LM, specifically the A3 format and the kaizen events. Also, authors illustrate an application to a carwash in the city of Celaya, Guanajuato, Mexico. Chapter 3 by Rodríguez-Borbón and Rodríguez-Medina and Chap. 4 by Meza-Jiménez et al. present a discussion related to Statistical Process Control (SPC) for quality assurance followed by its applications in the Maquiladora sector in Ciudad Juarez, Mexico, and the textile sector in Colima, Mexico.

In Chap. 5, Tlapa-Mendoza et al. discuss the six sigma approach that is applied in the maquiladora sector in the Mexican state of Baja California as an important tool applied for quality control. The six sigma concept evolved to lean-sigma concept have created synergy, which is discussed by Estrada-Orantes and Alba-Baena in Chap. 6. Also, in Chap. 6, some applications to maquiladora industry are discussed.

Usually, quality control requires the application of several advanced technologies for inspections, and Vergara-Villegas et al. in Chap. 7 discuss an elemental process for computer vision applied to a production system through a case study applied to apples selection. Since this inspection difficulty is based on repetitive inspection of products made by workers, total quality control is a philosophy that must be applied to enhance human labor. Rivera-Mojica and Rivera-Mojica in Chap. 8 discuss the Kaizen philosophy, in which they expose the current situation of Mexico in relation to other countries and report a structural equation model to identify critical success factors for the implementation of this philosophy.

However, high-quality manufactured products not only need people, they also require reliable equipment and the best operational conditions to satisfy the strictest specifications demanded by costumers. Therefore, in Chap. 9, Torres reports a Total Productive Maintenance approach and the key success factors for its implementation so that the integration of human–machine duo can offer improved results. Nevertheless, it should be noted that one of the most important quality requirements is the delivery of products just-in-time in production systems. Thus, arrivals of raw material must be on time, as well as deliveries between production processes and the final product delivered to customers. Therefore, Rivera-Mojica and Rivera-Mojica discuss the Just-in-Time (JIT) philosophy in Chap. 10. Authors include JIT concepts and identify the key factors for its successful implementation in the maquiladora industry that is established in Ciudad Juarez, Mexico.

Meanwhile, in Chap. 11, Villanueva-Ponce et al. focused their study on the supplier's selection process by taking into account that JIT philosophy implementation implies the compliance and best performance of production programs. Authors also discuss in this chapter the main techniques and most commonly used attributes for supplier evaluation and expose an example by applying a multi-attribute technique called TOPSIS. Finally, in the last chapter of this part (Chap. 12), Lagarda-Leyva et al. discuss and analyze flow of materials from a more holistic view, proposing a mega planning approach for all the logistics in the company, and they illustrate some examples for the improvement of small and medium businesses located in the Mexican city of Hermosillo.

Part III: Human Factors and Ergonomics in Lean Manufacturing

It is well-known that the success of LM is mainly derived from the effort of workers, supervisors, and managers in production systems; therefore, Part III of this book presents applications of Human Factors and Ergonomics in a Lean Manufacturing environment. This part of the book includes seven chapters that promote ergonomic practices and describes new concepts and theoretical frameworks for the Lean culture.

First, in Chap. 13, Naranjo-Flores and Ramírez-Cárdenas discuss from a general point of view the human factors and its importance in LM, specifically the Lean Ergonomics concept. Prado, in Chap. 14, presents an epidemiologic review of studies related to low back pain treated as a musculoskeletal disorder; he thus emphasizes on the main occupational risk factors.

Lean Manufacturing has been contributing to the success of numerous companies all around the world, which are permanently making efforts to improve their decision-making processes to assure competitiveness. This improvement includes the evaluation and selection of Advanced Manufacturing Technology (AMT). However, human factors and ergonomics aspects have been neglected in the actual decision-making models for AMT. In this way, Maldonado-Macías et al. expose in Chap. 15 an innovative fuzzy axiomatic design approach for ergonomic evaluation of Advanced Manufacturing Technology (AMT) proposed under the theoretical framework of Lean-Six Sigma. In the same matter, in Chap. 16, Realyvásquez-Vargas proposes an expert system approach based on fuzzy if–then rules for ergonomic assessment of AMT for Lean environments.

Furthermore, with the inclusion of this technology, important changes in the levels of mental and physical effort performed by workers have been experimented. In Chap. 17, Hernández-Arellano et al. present the comparison of fatigue scores of two processes including (AMT). The former involves machining of steel pieces using Computerized Numerical Control (CNC) lathes, while the latter deals with Constant Velocity (CV) Joints assembly using hydraulic presses, this is a case study of the assessment of human fatigue in industrial environments. Human fatigue effects have generated a wide variety of approaches, including accident analysis. In this way, researchers have emphasized on the need to incorporate human factors in risk assessment, due to the increasing relevant role of humans in the modern work environment. Hence, in Chap. 18 Reyes-Martínez et al. develop a theoretical approach for accidents' causality identification in industrial environments.

Part IV: Alternative Methodologies for Lean Manufacturing

This part content was designed to expose modern alternative methodologies that have facilitated Lean Manufacturing implementation, and are explained in nine chapters in next paragraphs.

Chapter 19 concerns optimization problems in manufacturing processed by Alvarado-Iniesta et al. In this chapter, authors easily describe the genetic algorithms methods and present an example step by step. This example will surely be useful to readers wishing to know quickly the power of this tool. Meanwhile, Adarme-Jaimes et al. in Chap. 20 refer to a technique that is applied very frequently with the total preventive maintenance; however it is more a philosophy than a technique, as it depends entirely on persons and refers to the 5's, which help determine and identify waste in a production system to be optimized.

However, the application of techniques, philosophies, and tools in a production system focused on Lean Manufacturing should be monitored. Therefore, Rivera and Manotas in Chap. 21 propose a performance measurement in Lean Manufacturing environments so that managers can identify the tools or philosophies that best fit their production system, as well as those that must be modified.

Still in the context of alternative methodologies, one of the most used is the plant layout for material flow optimization, since the transport of materials is seen as a waste and the source of accidents. This issue is discussed by Blanco-Fernández et al. in Chap. 22. Authors analyze the different techniques used and solve an example as a case study. Another large waste that has been observed in production systems is the preparation of equipment for new batch production, usually called setup. Therefore, in Chap. 23, Carrizo-Moreira discusses the SMED (Single-Minute Exchange of Die) system and reports the cases of seven companies.

These production systems have generally required adjustments to the process, which may be due to mismatch of machines and its sensors; therefore, in Chap. 24, Molina-Arredondo presents a model permitting fast adjustment to process, having information feedback obtained from production processes. Such adjustments prevent the production of whole lots with defects. However, obtaining parameters in LM is very difficult since it is a very broad concept; thus, specific parameters are often obtained, such as supply chain indicators. For this reason, in Chap. 25, Avelar-Sosa et al. compile the tendencies of techniques and attributes for supply chain performance measurement, which can tell the manager what they can do to improve the flow of raw materials and information.

To end with this part, optimization methodologies such as Design of Experiments (DOE) and dynamic analysis are covered. In Chap. 26, Becerra-Rodríguez et al. present an optimization case for manufacturing using DOE, and Sánchez-Ramírez et al. expose a dynamic analysis of inventory policies for improving scheduling in Chap. 27.

Jorge Luis García-Alcaraz
Aidé Aracely Maldonado-Macías
Guillermo Cortes-Robles

Acknowledgments

The editing process for this book has invested more than a year of work, and has been done with the valuable support of many other persons who intervened at different times. Specially, we would like to thank Judith Hinterberg, Mayra Castro, and Petra Jantzen, as well as Springer publishing people, who facilitated our work with their large editorial advice and knowledge.

Likewise, we wish to thank all the authors who have entrusted their work to be published in this book. We know that they are busy people with many duties; still, they supported this project.

Finally, we also thank our families. As authors, we dedicated a great amount of time to this book, time that we would have liked to spend with our beloved ones. Therefore, we thank them for their comprehension and also apologize for not being there with them all the time; however, we cannot promise this will not happen again.

Contents

Part I
Introduction

Chapter 1
Lean Manufacturing in Production Process in the Automotive Industry

Jesús Salinas-Coronado, Julián Israel Aguilar-Duque, Diego Alfredo Tlapa-Mendoza and Guillermo Amaya-Parra

Abstract The industry in search of the optimization of the processes has resorted to the application of lean manufacturing to solve a variety of problems that are present in the manufacture of different products. The Design of systems and products requires an understanding of the causes that achieve desirable performance, the need for an efficient system is the principal cause of performance Lean Manufacturing strategies. This chapter has the aim to describe the implementation of Lean Manufacturing focus in optimize the manufacturing process of a product for the automotive industry, using Total Productive Maintenance (TPM). The case of study presented in this chapter describes a specific problem whit the equipment used in the adhesive application process. The methodology was focused in determinate the six big losses in critical equipment and the implementation of TPM, achieving a reduction the problem 30 %.

Keywords Lean manufacturing · Equipment problems · Total productive maintenance

1.1 Introduction

The industry's concern to provide a product or a service with excellence to clients has opened an innumerable list of actions that need to be directed to: reducing the wastes, improve the productivity, to train the personnel and create a culture of quality. Lean tools are frequently used in many Mexican companies that manufacture products to be exported mainly to the United States. Mass production is

J. Salinas-Coronado (✉) · J. I. Aguilar-Duque · D. A. Tlapa-Mendoza · G. Amaya-Parra
Faculty of Engineering, Architecture and Design, Autonomous University of Baja California, Campus Ensenada, Km. 103 Carretera Tijuana—Ensenada, C. P. 22860 Ensenada, Baja California, Mexico
e-mail: jesus.salinas.coronado@uabc.edu.mx

J. L. García-Alcaraz et al. (eds.), *Lean Manufacturing in the Developing World*,
DOI: 10.1007/978-3-319-04951-9_1, © Springer International Publishing Switzerland 2014

one of the characteristics of the manufacturing industry and it should focus on the performance of production processes, a distraction can result in a poor quality products. It is important to have senior management that cares in providing high quality products and world-class processes, these companies invest time and resources to update their staff on different methods or tools such as lean manufacturing (5's, Heinkunka, Jidoka, Kaikaku, Kaizen, Kanban, Manufacturing Cells, Poka-Yoke, Push Pull System, SMED, Standard Work, TPM, Visual Factory, Value Stream Mapping). Also on the administrative staff updates of time management, supervision, teamwork skills, communication between members of the organization and senior management.

The way to manage senior management has changed in recent years. In the past administration employees were not seen on the floors of production but it was effective to a certain point.r. Nowadays it is very common to see senior management making daily rounds in the production areas identifying opportunities for improvement, follow up and supporting the implementation of the continuous improvement. In the next paragraphs we present general information about Lean Manufacturing and the different tools used for continuous improving.

1.1.1 Lean Manufacturing

Today there is a new evolution in manufacturing and it's driven by two factors: (a) A sustained economic growth. (b) The old management styles are failing to work with employees without no multi-task training.

In an effort to be more productive, many organizations are adapting to lean manufacturing processes setting solid goals as:

- Manufacturing quality product
- Cost Reduction Projects
- Total Employee Involvement
- Cultural Approaches

In the beginning many manufactures had many questions about the importance of Lean Manufacturing; We describe some of these questions, and their answers.

1.1.2 What an Organization Seeks with Lean Manufacturing Process

What a company is looking to get through lean manufacturing can be summarized as follow:

- Increase annual profits
- Establish cost reduction projects

- Shorten delivery times
- Manufacturing high quality products
- Retain employees
- Promote continuous improvement
- Be willing to change

1.1.3 What is Lean Manufacturing?

Lean manufacturing can be define as a combination of multiple tools to help eliminate activities that do not add value to the product, service and/or process by increasing the value of each activity, aimed to eliminate or reduce waste and improve operations.

1.1.4 Lean Manufacturing Objectives

- Specify the value for the customer
- Identify all actions required to bring the product from being a concept to being launch, from an order to delivery, from raw material to the customer's hands and even his life
- Remove any action that does not add value and streaming align every action to add value as required by the customer
- Analyze the results and start the evaluation process again.

Lean Manufacturing defines waste as: "Anything other than the absolute minimum of time and resources to add value to the product and get the highest quality" (Martinez 2003a, b).

Lean manufacturing focuses on reorganizing the following processes Waste reduction, minimizing steps and thereby reducing costs. Several tools and techniques have been proposed in the context of lean manufacturing. Though lean manufacturing is widely regarded as a business strategy, few researchers have concentrated on the validation of its positive link with business performance (Vinodh Dino and Joy 2012). One of the ambiguous questions is that how a manufacturer can identify tools and techniques, and the relevant capabilities and abilities to become lean. There are several important issues that need to be addressed to understand how lean manufacturing and management can be achieved with clarity of purpose, focus and goals (Vinodh Dino and Joy 2012).

Lean Manufacturing is open to every researcher, according to the breadth of it, so below is a sequence of appearance or application in history and is in constant update (Meyer 2011). Table 1.1 presents the ***Lean Manufacturing history and timeline.***

Table 1.1 Lean manufacturing history and timeline

2000–2010	2007: Application lean manufacturing concepts in the industry mass production of parts for automotive; company located north of Mexico, concepts such as SMED (Single-Minute Exchange of Die), TPM (total productive maintenance), Value stream mapping, to name a few
	2004: Shingo Prize-winning "kaikaku" published by Norman Bodek, chronicling the history and personal philosophies of the key people that helped develop TPS(Toyota production system)
	2003: Shingo Prize-winning "Better Thinking, Better Results" published, case study and analysis of The Wiremold Company's enterprise-wide Lean transformation
In the 1990s	1996: Lean Thinking by Womack and Jones
	1990: The Machine That Changed the World by Womack and Jones
1980s	1988: Kaizen Institute of American holds kaizen seminars at Hartford Graduate Center (Hartford, Conn.) with TPS sessions taught by principals from Shingijutsu Co., Ltd
	1988: Shingo Prize for Manufacturing Excellence created by Norman Bodek and Professor Vern Buehler of Utah State University
1960s	1969: Start of Toyota operations management consulting division
	1965: Toyota wins Deming Prize for Quality
	1961: Start of Toyota corporate wide TQC (Total Quality Management) program
1950s	1956: Shigeo Shingo begins regular visits to teach "P-Course"
	1951: J.M. Juran publishes his seminal work The Quality Control Handbook
1940s	1946: Ford adopts GM management style and abandons lean manufacturing
	1943: Taiichi Ohno transfers from Toyoda Auto Loom to Toyota Motor Corporation
	1940: Deming develops statistical sampling methods for the
1930s	1939: Walter Shewhart publishes Statistical Methods from the Viewpoint of Quality Control. This book introduces his notion of the Shewhart improvement cycle Plan-Do-Study-Act. In the 1950s his colleague W Edwards Demming alters the term slightly to become the Plan-Do-Check-Act cycle
	1938: Just-in-time concept established at Koromo/Honsha plant by Kiichiro Toyoda. JIT wa later severely disrupted by World War II
1920s	1929: Sakichi Toyoda sells foreign rights to loom and Kiichiro Toyoda visits Ford and European companies to learn the automotive business
	1926: Henry Ford publishes Today and Tomorrow
1910s	1914: Ford creates the first moving assembly line, reducing chassis assembly time from over 12 h to less than 3 h
	1912: The Ford production system based on the principles of "accuracy, flow and precision" extends to assembly
1900s	1908: Ford introduces the Model T
	1906: Italian economist Vilfredo Pareto creates a mathematical formula to describe the unequal distribution of wealth in Italy. He notices that 80 % of the wealth is in the hands of 20 % of the population
	1905: Frank and Lillian Gilbreth investigate the notion of motion economy in the workplace. Studying the motions in work such as brick laying they develop a system of 18 basic elements that can depict basic motion

(continued)

Table 1.1 (continued)

1890s	1890: Sakichi Toyoda invents a wooden handloom
1850s	1850: All of the American armories were making standardized metal parts for standardized weapons, but only with enormous amounts of handwork to get each part to its correct specification. This was because the machine tools of that era could not work on hardened metal
1800s	1807: Marc Brunel in England devised equipment for making simple wooden items like rope blocks for the Royal Navy using 22 kinds of machines that produced identical items in process sequence one at a time
1790s	1799: Whitney perfects the concept of interchangeable parts when he took a contract from the U.S. Army for the manufacture of 10,000 muskets at the low price of $13.40 each
1760s	1760: French general Jean-Baptiste de Gribeauval had grasped the significance of standardized designs and interchangeable parts to facilitate battlefield repairs

It's important to define that "The Lean manufacturing system" has subsystems (tools), and these subsystems are used to remove or reduce waste in organizations, in Fig. 1.1 appears some tools of Lean manufacturing.

In order to describe the Lean Manufacturing, tools and application, it's important to describe generalities about them. The next paragraphs describe general information about the different tools.

1.1.5 5S's

5S's began in Toyota in 1960 with the aim to make workplaces better organized, more orderly and cleaner permanently to achieve higher productivity and a better working environment. We all face tough challenges Unabated Competitive pressures. Companies strive to reduce costs. Some look to improve technology. Some operators reduce template. Very few have become excellent operational. 5S's, was developed by Hiroyuki Hirano to improve the industry to improve this concept. One Answer is 5S's

Some companies beat the odds and encourage strong, positive cultures. Danaher and Toyota are two of the better known examples. The method of 5S' is one way to engage people and contribute to culture change. 5S' is a visually-oriented system of cleanliness, organization, and arrangement (Fig. 1.2) designed to facilitate greater productivity, safety, and quality (Fig. 1.3). It engages all employees and it is a foundation for more on the job self-discipline, better working environment and better products.

5S' is a foundation for more discipline actions. If workers cannot even put a tool back in its designated location, will they follow standards for production? It's visual nature that makes things that are out of place stick out like a sore thumb. And, when properly supported, it builds a culture of continuous improvement. The benefits of 5S' are:

Fig. 1.1 Tools of Lean Manufacturing

LEAN MANUFACTURING
- 5S´s
- JIDOKA
- KAIKAKU
- KAIZEN
- KANBAN
- MANUAFCTURING CELLS
- POKA YOKE
- PUSH PULL SISTEM
- SMED
- STANDARD WORK
- TPM
- VISUAL FACTORY
- VALUE STREAM MAPING

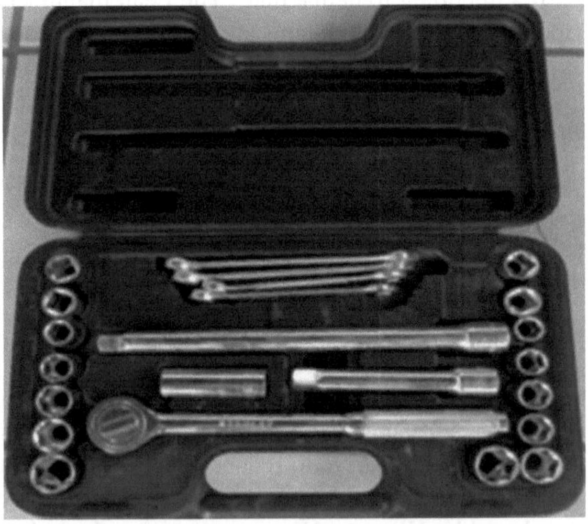

Fig. 1.2 5S's is visual oriented

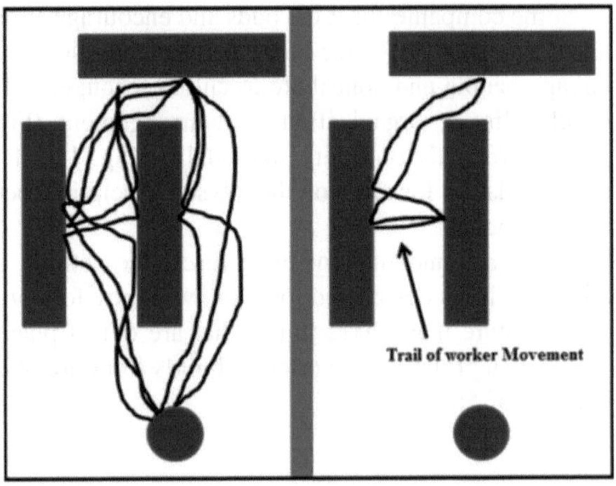

Fig. 1.3 5S' facilities greater productivity, safety and quality

- Cleaner and safer work areas: when a work area is clean and organized tripping hazards and other dangers are eliminated.
- Less wasted time through more workplace organization: when tools and materials are accessible and in order, workers need less time to "go get" this means less time to search for what they need.
- Less space: when unneeded items are eliminated and the needed ones are organized, the required floor space is dramatically reduced.
- Improve self-discipline: the 5S' system, especially its visual nature, makes abnormal conditions noticeable and makes ignoring standards more difficult.
- Improve culture: when 5S' is applied systematically, it encourages better teamwork and enthusiasm.

People like to work in a well-organized and clean environment. They feel better about themselves and better about their work, and they restore the self-discipline that is found in winning teams.

Markovitz (2012) said "5S is the foundation of lean". It's not just about "cleaning your room" or being faster at finding your stapler, with all the triviality that implies. In reality, the decisions that 5S's forces you to make—and the discipline it imposes—It is the basis for spotting waste, for creating systems that enable work to flow more efficiently and for helping to clarify "standard work" in the complex, highly variable office environment. To be sure, applying 5S' yields time savings from not having to search for information, but the more significant benefit comes from surfacing abnormalities and waste in processes so they can be fixed. Some people will claim that 5S's isn't really important for knowledge workers unless they're sharing an office space or a desk with someone else.

1.1.6 Jidoka

Who invented Jidoka? The original concept is very old and it goes back to the Toyoda Auto Loom Company. Mr. Sakichi Toyoda invented an automatic loom that would shut down as soon as a single thread broke. This saved a lot of wasted material and helped highlight problems as soon as one happened. That was the starting point.

Jidoka is one of the core principles of the Toyota Production System. It means applying the "human touch" to immediately address manufacturing problems at the moment they are detected. Jidoka is used at Toyota to empower every worker to stop the assembly line whenever a quality problem is detected. The worker pulls a red cord and the entire assembly line stops, idling every machine and every worker on that line until the problem is solved or a remedy is found to prevent a defect moving forward. When the line stops, fellow workers run over to the person who pulled the red cord to help them resolve the problem. In reality, the problem resolution often takes less than a minute and the line is again up and running. In the typical Toyota plant, the line is stopped dozens of times each day.

We discuss how Jidoka can be applied to the world outside of the factory. Dealers, sales people, and service technicians interact daily with customers and have countless opportunities to identify and react to problems before they spiral out of control. Like many aspects of the Toyota Production System, Jidoka is a simple common-sense methodology, with many powerful benefits:

- Increases trust—Powerfully conveys the Toyota principle of "Respect for People" that empowers and encourages people to report defects and problems without fear of blame.
- Improves communication—Provides clear notification of a problem to customers (downstream workers) and suppliers (upstream workers).
- Creates urgency—Signals an immediate and pressing need to solve the problem.
- Contains the problem—Limits the number of defects produced.
- Involves others—Calls on the supervisor, customers and suppliers (downstream and upstream workers) to help solve the problem.
- Drives prevention—Requires the identification of the root cause to keep the problem from recurring.
- Changes the culture from "blame" to "blameless."

Jidoka will improve the quality and safety of automobiles, and in the long term, decrease accidents, reduce costs, and restore customer confidence (Bodek 2011).

1.1.7 Kaikaku

Bodek (2004) said that Kaikaku have origin in Toyota Production System; Kaikaku means a radical change, during a short time, of a production system. Means "that an entire business is changed radically, normally always in the form of a project". Kaikaku is most often initiated by management, since the change as such and the result will significantly impact business. Kaikaku is about introducing new knowledge, new strategies, new forms thinking, lot communication between staff and all levels employees, new approaches, new production techniques or new equipment. Kaikaku can be initiated by external factors, e.g. transfer new products, new technology, launch new production lines. Kaikaku can also be initiated when management see that ongoing Kaizen work is beginning to stagnate and no longer provides adequate results in relation to the effort.

Kaikaku, which translates to "radical improvement or change", is a more transformational process. It stars with customers, priorities and links directly to the business strategy. Correct application of Kaikaku can help an organization move ahead of competitors by dramatically reducing the time required for major improvement in quality, cost and delivery. It is suited to companies facing merger and integration, or other cases demanding an enterprise-level transformation. (Seeliger 2006).

1.1.8 Kaizen

> The technique we called Toyota Production System (TPS) was born through our various efforts to catch up with the automotive industries of western advanced nations after the end of World War II, without the benefit of founds or splendid facilities (Monden 2012).

The history of Kaizen has developed of the following way; in 1950 Toyota implemented circles leading to development "Toyota Production System" focus in continuous improvement in quality, technology, processes, company, culture, productivity, safety and leadership. These continual small improvements (Kaizen) add up to major benefits, resulting; faster delivery, lower costs, and greater customer satisfaction (Imai 2012).

The 10 principles of Kaizen are:

- Say no to status quo, implement new methods and assume they will work.
- If something is wrong, correct it.
- Accept no excuses and make things happen.
- Improve everything continuously.
- Abolish old, traditional concepts.
- Be economical. Save money through small improvements and spend the saved money on further improvements.
- Empower everyone to take part in problems solving.
- Before making decisions, ask "why" five times to get to the root cause.
- Get information and opinions from multiple people.
- Remember that improvement has no limits. Never stop trying to improve.

The kaizen umbrella covers many elements, the next Fig. 1.4 show's an example.

Generally Kaizen activities include the next basic pattern:

- Discover the Improvement Potential
- Analyze the current Methods
- Generate Original Ideas
- Implement the Plan
- Evaluate the new Method.

These five steps include activities which the multidisciplinary groups join with an only one goal "the improvement continues". (Kato and Art 2011).

At the present, the world is so intensive that the companies that achieve the success in his business are the most flexible to change, having major capacity and speed of adjustment. This ability is the result of implement Kaizen, focus in the specific philosophy of continuous improvement (Maurer 2013).

Fig. 1.4 Umbrella Kaizen

1.1.9 Kanban

Kanban is a Japanese method of process control based only on actual consumption of materials. Kanban can reduce inventories in the production hall, which are required for optimized production activities. The aim of the system is to control and reduce the cost of storage of materials, increasing the availability of materials and eliminating production downtime due to lack of required materials (Gross and Mcinnis 2003).

The original Kanban system was founded in 1947 by Taiichi Ohno, an employee of a Japanese company Toyota Motor Corporation. One of the reasons was insufficient productivity of the company in relation to its competitors in the U.S. Ohno described the concept as follows: "It should be possible to organize the flow of materials in production, according to the principle of a supermarket, the consumer takes a certain amount of off the shelf product, and noticed the gap is immediately replenished" (Komentarzy 2013).

The increase in customer expectations has forced the company to increase the availability, resulted in a significant increase in storage. It was necessary to find the optimal solution, in particular, a narrow island that is Japan, would reduce the size of costly stocks of raw materials and semi-finished products. To minimize these costs, we need a flexible and efficient production system.

Komentarzy (2013) said an example of Toyota, "High motivation to implement the method for Japanese Kanban was after 3 years of the introduction of the Kanban, there was achieved the following results: 75 % reduction in any stocks,

95 % reduction in defects, 25 % increase in production, 10 % reduction in storage space, 10 % reduction of employees working in the warehouse."

One of the most important parts in Kanban is the visual aids, which are used to show when a specific part is needed. These visual aids are usually named "Kanban Cards". That future the part's name, picture, code, and the quantity needed. Kanban cards are often attached to a container where they become visible as soon as the quantity reaches a point where more should be ordered.

Some typical elements of most Kanban—controlled systems include:

- Standard lot sizes or standard containers
- A single card per container
- Containers returning empty, usually as a signal that further material has been authorized for movement.

1.1.10 Manufacturing Cells

The most common definition for "Manufacturing cells" it's "Sets of machines that are grouped by the products or parts they produce in a lean manufacturing environment". This system is used in the cellular manufacturing concept, which is distinct from the traditional functional manufacturing system in which all similar machines are grouped together. (Black and Hunter 2003).

Adenzo-Diaz and Lozano (2008) said, "The change in power sharing in the market could no less affect the organization of workshops. Early layouts were fundamentally of two kinds:

- Functionally centered layouts (Typically in job-shops, of the engineer-to-order type with low production volumes per product) sought (for highly variable flows) greater flexibility for the assigning of jobs to resources and the exploitation of scale economies, in change for certain inefficiencies as a result of heterogeneity (greater difficulty in scheduling, higher WIP, lead times and material costs).
- Product oriented layouts (typical in cases such as make-to-stock, with large production volumes) aimed are reducing flow-associated costs, at the expense of replicating machines, situated in flow line in accordance with the operational needs of products.

For the design of a manufacturing cell it's important to follow the next steps:

1. Define the manufacturing system (Manufacturing vs Production System) this step it will help to design the manufacturing system
2. Determinate the Functional Requirements
3. Determinate the manufacturing and Assembly Cells
4. Analyze the opportunity for setup reduction

5. Consider the Integrated Quality Control
6. Integrated Reliability
7. Think in Lean Production and Inventory Control
8. Design considering ergonomics

1.1.11 Poka Yoke

The term Poka-Yoke was applied by Shigeo Shingo in the 1960s to industrial processes designed to prevent human errors. Shingo redesigned a process in which factory workers, while assembling a small switch, would often forget to insert the required spring under one of the switch buttons. In the redesigned process, the worker would perform the task in two steps, first preparing the two required springs and placing them in a placeholder, then inserting the springs from the placeholder into the switch. When a spring remained in the placeholder, the workers knew that they had forgotten to insert it and could correct the mistake effortlessly. Shingo distinguished between the concepts of inevitable human mistakes and defects in the production. Defects occur when the mistakes are allowed to reach the customer. The aim of Poka-Yoke is to design the process so that mistakes can be detected and corrected immediately, eliminating defects at the source. (Shingo 1986).

This system it's really easy to implement. Shingeo Shingo recognized three types of Poka-Yoke for detecting and preventing errors in a mass production system:

- The contact method identifies product defects by testing the product's shape, size, color, or other physical attributes.
- The fixed-value (or constant number) method alerts the operator in a certain number of movements are not made.
- The motion-step (or sequence) method determines whether the prescribed step of the process has been followed.

When to use it? The Portal Quality said: "Poka Yoke can be used wherever something can go wrong or an error can be made. It is a technique, a tool that can be applied to any type of process be it in manufacturing or the service industry". Errors are many types:

- Processing error: Process operation missed or not performed per the standard operating procedure.
- Setup error: Using the wrong tooling or setting machine adjustments incorrectly.
- Missing part: Not all parts included in the assembly, welding, or other processes.
- Improper part/item: Wrong part used in the process.
- Operations error: Carrying out an operation incorrectly; having the incorrect version of the specification.
- Measurement error: Errors in machine adjustment, test measurement or dimensions of a part coming in from a supplier.

Robinson (1997) describes two categories of Poka-Yoke "Prevention and detection", they were defined as:

- A prevention device engineers the process so that it is impossible to make a mistake at all. Prevention devices remove the need to correct a mistake, since the user cannot make the mistake in the first place.
- A detection device signals the user when a mistake has been made, so that the user can quickly correct the problem. The small dish used at the Yamada Electric plant was a detection device; it alerted the worker when a spring had been forgotten. Detection devices typically warn the user of a problem, but they do not enforce the correction.

Finally, o good Poka-Yoke device is:

- Simple and Cheap
- Part of the process
- Easy to place close to where the mistakes occur, providing quick feedback to the workers.

1.1.12 Push Pull System

Push or Pull system are two different ways for the production systems. In a push system, production planning is largely driven by the production capacity with the objective to achieve high capacity utilization. The traditional Material Requirement Planning (MRP) method usually provides a good planning solution when demand is relatively stable. (Cheng et al. 2012). With highly uncertain demand, it is difficult for a push system to react quickly to match supply chain to be highly flexible in order to be responsive to change in demands.

The Fig. 1.5 Present a diagram process of the Push system versus Pull system.

Hopp and Spearman (1996), define push systems as a schedule work releases based on demand inherently due-date driven, control release rate, observe WIP level and pull systems: authorize work releases based on system status, inherently rate driven, control WIP level, observe throughput.

Hopp and Spearman (1996) remarks: "Push type" means make to Stock in which the production is not based on actual demand. "Pull type" means make To Order in which the production is based on actual demand. In supply chain management, it is important to carry out processes halfway between push type and pull type or by a combination of push type and pull type.

Why Push and Pull? Lindeke (2005) said: MRP is the classic *push* system. The MRP system computes production schedules for all levels based on forecasts of sales of end items. Once produced, subassemblies are *pushed* to next level whether needed or not. JIT is the classic *pull* system. The basic mechanism is that production at one level only happens when initiated by a request at the higher level.

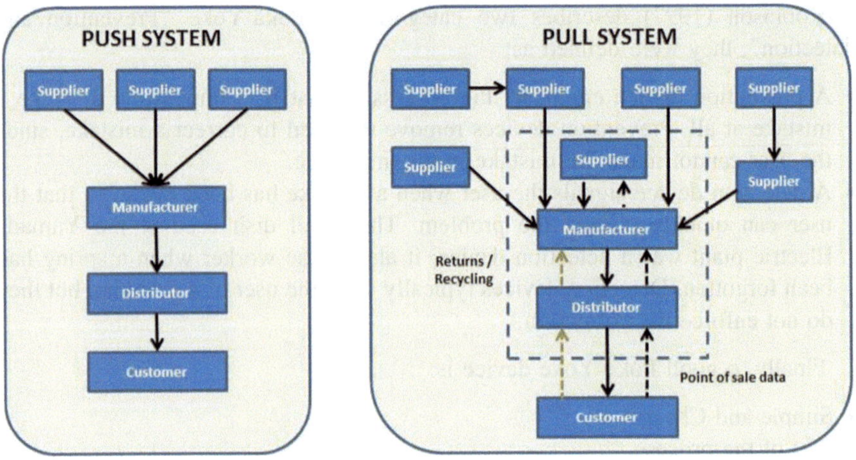

Fig. 1.5 Diagram of push system versus pull system

That is, units are *pulled* through the system by request. Continuing whit the comparisons, Lindke (2005) present the Table 1.2:

Finally, both systems have their advantages and disadvantages. The push system is more effective in dealing with fluctuating demand. Producers can store finished products in anticipation of demand, even though this incurs an inventory cost, or they can create a new demand by supplying products in the finished goods inventory, which means an overstocked sale.

In a push system, the producers control the pace of product development. Design changes are made infrequently, only when the current design becomes completely obsolete. But this system promotes the producer's control over the product and risks dissatisfying consumers.

On the other hand, the pull system forces producers to invest heavily into research and development to meet ever-changing customer requirements, which increases product cost. But customers are also more satisfied.

Recent research suggests using neither a pure push nor pure pull strategy, especially if you are producing multiple products. The pure pull system was initially designed for manufacturing environments producing repetitive products with stable demands, and requires at least a minimum inventory of each product. This may make it impractical for lines manufacturing a large variety of custom products.

1.1.13 Single Minute Exchange of Die (SMED)

The technique Single Minute Exchange of Die (SMED), by (Cuatrocasas 2012) comes from a study by engineer Shigeo Shingo, where TOYOTA asks the engineer to investigate as Volkswagen had managed to change in two hours the matrix of a

Table 1.2 Advantages and disadvantages of pull and push systems

Advantages	Disadvantages
JIT = PULL	
Limited and known Final Inventory	Every job is a 'High Stress' Rush order
Workers only consume their time and Raw Materials on what is actually needed	Balanced systems MUST be in place
Quality MUST be High—each piece has a definite place to go—else immediate feedback is given	Setup times will greatly impact throughput
	Any problem will lead to unhappy customers (either internal or external)
MRP = PUSH	
Allows Managers to manage—that is, plan and control things	Can lead to large inventories
Requires intricate knowledge of Production Times and Product Flow	Can generate large quantities of scrap before errors are discovered
Can lead to economies of scale in purchasing and production	Requires diligence to maintain effective product flow

press that weighed 1,000 tons, when they required four. At 6 months, the team had already lowered Shingo 90 min this time, with a whole new approach, eventually giving rise to a completely new change happened three minutes, the SMED born with this.

By De la Fuente et al. (2006) SMED system born as a set of concepts and techniques intended to reduce the preparation time to time to state them in less than 10 min. SMED is a considerably reducing set up times, avoiding the need to work with large batch production.

System Benefits SMED:

- Reduced batch size, the production time and inventory level.
- Greater flexibility to the company to adapt to fluctuations and changes in demand.
- Increased utilization rate of equipment and productivity by reducing downtime of the changes.
- By allowing manufacturing and delivery time very short, the company may discontinue to store and adapt their production to actual customer orders.
- When working with smaller batches, quality problems are detected quickly and affect fewer parts.

Espin (2013) indicates that the steps to work with SMED are different according to each author but essentially the methodology is the same and you get the same result correctly implementing the proposed technique thus the few phases to perform SMED see Fig. 1.6.

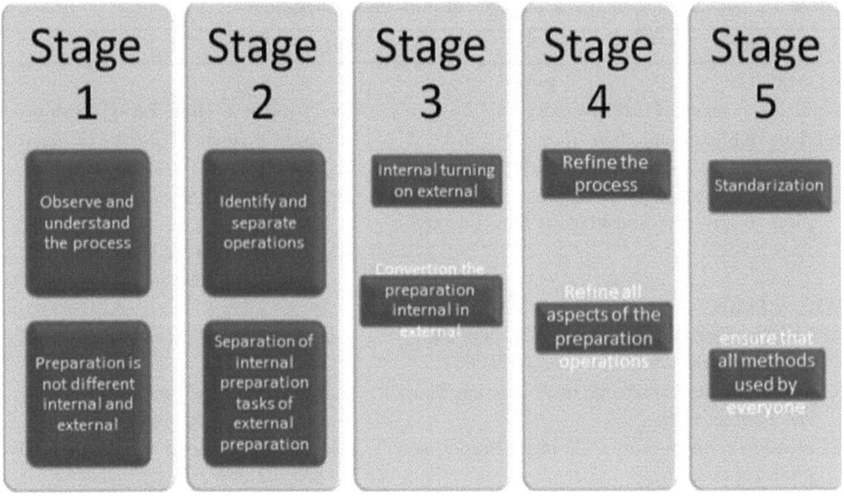

Fig. 1.6 Stage of SMED

1.1.14 Standard Work

Standard work is a valuable tool to use for any improvement project and consists of seven steps: (1) document reality, (2) identify the waste, (3) plan countermeasures, (4) implement changes, (5) verify changes, (6) quantify changes, and (7) standardize changes (Zidel 2006).

- First step. Document reality: go out to the area, observe the job being done, and document it. Do not attempt to do this from memory.
- Second step. Identify the waste in the process. Study your documentation and categorize each step as value added, type 1 non–value added, or type 2 non–value added.
- Third step. Plan countermeasures. Brainstorm ideas to eliminate the waste and make the process more efficient.
- Fourth step. Then implement the changes. Be sure to communicate with the people actually doing the work. Do not, under any circumstances, make changes without their knowledge.
- Fifth step. Verify that the changes do indeed make things better. It may be worthwhile to do a pilot run for a certain period to verify that the changes have been made.
- Sixth step. Quantify the benefits either monetarily, in time savings, or in workforce reduction.
- Seventh step. Make the changes standard. Make sure that all understand what they need to do, and write a policy, if necessary.

1.1.15 Total Productive Maintenance (TPM)

TPM is innovative Japanese concept. The origin of TPM can be traced back to 1951. Nakajima S. is a done pioneering work and has given basic definition of TPM. TPM was born to achieve the following objectives. The important ones are listed below (Venkatesh 2005).

- Avoid wastage in a quickly changing economic environment.
- Producing goods without reducing product quality.
- Reduce cost.
- Produce a low batch quantity at the earliest possible time.
- Goods send to the customers must be non-defective.

The fundamental processes underlying the TPM are also called "pillars" that support for the construction of a production system ordered, and implanted following a disciplined and effective methodology. Therefore, we can say that the implementation of these pillars is based philosophy Total Productive Maintenance. (Zambrano and Leal 2005). See the next points:

- Scheduled Maintenance, to unify criteria according to the types of maintenance employees and do some planning, programming and maintenance control.
- Individual improvements in equipment are activities performed by a group of workers who seek to eliminate losses and devices processes.
- Projects MP/LCC (Preventive Maintenance/Life Cycle Cost) to get the highest availability systems through analysis costs.
- Education and Training, to be multifunctional staff is essential maintain ongoing training to obtain a operator mentor.
- Maintaining the quality, look for a link between defects in the product and any process inputs, labor, machines, methods and materials, in order to establish parameters which may determine the conditions and process and implement actions to prevent future defects.
- Administrative controls to achieve the best ways to control areas maintenance related some of these forms of control are: Five S, Brainstorming among others.
- Environment, Health and Safety, optimization studies the same.

Morales (2012) explain the 6 large losses and indicates that these factors impeding achieve overall equipment efficiency, these 6 major losses are grouped into three categories according to the type of effects that cause the performance of production systems, as can be seen in more detail in Table 1.3.

1.1.16 Visual Factory

Visual Factory is a combination of signs, charts and other visual representations of information that enable the quick dissemination of data within a lean manufacturing process. The visual factory attempts to reduce the time and resources

Table 1.3 Large losses

Wastage	losses	Type and characteristics	Target
Time dead and vacuum	1. Breakdowns	Process downtime by failures, errors or failures, occasional or chronic	Remove
	2. Times preparation and settings equipment	Process downtime by preparation of machines or tools necessary for its implementation	Minimizing
	3. Operation at reduced speed	Difference between the actual speed and equipment design as capable. You can also see other improvements in the team to overcome design speed	Cancel or negative to the difference with the design
Losses speed on the process	4. Time load and short stops	Time intervals in which the equipment is expected to continue. Stops short by multiple mismatches	Remove
	5. Quality defects and rework	Production with chronic or occasional flaws in the resulting product and consequently, in the mode of development of their processes	Remove processes outside tolerances
Processes defective	6. Commissioning	Yield losses during the startup phase of the process, which can be derived from technical requirements	Eliminate or minimize as technical requirements

required to communicate the same information verbally or in written form, as both are viewed as a "waste" within the framework of a lean manufacturing process.

1.1.17 Value Stream Map (VSM)

Value stream is a collection of all actions (value-added as well as non-value-added) that are required to bring a product (or a group of products that use the same resources) through the main flows, starting with raw material and ending with the customer (Abdulmalek and Rajgopal 2007). But Cabrera Calva (2013) defines the value stream mapping as tool used to view and understand a process and identify waste, also to detect sources of competitive advantage, helps to establish a common language with everyone besides communicate ideas for improvement, focused on what the customer values, including materials, information and processes that contribute to get what the client is interested in purchasing. Is one a technique used to draw a map or flow chart showing the materials and information in the form flowing, looking to cut waste, is useful for the management of change. Some authors propose four stages in order to develop a VSM:

1st. Selection and training of the group and identify product family
2nd. Current State Diagram
3rd. Future state mapping
4th. Future State

Taking into account the symbolism and process, which is carefully managing the process and what are the requirements of the clients.

After describe of the different tools that are used normally by Lean manufacturing, we describe the case of study in the next section.

1.2 Case Study

1.2.1 Problem Description

Lean manufacturing, it's the strategy that had been used in the present research. The implementation of this strategy was developed in a north Mexico factory. This organization had eight production lines, six lines in two shifts and two lines in three shifts; this work was developed in line two. The main problem was two equipment's machinery: one to apply a stamp on the product, and another that applies an adhesive for join parts in the manufacturing process for automotive speakers. This equipment's were the principal cause of a long dead time. For this reason the team decided to use the methodology of total productive maintenance focus in Lean Manufacturing.

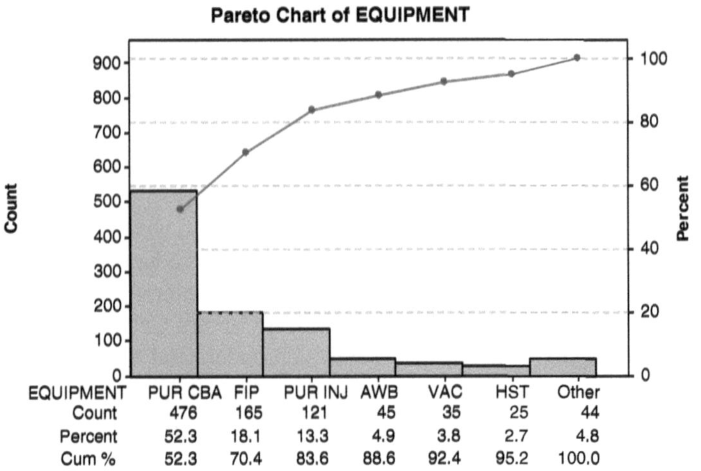

Fig. 1.7 Pareto Chart indicates equipment with more down time for month

1.2.2 Methodology

We worked with a multidisciplinary team, with representatives from the departments of Maintenance, Quality, Production, Materials, Engineering and Purchasing. We started to analyze the six big losses in critical equipment and the implementation of TPM. We found principally lost by impairments and lost for stop machine.

Was the lean manufacturing relying on total productive maintenance, we proceeded to perform measurements at high detecting process downtime FIP equipment (foam in place) which injected a seal to the product in the first workstations another machine was PUR (polyurethane reactive hot melt adhesive) which given an injection of adhesive to bond parts of the product. In Fig. 1.7 we present a Pareto Chart of Equipment downtime, before applying lean manufacturing. PUR CBA (PUR Change Bottle Adhesive), PUR INJ (PUR Injection), ABW (Adhesive Water Base), VAC (Vacuum), HST (Heat Stake), MAG (Magnetizing).

According whit the Parte analysis and the application of TPM, the engineers in the factory design different fixtures whit the goal to improve de system and reduce the PUR that represent almost the 60 %. In the Fig. 1.8 it shows an arraign.

After implementing improve. The engineers analyze the system, and the new results obtained are showed in the next Pareto (Fig. 1.9).

Figure 1.10 shows the result of the application of Lean Manufacturing in the system.

Fig. 1.8 Devises for fine adjust of equipment in objective application FIP

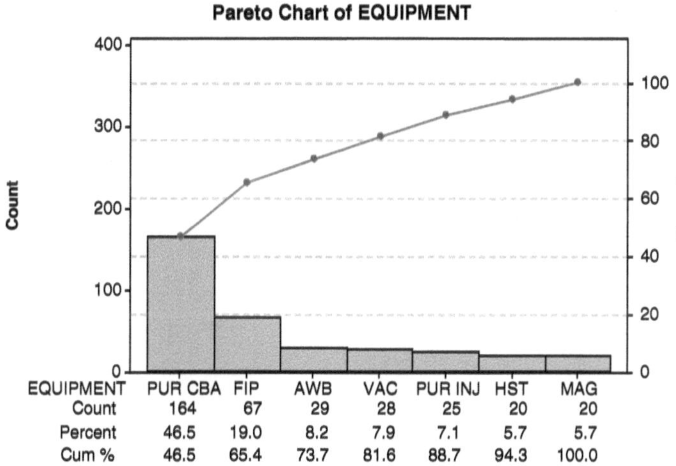

Fig. 1.9 Pareto Chart indicates equipment after apply lean manufacturing

1.2.3 Conclusions and Recommendations

It achieved a 67 % reduction in dead time which means a saving of $13,359 us per month, the commitment is to continue working on this and other projects to standardize these experiences in other lines of production. This type of teamwork brings much knowledge to all members as they interact with various participants and generate valuable information for both those who do and do not experience.

Fig. 1.10 Before and after results of applications of Lean Manufacturing

Information before of apply Lean Manufacturing			Cost USD
EQUIPMENT	DT	Production line cost, 40 operators, 1 supervisor, 2 technicals	465
PUR CBA	476		
PUR INJ	121		
FIP	165	Cost to stop production. The line produces 500 pcs/hr. Cost unproduced part 3 usd.	19950
Tot. Min	762		
Tot. Hrs	12.7		
		Total	20415

Information after of apply Lean Manufacturing			Cost USD
PUR CBA	DT	Production line cost, 40 operators, 1 supervisor, 2 technicals	156
PUR INJ	164		
FIP	25		
Tot. Min	67	Cost to stop production. The line produces 500 pcs/hr. Cost unproduced part 3 usd.	6900
Tot. Hrs	256		
Tot. Hrs	4.27		
		Total	7056

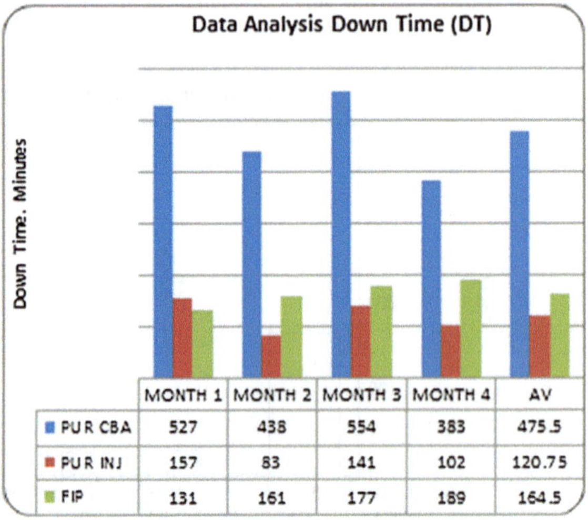

Fig. 1.11 Data analysis before application of Lean Manufacturing

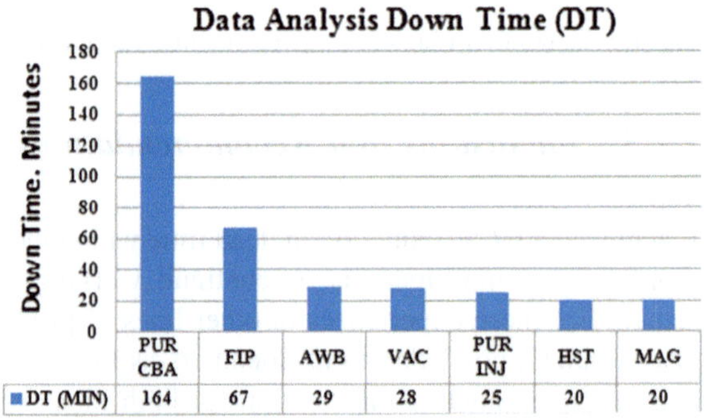

Fig. 1.12 After results application of Lean Manufacturing

The recommendation would be to continue to prepare the different teams to learn new tools to improve application processes already in full production and new products coming month after month or year after year with engineering changes (Figs. 1.11 and 1.12).

We present a comparison of before and after on the statistical information of Figs. 1.7 and 1.6 teams.

References

Abdulmalek, F., & Rajgopal, J. (2007). Analyzing the benefits of lean manufacturing and value stream mapping via simulation: A process sector case study. *International Journal of Production Economics, 107*(1), 223–236.

Adenzo, B. B., & Lozano, S. S. (2008). A model for the design of dedicated manufacturing cells. *International Journal of Production Research, 46*(2), 301–309.

Black, J. T., & Hunter, S. L. (2003). *Lean manufacturing systems and cell design*. United States of America: Society of Manufacturing Engineers.

Bodek, N. (2004). *10 Commandments of Kaikaku*. Lean CEO.

Bodek, N. (2011). *Jidoka simple tool a complex problem*. Morro Bay: Strategies Group LLC.

Cabrera Calva, R. (5 de 05 de 2013). *wordpress*. (13 de octubre de 2013). http://eddymercado.files.wordpress.com/2013/05/analisis-del-mapeo-de-la-cadena-de-valor.pdf.

Cheng, F., Ettl, M., Lu, Y., & Yao, D. (2012). A production-inventory model for push-pull manufacturing system with capacity and service level constrains. *Production and Operations Management, 21*(4), 668–681.

Cuatrocasas, L. (2012). *Gestión de la Producción, Modelos Lean Management*. Madrid: Diaz de Santos.

De la Fuente, D., García, N., Gómez, A., & Puente, J. (2006). *Organización de la Producción en Ingenierías*. Asturias: Universidad de Oviedo.

Espin, F. (2013). Tecnica SMED. Reducción del Tiempo de Preparación. *Revista de Investigación, 22*(1): 1–11.

Gross, J. M., & Mcinnis, K. R. (2003). *Kanban made simple: demystifying and applying toyota's legendary manufacturing process*. New York: AMACOM.

Hopp, W. J., & Spearman, M. L. (1996). The key difference between push and pull. Recovered 14-10-2013, of Factory-physics.com: http://www2.isye.gatech.edu/~jswann/teaching/6201/6201_Ch10_Stud_S03_6.pdf.

Imai, M. (2012). *Gemba Kaizen: A commonsense approach to a continuous improvement strategy 2/E*. United States of America: McGrawHill.

Kato, I., & Art, S. (2011). *Toyota Kaizen methods: Six steps to improvement*. New York: Productivity Press.

Komentarzy, B. (17 de 01 de 2013). *Kanban*. Obtenido de Dexription of the system, implemenation, marketing material, lean managmenet: http://en.system-kanban.pl/kanban/.

Lindeke, R. (2005). *Push vs. Pull Process Control*. IE 3265 POM.

Markovitz, D. (2012). Information 5S. *Management Sevices, 56*(1), 8–11.

Martinez, N. (2003a). *Curso Lean Manufacturing*. Ciudad Juarez: ITCJ.

Martinez, N. (2003b). *Lean manufacturing Course*. Mexico: ITCJ.

Maurer, R. (2013). *The spirit of kaizen creating lasting excellence one small step at a time*. United States of America: McGraw-Hill.

Meyer, K. (2011). *Lean manufacturing history & timeline*. Morro Bay: Kanso Strategies.

Monden, Y. (2012). *Toyota production system: An integrated approach to just-in-time*. Boca Raton: CRC Press.

Mora Gutierrez, A. (2009). *Mantenimiento, Planeación, Ejecución y Control*. Mexico: Alfaomega.

Morales, J. (2012). *implantación de un programa de mantenimiento Productivo total (tpm) al taller automotriz del Municipio de Riobamba (imr)*. Ecuador: Superior Politécnica.

Robinson, H. (1997). http://facultyweb.berry.edu/: http://facultyweb.berry.edu/jgrout/pokasoft.html.

Seeliger, J. (2006). *So you want to get Lean Kaizen or Kaikaku?*. Aviation, Aerospace and Defense: Oliver Wymar.

Shingo, S. (1986). *Zero Quality Control: Source Inspection and the Poka-Yoke System*. Stanford: Productivity Press.

Venkatesh, J. (2005). An Introduction to Total Productive Maintenance (TPM). *Plant-Maintenance, 18*, 18–25.

Vinodh Dino, S., & Joy, D. (2012). Structural equation modelling of lean manufacturing prectices. *International Journals of Production Research, 50*(6), 1598–1607.

Zambrano, S., & Leal, S. (2005). *Proceso de Implementación de las nuevas Tendencias del Mantenimiento en Procesos Productivos*. Venezuela: UNET.

Zidel, T. (2006). A Lean Toolbox—Using Lean Principles and Techniques in Healthcare. *JHQ WEB EXCLUSIVE, 28*(1), W1-7–W1-15.

Part II
Main Lean Manufacturing Tools and Techniques

Chapter 2
Troubleshooting a Lean Environment

Moisés Tapia-Esquivias, Manuel Darío Hernández-Ripalda, José Antonio Vázquez-López and Alicia Luna-González

Abstract Due to the necessary of having a systematic and disciplined approach to attack the root and not just to manage symptoms; methodologies have been developed from the vast literature of tools for the solution of problems. This chapter will focus on assessing the troubleshooting methodology on a Lean environment, especially with the use of A3 Problem Report. The Lean methodologies approach shall be submitted in the first section of this chapter, which allows characterizing each of them, in the second section it takes the self-assessment proposed by the ISO 9004 standard and a case is presented in the third section.

Keywords Lean manufacturing · Format A3 · Problems solving

2.1 The Troubleshooting Methodologies Lean Environment

The methodology for troubleshooting Lean environment is presented as a series of activities including a toolset alternatively described in terms of an essential logic of each of its steps.

A troubleshooting methodology is an improvement strategy (De Mast et al. 2000) it helps to find both the causes and the atypical situation, as the variation of the components within the process performance.

M. Tapia-Esquivias (✉) · M. D. Hernández-Ripalda ·
J. A. Vázquez-López · A. Luna-González
Department of Industrial Engineering, Instituto Tecnológico de Celaya,
Av. Tecnológico y A. García Cubas s/n., Celaya, GTO, México
e-mail: moises.tapia@itcelaya.edu.mx

J. L. García-Alcaraz et al. (eds.), *Lean Manufacturing in the Developing World*,
DOI: 10.1007/978-3-319-04951-9_2,

Under the premise that the methodologies are improvement strategies that are required to be placed in a context which is characterized in three parts: structure, deployment and method (Pozos et al. 2012), same as described below:

- Structure: The people involved in the process of troubleshooting methodology.
- Deployment: Objective of the methodology for troubleshooting.
- Methodology: As the name implies, is the method, steps or tools that takes place in the methodology for troubleshooting.

2.2 Lean Environment and Evolution

Lean Environment is defined as troubleshooting to the developed environment around the methods generated from best practices to improve processes of Japanese auto companies.

These practices range from the formation of the mentality in the general management of the organization, to the development of routine practices at all levels.

Lean environment now includes several definitions that reflect the historic step of the concept, ranging from; Lean is what makes Toyota, passing tools to attack the waste in processes, to a philosophy and administrative culture.

Unlike systems linked to quality improvement, it starts in the attack on the variability or other cost reduction. As discussed in the Shingo Prize, part of identifying the value demanded and then flowed into the required amount. However, it shares with quality initiatives based on the concept PDCA (Plan, Do, Check, and Act) and have influenced each other. For an academic discussion on the definitions of Lean, the reader may refer to Bhasin and Burcher (2006), Pettersen (2009).

We recognize that each organization must actively choose to adapt the elements that best suit their needs. Through this process of recursive adaptation of the elements that help you improve and learn how, the organization increases its ability to have a predictable and successful implementation.

The original purpose of the studies of Japanese auto companies was to increase the performance of equivalent organizations in the United States of America, especially getting shorter times, lower costs and better-quality to produce and develop their products Cusumano (1985), Krafcik (1988), Womack et al. (1990).

The first studies on the best practices of Japanese auto production Cusumano (1988) reported as findings include:

- Production on small lots and Just in Time
- Inventories minimum process
- Processes pull production processes
- Production level
- Rapid changes in models
- Streamlining of machinery and line

- Standardization of work
- Poka Yoke Devices
- Workers with multiple skills
- High levels of subcontracting
- Mechanisms for continued incremental improvement
- Selective use of automation
- Replacement and rapid expansion of new models
- shorter Phases in product development
- Supply Engineering high level
- Project Managers with full authority and expertise.

Thus Lean started as the study of manufacturing processes. It was later on when the development of Japanese automotive companies began to be studied (Kamath and Liker 1994), Sobek et al. (1999), leading to better results in the launch of new products.

American producers also observed in Japanese producers, obsessive process for improvement, "kaizen" which involved direct work with continuous improvement tasks. Besides speed of material flow rate was observed to find and solve problems.

It began to be discovered so that not only the production process should be "Lean", how to run the organization was also important, that is, the internal environment can be formed for positive effects.

Competitors of the Japanese in North America began to improve their quality and manufacturing efficiency, but Japanese firms had increased further as they created new technology and new brands. Suddenly they realized that only imitating the leader's job at one point in time and space would have better results.

The first studies about the Japanese automotive production methods had only studied the results of a self-improvement mechanism, therefore a study of how they think when designing or improving a process was began, which involved not only production processes but also training people processes, product design, strengthening administrative capacity and maintenance.

2.3 The DNA of Toyota's Production System

In 1999, the work of Spear and Bowen on Toyota System DNA appears subsequently Spear's work applies in other non-automotive especially in hospitals, creating a new application and development.

Spear and Bowen find that organizations are places not only to produce but also are places to learn how to produce and keep learning. In the activities of organizations seems to be the possibility of losing what they have learned to focus on the tools and forget the development of a culture.

The culture that Spear and Bowen (1999) propose in their industrial application, is a culture that they identify as a scientific method, since when it is going to specify something, is done through a rigorous process based on a number of

Table 2.1 The four rules of DNA

Rules	Sign of Problem
Rule 1, How people work: all work must be completely specified in content, sequence, timing and outcome	The activity is not made as specified The result is defective
Rule 2, How the connections between the People are: all client-provider connection must be direct, in a way yes-or-no unambiguous for sending requests and receiving responses	The responses are not keeping pace with the requirements The provider is idle, waiting for requests
Rule 3, How to build the production line: the path for every product and service must be simple and direct	A person or machine is not really needed An unspecified provider catered for intermediate goods or services
Rule 4, Improving: any improvement must be made in accordance with the scientific method, under the guidance of a teacher, at the lowest level possible in the organization	The actual result differs from expected result

Source Based on Spear and Bowen (1999)

assumptions that have to be tested, and to make any changes using a rigorous process of solving problems that require detailed assessment of the current state of the facts and a plan to improve it, and for this purpose an experimental test of the proposed change.

This culture has a method that is based on four rules; all rules require that activities, connections and flow paths have built tests to signal problems automatically. The continuing response to the problems makes a seemingly rigid system to remain flexible and adaptable to changing circumstances. Rules and issue signals are as shown in Table 2.1.

Spear and Bowen (1999) report that when the first rule is taught by a supervisor, the person is asked a series of questions that help him/her to understand and discover.

How do you do this work, how do you know that you are doing it correctly? How do you know that the outcome is free of defects, what do you do if you have a problem? This recalls Juran's principles of self-control, as seen in Defeo and Juran (2010), and the Shewhart-Deming cycle of Plan, Do, Check and Act.

We also found that there is a teaching-learning path that will cascade from the highest administrative levels to workers. The needs of people in direct contact with the work determine assistance, problem solving and higher activities. Very different to who works for whom in the traditional command and control, where orders diffuse downward and upward reporting job status.

In brief, the guide is to specify all design, test it with every use and improve as close in time, place and person to the occurrence of any problem. If the company does consistently is showing through action, that when people come to work, they are entitled to expect to achieve something of value to another person. If they cannot, they are entitled to know immediately that they did not, and have the right to expect to be involved in creating a solution that makes the achievement more likely next time. If a person cannot subscribe to these ideas, either in words or actions, it is unlikely that they can lead effectively in this system.

These rules were translated to implement them to an environment of health care, Spear (2005) presents the "four basic organizational capabilities in operations excellence," same as illustrated below:

1. People at all levels of the organization are trained to become experimentalists.
2. Solutions are disseminated adaptively through collaborative experimentation.
3. Problems are addressed immediately through quick experimentation.
4. The work is designed as a series of ongoing experiments that immediately reveal problems.

2.4 The First Kaizen Event

The application of Lean tools and especially A3 Format used to document a summary of the experiences to confront the problems of the organization, in resolving problems as a result of a Kaizen event, in the Kaizen event teams of people directly involved in a workplace to bring about change that generates waste disposal and performance enhancements.

Norman Bodek (2004) narrates the first Kaizen event held in North America in the Jake Brake Danaher plant in Bloomfield, Connecticut. This happened in 1988. The event was conducted by Iwata and Nakao from Shingijitsu consulting firm. It was called "five days and one night" and describes the program implemented as follows:

1. Monday: Principles of Toyota Production System.
2. Tuesday: Five teams of 10 people with the goal of forming manufacturing cells, studying processes observe and estimate cycle time, takt time, studying how to fill out standard worksheets, look for wastes and how to eliminate them.
3. Wednesday: Finish the value stream map showing value added, cycle time, takt time, inventory, and opportunities for improvement by removing waste in operations, in addition to the standard worksheets.
4. On Wednesday night, machines are moved to form 5 cells. A list of pending projects is produced that generate post-event projects to enhance the implementation.
5. On Thursday operators are instructed (those who had not participated) in the new process, relying on standard worksheets.
6. On Friday morning the five groups present their case reports.

Bodek recounts the psychological impact on workers and suggests that workers should have been part of the teams that made the change, also narrated that the job change, elimination of inventories and the new responsibilities of people working on the floor taking 3 months to eliminate the problems that at the time became visible. Spear (2009) notes that Kaizen events typically do not ensure increased capacity to design operate and improve daily working people in the process; additional Kaizen event within the Lean environment, there are also systems suggestions for improvement, self-study groups to increase learning ability and Kaizen projects among others.

2.5 Shigeo Shingo Prize

In 1988 the Utah State University created the awarded "Shigeo Shingo" to honor the engineer who developed at Toyota, along with Taiichi Ohno, the changes and the necessary tools for a production system that was not dependent on the mass production.

The award aims to encourage the creation of enhancement in organizations systems and create a canon against which you can compare how close or far is an organization in their efforts to improve, especially at Lean environment.

The award is a qualification that gives 1,000 points distributed in four dimensions, specifically, the second dimension called "Continuous Improvement Process" account for 350 points and must describe the philosophy of the organization to the principles and concepts of Lean, reviews several principles of the prize compliance, among which may be mentioned: seek perfection, quality assurance at the source, necessary level flow value, take scientific thinking and focus on the process.

In the dimension of continuous improvement were 18 examples of systems, one of which is the troubleshooting system, which in turn points 3 options PDCA, DMAIC and "A3 Thinking" (The Shingo Prize 2013).

A3 thinking refers to the use of A3 Format to achieve a disciplined way to report on the problems and in turn encourages a disciplined way to solve guided primarily in the application of PDCA, documenting the findings and enabling learning and improvement process learning thereof by applying it recursively.

The Lean environment has evolved from a competitive comparison of systems of automotive production to an administrative system that involves the whole structure of the organization in planning the work, checking if it is good, acting immediately if not well, learning and making explicit what is found. The Lean method goes into an experimental approach that can work as an experiment to learn and perform a show based on the PDCA cycle ensuring organizational learning, which allows you to convert the extraordinary into standard.

Turn into explicit what was found to confront and solve problems involving a system of documentation and at the base of this task is A3 format that enables to leave explicitly what was learned.

Table 2.2 A3 format characteristics as methodology

Methodology	Structure	Display	Method
A3 for problem solving	Multifunction team in the workplace	Eliminating special causes, maintaining common causes conformance to the customer and getting knowledge	Plan, Do, Check and Act

2.6 Format A3

Here is characterized the tool called A3 at Lean environment as the tool used to manage and document the solution of problems, as shown in Table 2.2.

The standard ISO 216 defines a size of paper called "A3," which corresponds to a rectangle of 297 × 420 mm (11.7 × 16.5 inches) and the area is close to an eighth of a square meter, which is similar in size to the American standard called "tabloid" of 279 × 432 mm (11 × 17 inches), which in turn corresponds to twice the size chart (letter) American (215.9 × 279.4 mm or 8.5 × 11 inches).

In the transformation initiatives "Lean" in organizations, A3 refers to information concerning a difficulty encountered in the course of business in a single sheet of paper. A3 therefore relates to a summary of the experiences to confront the problems of the organization.

The use of A3 emerged from Toyota to perform two administrative processes: Hoshin Kanri (Strategy Management) and the solution of problems.

A3 is used as a tool to solve problems, make improvements and get things done. A3 ensures rapid reporting thought necessary for a team facing a problem; encouraging to take a learning management process to solve problems and make decisions, and encourages the formation of a team of people learning how to do their job, if well and if it is not correct it by continuously improving operations and results.

There is no unique A3 Format, as each organization adopts its own style, however, the use of the experiences in Japanese car company Toyota, and formats found are generally derived from the definitions of Toyota. We present versions of A3 format elements in Table 2.3.

A3 format elements must have a logical and natural sequence, which allows you to bind the problem, its root causes, the goal, the actions to achieve the goal and the means to judge the success in a clear and easy way to be understood. The format should allow participants in the care of an issue or problem follow the thought through the PDCA cycle (Plan, Do, Check, Act).

Incorporating A3 in the activities of the teams, the organizations learn to face problems, and begin to recognize problems as opportunities to learn and improve. Leaders in Lean initiatives direct preferably working groups based on knowledge, based on the facts, strong-willed yet flexible. Media are administered, the same process that actually leads to the results. An A3 process directly identifies the owner's responsibility—Direct author of the A3 process. This person may not have direct authority over every aspect of the proposal, but the owner is clearly

Table 2.3 A3 format elements versions

A3 components	Shook	Jackson
Theme	Title: name of problem, difficulty or theme	Statement of the problem or issue at the beginning of the format
Problem statement	Background: the business context and the extent of it Conditions in this: what is known today about the issue or problem?	Statement of the problem—the reason for the project definition (including the present initial state)
Goals and objectives	Goals and objectives: what outcomes re desired?	Goal statement—(future state) defining project scope
Methodological environment	PDCA	PDCA, 8-Ds, CEDAC, DMAIC
Systematic analysis	Analysis: method of analysis and the identified causes that produce the difference between the present situation and the desired outcome	Systematic analysis (5 whys, cause and effect diagram, design of experiments, etc.)
Proposed solution	Countermeasures proposed: proposed corrective actions or countermeasures to address the problem, reducing the difference or achieving a goal	Proposed solution (including coordination of functional resources)
Timeline for implementation	Plan: prescribe an action plan indicating who will do what at what time to reach the goal Tracking: a process is created for reviewing and monitoring anticipated learning difficulties	Implementation time (with actions, responsible, and dates for the deliverables of the actions)
Graphical illustrations		Graphical illustrations to convey information at a glance
Date and reporting unit or owner	Date: the date of the most recent done Owner: who is the "owner" of the process where the problem or issue is identified	Date and reporting unit or owner the end of the format individual or team responsible for this matter A3

Source Shook (2008), Jackson (2006)

identified as the person who has taken or accepted the responsibility to ensure that decisions are taken and implemented.

The use of Toyota A3 format emerged at Toyota to perform two administrative processes: Hoshin Kanri (Strategy Management) and the solution of problems.

On a macro level of the organization, Hoshin Kanri aligns the goals and objectives of the organization with the operations and activities, the solution of formalized problems creates micro-level organizational learning. A3 process combines and incorporates both. A3 is a means to propose projects, take initiatives, show responsibility, sell ideas, gain agreement and learn. Managers can use A3 to guide and teach, to clearly assign responsibility, empowerment and accountability, to get good plans of their subordinates and encourage learning.

Jackson (2006) reports, for example, six different types of A3 formats, one for Trouble Reporting, five related to Hoshin Kanri process: 1. Intelligence Report, 2. Matrix X, 3. Team Charter, 4. Status Report, and 5. Summary Status Reports.

Matrix X is a tool that can generate an action plan in about a year to develop new capabilities and maintain paths aligned organizational operations within the broader strategy. Link through relationship matrices attempted strategy, tactical actions, outcomes and operational teams.

A3 form for Reporting Problems is associated with problem solving immediate action to address the special causes that arise during the daily standard work or to take advantage of identified opportunities for improvement. It is usually associated with a Kaizen event conducted by a team to address a problem or seize an opportunity for improvement in the workplace.

A3 form is a structured process to create problem solvers at the same time it is a troubleshooting tool; A3 format helps search and spread structured knowledge, allowing participation in decisions in an environment of critical discussion, forces individuals to observe reality, present data, propose countermeasures designed to achieve a stated goal and follows a process of checking and adjusting for actual results.

An organization using A3 thinking, achieves that: decisions taken to achieve goals and get things done, guiding individuals and teams along common goals and learn to get effectiveness, efficiency and improvement.

2.7 Self-Evaluation in ISO 9004:2009

Since February 1947 there is an international organization called ISO, International Organization for Standardization (2013), whose function is to develop, publish and promote the use of applicable standards or international standards, to assist the development between industry and trade. Within these rules, the ISO organization has launched a system called ISO 9000 standards of quality management, the rules guiding the organizations and businesses in meeting the requirements of its customers and consistent improvement in quality. The rule is explained based on eight principles of quality management, defined by a technical committee.

The system of quality management ISO has many rules, the main one is the ISO 9001:2008 standard that establishes the requirements for a quality management system, is the only standard that can be certified from all the other dedicated to the administration of quality. Another is ISO 9004:2009 guiding continuous achieving better efficiency in the organization or company, this rule does not certify or have regulatory or contractual use.

The international standard ISO 9004:2009 quality management complements the ISO 9001:2008 international standard providing guidance on continual improvement of quality management.

An important tool in ISO 9004:2009 is the self-assessment that allows management to know the points to act on their own system. Although ISO 9004:2009 and ISO 9001:2008 complement, ISO 9004:2009 can be used independently.

The self-assessment is a comprehensive and systematic review of the activities and results of an organization with respect to a selected maturity level. In the context of the methods of solving problems provides an overview of the performance, which identifies priorities in each of the methodologies, which allow a better understanding of each is steps and/or stages.

The way in which ISO 9004 proposes self-evaluation is based on five levels of maturity, maturity levels describe these six elements that indicate the attachment to the effective management of quality in the organization or company, the maturity levels ranging from basic (Level 1) to the development of best practice (level 5).

These maturity levels are grouped into six elements which are:

- Managing for the sustained success of an organization
- Strategy and Policy
- Resource Management
- Process Management
- Monitoring, measurement, analysis and review
- Improvement, innovation and learning.

In order to carry out self-assessment, it is necessary to define what features are most relevant in each of the maturity levels; the maturity levels suggested by ISO 9004 are shown below as an important tool to review the level of maturity of the organization. An organization may be at different levels of maturity between different elements.

Management maturity levels for the sustained success of an organization:

1. The focus is on products, shareholders and some customers with ad hoc reactions to changes, problems and opportunities.
2. The focus is on customers and statutory/regulatory requirements with some structured reaction to problems and opportunities.
3. Processes are defined and implemented for reacting to problems and opportunities.

4. Continual improvement is a as part of the organization's focus.
5. The focus is on the balancing the needs of emerging interested parties.

Maturity levels for policy and strategy:

1. Decisions are based on informal inputs from the market and other sources.
2. Decisions are based on customer needs and expectations.
3. Decisions are based on the strategy and linked to the needs and expectations of interested parties.
4. Decisions are based on the deployment of the strategy, into operational needs and processes.
5. Decisions are based on the need for flexibility, agility, and sustained performance.

Maturity levels for process management:

1. There is non- systematic approach to the organization of activities, with only some basic working procedures or instructions in place.
2. Activities are organized by function, with a basic quality management system in place.
3. Activities are organized in a process-bases quality management system that is effective and efficient, and which enables flexibility.
4. There is a quality management system that is effective and efficient, with good interactions between processes and which supports agility and improvement. The process addresses the needs of identified interested parties.
5. There is a quality management system that supports innovation and benchmarking, and which addresses the needs and expectations of emerging, as well as identified, interested parties.

Maturity levels for monitoring, measurement and analysis:

1. Results are achieved in a random manner. The commercial and financial indicators of productivity are implemented.
2. Corrective and preventive actions are performed in a systematic way.
3. It keeps track of the satisfaction of people of the organization and its stakeholders.
4. The key performance indicators aligned with the strategy of the organization are used to keep track.
5. The achieved results are above the sector average for the organization, and are maintained long-term.

Maturity levels for improvement, innovation and apprenticeship

1. Improvement priorities are based on errors, complaints or financial criteria.
2. Improvement priorities are based on customer satisfaction data or corrective and preventive actions.

3. Improvement priorities are based on the needs and expectations of some interested parties as well as those of suppliers and the organization's people.
4. Improvement priorities are based on trends and inputs from other interested parties.
5. Improvement priorities are based on inputs from emerging interested parties.

2.8 Key Elements and Estimated Maturity Level

Sustained success management: 2, the focus is on the costumers and the statutory/ regulatory requirements with some structured reaction to problems and opportunities.

Strategy and Policy 2, decisions are based on the needs and expectations of customers.

Resource Management: 3, resources are managed efficiently.

Process Management: 3, activities are organized in a process-based quality management system that is effective and efficient, and which enables flexibility.

Monitoring, measurement and analysis: 3, it keeps track of the satisfaction of people of the organization and its stakeholders.

Improvement, innovation and learning: 2, improvement priorities are based on customer satisfaction data or corrective and preventive actions.

The maturity level stated in each of the elements, notes a type of practice in organizations and companies that meet the immediate requirements of everyday life, without greater involvement of senior management, or developing recursive learning mechanisms.

A3 thinking has much potential to help in a Lean environment to these actions and learning strategies, but as seen in the Shingo Prize, A3 Format is considered just one option among several possibilities, within a subsection. On the other hand, Liker and Rother (2013) report on a survey conducted on November 2007 by Industry Week finds that the two percent of the companies have a Lean program that has achieved the anticipated results, Liker and Rother also reported that a review by the committee that awards the prize Shigeo Shingo at the same time, have found that many of the winners had not maintained or increased their level of performance after winning the award, a large percentage of those evaluated in the award were found to be experts in implementing Lean tools but did not have them deeply embedded into their culture. The presented levels reflect regulatory compliance but not growth in learning and strategy.

A methodology may be at different levels of maturity for the different elements. Recalling that in the implementation of Lean initiatives A3 format role runs from a simple format for recording to "A3 Thinking" recursive learning, improvement and action.

2.9 Enforcement Case A3: Case of Study

2.9.1 Problem Description and Objective

The following case study is an application of the solution of a problem by following the steps and documentation required by A3 format. The case is presented in a car wash business and consists in an increase in service demand causing the installed capacity to be exceeded since 2009. From vehicles entering the car wash, 26 % are not treated and as a result, complaints increased having as its main complaint the poor quality or lack of cleaning or service, as observed in Fig. 2.1:

The problem is assigned to a project team consisting of 4 employees of the business. The team is responsible for resolving the problem by following the steps in A3 format.

2.9.2 Methodology

In the understanding of the situation the following is discovered.

The distribution of business is designed on 2 levels, on the ground floor the mechanical service is performed and in the first level the washing is done, with a dead time of 10 min to raise and lower the vehicle. Furthermore, KPI's has not been defined (key process inputs) as the washing area and there is no standard working method, the washing time is 46 min with 4 people. Benchmarking was conducted in a different car wash, finding an average time of 27 min wash. Currently 74 % of cars entering the business are washed.

When decomposing the problem to find the root cause it is determined that 26 % of vehicles entering the service facility are not washed because there is no suitable distribution facilities. From the previous discovery, the goal is assigned to be achieved by the end of March 2010, 100 % of clients attending the accommodation cleaning service should be offered in a timely manner see the plan activities in Table 2.4.

Do

A plan is made to ensure an adequate distribution for washing cars, which consists of the following:

2.9.3 Results

These actions resulted in a need to redesign the distribution to implement the activities required for washing (see Fig. 2.2). It was also implemented a standard process considering ergonomics work and establishing a control board to level workloads of operators, thus washing the cars efficiently and effectively according to the needs and expectations of customers.

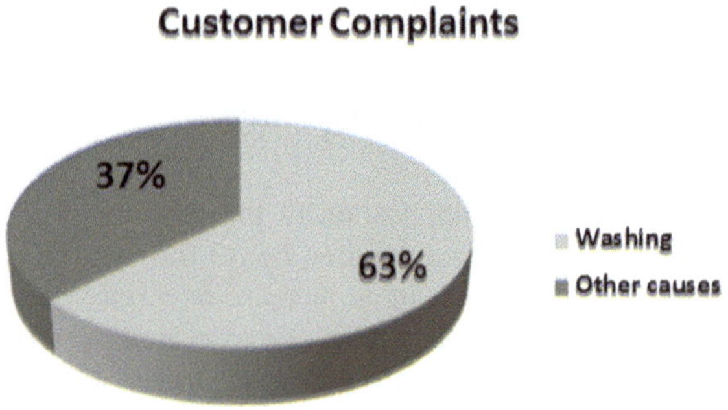

Fig. 2.1 Customer complaints about the poor quality or lack of car wash

Table 2.4 Plan activities

	Activities		Dates	
	What	Who	Onset	Term
1	Assessment of the architectural project (costs, space, materials)	General Manager	01/10/2009	15/10/2009
2	Benchmarking with other washing services	Service Manager	19/10/2009	24/10/2009
3	Design of the layout according to the space	Service Manager	26/10/2009	28/11/2009
4	Taking motion and time	Practitioners	26/10/2009	18/12/2009
5	Execution of civil works	General Manager-Architect	31/11/2009	01/02/2010
6	Stations equipment	General Manager	18/01/2010	30/01/2010
7	Standardization of the process with board	Kaizen Promoter	01/02/2010	27/02/2010

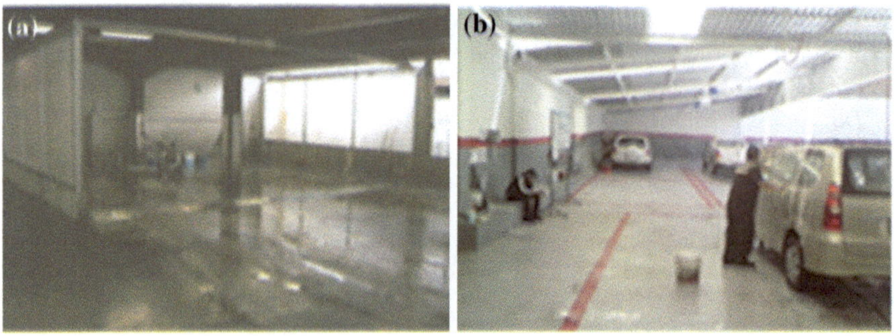

Fig. 2.2 Layout before and after actions. a Prior distribution, b Improved distribution

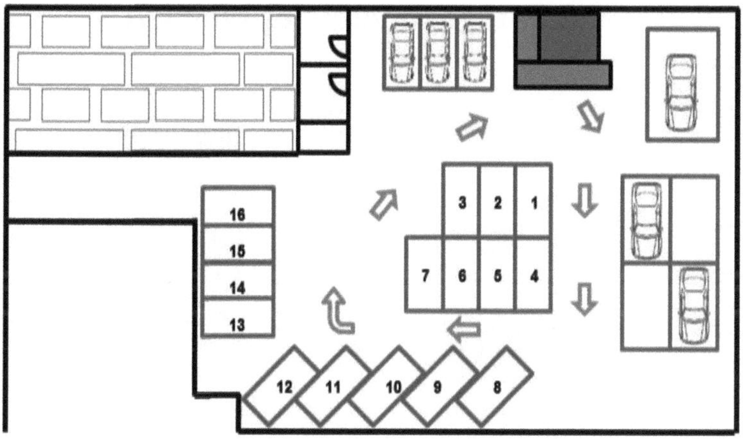

Fig. 2.3 Layout redesigned for car wash

The resulting distribution is immediately shown in schematic form (see Fig. 2.3), where the cars are placed in specific areas waiting to be served once the cars advance they are placed in different stations to be processed then they are put in a waiting place to be delivered to their owners.

2.9.4 Verification of the Results

With operations previously carried out, after 4 months, the results achieved by the implemented actions are verified quantitatively. For the percentage of vehicles washed after the implemented actions the 100 % is achieved as shown in the Fig. 2.4 below, thus achieving the goal.

The car wash time improves from 46 to 30 min and washes complaints decrease from 60 to 34.8 %. To keep improving, we outline a series of recommendations that can be made in the future such as acquire a foaming machine and implement flexible workforce.

As seen in the previous case, we can conclude that everything can be improved; hence the importance of adopting continuous improvement as a life philosophy and document improvements in a logical format and orderly as is the case of A3 format. The following shows the documentation of the previous case in A3 format (see Fig. 2.5).

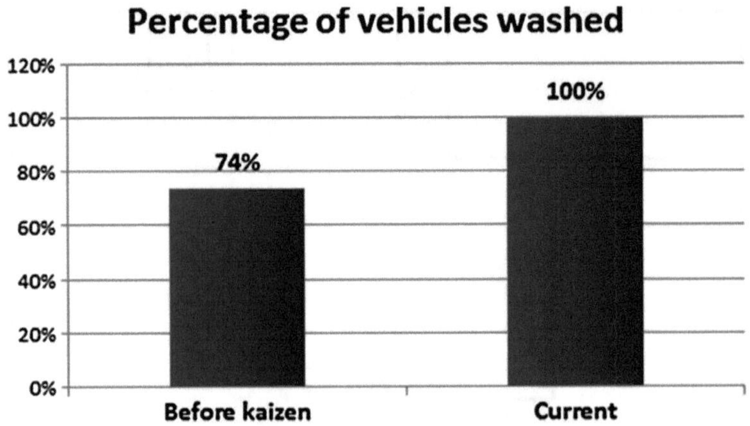

Fig. 2.4 Percentage of vehicles washed after the implemented actions

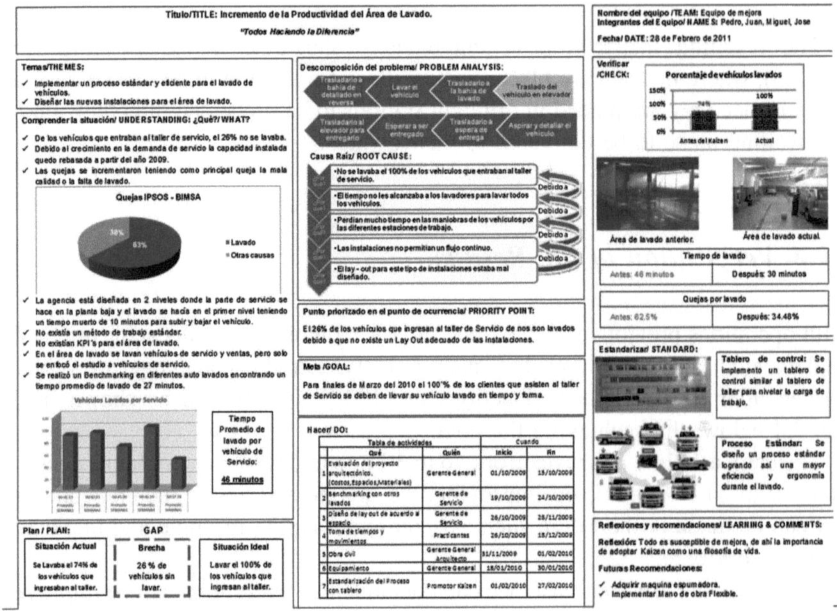

Fig. 2.5 A3 format for the case described above

References

Bhasin, S., & Burcher, P. (2006). Lean viewed as a philosophy. *Journal of Manufacturing Technology Management, 17*(1), 56–72.

Bodek, N. (2004). *Kaikaku: The Power and Magic of Lean.* Vancouver, WA: PCS Press.

Cusumano, M. A. (1985). *The Japanese automobile industry: Technology and management at Nissan and Toyota.* Cambridge, MA: Harvard University Press.

Cusumano, M. A. (1988). Manufacturing innovation: Lessons from the Japanese autoindustry. *Sloan Management Review, 30*(1), 29–39.
De Mast, J., Schippers, W., Does, R., & Van de Heuvel, E. (2000). Steps and strategies in process improvement. *Quality and Reliability Engineering International, 16*, 301–3011.
Defeo, J., & Juran, J. M. (2010). *Juran's quality handbook: The complete guide to performance excellence* (6th ed.). New York: McGraw-Hill.
International Organization for Standardization. (2013). Home. Retrieved September 20, 2013 from www.iso.org
Jackson, T. L. (2006). *Hoshin Kanri for the lean enterprise: Developing competitive capabilities and managing profit.* London, England: Productivity Press.
Kamath, R. R., & Liker, J. K. (1994). A second look at Japanese product development. *Harvard Business Review, 72*, 154–173.
Krafcik, J. F. (1988). Triumph of the lean production system. *Sloan Management Review, 30*(1), 41–52.
Liker, J., & Rother, M. (2013). *Why lean programs fail.* Retrieved September 20, 2013 from http://www.lean.org/admin/km/documents/A4FF50A9-028A-49FD-BB1F-CB93D52E1878-Liker-Rother%20Article%20v3_5_CM.pdf
Pettersen, J. (2009). Defining lean production: Some conceptual and practical issues, *The TQM Journal, 21*(2), 127–142.
Pozos, T., Tapia, M., Hernández, D., & Luna, A. (2012). ISO 9004-2009 y las Metodologías de Solución de Problemas. *Congreso internacional de investigacion, 4,* 2396–2400. Celaya, Guanajuato, Mexico.
Shook, J. (2008). *Managing to learn: Using the A3 management process to solve problems, gain agreement, mentor and lead.* Cambridge, MA: Lean Enterprise Institute.
Sobek, D. K., Ward, A. C., & Liker, J. K. (1999). Toyota's principles of set-based concurrent engineering. *Sloan Management Review, 40*(2), 67–82.
Spear, S. (2005). Fixing health care from the inside, today. *Harvard Business Review, 83*(9), 1–15.
Spear, S. J. (2009). *The high-velocity edge: How market leaders leverage operational excellence to beat the competition.* New York, NY: McGraw-Hill.
Spear, S. J., & Bowen, H. K. (1999). Decoding the DNA of the Toyota production system. *Harvard Business Review, 77*(5), 96–108.
The Shingo Prize. (2013). Application Guidelines. Retrieved September 20, 2013 from www.shingoprize.org
Womack, J. P., Jones, D. T., & Roos, D. (1990). *The machine that changed the world.* New York, NY: Rawson/MacMillan.

Chapter 3
Statistical Process Control

Manuel Iván Rodríguez-Borbón
and Manuel Arnoldo Rodríguez-Medina

Abstract In this chapter appears a description for the main concepts of statistical process controls, its graphs and common interpretation in manufacturing industries. Finally, appears a real application for statistical process control.

Keywords SPC · Manufacturing industry · Lean process · Quality control

3.1 Introduction

Statistical Process Control is a powerful objective tool helpful in decision-making and consequently the continuous improvement of products and processes.

It is mainly intended to analyze the statistical tools that detect variability, consistency, control and process improvement, specifically the control charts.

These are used to monitor, detect and prevent the behavior of the process through their vital steps. In designing and implementing control charts, the main objective is to understand the changes in the production processes, the nature of these changes, and time periods in which they occur. In general, we want to detect the change and identify potential root causes, later establishing possible corrective actions.

The construction of the control chart starts with the measurement process, taking into account that the quality of data measurement instrument depends, starting with the resolution, accuracy, and measure, including environmental conditions.

M. I. Rodríguez-Borbón (✉)
Department of Industrial and Manufacturing Engineering, Autonomous University of Ciudad Juarez, Av. Del Charro 450 N, Ciudad Juárez, Chihuahua, Mexico
e-mail: ivan.rodriguez@uacj.mx

M. A. Rodríguez-Medina
Graduate and Research Programs, Ciudad Juarez Institute of Technology, Av. Tecnologico 1340, Ciudad Juárez, Chihuahua, Mexico

J. L. García-Alcaraz et al. (eds.), *Lean Manufacturing in the Developing World*,
DOI: 10.1007/978-3-319-04951-9_3, © Springer International Publishing Switzerland 2014

3.2 Type of Control Charts

In order to construct control charts it is important to define the different types of data to be plotted, that is, whether data are continuous or discrete, since the control methods depend on this.

The graphs for continuous variables depend largely on the shape of monitoring the variance of the process i.e. by using ranges or sample variances.

There are some important points to consider before making control charts, such as:

- The variable type
- The sample size
- The sampling frequency
- The type of test to be performed
- Staff training

Another important aspect is the size of the production batches, which leads to a different type of charts: the chart for short production runs used when lots are small and different models are similar in the same production facility.

Within this type of graphics can be mentioned those considering deviations from the target value and constant variance, which could consider some unavoidable errors of the measurement system.

Wise and Fair (1998) mention that quality engineers are currently facing problems of controlling processes where control techniques are different from traditional control forms, facing situations that lead to:

- similar features with different dimensions
- small batch sizes
- high product mix with lower production volumes

When these conditions are specific, quality engineers faces a number of important decisions, such as selecting the sample sizes, the type of chart to use, verification of assumptions, such as normality of the data, independence, and adding the further homoscedasticity somehow measurement errors.

Thus, for the situations mentioned, Wise and Fair (1998) recommend the use of the following control charts:

- Control Charts with individual objective values and moving range
- Control Charts with objective values and sample sizes $n > 2$ (using ranges)
- Control Charts with values and sample sizes $n > 2$ using S
- (Standard deviations)

All these charts are recommended for production runs with small batches, i.e. very limited amount of data. The charts are coded according to the mean of the process and the desired value, thereby maintaining their original units. This allows the use of common control limits making it possible to use multiple control limits for part numbers or different characteristics.

The basis required for the construction of both short runs and traditional control charts are the same. Practically, the calculated parameters and the assumption are the same in both types of graphic control charts.

In industrial plants is very common to have a lot of processes that have short production runs. The great diversity of characteristics of components regularly causes assuming equal variances (homoscedasticity) is not met, as the assumption of normality of data. Another important factor in establishing controls for this type of process, is the measurement error, which even if it is small there, and necessarily, the plot must detect when this source of variation is present.

The basic reason for the implementation of statistical process control is that the industry is always trying to implement controls to increasingly close tolerance processes. The variables to control industrial processes may be continuous and discrete.

If the case is to control continuous variables, such as diameters, pressures, temperatures, could be considered unilateral and bilateral limits. In the case of bilateral limits the target will have the quality characteristic variable or as close to the nominal value, so that there is less likelihood that there are parts out of specification.

For discrete variables, consider the case of non- conforming parts in industrial processes. Here you can generate two different conditions: damaged parts and defective parts. In the case of defective parts may be provided out of specification parts, or parts that do not pass a test of functionality. For parts with defects include visual inspection results established for the verification of defects in parts.

Otherwise within the need for controls is destructive testing, for example, parts requiring minimal resistance to stress, accelerated life testing, testing of degradation conditions in which the drawing of inferences is somewhat different.

Generally, it is said that a process is in statistical control when they have managed to eliminate, or minimize special causes of variation. Detecting the occurrence of these special or assignable cause variations is accomplished by a control chart. These allow us, first detect and then predict behavior when the process is in control.

The form of process control necessarily depends on the type of variable to be controlled, so we can classify them into discrete variables and continuous variables. For discrete variables, subdivided these defective parts and parts with defects. The assumed distributions are the binomial and Poisson respectively.

For continuous variables, the probability density function is assumed normal distribution, which is based on monitoring and analyzing the quality of the process.

3.3 Traditional Control Charts

Montgomery (2003) and Brunk (1979) mentions that one of the critical decisions that must be made in the design of a control chart is the specified limits. If you move the control limits from the center line separating them, it decreases the risk of type I error, which indicates an out of control when there is no assignable cause. However, widening the control limits also increase the risk of committing a Type II error, i.e., the risk that a point appears within the control limits when the process is out of control.

Control limits are generally a Shewhart chart placed at a distance of plus or minus three standard deviations of the variable plotted on the center line of the graph, which are called 3-sigma control boundary.

The average run length (ARL) is a way of assessing decisions about sample size and the sampling frequency control chart. The ARL is, essentially, the average number of points that must be plotted before a point indicates an out of control condition.

Montgomery (2003) states that for any Shewhart control chart, the ARL can be calculated from the mean of a geometric random variable. Suppose that p is the probability that any point exceed the control limits. Then:

$$ARL = \frac{1}{p} \tag{3.1}$$

Thus, for a graph limits 3, p = 0.0027 is the probability that a point appears outside the boundaries when the process is in control, and

$$ARL = \frac{1}{p} = \frac{1}{0.0027} \cong 370$$

This is the average run length of the chart when the process is under control. That is, even when the process remains in control signal out of control condition is generated every 370 points, on average.

The Long Run Average is an indicator that allows us to compare the quality of control charts designed, because a good control chart will be the one that displays a false signal as far as possible and actual signal as close as possible.

3.4 The Recalculation of Control Limits

Generally, the responsible persons of the interpretation and maintenance of control charts have problems when recalculating the limits of the control charts. This should be considered when all the following conditions are true:

- The process has changed because it had an assignable cause variation.
- The cause of change of the process is known.

- It is expected a continue change in the process.
- There are insufficient data to recalculate the control limits (15–25 points plotted).

Just as it is important to know when to recalculate the control limits, it is equally important to know when it should be recalculated the control limits. For traditional control charts, control limits should not be recalculated.

When less than 15 data subsets. Because the control limits are predictions of future behavior of the process, some people might conclude that two subgroups are not enough data to make a good and reliable production.

With 40 subgroups of data. The first instance represents a situation in which there are insufficient data, but this is a situation of too much data. Using more than 40 subgroups to calculate control limits could be a very long period of time in a process. The process changes that may occur within forty subgroups in time may unnaturally inflating estimates of control limits.

With each new subset of data. The recalculation of control limits after each additional subgroup is obtained could mark assignable causes in a process.

The management of the control charts consists mainly in identifying assignable causes and reduction or elimination of them if they cause a detriment on the performance of the process, or into the process if they improve performance. This is the true sense of the implementation of controls in the process. Control charts are the tools to predict and control the performance of process characteristics.

3.5 The Baseline and Control Limits

The first thing you must do to build control charts is to choose what is going to work, identify the appropriate graph for the data type, define the size of the sample, and start collecting data. The control limits are calculated are collected when 15–25 samples. Control of the process really is to observe whether the graph shows a lack of control condition, analyze the special cause variation and eliminate a definitive way. The procedure must be iterative until all assignable causes have been eliminated and the control chart shows stability and this on the target value and with an acceptable variation. This will be the baseline of the process.

3.6 Maintenance of Control Charts

After establishing the baseline, the next step is maintenance. The established baseline will serve to extend the control limits for the control in the future. This means that you should take more data and verify lack of control conditions, changes in the mean and standard deviation. If there is evidence that the process has improved significantly, it is likely to have to get a recalculation of the baseline.

That is, we must establish new control limits for the process, but when it does not recalculate re-calculate control limits is not always well defined. The idea is advisable to recalculate limits only when the current limits do not represent roughly the current operational status of the process.

3.7 Control Charts for Short Runs (DNOM)

When carrying out short production runs, few parts are of different sizes so that the traditional control is not significant. Thus, instead of evaluating the variation in each measurement, it is used the variation of the parts with a target value. These types of charts are known as Deviation from Nominal Charts (dnom). The assumption here is that the variation ideally remains stable even when the samples vary in size.

Then, if we are doing a few parts at a time, then it makes sense to have a specific treatment to manage statistical process control for short runs. The graphs that we build are:

- Control Charts with Central Line based on the average (the average use of your data for the centerline)
- Control Charts with mobile data and Limits of Control (Graph most recent data as you define)
- Control Charts for Short Runs (compare your data against a target for each point). When making use only few parts of different sizes
- Horizontal Control Charts (for data organized in rows against columns)
- Control charts with fixed control limits (for data organized in rows against columns)
- Control charts with Central Line based on the median (graphics x only)

Hawkins and Olwell (1997) consider very important calibration of control charts and of course the parameter estimation, in which as much as possible to reduce the uncertainty in the estimate. To calibrate a chart, in general, take a sample of size m and calculate its mean and standard deviation which will be the estimators of μ and σ, which means that the true mean and the true standard deviation in the errors.

$$\mu - \bar{X} \text{ and } \frac{\sigma}{s} \tag{3.2}$$

Assume that a sample of size m = 50 is used for the calibration graph. Then the standard error is

$$se_{\bar{X}} = \frac{1}{\sqrt{m}} = \frac{1}{\sqrt{50}} = 0.141 \tag{3.3}$$

and the standard error of S is approximately

$$se_s = \frac{1}{\sqrt{2m}} = \frac{1}{\sqrt{2(50)}} = 0.1 \tag{3.4}$$

So, there are 2/3 chance that will be within one standard error of μ, and there is 1/6 of opportunity that is more than one standard error above μ, and 1/6 of a chance to be more than one error μ below standard. Then, an error of more than one standard deviation is quite likely. From this it concludes that what is more important is the sample size, since the type II error would always be present, i.e. no change in the parameters of the process when in fact if any.

3.7.1 Control Charts for Similar Features

Wise and Fair (1998) describe single charts and R to control the variation.

Chart IX objective is used to monitor and detect changes in measurements multiple product or process characteristics on the same graph. It is also possible to plot process parameters which have changes in their target values, separating the process variation due to the specific product. Tables 3.1 and 3.2 illustrate the calculations for the points to graph and control limits of the charts IX and MR.

3.7.2 Chart of Moving Range of Short Runs

The short-run chart is used to monitor and detect changes in the averages between multiple features of any kind. These characteristics may have different ratings, different units of measurement and different standard deviations, only that there must be a good relationship between features so they can be plotted together.

The graph R is using short runs to monitor and detect changes in the standard deviation between multiple characteristics Table 3.3 llustrates the formulations to calculate the points to graph for graphics and short runs ranges. Table 3.4 shows the control limits for both charts.

3.8 Short Run Control Charts for Variability

Short run charts are designed to monitor characteristics with different sizes, different units of measurement and standard deviations, all on the same control chart. Like the graphics dnom, short runs require mathematically encoded data. The short-run graphics require encoded values of averages and ranges.

Table 3.1 Equations to graph points for short run charts

Chart	Point to graph	Equation
Short run individual values	IX coded	$\frac{IX - IX \text{ objective}}{MR \text{ objective}}$
MR Short runs	MR objective	Absolute difference between two consecutive values IX encoded in the same part or location

Table 3.2 Control limits for short run charts

Chart	Upper control limit	Lower control limit
Short run individual values	$+A_2$	$-A_2$
MR Short runs	D_4	D_3

Table 3.3 Equations for $\bar{X}$ and R short run charts

Table	Point to Graph	Equation
Short run $\bar{X}$	encoded $\bar{X}$	$\frac{\bar{X} - \bar{X} \text{ objective}}{\bar{R} \text{ objectve}}$
Short run R	Encoded R	$\frac{R}{\bar{R} \text{ objective}}$

Table 3.4 Limits for $\bar{X}$ and R short run charts

Table	Upper control limit	Lower control limit
Short run$\bar{X}$	$+A_2$	$-A_2$
Short run R	D_4	D_3

3.8.1 Points to Graph for the R Chart

The plot points are based on traditional point averages, ranges and sample standard deviations. In order to monitor different characteristics on the same graph, the points must be encoded, which allows different units and different product characteristics can be plotted on the same graph. The chart control limits ranges are given by:

$$LCS_R = D_4\bar{R} \text{ and } LCI_R = D_3\bar{R} \tag{3.5}$$

A point is plotted on a graph R control when it appears between the control limits given

$$\mathrm{LCS_R} < \mathrm{R} < \mathrm{LCI_R} \tag{3.6}$$

or

$$D_3\bar{R} < R < D4\bar{R} \tag{3.7}$$

where R is the current range value in the subgroup.

To make the plotted points unitless reasons, the R should be disposed of inequality. To eliminate inequality unchanged, just divided between the three terms,

$$\frac{D_4\bar{R}}{\bar{R}} > \frac{R}{\bar{R}} > \frac{D_3\bar{R}}{\bar{R}} \tag{3.8}$$

If we cancel $\bar{R}$ we get

$$D_4 > \frac{R}{\bar{R}} > D_3 \tag{3.9}$$

$\bar{R}$ for a given process the average range is expected objective can call. Therefore, the plotted point is

$$\frac{R}{\bar{R}\text{ objective}} \tag{3.10}$$

The objective $\bar{R}$ is the most important part of the run short chart, which represents an expected range or estimated range.

3.8.2 Graphical Points Under S Chart for Short Runs

The control limits of traditional s charts are given by

$$LCS = B_4\bar{s} \quad LCI = B_3\bar{s} \tag{3.11}$$

Then we can write as:

$$\text{LCS} > \text{s} > \text{LCI} \tag{3.12}$$

or

$$B_4\bar{s} > s > B_3\bar{s} \tag{3.13}$$

Trying to eliminate $\bar{s}$ from last equation we have

$$\frac{B_4\bar{s}}{\bar{s}} > \frac{s}{\bar{s}} > \frac{B_3\bar{s}}{\bar{s}} \text{ and } B_4 > \frac{s}{\bar{s}} > B_3 \tag{3.14}$$

We conclude that the graphical points are $\frac{s}{\bar{s}}$. Then the point in t chart will be

$$\frac{s}{\bar{s}\text{ objective}} \tag{3.15}$$

3.9 Control Charts for Short Runs with Non-Constant Process and Measurement Error

Farnun (1992) presents a work for the construction of control charts deviations from the nominal value when the equal variance assumption is not met. The disadvantage of handling graphics of control when this assumption is not fulfilled is estimated standard deviations for each of the subgroups. Then, the alternative is to generate a control procedure that includes reasons subgroup mean nominal values.

This author (Farnum) recommends that for production should be used:

(1) Use birth deviations control increased power detection small changes early process.
(2) Use charts based on individual data to monitor controllable variables instead of controlling variables of the product, or
(3) Deviations from nominal values plot (dnom) where several different parts are run through the same process.

The overall of the foundation of dnom chart is that when small batches of different types of parts are run through an established process, the measured differences of these parts of their face value (or target values) can be plotted on the same chart control. Then if Xij, j = 1,2,…,n represents the j-th dimension of a process as a target value Ti, then the average difference is plotted for each data set. Each group consists of parts of the same type and each type of part occur any number of times in the sequence of subgroups. Thus, the combinations of deviations over all types of parts, the chart covers dnom data constraints that may typically present the use of a control chart for a single part. A dnom chart is restricted to various types of part running in the same process.

The reason for the restriction to a single process graph is that plotted deviations should be measured in the same units, but the most important reason is that the variances of the terms remain a better chance to have approximately the same magnitude if only one process is involved. That is, the homoscedasticity assumption is critical to the usual procedure dnom. This not only ensures constant control limits, but in some cases equal variances is the only justification you have to accumulate data from different parts in order to build the chart. If variances are not constant across the different types of parts, it is possible to use a standardized dnom graphic, which plot the statistical σ standard deviation of the measured values of the i-th subgroup. It should be noted that estimates of the σ i's are commonly estimated graphics information from individual parts control.

Farnum (1992) mentions that there are many processes for which non-constant variances are probably the rule rather than the exception. It is possible for both, process variation and the variation of measurement error, depend on the particular nominal dimension Ti .

It is also possible that the measurement error itself can cause changes in variance. In the analysis of the measurement system, measurement errors are specified in either of two ways: in absolute terms (as the maximum possible error on the full scale of the instrument) or in relative terms (as a percentage of error true reading. A common case is the analysis of bias for a certain part, which could be significant, and linearity analysis could show that there are different types of parties with significant biased values which will influence the calculation of the differences.

The main problem of handling the dnom chart is mainly to have a suitable model for the presence or absence of variability (i.e. homoscedasticity or heteroscedasticity). The establishment of appropriate model will lead to correct inferences about the parameters of the process.

The approach recommended by Farnum includes the usual dnom chart, which is the one that considers the process and measurement model as follows:

$$X_m = X + \varepsilon \tag{3.16}$$

where Xm is the measured value, X is the "true" value of the measured characteristic and "ε" is the measurement error. The X's are seen as generated by a process with a target value or nominal dimension of Ti and depending on the particular process and model of measurement error this equation can be used to generate estimates of σ i's required for the dnom standardized chart. Specifically

$$\sigma_i^2 = Var\left(\frac{X_m}{T_i}\right) = Var\left(\frac{X}{T_i}\right) + E\left(Var\left(\frac{\varepsilon}{X}\right)\right) \tag{3.17}$$

which remain under the considerations that

$$E\left(\frac{X}{T_i}\right) = T_i \text{ and } E\left(\frac{\varepsilon}{X}\right) = 0 \tag{3.18}$$

The estimated values of σ_i could be applied now to DNOM:

Model I

This model assumes that the process variation does not change even though the nominal values change, coupled with the consideration that errors are independent of the magnitude being measured.

$$E\left(\frac{X}{T_i}\right) = T_i \text{ and } Var\left(\frac{X}{T_i}\right) = \sigma^2 \tag{3.19}$$

For any X and any nominal value T_i, where σ^2 is constant and for the measurement $E\left(\frac{\varepsilon}{X}\right) = 0$ and $Var\left(\frac{\varepsilon}{X}\right) = \sigma_\varepsilon^2$, for any X_i and the variance error does not depend on the value X.

Model II

In this model it is considered that there are both measurement error and variation in the process. The standard deviation of measurement error is considered proportional to X, the measured quantity, while the standard deviation of the process is considered to be proportional to the nominal value Ti

$$E\left(\frac{X}{T_i}\right) = T_i \text{ and } Var\left(\frac{X}{T_i}\right) = K^2T_i^2 \tag{3.20}$$

For any X and any nominal value of the process T_i^2, and constant K (That means, the standard deviation of the process is proportional to the nominal value). For the measurements we have:

$$E\left(\frac{\varepsilon}{X}\right) = 0 \text{ and } Var\left(\frac{\varepsilon}{X}\right) = k^2X^2 \tag{3.21}$$

For any X and constant k (That means, The standard deviation of the measurement error is proportional to the value of the characteristic)

K can obtain the measurement system of the manufacturer. Consider for example that the system has the nearest 5 % of the true value X being manufactured and measured, so for normal errors, 0.05x would be approximately equal to $3\sigma\varepsilon$, which leads to k = 0.05/3 = 0.017.

The estimated value of K can be obtained by solving

$$\mathrm{s}^2 = \mathrm{K}^2 + \mathrm{k}^2\mathrm{K}^2 + \mathrm{k}^2 \tag{3.22}$$

where

$$\mathrm{K}^2 = \left(\mathrm{s}^2 - \mathrm{k}^2\right)/\left(1 + \mathrm{k}^2\right) \tag{3.23}$$

3.10 Analysis Process Capability

A process is a set of operations, machines, tools, materials, methods, all these combined to achieve a measurable objective. General processes have inherent statistical variability, which can be measured by means of ranges and/or standard deviations, which in a way, measure the performance of the process.

The ability of the process (or the ability to make good parts) is a measurable property of the process can be evaluated by means of the process capability indices (Cp and/or Cpk) or process performance indices (Pp and/or Ppk). The process capability indices are defined for stable processes with normal data, while the performance indices are defined for unstable processes and non-normal data.

It can be considered that the analysis of capacity or performance of the process may include the following:

1. The analysis of the measuring system, considering linearity studies and R & R. also checking the normality of the data.
2. Sampling suitable for the determination of process parameters.
3. Verification of the stability of the process by a control chart.

4. Calculation of the rates of capacity or performance of the process.
5. Compare the ability of the process with a given tolerance or specification.

The equations for calculating the measurable process capability are given as follows:

The Process Potential Capacity will be given by

$$Cp = \frac{USL - LSL}{SNTL - NLTL} = \frac{USL - LSL}{\mu + 3\sigma - (\mu - 3\sigma)} = \frac{USL - LSL}{6\sigma} \tag{3.24}$$

where:

SNTL: Superior Natural Tolerance Limit Process
NLTL: Natural Lower Tolerance Limit Process
USL: Upper Specification Limit
USL: Lower Specification Limit

Process Real Capability is obtained by

$$Cpk = \min\left\{\frac{USL - \mu}{3\sigma}, \frac{\mu - LSL}{3\sigma}\right\} \tag{3.25}$$

where μ and σ are obtained by:

$$\hat{\mu} = \bar{\bar{X}} \quad \hat{\sigma} = \frac{\bar{R}}{d_2} \quad \text{or} \quad \hat{\sigma} = \frac{\bar{S}}{c_4} \quad \text{for capability indexes and}$$

$$\hat{\mu} = \bar{\bar{X}} \quad \hat{\sigma}^2 = s^2 = \frac{1}{n}\sum_{i=1}^{n}\left(x_i - \bar{\bar{X}}\right)^2 \quad \text{for performance indexes.}$$

There are some standard values given for comparison with the values obtained in the process, such as the automotive industry sets a Cpk = 1.67 and an acceptable level of quality, which may generally be considered as a reference value.

However, considering that the Cpk is a measure of the centering of the process, what might be considered somewhat inadequate, since for any fixed value of process average in the range given by the specification limits, depends inversely on the standard deviation grows as it tends to zero. Given this, the Cpk can be an incorrect measure process capability Cpm proposing the capacity as an index for measuring fittest process focused.

$$Cpm = \frac{Cp}{\sqrt{1 + \xi^2}} \quad \text{where} \quad \xi = \frac{\mu - T}{\sigma} \tag{3.26}$$

The methodology for the construction of control charts when the variance rate of the process for the different parts is constant and there is no measurement error, specifically linearity and repeatability and reproducibility is as follows:

1. Determine the parts that could be made (or are making) in the process.
2. Make a study of linearity for the team (or measuring equipment) used in the process.
3. Analyzing the values of bias and linearity to first determine the significance of the existence of linearity (slope of the regression line) and the significance of the bias values for each of the parties.
4. Constructing a scatterplot part measurements against their nominal values, confirming the line graph of regression in linearity study.
5. Calculate averages and standard deviations of each of the samples.
6. Calculate $\frac{\bar{X}_i}{T_i}$, $\frac{s_i}{T_i}$ and $\bar{X}_i - T_i$.
7. Calculating variances for each of the cases: dnom chart with constant variances and variances dnom chart with constant and measurement errors in the bias and linearity in addition to errors in repeatability and reproducibility (Percentage of R & R).
8. Calculate the control limits in each case.
9. Interpret graphs and analyze the importance of the use of each.

3.10.1 The Assumption of Equal Variances

When trying to control the arithmetic mean of a process using control charts, one of the most important assumptions is the one of equal variances. If the assumption of equal variances is not met, the inferences made in relation to the average population might not be reliable in addition to the process capability studies.

When plotting several parts in one control chart for the arithmetic mean, verification of behavior depends largely on this assumption is met. So if there is heteroscedasticity, the control chart should be able to detect it. Some data in four different lengths to control in the same short production run chart is analyzed. These data were obtained before behaviors, such as behavior scatterplot standard deviation against the target values of each.

3.11 Application

As an example, an analysis shows a windshield wiper manufacturing plant, where most of the processes, such as the rubber extrusion process, the cutting and bending of the rods, and the painting process (where measured thickness of the paint layer, etc.) run different models on the same turn. In different plant processes are used graphic control $\bar{X} - R$, of which is taken for analyzing information processing capacity.

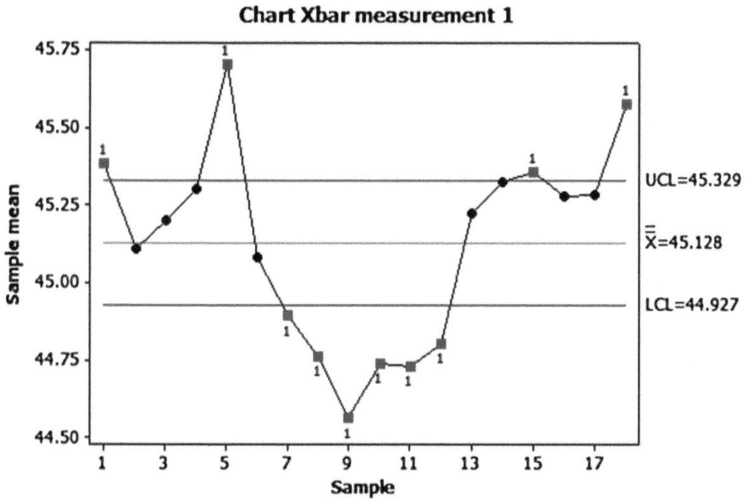

Fig. 3.1 Control Chart for length 1

Figure 3.1 shows a process capability analysis for the Length 1, where Cpk = 0.83 and Cp = 1.11 demonstrating variability and problems in location, i.e. not centered process.

You need to consider that regularly plants have severe time constraints, equipment and personnel, which obligate them to look for alternative solutions to problems that are regularly solved with the purchase of more equipment and hiring more staff.

It is possible to consider that the problems in the measurement system are the cause of the behavior of processes, mainly because some computers are used as calibrators of others, and the data to calculate capability indices are taken from control charts.

Figure 3.2 shows a graph for the length 1, where the parameters for the graphs are calculated with the sample.

So, what is intended is to implement graphs in which parties can plot multiple batch production whose parts are run in the same process, where mostly the assumption of equal variances is violated, together with the possible presence of measurement errors.

It is important to note that a manufacturing plant has several production cells which run different models, but is often considered as a single cell in order to show what happens when trying to control incorrect models. Furthermore, this project began with the intention of improving the measurement system, which is fully achieved. It is well known that the average control of a first control process necessarily involves the variability of this, because the assumption is that there may be changes in the mean, but not in the variance. If a machine is going to run various different lengths, thicknesses, weights, etc., It is almost certain that this

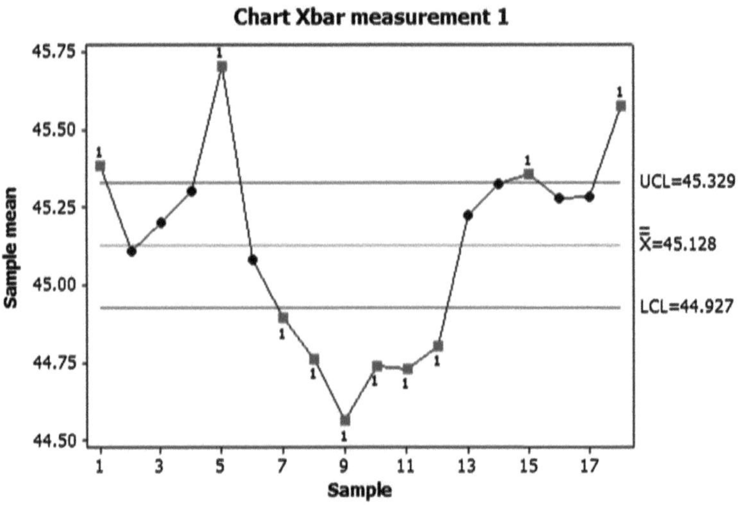

Fig. 3.2 DNOM Chart of the ratio Prom/T with a situation out of control

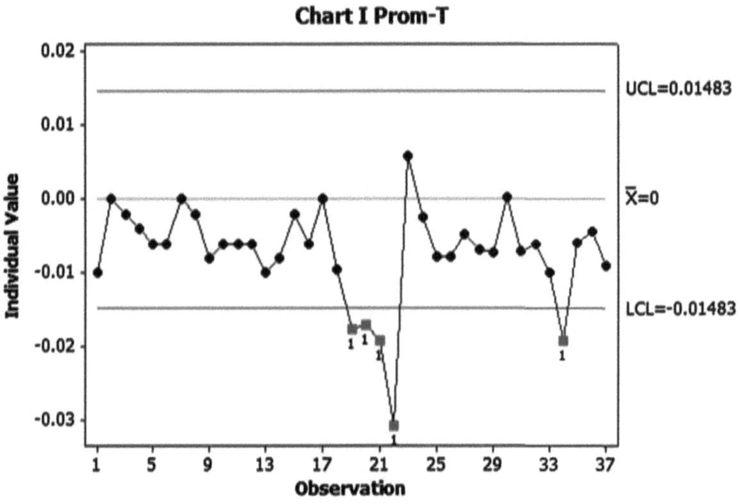

Fig. 3.3 DNOM Chart Prom-T with false out of control situation

course, insurance is not fulfilled. Then data and graphs are shown for testing this condition.

Figure 3.3 shows the graph dnom of reason $\bar{X}/T$. This considers the detection of heteroscedasticity and the graph shows no runaway situation. Figure 1.3 shows the graph of the difference $\bar{X} - T$ dnom. This was not designed to detect non-constant variance.

3.12 Conclusions

One of the common problems that arise in business is the resistance to change by staff, and as a result, resistance to providing information, the vast majority of managers do not provide the information of what happens in the process causing inadequate improvement actions.

The first objective of any improvement of the quality must be the training of personnel in the Measurement System Analysis (MSA) and Statistical Process Control (SPC), with a commitment to first analyze your area and really the situations. More importantly, it must be performed bias studies and linearity in addition to the Reproducibility and Repeatability of the measuring system. Then, control charts could be produced, depending on the actual situation that occurs and, of course, select which type of chart to implement and finally the process interpretation.

References

Brunk, H. D. (1979). *Introduction to Mathematical Statistics*. Mexico: Editorial Trillas (in Spanish).

Farnun, N. R. (1992). Control charts for short runs: Noncostant process and measurement error. *Journal of Quality Technology*, *24*(3), 138–144.

Hawkins, D. M. & Olwell, D. H. (1997). *Statistics for engineers and physical science*. New York: Springer.

Montgomery, D. C. (2003). *Applied Statistics and Probability for Engineers* (3rd ed.). New York: Wiley.

Wise, S. A., & Fair, D. C. (1998). *Innovative control charting: Practical SPC Solutions for Today's Manufacturing Environment*. Milwaukee: American Society for Quality.

Chapter 4
Statistical Process Control: A Vital Tool for Quality Assurance

Jorge Meza-Jiménez, Miguel Escamilla-López and Ricardo Llamas-Cabello

Abstract This chapter reviews the main graphs that are usually used in Statistical Process Control in a continuous manufacturing production system. The chapter begins with some general descriptions and concepts related to SPC, there are a brews description related to SPC evolutions and tendencies appears some examples related to each graph type and finally, appears an application.

Keywords Quality charts · SPC · Textile industry

4.1 Introduction

Production systems are started when people are grouped to produce a product, historically, it been saw the need to control the production, seeking to ensure that control the production, generated scientists dedicated to tasks seek tools that would allow them to be more productive and a consequence become more competitive in their markets. However, in the course of years it was observed that the work required of more effective ways to couple a large number of employees with a large number of machines, additionally, of huge quantities of information coming from the production process.

The growth of quality control as we know it today was developed in the last century, where significant changes to the focus of work have occurred on average every 20 years, as stated (Feigenbaum 2004).

The activity of the control of product quality in the last century, was centered on the worker, in this context it is established that at least one or a very small group of workers they were responsible for manufacturing (manufacture) Product

J. Meza-Jiménez (✉) · M. Escamilla-López · R. Llamas-Cabello
Department of Industrial Engineering, Instituto Tecnológico de Colima,
Av. Tecnológico No. 1, Villa de Alvarez, Z. P. 28976 Colima, Colonia Liberación, México
e-mail: jorge.meza@itcolima.edu.mx

J. L. García-Alcaraz et al. (eds.), *Lean Manufacturing in the Developing World*,
DOI: 10.1007/978-3-319-04951-9_4, © Springer International Publishing Switzerland 2014

Full and therefore, each worker controlled the quality of its products. In the early 1900s, there is noticeable progress in production systems, appearing the quality control supervisors, here starts the period of development of modern factories, in which a group of workers with similar tasks were controlled by a person called supervisor, who, took full responsibility for the quality of work and in consequence of the production. As a result of this activity and the strong production growth driven by the First World War, first appear full-time inspectors and kicks off what is known as *quality control by inspection.*

Accordingly, during the decade of 1920–1930 were established large inspection organizations, separate from production systems and led by a figure who was named superintendent. The program remained in office until the need for mass production required by World War II forced them to develop a more efficient control technique that was very directly related to the production, emerging concept Statistical Quality Control.

Inspectors to apply statistical quality control were provided with basic statistical tools such as: sampling and control charts with these tools is achieved most important contribution in Statistical Quality Control with the concept of sampling inspection, instead of 100 % inspection. No clutch, the concept was anchor in the production process and its development and growth was relatively slow, in this decade and on time in 1924 Dr. Shewhart developed the first control charts, which is established with your implementation, *the quality of product as manufactured is subject to some variation due to chance* (Grant and Leavenworth 1987).

The slow growth of quality control had to do with the development of the ideas and statistical techniques; well as, the lack of tools for adequate and prompt manipulation of the data obtained from the process operations, other impediments were, the will or ability of business organizations and governmental organizations, to take appropriate action relating to the discoveries of statistical technical work.

It was in the decade of 60s when the business community takes up the statistical concept for application in production operations, based on the beginning of the automation of industrial processes with Numerical Control (NC) and later with Computer Numerical Control (CNC), likewise, takes up the concepts of Shewhart and more emphasis on production sampling to measure the quality of manufactured products.

In the decade of 80s, the emphasis is on total quality systems as a way to ensure the survival of the business sector to competition it was experienced at that time; is started the development of computer systems with more speed, starts production of the first desktop computers and born the decade of the Pentium. With Pentium computers can process more information and develop the first software for statistical analysis of data such as; STATISTICS and STAT GRAPHICS, with these advances the business sector begins to establish the first rules for statistical control of quality and thus formalized statistical process control; also, it working in the training of engineers with high level of expertise in these techniques.

In the recent decade from 2000 to 2010, the analysis has taken some important processes in the productive sectors; are implemented as 6σ tools and computer systems, as well as the developed software for data management (MINITAB,

STAT GRAPHICS, SPSS, and STATISTICS), the computer systems become more fast and high capacity for handling the information. With this premise, statistical process control is essential to control the quality of products manufactured, regardless of the company's turnover, considering continuous production, discrete or services.

The historical development of statistical process control in the last century and today so far, has made great contributions and has been supported by the contributions of scientists and companies to improve production processes. Table 4.1 shows the most important contributions in this area of study.

4.2 What is Statistical Process Control?

It is defined as a process, the set of interrelated activities that combine machines, tools, methods, materials and workmanship to carry out the transformation of an input (raw material) into a product or service with a certain quality. Figure 4.2 shows schematically the relationship between input and output process.

Order for a product meets quality characteristics demanded by the client, you must have control of the variables that are used in the process with due care that inputs have to meet the quality characteristics that the process required. By natural law, all processes have variations and are the property of the variability, so that the products can't leave the same, since there are factors that change the outcome of actions (Guajardo 1996). These factors or causes in most cases are inherent in the production process and sometimes occur by external eventualities; however, it is important to perform measurements that will control these variations in order to meet the quality characteristics of the product manufactured. Statistical process control is the mathematical language by which we can understand the process and distinguish the differences they experience in the course of time, as well as the causes of their variation.

When it is possible to predict statistically they rank will present the variation of a process, manufacturing staff can plan their work to meet the required quality. Generally, a production process has many sources or causes of variation, these causes can be divided into two types: the causes provoked by process inputs and the causes assigned by the characteristics of the process. The characteristics of the process include: equipment, work procedures and environmental conditions on the ground, in both cases the combination of the lead causes of process variability.

When the causes of the variation inherent in the process can lead to chronic problems, is commonly called the common causes, the effect of the common causes, can be described by means of continuous or discrete probability distributions in both cases are mathematical models linking the value of the variable under study with the probability of occurrence (Fig. 4.1).

The data presented in statistical process control usually come from two sources: from measurements and counts.

Table 4.1 Most notable contributions to the development of statistical process control

Year	Personage	Contribution
1916	Ford motor company	Developed the materials handling system, lay out factory and final inspection
1917	G. S. Radford	Publishes an article first introduces the term quality control
1922	G. S. Radford	Publishes the first book on quality control (The control of quality of manufacturing)
1924	Walter A. Shewarrt	Develop the concept of control charts and was named the father of statistical control
1925	Harold F. Dodge	Develops concepts and terminology for acceptance sampling of production lots
1950	Joseph M. Juran y W. Edwards Deming	Teaching statistical methods and statistical control to the Japanese
1950	Genichi Taguchi	Emphasized in product specifications and its relation to the quality, design and strategies for reducing variation, robust design and design of experiments
1951	Shingeo Shingo	Developed the concept of zero quality control of using inspection at the source, created the concept Poka-Yoke
1951	Joseph M. Juran	Publishes the first edition of quality control handbook
1960–1963	Kaoru Ishikawa	Integrated the statistical quality control concepts development total quality, continuous improvement and customer service; he devised using the seven basic tools and developed the cause and effect diagram
1970s	Industrial sector	The industry focus is given to continuous improvement and employee involvement
1980s	Industrial sector	Emphasis on quality of design for manufacturing, computers are widely used in all aspects of quality
1985	Bill Smith	Introduces the concept of six sigma to standardize the way defects are found, from design to delivery of the product to the customer, taking into account all the processes of the organization
1987	ISO	Outputs the ISO 9000 series of standards for quality systems
1987	USA	Sets the national quality award (Malcolm Baldrige) by agreement of congress
1990s	Industrial sector	The concept of quality extends service companies, the emphasis on total quality management (TQM) and customer satisfaction
1994	ISO	Outputs the revised version of the ISO 9000 series. The series was renamed ANSI/ASQC Q9 BY QP9000
2008	ISO	It issues an update of the 2000 version of the ISO 9000 series of quality systems based on 8 principles approach to the process and a management model based on 5 requirements

In the case of data coming from the measurements, can be given to the statistical normal probability distribution, this distribution is in the form of smooth curve to the area under the curve equal to the probability as shown in Fig. 4.2, so

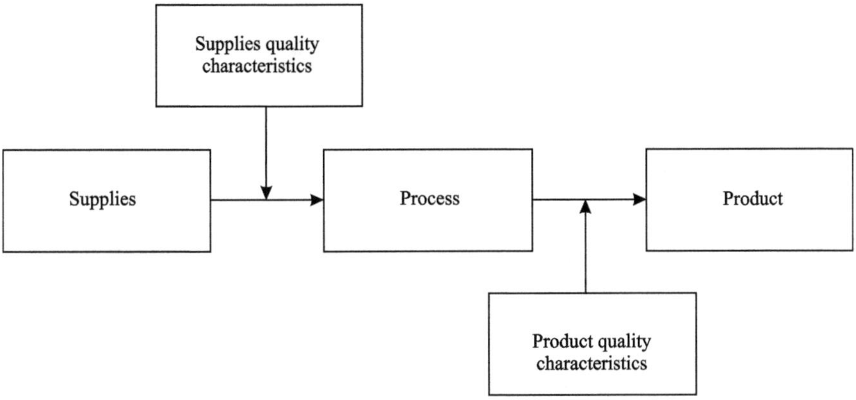

Fig. 4.1 Input process product relationships

that the occurrence probability of a value x is in the range defined by a–b as shown in Eq. 4.1:

$$P\{a \leq x \leq b\} = \int_a^b f(x)dx \tag{4.1}$$

The normal distribution is one of the most important theoretical and practical applications of statistics. The definition of the distribution is based on:

If x is a normal random variable, then the probability distribution is:

$$f(x) = \frac{1}{\sigma\sqrt{2\pi}} e^{-\frac{1}{2}\left(\frac{x-\mu}{\sigma}\right)} \text{ for } -\infty \leq x \leq \infty \tag{4.2}$$

The distribution parameters are the mean $\mu\,(-\infty \leq \mathrm{x} \leq \infty)$ and variance $\sigma^2 > 0$.

The formality of the analysis of the equation that defines the normal distribution is posed widely in Mendenhall (1986), Jonshon (1997). It is important to point that there are more important Continuous probability distributions as are: the Exponential Distribution, Gamma Distribution, Weibull Distribution, widely reported (Montgomery and Valckx 1991). However, in this chapter, only the normal distribution is used as a basic statistical tool for analysis in statistical process control.

When the data come from a counting process are discrete data; in this case it speak of fraction defective or defects to proportions and treatment for the analysis is performed from discrete probability distributions, among which are the Hypergeometric Distribution, Binomial Distribution, Poisson Distribution, and Pascal distribution among others. In this context, the phase of the statistical control is based only on the manufacturing process, which implies that the process has to be stable and able to operate virtually, so that the manufactured products meet design specifications and process.

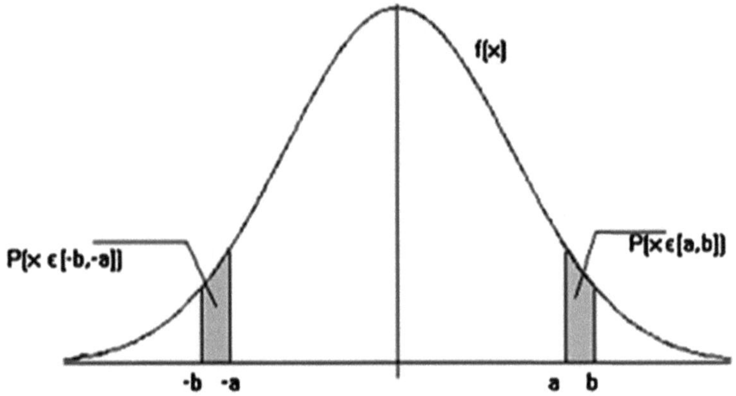

Fig. 4.2 Graphical representation of the normal distribution

4.3 Tools for Statistical Process Control

Montgomery and Valckx (1991) provides: it is impossible to incorporate the product quality by inspection or testing. In this case, statistical process control is the easiest way to monitor the manufacture of the product from the start to the end of operations; the tools that help to have this type of control are statistical control charts.

Montgomery and Valckx (1991) points out, that the fundamental uses of control charts is for tracking and monitoring of a process, reducing the variability of a process and estimate the parameters of a product or process.

Control charts are broadly graphical representations of a quality characteristic of the product; the quality characteristic is measured or computed over a sample. Statistical their livelihood is on the average μ and standard deviation σ of the sample or population, in addition, depends on the methodology used to obtain the data required to analyze. For the analysis of control chart is generally set as the average μ center line representing the same average value of the quality characteristic measure, or random causes indicating the status of the process under control. Moreover, in the same chart, are sets two lines placed with the same value range of variation, they represent the natural limits of the process also known, upper control limit (LSC) and lower control limit (LIC) respectively. The limits are chosen according to the variation having the process and can be as wide or narrow as required for the process analysis and standard deviation σ.

Control Charts can be classified into two types: the first is based on the measurement of the quality characteristic (continuous data) in this case, should be describe the quality characteristic by means of a measure of central tendency and variability; The second concept refers to the data that in the process can measure the conformity or unconformity of the product that it has certain attribute or not, or

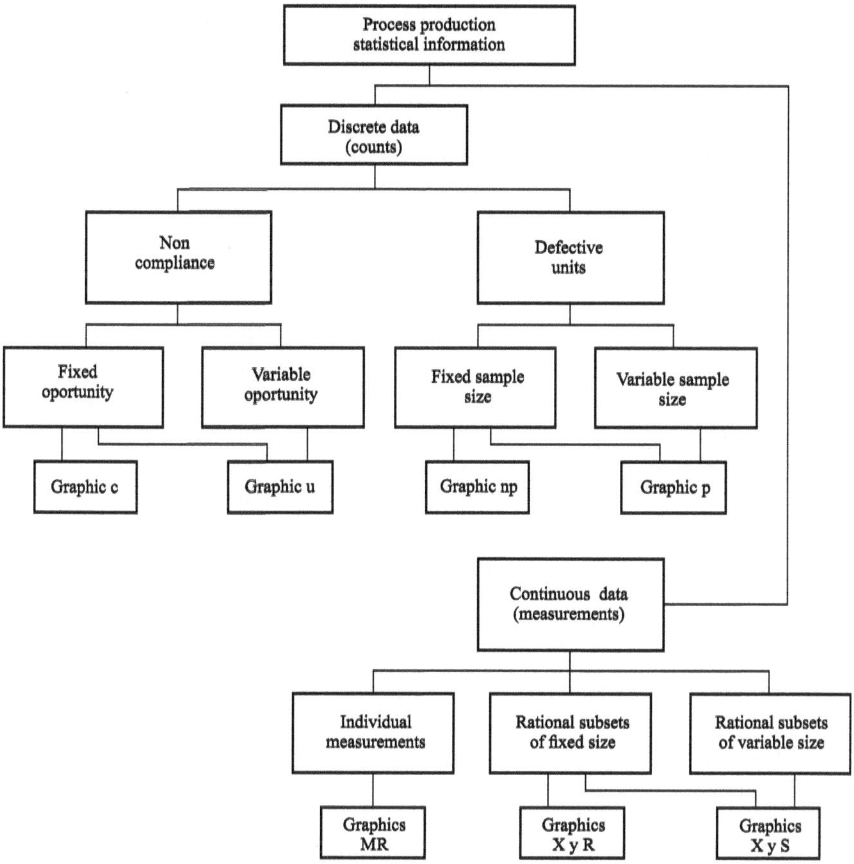

Fig. 4.3 Classification of control chart

you can count the number of defects that appear in a product unit (discrete data). The Fig. 4.3 shows the classification of control charts discussed in this chapter.

As has been discussed above, the basic idea of a control chart is analyze the performance of a process in order to distinguish the existing variation due to common causes caused by special causes. This will allow us to detect important changes in trends in the process (Gutiérrez and De la Vara 2005).

4.4 Analysis of Control Charts

For the analysis of the control chart, the start is set to with fundamental understanding of the process, which determines the type of data that the process provides for the analysis (Fig. 4.3). Subsequently, we establish the conditions for the elaboration of process control chart calculating the basic parameters such as: sample mean or population, variation and natural control limits of the process.

4.4.1 $\overline{X}$–R Chart

This type of control chart applies where the quantities or production batches are big and the quality characteristic of interest provides continuous data; for example; for water pump shaft, in this machine element, the feature may be defined by the inner diameter, the finish (the relative roughness), the thickness and hardness of steel. Another basic example is the production of special screws which can be important features: the degree of hardness, the screw diameter and thread pitch. In these cases, chart ($\bar{x}$ or μ), analyze of the behavior the mean of the quality characteristic, whereby, information is obtained from the central tendency of the process, is also contemplated in this chart, the calculation to process variation across the range (R) and standard deviation (σ or s) as measures of dispersion. Counting with these parameters can be determined statistical control limits of the process as shown in Eqs. 4.3 and 4.4

The location of the control limits on a control chart, is fundamental because if they are located too far from the mean, will be more difficult detect changes in the process, however, if they are very strait, or very close to the mean, it can increase the type I error (declaring a change when there is not). For calculating limits proceed as follows:

Under statistical control: the statistic, which is plotted on the chart, has a high probability of falling within the limits.

Finding the statistical probability distribution, estimating its parameters, and locate the limits so that a high percentage of this distribution within them (Duncan 1999).

Determining the relationship between the mean μ and the standard deviation σ for the study quality characteristic.

For the case where the variable has the characteristics described in paragraph 3 and is under statistical control conditions, the limits are determined by:

$$\text{LIC} = \mu - 3\sigma \tag{4.3}$$

$$\text{LSC} = \mu + 3\sigma \tag{4.4}$$

In Table 4.2, 100 measurements are shown the inner diameter of the sleeve machined to pump shaft 8 × 6 in. The quality characteristic of interest is the inner diameter of the sleeves, measurements were grouped into 25 subgroups with 4 data each, in order to obtain the condition that the process has with respect to characteristic referred. The diameter is limited to 3250 ± 002 mils according to the manufacturing drawing.

Table 4.2 Measurements of the diameter of the pump shaft sleeves for 8 × 6 in.

Subgroup	Measurements of the diameter of the pump shaft sleeves 8 × 6 in.				Mean	Rank
1	3248.888	3251.48	3248.264	3251.628	3250.07	3.364
2	3248.436	3248.116	3250.116	3250.116	3249.2	2
3	3250.04	3250.752	3248.684	3251.736	3250.3	3.052
4	3250.92	3251.376	3251.744	3249.996	3250.76	2.744
5	3250.084	3249.672	3250.136	3249.924	3249.95	0.464
6	3251.964	3250.472	3249.128	3248.3	3249.97	3.664
7	3251.253	3248.916	3250.024	3250.912	3250.28	2.337
8	3250.932	3249.084	3250.604	3251.06	3250.42	1.976
9	3251.504	3249.88	3251.908	3249.388	3250.67	2.52
10	3248.64	3250.804	3249.896	3251.936	3250.32	3.296
11	3250.792	3248.568	3251.636	3250.36	3250.34	3.068
12	3248.288	3249.08	3251.628	3250.388	3249.85	3.34
13	3250.672	3248.884	3251.468	3250.924	3250.49	2.584
14	3249.52	3248.34	3251.808	3251.304	3250.24	3.468
15	3251.004	3251.84	3251.588	3248.044	3250.62	3.796
16	3249.172	3248.348	3250.54	3249.052	3249.28	2.192
17	3250.956	3248.076	3250.428	3248.236	3249.42	2.88
18	3248	3249.812	3251.928	3249.44	3249.8	3.928
19	3248.04	3249.676	3250.376	3249.296	3249.35	2.336
20	3248.992	3250.576	3250.7	3249.816	3250.02	1.708
21	3250.896	3251.272	3250.16	3249.852	3250.55	1.42
22	3250.636	3251.292	3249.732	3250.604	3250.57	1.56
23	3251.968	3251.032	3248.464	3251.368	3250.71	3.504
24	3251.392	3249.744	3250.696	3249.508	3250.34	1.884
25	3251.8	3248.174	3250.068	3250.704	3250.19	3.626

The steps are performed for the elaboration of a control chart $\overline{X}$–R are defined by the following points:

1. Select the quality characteristic.
2. Choosing the sample size or subgroup.
3. Collect data quality characteristic.
4. Determine the control limits.

Once you have determined the sample of the quality characteristic, the most concrete way of choosing the control chart is based on two basic procedures: method instantly and the period method.

The first one, is used more frequently, because with this procedure homogeneous samples is achieved and provides a reference time, which is useful for locating the causes of variation and can react more timeliness and accuracy.

For the second method, it provides more information about the product quality and less on variability, so that the sample size for a graph $\overline{X}$–R must be determined from the changes to be detected.

As mentioned previously, the control limits Shewhart Chart are determined by the mean and standard deviation.

For the case: the average of the data that are plotted in the diagram can be represented as:

$$\mu_{\varpi} \pm 3\sigma_{\varpi} \tag{4.5}$$

In Chart $\overline{X}$, and ω represents the statistical mean of the samples, so the form of to estimate the mean to $\omega\overline{x}$ is given by:

$$\mu\omega = \mu\overline{x} \approx \overline{\overline{X}} \tag{4.6}$$

where $\overline{X}$ = is the mean of the samples

Thus: the value of the standard deviation of the mean of the subgroups is given by:

$$\sigma_{\omega} = \sigma_{\bar{x}} = \frac{\sigma}{\sqrt{n}} \tag{4.7}$$

In most early studies σ is unknown, so they are required estimated from the sample data for that, an alternative computation is to determine it from the standard deviation (S) of the measurements of the inner diameters of sleeves for pump shaft 8 $\times$ 6 in. However, you should not lose sight, to do it that way includes the variability between and within these while. For the graph $\overline{X}$, is more appropriate to include only the variability within samples (Duncan 1976). Then, the alternative to estimate the short-term variation is to estimate σ from the mean of the ranks $\overline{R}$ as follows.

$$\sigma \approx \frac{\overline{R}}{d_2} \tag{4.8}$$

where d_2 is a constant that depends on the sample size. In this case the control limits are obtained by the following equations.

$$\text{LSC} = \overline{\overline{X}} + A_2\overline{R} \tag{4.9}$$

$$\text{LIC} = \overline{\overline{X}} - A_2\overline{R} \tag{4.10}$$

where $\overline{\overline{X}}$ = Centerline of control chart, and for $\overline{R}$

$$A_2\overline{R} = 3\sigma_x = 3\left(\frac{\sigma}{\sqrt{n}}\right) \tag{4.11}$$

$$A_2\overline{R} = \frac{3}{d_2\sqrt{n}}\overline{R} \tag{4.12}$$

Considering different types of samples, if the equation where d_2 intervenes and considering that represents the average relative range W = R/σ, which is a random variable, which sets the relationship between the range of sample the normal distribution and standard deviation of the same (Gutiérrez 2005).

In the case of the control limits for the inside diameter of the sleeves of shaft pump 8 × 6 in., the boundaries are given by:

LSC = 3250.207 + (0.729 × 2.618) = 3252.11 mpulg.
LIC = 3250.207 − (0.729 × 2.618) = 3248.30 mpulg.

The boundaries are represented in Fig. 4.4. This graph shows the average of the averages of each group has a value of 3250.207 mils. For example, the process is in statistical control, as the chart no cumshots or abnormal patterns, so that might indicate that the machining of the parts being out of control or have to make any revisions to the process. The graph can also be observed that there is average amplitude of 2.618 mils, indicating that there may be problems in machining because the maximum permitted in the inner diameter of the sleeve is 2 mils. With these values represented can affirm that there is stability in the process, since the inner diameter of the sleeve varies around 2350.207 mils, with a target value of 2350 mils.

According to the control limits calculated for the graph of means, it follows, from the assumption that there is statistical normality. *If the quality characteristic (ID) does not follow a normal distribution, the graph* $\overline{X}$ *is still valid to detect a significant change in the central tendency of the quality characteristic.*

With respect to graph R (range) for the measurements of the inner diameter of the sleeve to pump shaft 8 × 6 in. The graph reflects the variability of the quality characteristic *Di*, and the behavior of the range at each sample or subgroup. Control limits are shown in Fig. 4.5, were established considering the average plus or minus three standard deviations, so the equation is representative.

$$\mu_R \pm 3\sigma_R \tag{4.13}$$

where the estimated average μ_R ranges is determined from the average range $\overline{R}$ and $\sigma_R\sigma_{\mathrm{R}}$ deviation ranges, which is obtained as follows.

$$\sigma_R = d_3\sigma = d_3(\frac{\overline{R}}{d_2}) \tag{4.14}$$

where d_3 is a constant that depends on the sample size. Then:

$$\mathrm{LSC} = D_4\overline{R} \tag{4.15}$$

$$\mathrm{LIC} = D_3\overline{R} \tag{4.16}$$

The center line $= \overline{R}$ and D_4, D_3 depend on the size of the sample, so the calculated limits for the graph are at. LSC = 5.972, LIC = 0, $\overline{R} = 2.618$

When it has worked with $\overline{X}$-R ó $\overline{X}$–S control charts as shown in Fig. 4.4 it is important to give a proper interpretation, since what is seen is a variable behavior in study quality, about it, there may be different ways in the diagram, which have a particular meaning to those configurations are known as behavior patterns quality characteristic analysis, these patterns can be defined in general as the following shows:

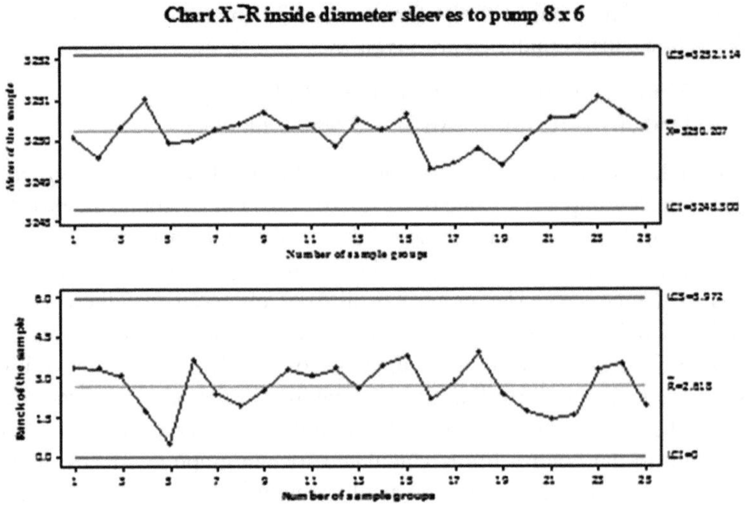

Fig. 4.4 Graph of means and ranges of the sample for sleeves diameter

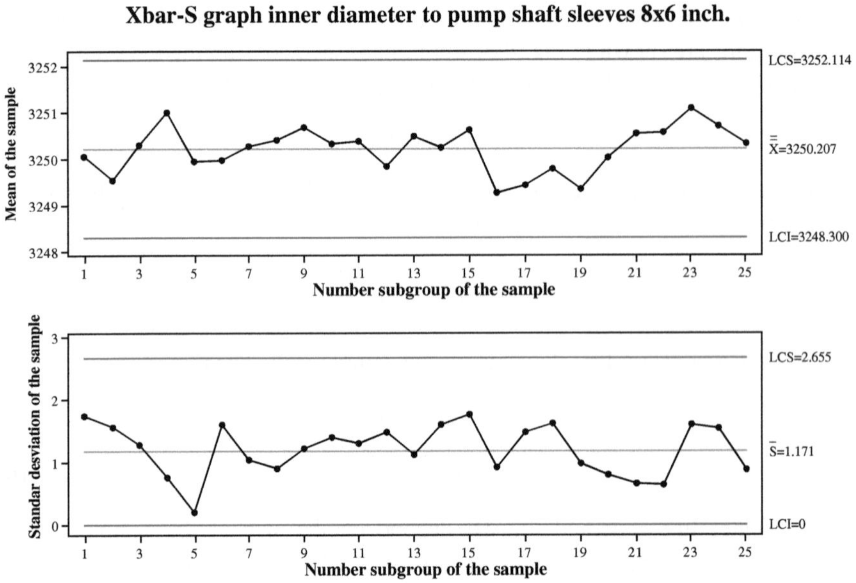

Fig. 4.5 Graph of means and standard deviations for the sample diameter sleeves

It says that it has detected a signal variation or a special cause of variation when:

1. There are changes or leap in process or one or more points fall outside the control limits.
2. Trends exist in the process level.
3. There are recurring cycles or time in the process.
4. There is a lot of variability in the process.
5. There is a lack of variability in the process.

4.4.1.1 When Changes or Leap in Process or One or More Points Fall Outside the Control Limits

In the example of the sleeves of the pump shaft presented in Table 4.2, are changed measurements obtained in some of the samples, with this pattern is observed which is again in the graph $\overline{X} - R$ and re-plot the samples presents a point out of natural control limits the process. Also, it has a pattern of 5 consecutive points above the average, so that in Fig. 4.6 clearly identifies that there jumps or special changes in the manufacturing process of the shaft pump sleeves. About changes observing in the level of the process requires verify the following tests.

1. Identify if there is a point outside the control limits.
2. Two or three consecutive points are between 2 and 3 standard deviations.
3. Four or five consecutive points are between 1 and 2 standard deviations.
4. There are 8 consecutive points on one side of the centerline.

4.4.2 Definition of Control Chart Patterns

As can be seen in Fig. 4.6, the subgroup sample 15 represents a point outside the lower control limit, presenting an abnormality in the process and it is observed that samples 7, 8, 9, 10 and 11 represent to test 3 says: 4 or 5 consecutive points are between 1 and 2 standard deviations. Compliance with these 2 tests is indicative which came a special cause in the process and makes it operate on another level, where change could give gradually until it reached considerable size and the control chart recorded it. Special causes why this behavior occurred might be:

- Originated by cutting tools.
- Inaccuracies in calculating machining parameters.
- Imbalances in the machine (lathe).
- The worker has little expertise or training to conduct lacks the machining process of the sleeves.
- The physical properties of the steel used the recommended (material suitable for manufacture).
- Errors in product measurements measuring instruments.
- Errors in product measurements activity personnel.

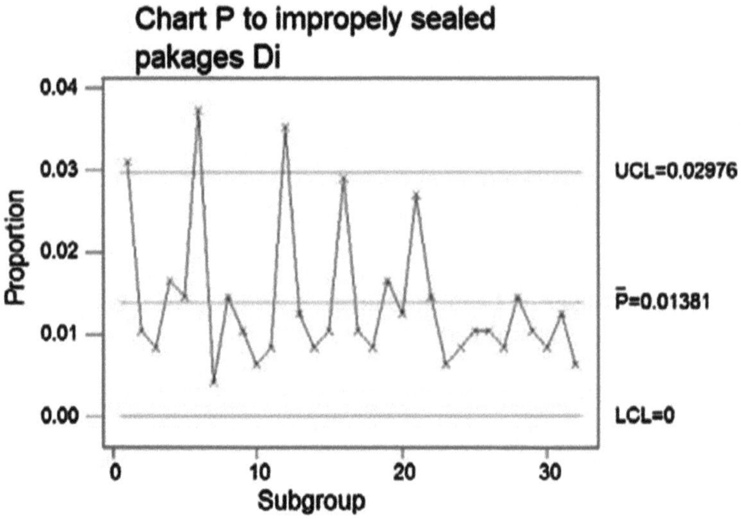

Fig. 4.6 Chart *P* to improperly sealed packages *Di*

4.4.2.1 When There are Tendencies in the Process Level

In the process presents trends to increase or decrease the values in the control chart, the test to be observed is:

Six consecutive points ascending or descending shown in the control chart.

4.4.2.2 When There are Recurrent Cycles or Periodicity in the Process

It is a non-random pattern, which is presented in the control charts and their behavior is cyclic, i.e. several points on the graph show a tendency to grow and immediately change decreasing trend, repeating this feature cyclically, in this case, the test can be defined as follows:

Fourteen consecutive numbers alternating above and below the mean.

4.4.2.3 When There is Much Variability in the Process

This pattern occurs when there is a high proportion of points near the control limits, on both sides of the graph on both sides of the center line and very little or none at the center line, consequently it is said that a signal that in the process there are special causes of variation, the test to observe this phenomenon, is summarized in:

Eight consecutive points on either side of the center line with none in area with one standard deviation.

4.4.2.4 When There is Lack of Variability in the Process

Occurs if the points are concentrated around the mean or center of the diagram and reflect little variability, this indicates a sign of abnormality in the process, which in most cases is due to:

The natural limits of the process were not calculated properly.

4.4.3 Charts for Attributes

Referring to Fig. 4.3, control charts for attributes, are used to determine the control state, which keeps a process when the quality characteristic being monitored is of the form, it passes not pass, i.e. when the product is analyzed and presents a value that comes from a measurement (continuous data), in this case have been developed several charts it possible to assess clearly and simple if the product meets any requirements or expected quality.

The most common graphics for statistical attribute handling are:

- Chart P and NP (for bad)
- Chart C and U (for defects).

The graph P shows the proportion of defectives per sample or subgroup, this type of graph is used to evaluate the performance of a particular product, or of a process or a part of the process, taking utmost account variability to detect causes or special changes. In this case, the calculation of the natural limits of control and the representation of the points on the graph is underpinned by the characteristics of the Binomial Distribution. For the generation of the control chart for attributes P, it must follow the following procedure:

Step 1: Identify the production batch or shipment you want to test
Step 2: Take a sample batch N_i items selected production or shipment
Step 3: Inspect N_i articles to determine the quality characteristic to be evaluated by the criterion fails passes
Step 4: Determine the number of defects D_i and making the graph of the D_i defective items. Consequently calculating the proportion of defective parts is given by:

$$P_i = \frac{d_i}{n_i} \tag{4.17}$$

where P_i = proportion of n defective units is equal to the subgroup.
d_i = Defects in the subgroup.
n_i = Number of items sampled in the subgroup.

To calculate control limits subgroup, it is assumed that the number of defective items follow the properties of the binomial distribution, such that the boundaries are given by $\mu_w \pm 3\sigma_w$ which indicates that is the mean plus or minus three standard deviations the statistic w that is plotted. Therefore:

If w $= P_i$ then:

$$\mu_{P_i} = \overline{P} \text{ y } \sigma_{P_i} = \sqrt{\frac{\overline{P}(1-\overline{P})}{n}} \tag{4.18}$$

where $n =$ Subgroup size $\overline{P} =$ Average proportion of defective items in the process. The control limits are:

$$\text{LSC} = \overline{P} + 3\sqrt{\frac{\overline{P}(1-\overline{P})}{n}} \tag{4.19}$$

$$\text{LIC} = \overline{P} - 3\sqrt{\frac{\overline{P}(1-\overline{P})}{n}} \tag{4.20}$$

In the following case, the problem posed in a small company that produces coconut candy, is related to the packing of sweets (coconut candies), in plastic bags. The problem that it has been found is that the bags are not properly sealed, this problem is detected by visual inspection, the bags, they are not well sealed are open and retrieved the candy to re-pack.

The attribute bad seal of the bag is important, because, if a poorly sealed package comes to market, the shelf life is shortened and can become contaminated quickly by the sugar content in sweet. This situation produces a wealth of dissatisfied customers and creates a bad image of the company producing sweets. In this context the packing machine operators, they should not pass let packet of sweets that is not well sealed, but this has not been enough and as it had not a record of the magnitude of the problem, not had quantifiable basis to detect changes in the performance of the machines, in addition, there was no way of knowing whether the measures employed to reduce the problem had been effective.

Based on this problematic was agreed with the company responsible for recording and analyzing data through a control chart in order to have information to make decisions with regard to what happens in the process. Therefore, it was decided to take sample every hour in the working shift for 4 days and record the number of defective packages with problems on the seal of the bag.

The counter of the machine was obtained the number of produced packets n_i and was obtained the number of packets with defect (*di*). The data are shown in Table 4.3.

4.4.3.1 Interpretation of the Data

It is clear that the problem of improperly sealed packages is the type, pass does not pass, therefore, to make an analysis of the variation is used control chart, type P, therefore, should be initiated from to the following points:

Table 4.3 Data for coconut candies packing process

Subgroup	Sampled packets n_i	Bad sealed packages d_i	Proportion P_i
1	490	15	0.0306
2	495	5	0.1010
3	507	4	0.0070
4	498	8	0.0160
5	504	7	0.0139
6	500	18	0.0360
7	490	2	0.0040
8	502	7	0.0139
9	500	5	1.0100
10	500	3	0.0060
11	497	4	0.0080
12	492	17	0.0345
13	496	16	0.0120
14	494	4	0.0080
15	498	5	0.0100
16	493	14	0.0284
17	492	5	0.0101
18	491	4	0.0081
19	500	8	0.0160
20	502	6	0.0119
21	494	13	0.0263
22	496	7	0.0141
23	493	3	0.0060
24	504	4	0.0079
25	500	5	0.0100
26	493	5	0.0101
27	495	4	0.0080
28	497	7	0.0140
29	506	5	0.0098
30	496	4	0.0080
31	495	6	0.0121
32	501	3	0.0059
Sum	15418	203	

1. Calculate the average ratio.
2. Calculate the average subgroup size n_i which corresponds to the number of coconut candies packaged per hour.
3. Calculate the control limits.

Therefore, the number of packets inspected was 15418 of which 203 were found defective, which imply that the calculation the mean of proportion is:

$$\overline{P} = \frac{Defective\ total}{Inspected\ total} = 203/15418 = 0.0138$$

The subgroup size for the problem is given by:

$$\bar{n} = \frac{\mathit{Inspected\ total}}{\mathit{subgroups\ total}} = 1541/32 = 481.8 \approx 482$$

The calculation of control limits is expressed as follows:

$$\mathrm{LSC} = 0.0138 + 3\sqrt{\frac{(0.0138)(1-0.0138)}{482}} = 0.029$$

$$\mathrm{LIC} = 0.0138 - 3\sqrt{\frac{(0.0138)(1-0.0138)}{482}} = -0.0024$$

As can be observed, calculating the lower control limit is negative, and the analysis that is performed at the proportions, the lower limit have to be greater or equal to zero, then, the lower control limit is taken as zero discarding the calculation result. The calculated limits are shown in the chart of Fig. 4.3.

When analyzing the $\overline{P}$ control chart shows.

1. For sealing the bags coconut sweets (coconut candy), the process worked with causes or situations that usually are not present in the process.
2. The special causes provoked, that the proportion defective in the process it was abnormally large.
3. In the control chart, the samples 23 through 32 show abnormality in the process because all found below the average proportion.
4. When doing research on the causes that provoked to the out of control points are determined.
5. At the start of the shift was observed that there was no standard procedure to start sealing the bags, sometimes was beginning to seal the bags without taking into account temperature on sealing the blades, so it was suggested draw up a standard start of operations to avoid these types of problems.

It was noted that ended up the to be sealed roll of bags and the change of it does not follow a standardized method, causing seal failure due to poor accommodation in the squeegee the film, in the same way developed a standard procedure for the arrangement of the roll when required change.

It was noted that exist sealing machine operators who do not have adequate training to run out its task. Was proposed and trained operators on the operation and maintenance of machines bag sealers.

After correcting the causes that led to the loss of control in the process of sealing coconut candy bags, we proceeded to calculate new limits on which the process must be working.

In this respect is was suggested that the data obtained in Table 4.2, were removed values that are out of control and even those that are very close to the upper limit.

Therefore new boundaries are calculated as follows:

$$\overline{n} = \frac{15{,}418 - 490 - 500 - 492 - 493 - 494}{32 - 5}$$

$$\overline{n} = 479.59 \cong 480$$

$$\overline{p} = \frac{\sum_{i=1}^{32} d_i - d_1 - d_6 - d_{12} - d_{16} - d_{21}}{\sum_{i=1}^{32} n_i - n_1 - n_6 - n_{12} - n_{16} - n_{21}}$$

$$\overline{p} = \frac{203 - 15 - 18 - 17 - 14 - 13}{15{,}418 - 450 - 500 - 492 - 493 - 494}$$

$$\overline{p} = \frac{126}{12{,}949} = 0.0097$$

The calculation of new control limits is:

$$\text{LSC} = 0.0097 + 3\sqrt{\frac{(0.0097)(1 - 0.0097)}{480}}$$

$$\text{LSC} = 0.0103$$

$$\text{Central line} = 0.0097$$

$$\text{LIC} = 0$$

The interpretation of the calculation of the new limits for coconut candy packaging described below:

In producing 480 packages cocada, a proportion is expected to sealing problems of the bags lies between 0 and 0010, with an average of 0.0097, that is, the percentage of poorly sealed bags varies between 0 and 1 % with an average of 0.97 %, this analysis shows the reality of the process, therefore, as the proportion is within the control limits and there is no special variation pattern is any indication that the process works.

As was mentioned in Fig. 4.3, exist in addition to the graphics P defective fraction, other graphics that may be useful, these control charts are used when the sample size has different characteristics than the P graph for this if we can consider the following Chart NP. This graph type is applicable when the sample size is constant in this case is plotted the number of defects, instead of using the graph of the proportion and control limits is determined by:

$$\mu_{di} = n\overline{p};\ \sigma_{di} = \sqrt{n\overline{p}(1 - \overline{p})} \tag{4.21}$$

$$\text{LSC} = n\overline{p} + 3\sqrt{n\overline{p}(1 - \overline{p})} \tag{4.22}$$

$$\text{Central line} = n\overline{p} \tag{4.23}$$

$$\text{LIC} = n\overline{p} - 3\sqrt{n\overline{p}(1 - \overline{p})} \tag{4.24}$$

where n = the size of subgroup $\overline{p}$ = the mean of proportion to defective products

Control charts C and U. These control charts help us to determine the characteristics of a product, when on inspection we find a defect and not simply that the product is defective or does not meet a single quality characteristic. That is, although defects are detected in an interim operation of a process unit being inspected can proceed to the next stage, which occurs otherwise control charts with P and NP. Therefore, to analyze the variability of the number of defects per subgroup, when the size remains constant, indicating C_i will graph the number of defects in the i-th subgroup and the control limits are obtained assuming that the statistical C_i continues a Poisson Distribution, so the estimators of the mean and standard deviation are given by.

$$\mu_{ci} = \overline{C} = \frac{Defective\ total}{subgroups\ total} \text{ and } \sigma_{ci} = \overline{C} \tag{4.25}$$

$$\text{LSC} = \overline{c} + 3\sqrt{\overline{c}} \tag{4.26}$$

$$\text{Center line} = \overline{c} \tag{4.27}$$

$$\text{LIC} = \overline{c} - 3\sqrt{\overline{c}} \tag{4.28}$$

With these relations we can see that the control limits in a graph C, reflect the variation expected for the number of defects per subgroup. Another control chart is important to know for people who are dedicated to monitor production processes or manufacturing, the graphics are also called U or number of defects per unit, this graph analyzes the variation in a process when the average number Flaw article size is from a subgroup in which this is not constant (Gutiérrez and De la Vara 2009). The parameters for the construction of the graph are calculated as expressed in the following equations.

$$U_i = \frac{C_i}{N_i} \tag{4.29}$$

$$\mu_{u_i} = \overline{U} = \frac{Defective\ total}{Inspected\ article\ total} \tag{4.30}$$

$$\sigma_{ui} = \sqrt{\frac{\overline{u}}{n}} \tag{4.31}$$

$$\text{LSC} = \overline{U} + 3\sqrt{\frac{\overline{u}}{n}} \tag{4.32}$$

$$\text{Center line} = \overline{U} \tag{4.33}$$

$$\text{LIC} = \overline{U} - 3\sqrt{\frac{\overline{u}}{n}} \tag{4.34}$$

The use of control charts is very useful as it can meet a perceived need by the person responsible of the production process (manufacturing) and will largely depend on how they are implemented.

Gutiérrez and De la Vara (2009) defines the important aspects that should be taken into ac-count for the implementation of control charts:

1. Describe the problem or current situation of the process.
2. Explain clearly the usefulness of the control chart to assess the situation found in 1 point.
3. Define the objectives of the control chart in concrete form. Select and list the variables of interest for analysis with control charts.
4. Make the selection of the variables listed in item 5.
5. Choose the variable to the previous analysis by the control chart.
6. Select the most appropriate graph according to the type of data to be handled.
7. Decide on the sub grouping of the data, in this context, the selection of the elements that make each subgroup must be in such a way that if special causes are present, these appear in difference between each subgroup.
8. Decide how subgroups collection, there are two ways about it, the instant method and the method of the period, they are clearly described (Gutiérrez and De la Vara 2009).
9. Choose the size and frequency of sampling are statistical techniques that can be checked for sample analyzes, these are described extensively in (Kenett and Zacks 2000).
10. Standardize the form of the data; it must clearly define the procedure for taking data, ensuring that the test instruments are adequate.
11. Determine the control limits and review future.
12. Recalculate the limits and monitor actions taken to implement new graphics to monitor these actions.

4.5 Conclusion

Since the late 1920s when control charts were created by Walter Andrew Shewhart, proved to be a powerful statistical tool for the analysis, control and improvement of production processes of goods and services, today more than 80 years of creation still are to be vastly used in manufacturing industries worldwide. Perhaps the secret of this is in its easy processing, simple calculations and its even easier interpretation.

Quality control is in control charts, in simple analysis tool and emblematic for statistical control of production processes as they have a preventive approach to be used to analyze real-time processes, in order to investigate the causes of variation, random or assignable and always with excellent results.

New digital technologies for information management and data processing have been to facilitate and enhance their use. Far from being in disuse and obsolescence, are more used today than in previous years, thanks to the availability is specialized statistical software such as Minitab, Statgraphics, Statistical and Mathlab among others. While it is true that today there are more tools for statistical process control, control charts are and will remain an effective tool against the poor quality of the products and services we use in our daily life.

References

Duncan, A. J. (1976). The economics design of *p* chart to maintain current control of a process: Some numerical results. *Technimetrics, 71*(274), 228–242.

Duncan, A. J. (1999). *Quality control and industrial statistics*. México: Alfaomega. (In Spanish).

Feigenbaum, A. (2004). *Total quality control*. México: CECSA. (In Spanish).

Grant, E. L. & Leavenwort, R. S. (1987). *Control estadisitico de calidad*. Compañia Editorial Continental. Mexico

Guajardo, E. (1996). *Total quality control*. México: Pax México. Second reimpresión (In Spanish).

Gutiérrez, H., & De la Vara, R. (2005). *Statistical quality control and six sigma*. México: Mc Graw-Hill. (In Spanish).

Gutiérrez, H. (2005). *Quality control and productivity*. México: Mc Graw-Hill. (In Spanish).

Gutiérrez, H., & De la Vara, R. (2009). *Control estadistico de calidad y seis sigma* (ed.). México: Mc Graw-Hill.

Jonshon, R. A. (1997). *Probability and statistics for engineers*. México: Prentice Hall. (In Spanish).

Kenett, S. R. & Zacks, S. (2000). *Modern industrial statistics*. México: Thomson. (In Spanish).

Mendenhall, W. (1986). *Estadística matemática con aplicaciones*. Mexico: Iberoamericana.

Montgomery, D., Valckx, V. D. (1991). Statistical quality control. México: Grupo Editorial Iberoamericano. (In Spanish).

Chapter 5
Process Improvement: The Six Sigma Approach

Diego Tlapa-Mendoza, Jorge Limón-Romero, Yolanda Báez-López and Jesús Salinas-Coronado

Abstract This chapter presents information for both practitioners and academicians about Six Sigma (SS) methodology. Main goal is to provide an overview of Six Sigma, from its beginning until nowadays, giving a description of its phases. A literature research was conducted to determine the Critical Success Factors (CSF) and Tools and Techniques (T&T) more frequently used by practitioners around the world. In order to illustrate the five phases development, a case study of six sigma implementation in an electronic manufacturing company is presented, obtaining a successful implementation taking into account defects rates and their impact on saving and customer satisfaction. Findings support congruence between theory and practice through the use of some CSFs and T&T. It is important to remark that the use of correct T&T plus a consideration of CSFs may considerable increase chances to obtain benefits implementing SS. Finally, is also mentioned the relevance of integrate methodologies for process improvement like six sigma and lean manufacturing, named Lean Six Sigma (LSS) in order to obtain optimal results.

Keywords Six sigma · Process improvement · Lean performance

5.1 Introduction

Today's organizations look for different strategies and working methods to increase the quality of their products, processes and services, and to reduce costs in order to remain competitive in a highly globalized market. In this sense high

D. Tlapa-Mendoza (✉) · J. Limón-Romero · Y. Báez-López · J. Salinas-Coronado
Faculty of Engineering, Architecture and Design, Autonomous University of Baja California, Campus Ensenada, Km. 103 Carretera Tijuana–Ensenada C.P. 22860 Ensenada, Baja California, Mexico
e-mail: diegotlapa@uabc.edu.mx

J. L. García-Alcaraz et al. (eds.), *Lean Manufacturing in the Developing World*,
DOI: 10.1007/978-3-319-04951-9_5,

quality level and reduced cost are competitive advantages for companies; then, these are important issues in long-term business success. One internationally accepted methodology is SS, which has been adopted by several companies to reduce variation of their processes and products.

Pioneers in SS application, are companies like Motorola and General Electric (GE), which reached financial gain and recognition of the quality of their products, surprisingly in the decade of the 80s and early 90s respectively. Following this, many other worldwide companies have implemented SS, some with great success, others with medium and sometimes without obtaining expected outcomes or even to failure, which results in dissatisfaction, distrust and disappointment among investors and workers toward the methodology.

SS is a business process that allows companies to drastically improve their bottom line by designing and monitoring everyday business activities in ways that minimize waste and resources while increasing customer satisfaction (Harry and Schroeder 2000). In a related way Montgomery and Woodall (2008) defined SS as a disciplined, project-oriented, statistically based approach for reducing variability, removing defects, and eliminating waste from products, processes, and transactions.

The goal of SS is to increase profits by eliminating variability, defects and waste that undermine customer loyalty. Then, SS is a rigorous and systematic methodology that utilizes information and statistical analysis to measure and reduce variation, improving an organization's operational performance, by identifying and preventing root causes of defects in manufacturing and service-related processes in order to exceed expectations of all stakeholders to accomplish effectiveness.

Six Sigma through time. It began officially at Motorola on January 15, 1987, when the CEO Bob Galvin launched a long-term quality program named *The Six Sigma Quality Program,* which was an aggressive corporate strategy to improve company performance thorough quality. However, was Bill Smith, a veteran engineer at Motorola and Mikel Harry a Ph.D. from Arizona State University, which developed a four stage problem solving approach: Measure, Analysis, Improve, and Control (MAIC). After implementing SS, in 1988, Motorola was among the first recipients of the Malcom Baldrige national quality award; then, companies like Allied Signal, Texas Instrument and GE, implemented also the methodology with great success, however, Jack Welch CEO of GE is recognized for making SS a central focus of his business strategy in 1995. Until this time, the framework used for improvement was MAIC. Later GE added the D, for define, in the late 1990s to create DMAIC. This was done because the existing framework, or theory, did not do an adequate job of defining the problem and its business context (Snee 2010). The DMAIC phases remain until now, so the vast majority of authors mention them (Tobias 1991; Harry and Schroeder 2000; Antony and Bañuelas 2002; Breyfogle 2003; Yam 2006; Brady and Allen 2006; McManus 2008; Tang et al. 2007).

Montgomery and Woodall (2008) indicated three generations of SS implementations. Generation I focused on defect elimination and basic variability reduction, primarily in manufacturing. Motorola is a classic exemplar of Generation I; In Generation II, the emphasis on variability reduction and defect elimination remained, but now there was a strong determination to tie these efforts to projects and activities that improve business performance through improved product design and cost reduction. General Electric is often cited as the leader of the Generation II. In Generation III, SS has the additional focus of creating value throughout the organization and for its stakeholders (owners, employees, customers, suppliers, and society at large). Creating value can take many forms, such as increasing stock prices and dividends, job retention or expansion, expanding markets for company products/services, developing new products/services that reach new and broader markets, and increasing the levels of customer satisfaction (perhaps by reducing cycle time or increasing throughput) throughout the range of products and services offered. In 2005 was added lean manufacturing to the SS methodology to allow improved process flow more effectively (Snee and Hoerl 2007).

5.2 What is Six Sigma?

SS can be seen and understood as a metric which describe a process performance, as a methodology entailing phases for problem solution, as a philosophy of driving behavior to improve business performance, as a strategy of cost cutting and boosting customer satisfaction, as a goal to be competitive and so on.

5.2.1 Six Sigma as a Metric

In a direct manner of seeing quality level in real life, 1 % of failure is not close of being satisfactory, for instance, Breyfogle (2003) indicated that a "goodness level" of 99 % equates to

- 20,000 lost articles of mail per hour
- Unsafe drinking water almost 15 min per day
- 5,000 incorrect surgical operations per week
- Short or long landing at most major airports each day
- 200,000 wrong drug prescriptions each year

Six Sigma i.e., "6σ" is a metric to measure process performance and product quality, where sigma "6σ" is a letter of the Greek alphabet used by statisticians to measure the variation (standard deviation). Because an aim of SS is to achieve specific target of critical to quality (CTQ) characteristic of product/process while reducing variability between specification limits, a way to measure this achievement is numbering the quantity of failure or defects, i.e. number of item overpassing this limits. Figure 5.1 shows a normal probability distribution for a critical to quality characteristic with specification limits on both sides of the target value at

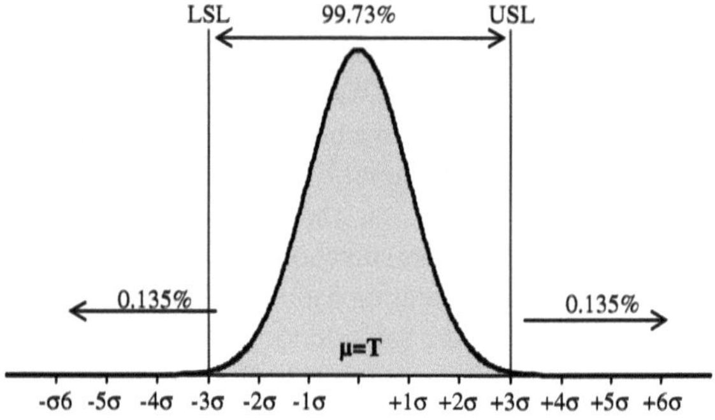

Fig. 5.1 Normal distribution of critical to quality characteristic with $\pm 3\sigma$ specification

Table 5.1 Number of standard deviations between specifications limits and process mean, when this is equal to target and its related ppm defective

Standard deviation between target and limits	Percent inside Specifications	PPM defective
$\pm 1\sigma$	68.27	317,300
$\pm 2\sigma$	95.45	45,500
$\pm 3\sigma$	99.73	2,700
$\pm 4\sigma$	99.9937	63
$\pm 5\sigma$	99.999943	0.5733
$\pm 6\sigma$	99.9999998	0.0019

three standard deviations. At this point, 99.73 % of the time, items would be conform to specifications, however around 2700 item will not per million of items.

SS as a metric implies that the specification limits are at least six standard deviations from the target, then 0.0019 failures would occur per million of items, this is near perfect output also known as short term capability of 6 Sigma. This level of near-perfect performance is significantly superior to that achieved by most organizations today (Black and McGlashan 2006), thereon Table 5.1 shows different scenarios of sigma levels.

The Motorola SS concept entails an assumption, that when the process reached the SS quality level, the process mean was still subject to disturbances that could cause it to shift by as much as 1.5 standard deviations off target (Montgomery and Woodall 2008). This corresponds to long-term capability of 4.5 Sigma, producing up to 3.4 parts per million (ppm) non-conforming to specifications (Fig. 5.2).

The drifting mean aspect of the Six Sigma metric has been a source of controversy. Some have argued that there is an inconsistency in that we can only make predictions about process performance when the process is stable. The 3.4 ppm metric, however, is increasingly recognized as primarily a distraction; it is the focus on reduction of variability about the target and the elimination of waste and defects that is the important feature of Six Sigma (Montgomery and Woodall 2008). Traditionally companies accept three or four sigma levels as a standard, although it represents between 6,200 and 67,000 defects per million opportunities (Pyzdek 2003a, b). A wider scenario suggests defects per million as observed in Table 5.2.

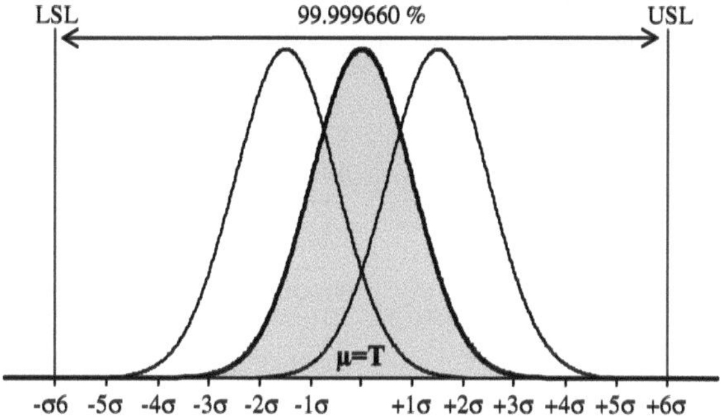

Fig. 5.2 Normal distribution of critical to quality characteristic with $\pm 6\sigma$ specification with the mean shifted by $\pm 1.5\sigma$ from target

Table 5.2 Number of standard deviations between specification limits and process mean, when this is shifted by $\pm 1.5\sigma$ from target and its related ppm defective

Standard deviations between target and spec. limits	Percent inside specifications	PPM defective
$\pm 1\sigma$	30.23	697700
$\pm 2\sigma$	69.13	608700
$\pm 3\sigma$	93.32	66810
$\pm 4\sigma$	99.3790	6210
$\pm 5\sigma$	99.97670	233
$\pm 6\sigma$	99.999660	3.4

5.2.2 Six Sigma as a Methodology

SS is based on the application of scientific method to the design and operation of management systems and processes that allow workers to deliver greater value to customers and partners (Pyzdek 2003a, b). It can be presented as a systematic structure for problem solving and process improvement with the configuration of various flexible tools and techniques between five phases: Define, Measure, Analyze, Improve, and Control (DMAIC). In turn, regarding a new process or product, Design for Six Sigma (DFSS) includes five phases also: Define, Measure, Analyze, Design, Verify (DMADV); they are mainly used for innovations in product development. Both approaches have good reputation, however DMAIC is more broadly used, and so, a brief description of its phases is presented.

Define. Main goal of this phase is to obtain a clear understanding of the problem to be addressed and define objectives in order to solve it. Here, some helpful questions are: Why is this Six Sigma project important? What are the key measures or performances indicators? What is the current performance level? Other purposes in this phase are to develop responsibilities, process steps, goals and milestone. The resulting project and its objectives usually are summarized in a project charter, this and others useful tools and techniques used in this phase are presented in Table 5.3.

Measure. In the second phase, project team maps out the current process to verify and assess all the variables that impact that process (inputs and outputs) in

Table 5.3 Tools and techniques frequently used in Define phase

Flow charts	Gantt Charts
Pareto charts	Project chart
Time plots	Communication plan
Control charts	Brainstorming
Histogram	Multi-vari chart

Table 5.4 Tools and techniques frequently used in Measure phase

Process flow diagram	Process failure mode and effect analysis (PFMEA)
Pareto charts	Measurement system analysis (MSA)
Box plot	Cause and effect matrix/diagram
Histogram	Capability analysis
Scatter plot	Control chart
Trend chart	Affinity diagram
Process map	Kano analysis
SIPOC diagram	Benchmarking
Function deployment matrix	

order to ensure that are working on the most critical to its operation (Big Y). Team evaluates the adequacy of existing measurement systems to accurately account for critical variables; findings are then used to determine such characteristics as stability, reproducibility, and repeatability of those system. Since project team focused on data analysis to determine a baseline and to observe for signs of root causes of the problem, then, the more reliable a measurement system is, the better the data are. At this point, project opportunities can begin to be highlighted with the help of tools and techniques frequently used in this phase (Table 5.4). Useful questions for project team are: where, when and how does the problem occur? How the processes currently perform? What gap is the project addressing? Is the measurement system reliable?

Analyze. In this phase, SS practitioners use tools and techniques to determine the relationships among the response variable or big Y, and the different variables affecting it, i.e., independent factors. Data are stratified and analyzed in order to make a diagnosis of the root causes of problem. Once determined, the root causes are validated through a statistical analysis of the significance of the X variables, at this time, important tools and techniques are used, such as design of experiments, regression analysis and others (Table 5.5). In this phase, Lynch and Cloutier (2003) indicated that helpful questions at this stage are: What source of variation is present? Which factors are affecting the response variable? Which factors are the most significant? How reliable is the concluded data?

Improve. The fourth phase involves elimination or reduction of root causes of the problem, which previously were validated in the analyze phase. Possible solutions are generated and evaluated by project team, employing tools and techniques like brainstorming, simulation, cost benefit analysis and similar used to this purpose (Table 5.6). Solutions should be clarified through different criteria

Table 5.5 Tools and techniques frequently used in Analyze phase

Design of experiments	Process failure mode and effect analysis (PFMEA)
Hypothesis test	Measurement system analysis (MSA)
Correlation and Regression analysis	Cause and effect matrices/diagram
Multi-vari studies	Capability analysis
Pareto analysis	Control chart
5 why's	

Table 5.6 Tools and techniques frequently used in Improve phase

Brainstorming	Process failure mode and effect analysis (PFMEA)
Benchmarking	Correlation and regression analysis
Cost-benefit analysis	Multi-vari studies
Stakeholders analysis	Pareto analysis
Design of experiments	Simulation
Hypothesis test	Pay-off matrix

Table 5.7 Tools and techniques commonly used in Control phase

Control plan	Check list
Statistical process control	Process failure mode and effect analysis (PFMEA)
Gauge control	Management chart
Poka yoke	Change management techniques

such as feasibility, capability, benefit-cost relation, sustainability, and safety among others. Important question to be addressed at this point are: what solutions can be implemented? What are the most important criteria to evaluate solutions? What are pros and cons about potential solution? How the solution will affect customers?

Control. The last phase of SS methodology is preventive in nature, and focuses on maintain the gains realized in the Improve phase. At this time, process owner take control of improved process, therefore, new process considerations are documented, implemented and monitored, so that the improvements could be institutionalized. Control plan and similar tools and techniques as seen in Table 5.7, are considered to achieve the above. Opportunities to replicate the improvements in other similar process are also identified; finally all lasting activities on the project should be completed as well by project team. Lynch and Cloutier (2003) suggested some questions for this phase: How is the improved process standardized and institutionalized? What can be done to ensure continuity in the event of changes in conditions, customers and technology? What action will be taken if the measurements are unsatisfactory? How will you transfer responsibility for ongoing monitoring to the process owner? To date, what progress has been achieved in process capability performance?

5.2.3 Six Sigma as a Philosophy

The difference between SS and similar programs is the degree of involvement of a company management in monitoring the results and achievements, a management system as SS, involves responsibility for results and on-going reviews to ensure the results (Pande and Holpp 2002). SS as a best practice is more than a set of tools for process improvement and problem solving based on metric. McCarty et al. (2005) stated that at the highest level, has been developed into a practical management system for continuous improvement based on four key areas:

- Understanding and managing customer requirements;
- Align key processes to achieve customer requirements;
- Using rigorous data analyzes to understand and minimize variation in key processes;
- Conduct a rapid and sustainable improvement of business processes.

5.2.4 Six Sigma as a Strategy

SS is a strategy and methodology for deploying statistical thinking and methods in an organization (Snee 2004). The statistical implications of a SS program go well beyond the qualitative eradication of customer-perceptible defects. It's a methodology that is well rooted in mathematics and statistics. The objective of SS Quality is to reduce process output variation so that on a long-term basis, which is the customer's aggregate experience with our process over time.

The role of the SS professional is to quantify the process performance (Short Term and Long Term capability) and based on the true process entitlement and process shift; establish the right strategy to reach the established performance objective.

5.2.5 Six Sigma as a Goal

As a goal seeks to reduce the degree of defects closer to zero, so that 6σ represents up to 3.4 defects per million opportunities. In this sense, while the process sigma value increases from zero to six, the variation of the process around the mean value decreases. With a high enough value of process sigma, the process approaches zero variation and is known as *zero defects*, then, if a company has a lower yield process, the goal look as close as possible to 6σ. In statistical terms means reaching 6σ achieve zero defects, but the real message beyond the statistics, is that it reaches a philosophy and total commitment to excellence, customer focus, process improvement, and the obligation to measure and not based on hunches or gut (Dirgo 2005). The implementation of SS allows the company to set new goals and, in effect, ask employees to cope with the change by thinking and acting differently, perform new tasks, and engage in new behavior.

5.3 Successful Implementation of Six Sigma

5.3.1 The Belt System

In addition to the tools that compose it, SS also defines a tactic for organizational structure that allows its implementation in a more reliable manner. In this regard, Motorola created the figure of the so-called black belt, i.e., specially trained individuals on statistical methods, quality, process improvement tools, and project management as well. Their primary purpose is to assure that root causes of problems in processes are investigated, analyzed and addressed so that customers are satisfied with the outcomes (Black and McGlashan 2006). Another hierarchical level called the green belt was added to the tactic, they are an extension of the black belt, who teach and assist them. Usually, green belts perform their daily work combined whit part time on SS projects.

Both SS approach (DMAIC and DMAIV) are executed by green belts and black belts, and are supervised by Master Black Belts when available. Small and medium enterprises (SMEs) applying SS usually lack this figure. A Master Black Belts (MBB) is a full time consultant who clarifies and teaches project members. MBB often write and develop training materials, are heavily involved in project definition and selection, and work closely with business leaders called Champions (Montgomery and Woodall 2008). Champions are Project sponsors, and at the same time, ensure that the right projects are being identified and worked on, that teams are making good progress, and that the resources required for successful project completion are in place. Today, many frameworks exist for implementing the methodology; in this sense some companies have incorporated the yellow belt and white belt to the rest of the staff as well. Kumar (2007) strongly argues for developing a White Belt system for SMEs instead of heavily investing in the Black Belt system.

5.3.2 Six Sigma Projects

Hahn (2005) argued that select initial projects result in expectations, which need to be met and met rapidly to maintain momentum, that's why projects need to be viable and doable in a short time, preferably less than 3 months. Commonly, projects are 4–6 months in duration; it depends of complexity, resource availability and others. When resource is limited, priorities should be defined in order to select a project. In this context Project selection is frequently the most important and difficult part, stated (Pande and Holpp 2002). Tkác and Lyócsa (2009) argued that exist a wealth of scientifically valid methodologies that could potentially be used to select SS projects; however, scientific publications of SS project evaluation and project selection are rare. In this sense, Yang and Hsieh (2009) proposed to use the Taiwan's quality award criteria for selecting a Project.

5.3.3 Six Sigma Success

Different measures are used as a metric to assess the success of a SS project; these include benefits in terms of hard and soft measures of organizational performance. Kumar (2007) noted the "hard measure" of organizational performance focuses on quality performance, operational and business performance indicators. The hard measure incorporates variables such as productivity, defect reduction, on-time delivery, warranty claim cost as a percentage of total sales, cost of quality as a percentage of total sales, profit improvement and sales improvement. On the other hand are "soft measures" such as employee satisfaction, customer satisfaction, organization commitment, job involvement, learning and work environment.

Despite a variety of measures to assess SS success, majority of its implementation reports are in term of savings, indistinctly of the area where was implemented, e.g., in medical services, Thibodaux Regional Medical Center reported a savings of more than US$475,000 per year (Stock 2002). In textile sector, HandV Floyd saved from 2002 to 2006 10 % of sales, but also increasing order-on-time deliveries to 95 %, improving raw material (Green et al. 2006). About safety, SS initiatives at Honeywell resulted in a 33 % improvement in global safety performance and $1.4 million in productivity improvements in 1999 (ReVelle 2004). Others reports in financial services, included the Bank Western Hemisphere reduced internal call backs by 85 % and reduced the credit process by 50 % (Perry and Barker 2006). Global Equipment Finance provides global financing and leasing services to Citibank customers, the credit decision cycle was reduced 67 %, from 3 days to one (Perry and Barker 2006).

5.3.4 Critical Success Factors for Six Sigma

Through SS, many companies have achieved billions of dollars in bottom-line benefits and improved customer relationships. However, not all organizations have experienced equal success (Breyfogle 2003). Despite the immense popularity and the wide spread adoption of SS, there is a rising concern regarding the failures of SS programs (Chakravorty 2009). According to Gray and Anantatmula (2009), in their study found that 67 % of the respondents have experienced SS project failure at least once; they identified several reasons of failure, which include: failure to identify and manage project stakeholders and their expectations, inadequate project selection process, inability to align projects with critical organizational priorities and others. In a similar sense, Kanani (2006) found that 144 from 181 SS projects implemented in a company were successful, this suggest a proportion of 20 % of non-successful projects. In a related way, Zimmerman and Weiss (2005) argued that less than 50 % of the survey respondents from aerospace companies were satisfied with their SS programs.

In this sense, to avoid failure it is important to know prior experiences. Cooke-Davis (2002) stated that learning from experience, is another Critical Success Factor (CSF). Organizations may have differing benchmarks of success for their SS projects as a result of diverging levels of maturity in the deployment of their initiatives (Shenhar et al. 1997). Thus, the term project success is used to depict the level to which desired results are achieved. This definition is applicable across different types of projects, and covers the domain of project success for organizations in varying stages of SS deployment (Anand et al. 2009). A common term in literature is CSF and this factor is critical to the success of any organization, in the sense that, if objectives associated with the factors are not achieved, the organization will fail, perhaps catastrophically (Rockart 1979). In the same way Antony and Bañuelas (2002) defined that in the context of SS project implementation, CSFs represent the essential ingredients without which a project stands little chance of success. Five of the most reported CSFs are shown below.

5.3.4.1 Top Management Involvement and Commitment

Literature refers to this CSF as *Top down commitment* (Tobias 1991), as *Total management commitment* (Hahn 2005) or *Management involvement and commitment* (Bañuelas and Antony 2002). However, all agree this CSF is the most important, and need to be considered. As stated by Montgomery and Woodall (2008), this CSF goes beyond just giving speeches at kick-off events; executives must devote considerable personal energy to ensure success. Top management cannot approve the SS implementation by just approving the budget for it without serious involvement and commitment (Goh et al. 2006).

5.3.4.2 Project Selection and Prioritization

Organizations should develop formal mechanisms to select SS projects. These mechanisms involve senior management to filter out SS project that do not have financial or strategic implication (Carnell 2003). From this perspective, the decision of initiating a project is allocated to senior management. Many publications revealed that project selection is the most critical and easily mishandled element during project implementation (Fundin and Cronemyr 2003).

5.3.4.3 Training and Education

Szeto and Tsang (2005) mentioned that quality improvement requires change, and change starts with people. People change when they understand the purpose and the skills to implement it. The implementation of SS allows the company to set new goals and, in effect, ask employees to cope with the change by thinking and acting differently, performing new tasks, and engaging in new behavior. In this sense, the curriculum of the belt program should reflect the organization's needs and requirements. It has to be customized to incorporate economical and managerial benefits. Training should also cover both qualitative and quantitative measures and metrics, leadership, and project management practices and skills (Kwak and Anbari 2006).

5.3.4.4 Understanding Method, Tools and Techniques

There is a direct relationship between this CSF and training and education, this implies not only the importance of receiving SS training but also verifies its understanding. A way to confirm real understanding of the methodology, is through the verification of savings in the implementation, and then could be suggested SS certification. However, this is not guaranteed, e.g., Moosa and Sajid (2010) observed that many training programs throughout the world which claim SS black/green belt certification are not capable enough to develop skills for the investigation of causal relations in complex systems through the use of these statistical techniques, resulting in qualified but incapable persons. In addition, DeRuntz and Meier (2009) stated that the Six Sigma Black Belt (SSBB) certification is granted by many organizations including industry and academia, each of these organizations have independently developed their own unique body of knowledge (BOK) by which their SSBB certification is granted.

5.3.4.5 Project Tracking and Review

Project needs to be reviewed in order to ensure that it is on track. By doing this, champions, BB, MBB and others can evaluate whether the project will be completed on schedule. If a deviation is present, resources, technical help and guidance could be provided. In the same way, Lynch and Cloutier (2003) expressed that the formal tollgate review process is a systematic review that takes place when a project reaches a milestone. This process is not intended to be a substitute for the daily and weekly project management controls already in place. Typically, a project should be reviewed after each phase of the DMAIC process is completed.

5.3.5 Tools and Techniques of Six Sigma

Nowadays, there has been a significant increase in the use and development of SS methodology in manufacturing and service industry; this is not only observed in a continuing increase of literature, but also in the increasing number of tools and techniques (T&T) used in the methodology, which makes difficult to be actualized on it, in addition, authors may differ about the important T&T to be used. McQuater et al. (1994) argued that T&T are essential process ingredients, and instrumental in success of a quality program. Many companies have used them without giving sufficient thought and have experienced barriers to progress. In general, T&T can be broadly defined as practical methods and skills applied to specific activities to enable improvement. A specific tool has a defined role and a technique may comprise the application of several such tools (Basu 2004). A single tool may be described as a device, which has a clear role and defined application. It is often narrow in its focus and usually, used on its own (Dale and McQuater 1998). Examples of tools are: cause and effect diagram, Pareto analysis, relationship diagram, and flow chart. A technique has a wider application than a tool. There is also a need for a greater intellectual thought process and more skill, knowledge, understanding and training in order to use them effectively (Dale and McQuater 1998). A technique may even be viewed as a collection of tools, for example, Statistical Process Control, Benchmarking and QFD.

A single T&T by itself will produce results in a limited area. It is the cumulative effect of a number of appropriate T&T that would create sustainable benefits for the whole organization (Basu 2004). That is why due to its holistic and tool-based approach, SS adds any other tool or method that may improve results (Hoerl 2004). However, T&T can be a double-edged sword, they are effective in the right hands and can be dangerous in the wrong ones (Basu 2004).

SS has management and technical components; the management focus is on identifying process metrics, setting goals, choosing projects, and assigning people to work on projects, while technical side is focused on enhancing process performance by reducing variation (Hu et al. 2005). In relation to SS training, it mainly involves three groups of T&T sets about team, process and statistic. Team and process tools are used to prepare the project leader with the required team building and leadership skills for implementation of the project (Szeto and Tsang 2005). These tools also help the project leader to create a shared need for the project as well as establish an extended project team.

Statistical tools help team members identify variables inherent to the process that may not be controlled adversely affecting the overall quality (McAdam and Evans 2004). Practitioners use data and statistical thinking as part of a disciplined improvement methodology. At this point, SS is a strategy and methodology for deploying statistical thinking and methods in an organization (Snee 2004). In this area, Yang (2004) stated that DOE and Regression analysis are among the most important *work horses* in SS movement and their applications achieved a great

Table 5.8 Tools and techniques more reported in literature from 2001 to 2010

Tool and technique	Presence	Percent (%)
Design of experiments (DOE)	86	60.10
Cause and effect diagram (C&E)	82	57.30
Failure mode and effect analysis (FMEA)	76	53.10
Statistical process control (SPC)	74	51.70
Process capability (CP)	73	51.00
Pareto chart	72	50.30
Process map	69	48.30
Hypothesis test	55	38.50
Brainstorming	48	33.60
Analysis of variance (ANOVA)	47	32.90
Regression analysis	47	32.90
Flow chart	45	31.50
Quality function deployment	45	31.50
SIPOC diagram	44	30.80
Repeatability and reproducibility	40	28.00
Multi-vari studies	37	25.90
Histogram	36	25.20
Measure system analysis	36	25.20
Process control plan	30	21.00
Benchmarking	27	18.90

deal in improving quality and profitability. In a similar way, Nonthaleerak and Hendry (2006) found that DOE is the most frequently studied tool in the literature.

Six Sigma's T&T are not new, there are plenty of statistical and data analysis tools which can be used for SS due to its quantitative approach, but, it also uses qualitative tools (Ricondo and Viles 2005). In the same line Thawani (2004), indicated that SS has been described as *old wine in a new bottle*, since most of the tools *packaged* in it have been around for several decades. Despite a large number of tools and techniques available, major subsets of these are especially important because of their frequent use (ReVelle 2004). In order to summarize the SS trend, a list of tools and techniques more reported from 2001 to 2010 was elaborated to separate the frequency of reports. A structured methodology was followed to narrow down the search from around 2000 papers mentioning SS to 143 mentioning T&T, this way Table 5.8 shows the 20 more frequently reported, where presence indicate the amount of times that each one is reported and it's percent respectively.

The most reported T&T are: *DOE, C&E diagram*, *FMEA*, *SPC*, *CP*, *Pareto Chart*, *Process map*, *Hypothesis test*, *Brainstorming* and *Anova*. Results suggest a balance between statistical and non-statistical T&T. This is consistent with a study conducted in China by Bañuelas et al. (2009), they found that *Pareto, FMEA*, *C&E diagram*, *SPC*, *DOE*, *hypothesis test*, *MSA*, *regression analysis*, *VSM*, *5'S and poke yoke* are the most utilized T&T. Other tools have been taken from other programs and disciplines e.g., discrete event simulation from operations research and mistake-proofing from Lean (Hoerl 2004). This suggests a benchmarking between different disciplines, a maturity level and practitioner's necessity to solve problems. In this context, organizations need to continuously learn and adapt the latest trends and techniques that are outside the SS domain that might be useful to complement the SS approach (Kwak and Anbari 2006). Nevertheless, Hoerl (2004) advises that the addition of tools should be selective so as not to slow down deployment.

Below is shown a case study reported by Báez et al. (2010), illustrating the application of the DMAIC methodology to solve a problem in a Mexican electronics company. For data analysis Minitab 14 was used.

5.4 Case Study

The present case study shows the application of Six Sigma methodology to tackle the problem of low strength in a Light Emitting Diode (LED) assembly when being welded on the customer printed circuit boards (PCBs) in the manufacture process of a cellular phone. Measurements of product strength is made by a pull test, which consists of a mechanism in which is placed the piece to be destroyed by a wedge that measures the strength used to separate the LED housing, thus obtaining the strength of the unit. The next section presents the Six Sigma's DMAIC methodology followed to face this problem.

5.4.1 Define Phase

In this initial phase, a complete analysis of company's information was conducted, including input and output process variables, key measures or performance indicators, CTQ, voice of customer (VOC), voice of business (VOB), defect rate, process capability (CPk), among others in order to determine the current performance level.

This LED represented 40 % of the company's sales; for this reason it is important to maintain or improve quality of this assembly. An analysis of this product indicated some recurrent defects, which are depicted in a Pareto chart (Fig. 5.3). It can be noticed that coplanarity is the most frequent defect; however, it was studied with an alternative successful project. The second most frequent defect was low strength, which is a CTQ characteristic. In addition, this defect is related to company's strategic objectives, and was detected that its Cpk index was below 1.33, which is the company target for strength pull test. Furthermore, engineering team expects that this situation could be corrected in three months; then in accordance with top management, it was defined as a SS project to be implemented.

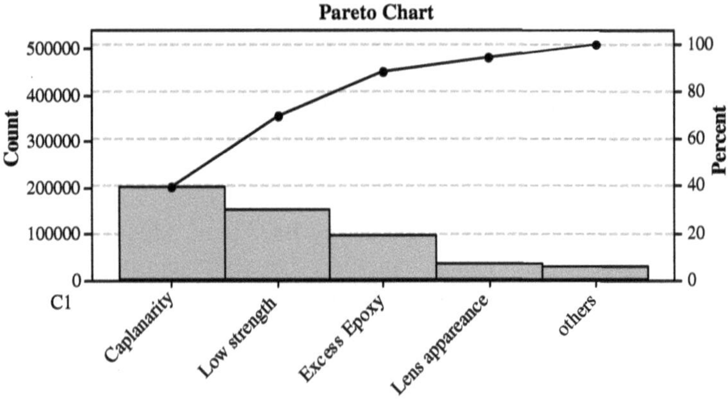

Count	202448	152612	95777	32623	27434
Percent	39.6	29.9	18.7	6.4	5.4
Cum %	39.6	69.5	88.2	94.6	100.0

Fig. 5.3 Pareto graph for defects in the production line

Table 5.9 Results obtained in the R&R study

Variation components	Components of variation	
	SD	Contribution to variation
Total gage R&R	0.18	4.06
Repeatability	0.18	4.06
Reproducibility	0	0
Part to part	4.27	95.94
Total variation	4.45	100
Distinct categories = 6		

5.4.2 Measure Phase

The next step was to determine the performance of the process according to the requirement of strength, being this is a CTQ for the assembly, so it was essential to establish the accuracy of the measurement system used to collect data. By the way, a study of repeatability and reproducibility (R&R) was conducted to identify the sources of variation that contribute to the total variation obtained by the measurement system, and as a consequence evaluate discrimination power (Henderson 2006). A measuring system is considered acceptable when its variability is less than 10 % of the total variability of the process, may be acceptable depending on the size or cost of the product if it represents between 10 and 30 % and if represents over 30 % is considered unacceptable (AIAG 2002). This R&R study showed that the measurement system represents only 4.06 % of total variation, which implies that it is acceptable. In addition, the number of distinct categories was 6, being 5 the lowest recommended (Table 5.9).

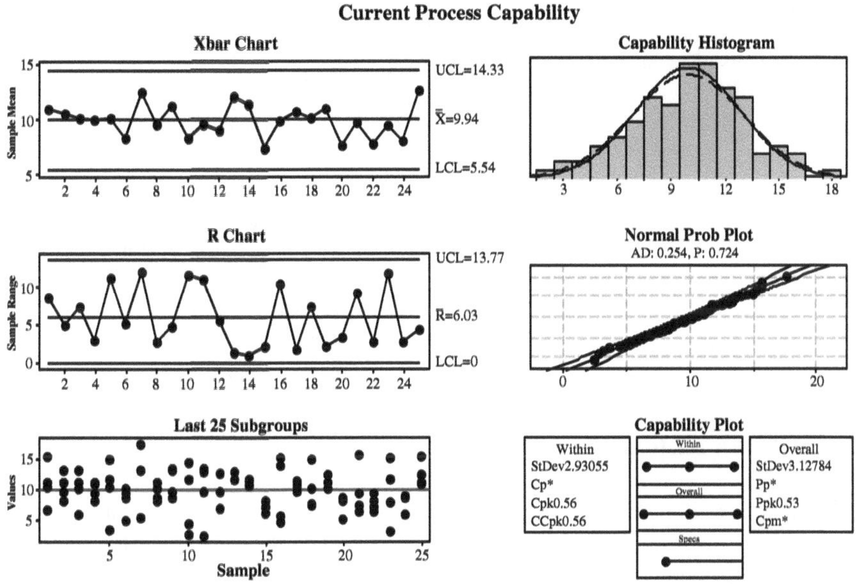

Fig. 5.4 Current process capability analysis

The baseline process capability (Cpk) is also obtained in this phase once that previously was determined that the measurement system was reliable. Thus the process capability study was conducted collecting 25 samples each one of size 4 during 5 days in different shifts (Fig. 5.4). It can be observed that the process is stable over the time, however it also has a really poor behavior with a Cpk = 0.56 which is too low considering that the quality characteristic is considered critical, so it is clearly evident the need to improve this process.

5.4.3 Analyze Phase

In this phase a comprehensive analysis of the process was realized through a brainstorming session considering the participation of different personnel involved with this LED assembly, generating a cause and effect diagram with potential variables affecting the strength to pull test of this product (Fig. 5.5). Further analysis with possible causes and previous tests helped to consider the variables in this study: epoxy load, injection pressure, height adjustment, ventilators, life of epoxy, mold wear, cavities and baking temperature.

Now, a brief description of these factors and why were considered in the study is presented. It is in this phase when it was decided to use Taguchi methods to optimize the process.

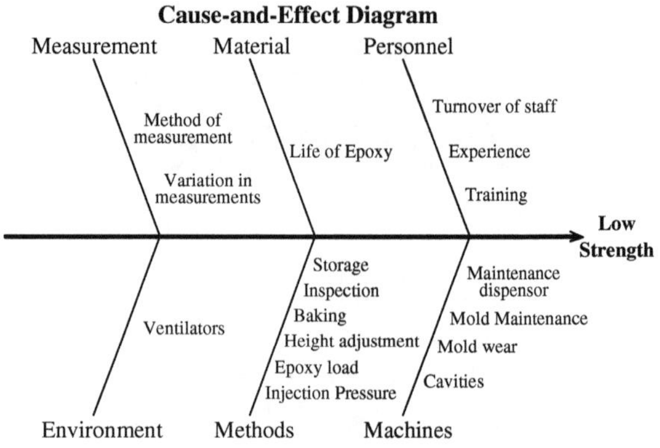

Fig. 5.5 Cause and effect diagram to determine possible factors affecting the pull test resistance

5.4.3.1 Noise Factors

Life of epoxy. It is known as maximum exposure time of epoxy resin at room temperature recommended by the supplier to ensure the substance maintain the appropriate features to be used in the process. During its 4 h of useful life, the resin will undergo small changes in its temperature, which will result in a viscosity increase, causing the dispenser's needles clogging, depositing different amounts of resin in the assemblies, provoking differences in strength of products to pull test.

Mold wear. This is an important factor since it may affect the assembly's strength, so the process should be set to be minimally affected by the mold wear during its useful life. The experiment was carried out in two molds, one in perfect conditions and other with considerable degree of wear.

Cavities. The differences among cavities of a same mold could affect the results obtained on the pull test. After preliminary tests, it was determined that for the experiment, cavities 3 and 12 would be analyzed.

It is necessary to include the above three parameters as noise factors because they could affect the strength of the assembly, so the intention is find out the process operating condition that is minimally affected by such factors. Besides these factors were considered as noise because pretend to control them is impractical or uneconomical.

5.4.3.2 Control Factors

Epoxy load. Was tested at four levels, 100, 75, 50 and 25 % of application because there was interest in knowing if less than 100 % of the epoxy application the assembly would reach the necessary strength.

Injection pressure. Was evaluated at two levels; the first level was the normal working condition (30–40 psi) and the second was a higher pressure (50–60 psi) tested previously with good results.

Table 5.10 Control factors and their levels

Control factors	Level 1	Level 2	Level 3	Level 4
(A) Epoxy load	100 %	75 %	50 %	25 %
(B) Injection pressure	30–40 psi	50–60 psi	–	–
(C) Height adjustment	High	Low	–	–
(D) Ventilators	On	Off	–	–
(E) Temperature	Level 1: 306, 300, 294, 170 and 70 °C			
	Level 2: 321, 315, 309, 185 and 85 °C			

Table 5.11 Noise factors and their levels

Noise factors	Level 1	Level 2
Life of epoxy	New	2 h
Mold wear	New	Old
Cavities	Cavity 3	Cavity 12

Height adjustment. Of epoxy dispenser was tested at two levels. The first level is high setting (which is normally used) and the second was in a lower setting. At this level dispensing needles are closer to the mold.

Ventilators. Are considered because the wind blowing cools the oven causing the epoxy does not cure properly in the assembly, and so the welding may not be adequate. To check its influence, this factor was studied at two levels: on and off.

Baking temperature. Refers to temperatures through which the product passes into the oven considering five stages. The first level was the baking process that is normally used, and for second level, the temperature was incremented 15 °C at each stage, considering that higher temperatures would give better results, according to the experience of team members.

5.4.4 Improve Phase

In this stage, the experiment was ran using the Taguchi methods as specified above, because this require a small number of experimental runs, since as explained by Cesarone (2001), only a small fraction of all possible factors combinations is tested to calculate the effects of all inputs on the outputs. For this reason, the factors were firstly divided as control and noise factors (Tables 5.10, 5.11).

To accommodate the factors it was necessary to use the L_8 and L_4 orthogonal arrays (OA) for inner and outer arrays respectively. For this experiment the inner array was modified to meet the requirements of the experiment, because the factor epoxy load was tested at four levels.

For noise factors was necessary an L4 OA, which allows studying up three factors at two levels. Table 5.12 shows the resulting design matrix, where the *Yij* symbol represents the results obtained when the experiment was run in each intersection between the different levels of the control and noise factors.

Table 5.12 Design matrix

						Outer array				
	Factors					3	1	2	2	1
						2	1	2	1	2
	Inner array					1	1	1	2	2
Run	A	B	C	D	E					
1	1	1	1	1	1					
2	1	2	2	2	2					
3	2	1	1	2	2					
4	2	2	2	1	1			Y_{ij}		
5	3	1	2	1	2					
6	3	2	1	2	1					
7	4	1	2	2	1					
8	4	2	1	1	2					

5.4.4.1 Conducting the Experiment

The experiment started being careful and watching every one of the details that could affect the outcome of the experiments and the runs were performed according to the order of the resulting array.

The next step of the experiment was to analyze the collected data to find which factors affect the variation and so, which of them were important and additionally at what levels they should be set, to establish the final configuration of the process. The variation is evaluated respect to the mean of the data and the signal-to-noise (S/N) statistic, which measures the performance of the robustness for each combination of control factors (Gutiérrez 2004). This study aims to maximize strength to pull test for this LED assembly, so that the response variable is of the type the bigger the best, corresponding to the following S/N transformation:

$$\eta dB = -10 \log \left[\frac{1}{n} \sum_{i=1}^{n} \frac{1}{y_i^2} \right] \tag{5.1}$$

Table 5.13 shows the complete results including the mean and S/N ratios for analyzing strength to pull test. With the results shown in this table, it was necessary to construct the factorial plots for the S/N ratio and mean, respectively (Figs. 5.6, 5.7). An analysis of these graphs, suggests as the best combination of control factors: A3 B2 C2 for the S/N ratio and A1 B2 C2 for means. The slopes close to zero for factors D and E suggest that the effect of these variables is not significant for both responses.

The ANOVA was performed for verifying the statistical effect for each factor in both, the S/N ratio and the mean response (Tables 5.14, 5.15). This analysis confirmed that the variables D and E are not significant and so were grouped at random error, resulting the grouped random error (Error) used in subsequent

Table 5.13 Data collected including calculations of means and S/N ratios

Factors						Outer array					
						1	2	2	1		
						1	2	1	2		
Inner array						1	1	2	2	Response	
Run	A	B	C	D	E					Mean	S/N
1	1	1	1	1	1	14.88	6.14	6.63	15.26	10.73	18.37
2	1	2	2	2	2	13.98	21.05	23.7	17.66	19.1	25.1
3	2	1	1	2	2	14.56	9.9	4.39	5.97	8.71	16.24
4	2	2	2	1	1	18.44	14.8	18.59	10.77	15.65	23.22
5	3	1	2	1	2	16.4	13.12	18.16	10.07	14.44	22.51
6	3	2	1	2	1	14.4	9.05	14	16.63	13.52	21.91
7	4	1	2	2	1	0.52	0.56	1.2	1.73	1	−2.96
8	4	2	1	1	2	5.08	0.93	2.74	0.3	2.26	−4.92

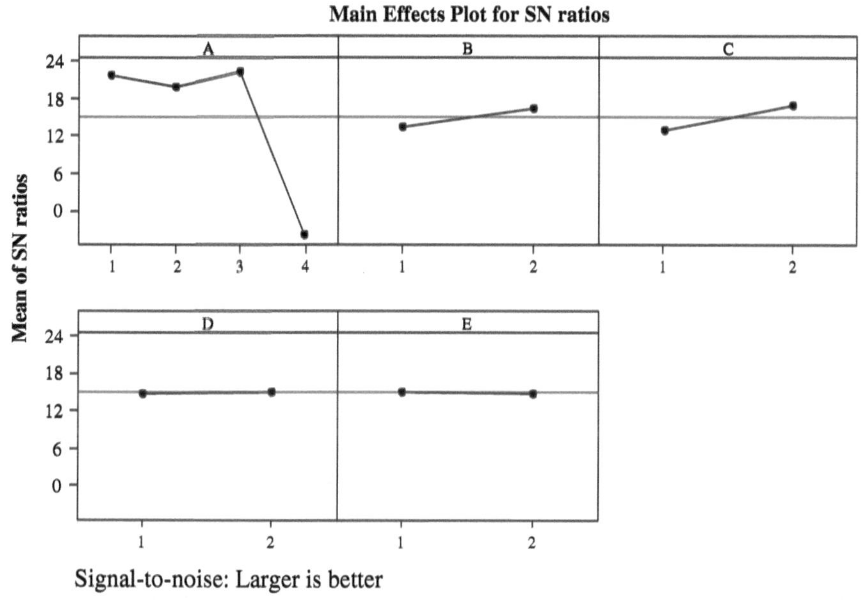

Fig. 5.6 Factorial plot for signal to noise ratio

calculations. This way ventilator and temperature best level could be established considering economical or practical issues.

Once the analysis of variance was realized, it was concluded that important factors for both response variables are: A, B and C. Later, observing factorial plots their suitable levels were defined in the following manner: Epoxy load = 100 %, Injection pressure = 50–60 psi and Height adjustment = low.

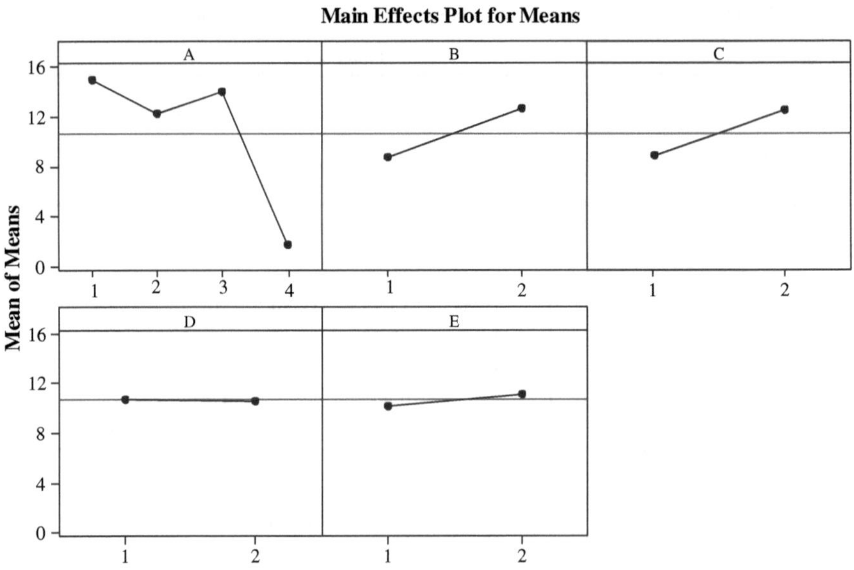

Fig. 5.7 Factorial plot for original data (mean)

Table 5.14 ANOVA for S/N ratio

Factors	Degrees of freedom	Sum of squares	Mean square	F	P
A	3	957.69	319.23	1329.05	0.001
B	1	15.5	15.5	64.52	0.015
C	1	33.06	33.06	137.65	0.007
*D	1	0.15	–	–	–
*E	1	0.33	–	–	–
Error	–	–	–	–	–
(Error)	2	0.48	0.24	–	–
Total	7	1006.73			

*No significant factors pooled into error

Table 5.15 ANOVA for natural response variable

Factors	Degrees of freedom	Sum of squares	Mean square	F	P
A	3	903.16	301.05	22.89	0
B	1	122.58	122.58	9.32	0.005
C	1	112.09	112.09	8.52	0.008
*D	1	0.28	0.28	0.02	0.885
*E	1	6.49	6.49	0.49	0.849
Error	24	315.61	13.15	–	–
(Error)	26	322.38	12.4	–	–
Total	31	1460.21			

*No significant factors pooled into error

Predicting the performance with optimal levels. The expected values for mean and S/N under the proposed process operating condition: A1, B2, C2, were calculated using Eq. (5.2):

$$\hat{Y} = \overline{y} + \left(\overline{A_1} - \overline{y}\right) + \left(\overline{B_2} - \overline{y}\right) + \left(\overline{C_2} - \overline{y}\right) \tag{5.2}$$

Prediction for the mean strength

$$\hat{Y} = 10.676 + (14.91 - 10.67) + (12.63 - 10.67) + (12.54 - 10.67)$$
$$\hat{Y} = 18.74 \text{ Psi}$$

Prediction for S/N

$$\hat{Y} = 14.93 + (21.74 - 14.93) + (16.32 - 14.93) + (16.97 - 14.93)$$
$$\hat{Y} = 25.16 \text{ dB}$$

It is important to mention that the factors and their indicated levels coincide with run number two of internal array and the predicted values are very close to those obtained in the experiment. Considering that optimal levels coincided with an experimental trial, it was not necessary to realize a confirmation run, however, a T-paired test was conducted to reinforce the decision resulting a 95 % confidence interval for differences (5.51, 8.61) demonstrating that the strength mean for proposed setting is higher statistically.

5.4.5 Control Phase

The great challenge of the SS methodology may not be the improvement of processes, but to maintain the optimum results for its daily operation. This requires standardization of working methods and monitoring and continuous control of operations. To achieve this, once validated the improvement, the new process operating conditions was established according to the optimal results found in the previous phase of the project and later a control chart was generated, for operators to take preventive action before this quality characteristic starts to go out of control with the consequent loss of quality (Fig. 5.8). For calculating the control limits 25 samples of size 4 each, were collected during 5 days in different shifts. Besides, with this information PPM's, Cpk index and process sigma level were calculated to allow comparison with the initial process.

5.4.6 Project Results

To determine the savings generated by the project only were considered the cost generated by a defective unit detected within the company. Table 5.16 resumes the

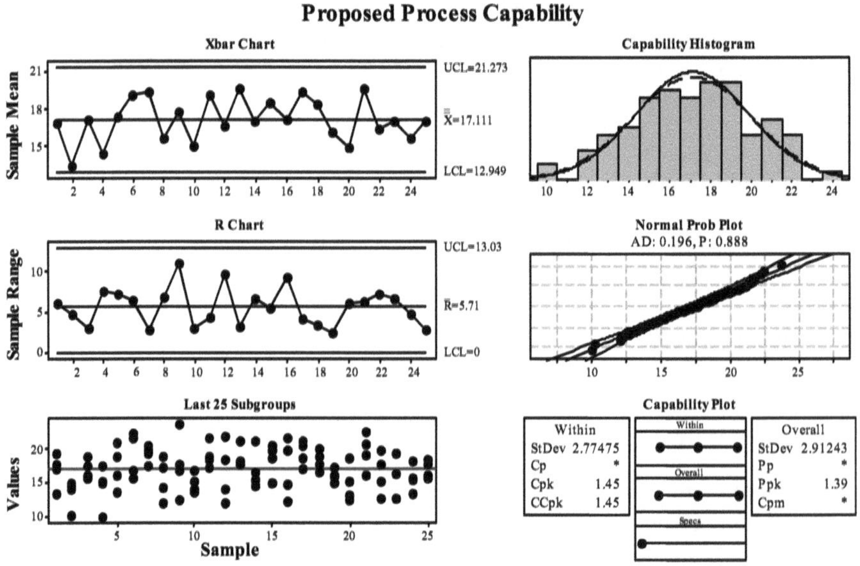

Fig. 5.8 Proposed process capability analysis

Table 5.16 Savings achieved by reducing the number of defects

	Before project	After project
Monthly production	1,170,000.00	1,170,000.00
Annual production	14,040,000.00	14,040,000.00
PPM	46,046.15	6.36
Annual defects	646,487.95	89.29
Annual cost for low strength	129,297.59	17.86
Savings after six sigma project	US$ 129,279.73	

savings generated by the project. These calculations were made considering the parts per million of defective products before and after this improvement project, obtained with their respective process capability studies.

Through the development of this project, the company was able to demonstrate the importance of undertaking improvements in their processes based on statistical techniques, in this case following the Six Sigma methodology. Before this project, improvements were made by trial and error or by collecting a large number of samples and analyzing only the mean and standard deviation to make a decision, without any statistical test, so that the conclusions were not adequate and also the costs of these tests were too high because sample sizes used even up to two thousand units. Through the development of the project, it was possible to achieve a drastic improvement with the consequent cost reduction, estimated around

Table 5.17 Comparisons of indexes for each analyzed process

Index	Proposed process	Current process
Mean	17.11	9.94
Standard deviation	2.77	2.93
Cpk	1.45	0.56
PPM	6.36	46,046.15
Sigma level	5.86	3.18

$ 129,279.73 dollars. General results are presented in Table 5.17. It is important to mention that after this successful implementation, was obtained an increased customer satisfaction of this subassembly reducing practically to zero their complaints for this problem.

5.5 Lean Six Sigma

Lean Manufacturing (LM) originated from the Japanese Toyota Production System (TPS), focuses on maximizing process velocity and separation of "value-added" from "non-value-added" activities to eliminate the root causes of non-valued activities. In this context, the essential idea is to identify and eliminate "waste" in its different presentations:

- *Transportation.* Moving products or materials.
- *Inventory.* Products or components waiting to be processed.
- *Motion.* Unnecessary human or machine movement.
- *Waiting Time.* People waiting for material, machine or operation.
- *Over processing.* Inappropriate operations and oversize equipment.
- *Overproduction.* Products ahead of demand (too much/too early).
- *Defects.* Nonconforming products or non-working properly.

Additional waste could be human talent, i.e., take little notice of what workers can contribute. Tools, techniques and strategies commonly included are: 5S, TPM, Cellular/Flow, Pull system, Kan-ban, Quick Changeover, work standardization, value stream mapping among others. LM and SS business improvement programs are often implemented separately. However, a growing number of companies have realized these programs are a dynamic, synergistic force rather than two separate and competing initiatives (Pickrell et al. 2005). Therefore, Gibbons (2006) argued the fusion of LM and SS is required because LM cannot bring a process under statistical control and SS alone cannot dramatically improve process speed or reduced invested capital. In this sense Su et al. (2006) stated that a pure SS approach lacks three desirable LM characteristics:

- No emphasis on improving the speed of a process
- Lack of direct attention to reduce the amount of investment in inventory
- No realization of quick financial gains owing to the inordinate amount of time required for the learning and applying its methods and tools for data collection and analysis.

Deficiencies of a pure lean improvement effort include:

- The processes are not brought under statistical control
- There is no strategy for evaluating variations in measurement systems used for decisions
- The process improvement practices do not link quality and advanced mathematical tools to diagnose and solve process problems that remain once the obvious waste has been removed.

In relation to Lean Six Sigma (LSS) implementation, Hoerl (2004) suggested that the incorporation of both methodologies should be leaded by SS methodology. Utilizing SS as the lead initiative allows the formal infrastructure of Six Sigma to remain in the LSS implementation (Pickrell et al. 2005). In this scenario, Montgomery and Woodall (2008) found that SS combined with lean principles and DFSS are entirely consistent with Deming's system of profound knowledge. In other words, what distinguish LSS from their solely implementation, is the acknowledgment that you cannot improve quality or improve flow velocity in an isolated mode. In order to improve quality as defined by VOC, in time and quantity among other strategic company's goals, a balanced process should be accomplished. At this time LSS can help to achieve the fastest rate of improvement in customer satisfaction, cost, quality, process speed, and invested capital.

5.6 Conclusions

In this chapter a literature review of SS is presented, remarking the most reported T&T as well as CSFs in literature for a successful implementation. Secondly can be noticed congruence between theory and practice through a case study, where four of the five CSF mentioned above were used, and about T&T, 10 of 20 more reported in literature were deployed, obtaining a successful implementation taking into account defects rates and their impact on saving and customer satisfaction.

It is important to manifest that not all T&T together with CSF apply in the same way for all kind of organizations, for some of them, a certain array may result better, but for others a different set could works better, and then it is important to know them. It is imperative to highlight that the use of correct T&T plus a consideration of success factor may increase chances to obtain benefits implementing SS. Finally, is also mentioned the relevance of integrate methodologies for process improvement like six sigma and lean manufacturing in order to obtain optimal results.

References

Anand, G., Ward, P., & Tatikonda, M. (2009). Role of explicit and tacit knowledge in six sigma projects: An empirical examination of differential project success. *Journal of Operations Management, 28*(4), 303–315.

Antony, J., & Bañuelas, R. (2002). Key ingredients for the effective implementation of six sigma program. *Measuring Business Excellence, 6*(4), 20–27.

Báez, Y., Limón, J., Tlapa, D., & Rodriguez, M. (2010). Implementing six sigma and Taguchi methods to obtain an increased resistance on a pull test of a light emitting diode. *Información Tecnológica, 21*(1), 63–76.

Bañuelas, R., & Antony, J. (2002). Critical sucess factors for the sucessful implementation of six sigma projects in organization. *The TQM Magazine, 14*(2), 92–99.

Bañuelas, R., Johnson, M., Virakul, T., & Antony, J. (2009). Assessing the role of six sigma in a supply chain: An exploratory study in the UK manufacturing organizations. *International Journal of Six Sigma and Competitive Advantage, 5*(4), 380–397.

Basu, R. (2004). Six-sigma to operational excellence: Role of tools and techniques. *International Journal of Six Sigma and Competitive Advantage, 1*(1), 44–64.

Black, K., & McGlashan, R. (2006). Essential characteristics of six sigma black belt candidates: A study of US companies. *International Journal of Six Sigma and Competitive Advantage, 2*(3), 301–312.

Brady, J. E., & Allen, T. T. (2006). Six sigma literature: A review and agenda for future research. *Quality and Reliability Engineering International, 22*(1), 335–367.

Breyfogle, F. (2003). *Implementing six sigma: Smarter solutions using statistical methods*. New York: Wiley.

Carnell, M. (2003). Gathering customer feedback. *Quality Progress, 36*(1), 60–61.

Cesarone, J. (2001). The power of Taguchi. *Institute of Industrial Engineers Solutions, 33*(11), 36–40.

Chakravorty, S. (2009). Six sigma failure: An escalation model. *Operations Management Research, 2*(1), 44–55.

Cooke-Davis, T. (2002). The real success factors on projects. *International Journal of Project Management, 20*(1), 185–190.

Dale, B., & McQuater, R. (1998). *Managing business imprevement and quality*. Oxford: Wiley-Blackwell.

DeRuntz, B., & Meier, R. (2009). Trainers' perceptions of the relative importance of the ten topics included in the American Society for quality's six sigma black belt certification. *Journal of Industrial Technology, 25*(3), 2–13.

Dirgo, R. (2005). *Look forward: Beyond lean and six sigma*. Boca Raton, FL: J Ross Publishing.

Fundin, A., & Cronemyr, P. (2003). Use customer feedback to choose six sigma projects. *ASQ Six Sigma Forum Magazine, 3*, 17–21.

Gibbons, P. M. (2006). Improving overall equipment efficiency using a lean six sigma approach. *International Journal of Six Sigma and Competitive Advantage, 2*(2), 207–232.

Goh, T. N., Tang, L. C., & Lam, S. W. (2006). Six sigma: A SWOT analysis. *International Journal of Six Sigma and Competitive Advantage, 2*(3), 233–242.

Gray, J., & Anantatmula, V. (2009). Managing six sigma projects through the integration of six sigma project management process. *International Journal of Six Sigma and Competitive Advantage, 5*(2), 127–143.

Green, F., Barbee, J., Cox, S., & Rowlett, C. (2006). Green belt six sigma at a small company. *International Journal of Six Sigma and Competitive Advantage, 2*(2), 179–189.

Gutiérrez, H. & De La Vara, R. (2004). Análisis y diseño de experimentos. México: McGraw Hill

Hahn, G. J. (2005). Six sigma: 20 key lessons learned. *Quality and Reliability Engineering International, 21*(3), 225–233.

Harry, M., & Schroeder, R. (2000). Six sigma: The breakthrough management strategy revolutionizing the world's top corporations. *Currency*. New York.

Henderson, G. R. (2006). *Six sigma: Quality Improvement with MINITAB* (1st ed.). Chichester: Wiley.

Hoerl, R. (2004). One perspective on the future of six-sigma. *International Journal of Six Sigma and Competitive Advantage, 1*(1), 112–119.

Hu, M., Bart, B., & Sears, R. (2005). Leveraging six sigma disciplines to drive improvement. *International Journal of Six Sigma and Competitive Advantage, 1*(2), 121–133.

Kanani, Y. (2006). Study and analysis of control phase role for increasing the success of six sigma projects. In *International Conference on Management of Innovation and Technology IEEE* (pp. 826–829).

Kumar, M. (2007). Critical success factors and hurdles to six sigma implementation: The case of a UK manufacturing SME. *International Journal of Six Sigma and Competitive Advantage, 3*(4), 333–351.

Kwak, Y., & Anbari, F. (2006). Benefits, obstacles, and future of six sigma approach. *Technovation, 26*, 708–715.

Lynch, D., & Cloutier, E. (2003). 5 steps to success. *Six Sigma Forum Magazine, 2*(2), 27–33.

McAdam, R., & Evans, A. (2004). The organizational contextual factors affecting the implementation of six-sigma in a high technology mass-manufacturing environment. *International Journal of Six Sigma and Competitive Advantage, 1*, 29–43.

McCarty, T., Bremer, M., Daniels, L., & Gupta, P. (2005). *The black belt handbook*. Schaumburg: McGraw-Hill.

McManus. (2008). So long six sigma? *Industrial Engineer, 40*, 18.

McQuater, R., Dale, B., Wilcox, M., & Booden, R. (1994). The effectiveness of quality management techniques and tools in the continuous improvement process. *Proceedings of Factory 2000 IEE, 14*(10), 574–580.

Montgomery, D., & Woodall, W. (2008). An overview of six sigma. *International Statistical Review, 76*(3), 329–346.

Moosa, K., & Sajid, A. (2010). Critical analysis of six sigma implementation. *Total Quality Management, 21*(7), 745–759.

Nonthaleerak, P., & Hendry, L. (2006). Six sigma: Literature review and key future research. *International Journal of Six Sigma and Competitive Advantage, 2*(2), 105–161.

Pande, P., & Holpp, L. (2002). *What is six sigma?*. New York: McGraw Hill.

Perry, L., & Barker, N. B. (2006). Six Sigma in the service sector: A focus on non-normal data. *International Journal of Six Sigma and Competitive Advantage, 8*(3), 313–333.

Pickrell, G., Lyons, H., & Shaver, J. (2005). Lean six sigma implementation case studies. *International Journal of Six Sigma and Competitive Advantage, 1*(4), 369–379.

Pyzdek, T. (2003a). *Six sigma handbook revised and expanded*. New York: McGraw Hill.

Pyzdek, T. (2003b). *The six sigma project planner*. McGraw Hill, New York.

ReVelle, J. (2004). Continuous improvement six sigma. *Professional Safety, 49*(10), 38–46.

Ricondo, I., & Viles, E. (2005). Six sigma and its link to TQM, BPR, lean and the learning organization. *International Journal of Six Sigma and Competitive Advantage, 1*(3), 323–354.

Rockart, J. (1979). Chief executives define their own data needs. *Harvard Business Review, 57*(2), 238–241.

Shenhar, A., Levy, O., & Dvir, D. (1997). Mapping the dimensions of project success. *Project Management Journal, 28*(2), 5–13.

Snee, R. (2004). Six sigma: The evolution of 100 years of business improvement methodology. *International Journal of Six Sigma and Competitive Advantage, 1*(1), 4–20.

Snee, R. (2010). Lean six sigma getting better all the time. *International Journal of Lean Six Sigma, 1*(1), 9–29.

Snee, R., & Hoerl, R. (2007). Integrating lean and six sigma a holistic approach. *Six Sigma Forum Magazine, 6*(3), 15–21.

Stock, G. (2002). Taking performance to a higher level. *Six Sigma Forum Magazine, 1*(3), 23–26.

Su, C. T., Chiang, T. L., & Chnag, C. M. (2006). Improving service quality by captalising on an integrated Lean six Sigma methodology. *International Journal of Six Sigma and Competitive Advantage, 2*(1), 1–22.

Szeto, A., & Tsang, A. (2005). Antecedents to successful implementation of six sigma. *International Journal of Six Sigma and Competitive Advantage, 1*(3), 307–322.

Tang, L., Goh, T., Lam, S., & Zhang, C. (2007). Fortification of six sigma: Expanding the DMAIC toolset. *Quality and Reliability Engineering International, 23*(1), 3–18.

Thawani, S. (2004). Six Sigma: Strategy for organizational excellence. *Total Quality Management, 15*(5), 655–664.

Tkáč, M., & Lyócsa, Š. (2009). On the evaluation of six sigma projects. *Quality and Reliability Engineering International, 26*, 115–124.

Tobias, P. (1991). A six sigma program implementation. In *IEEE 1991 Custom Integrated Circuits Conference IEEE* (pp. 2911–2914).

Yam, H. Y. (2006). Six sigma: Past, present and future. In H. Yam, & T. Yoap (Eds.). *Six sigma: Advanced tools for black lelt and master black belt* (pp. 2–17). Wiley, New York.

Yang, K. (2004). Multivariate statistical methods and Six-Sigma. *International Journal of Six Sigma and Competitive Advantage, 1*(1), 76–96.

Yang, T., & Hsieh, C. H. (2009). Six-sigma project selection using national quality award criteria and Delphy fuzzy multiple criteria decision-making method. *Experts Systems with Applications, 36*(4), 7594–7603.

Zimmerman, J., & Weiss, J. (2005). Six sigma's seven deadly sins. *Quality, 7*(10), 62–67.

Chapter 6
Creating the Lean-Sigma Synergy

Francisco J. Estrada-Orantes and Noé G. Alba-Baena

Abstract Applying Continuous Improvement techniques in a manufacturing environment has become an issue when trying to improve a situation known to cause problems to customers, which needs to be stopped immediately. Typically, continuous improvement activities are organized around two major methodologies: Lean Manufacturing and Six Sigma. While Lean Manufacturing tends to be a perfect approach for most quick fixes, Six Sigma tends to work better when there is plenty of time to conduct deeper statistical analysis. This chapter explores a proposal to combine the Lean Manufacturing speed with the Six Sigma power of analysis, to create synergy amongst the two methodologies. The results presented in this chapter show that the use of a Lean-Sigma approach may lead to improvements in overall performance of up to five sigma levels, and achieve a fast fix of the undesirable situation in a short period of time.

Keywords Six sigma · Lean sigma · Project improvement

6.1 Introduction

Henry Ford's vision of a continuous mass production line was based on the principle of keeping everything in motion taking the work to the man, and not the man to the work. According to Ford, the speed of a continuous assembly line was critical to maintain a high throughput and a low cost. Everything had to keep moving in a steady and continuous pace without interruptions: raw materials, work-in-process, and finished products. To keep everything on the same pace, he

F. J. Estrada-Orantes (✉) · N. G. Alba-Baena
Department of Industrial and Manufacturing Engineering, Institute of Engineering and Technology, Autonomous University of Ciudad Juarez, Av. Del Charro 450 Norte, Col. Partido Romero, Ciudad Juárez, CHIH, México
e-mail: frestrad@uacj.mx

J. L. García-Alcaraz et al. (eds.), *Lean Manufacturing in the Developing World*,
DOI: 10.1007/978-3-319-04951-9_6, © Springer International Publishing Switzerland 2014

had to pay attention to all kind of details, from the individual tasks and tools used in the assembly process, to the large cranes and conveyors required at different stages of the operation (Hopp and Spearman 2000). As mass production gained popularity in the manufacturing arena, and the scientific management approach made possible to standardize individual tasks, then it was a matter of keeping the worker motivated to achieved the desired pace. Besides and after Elton Mayo, different motivational approaches were used to keep workers concentrated on achieving a specified throughput rate.

For a century the efforts were focused in reaching high production rates up to the late 1970s when many companies in the USA faced this scenario (in the late 1970s and early 1980s) during the economic crisis that forced them to look into and follow the continuous improvement paths. In general, improvement programs are not in the daily agenda, but, when the burden of a company becomes higher than the expected, and the financial health of the company is in risk, the continuous improvement initiatives arises as the most important programs in the organization. Learning from this experience US companies overcome critical situations using mainly the "quality path" or the "Just-in-Time (JIT) production path (Schonberger 1986). These paths represent the extremes of the series of improvement methodologies. Just in time ideas were firstly introduced by Taiichi Ohno in 1988, and are based on the Toyota Production System (TPS), to be later known as Lean Manufacturing. A more efficient use of this path was proposed and developed after Womack's (1990) research. Womack's ideas redefine Lean to be oriented in achieving a rapid and continuous improvement in the production systems. And today the "Lean" path is identified as the "just do it" and "keep it simple" strategy. On the other hand, since 1982 (Deming 1982) the "Quality" path is based in Deming's statistical approach (Deming 2000) to quality control and efficiency improvement. The main goal here is to identify and eliminate the root cause of problems through an effective in-depth statistical analysis, causing an improvement chain reaction.

6.1.1 Lean Manufacturing

To find the synergetic elements between these different strategies and be able to visualize the possibility for merging them and create a synergetic relationship, it is necessary to understand the basic concepts of these paths for improvement (Lean and Six Sigma), it is necessary to understand their principles, tools and methods. According to Womack and Jones (1996) the Lean Manufacturing methodology is based on five principles:

1. Specify the Value added of the product. Provide exactly what the customer wants, at the right time, and at the right price.
2. Identify the main Value Stream. From the customer's perspective, identify all the activities in the production process that add value to the product.

3. Develop Flow. Make sure the process flows without interruptions, delays or accumulations.
4. Implementing Pull production scheduling. The Production should be customer driven, or oriented to fulfill the customer demands. Note: Here is when just-in-time is needed.
5. Strive for Perfection. Continue looking for perfection by eliminating the different types of waste.

There are mainly eight types of wastes identified in the Lean manufacturing methodology as:

1. Defective product. Parts or products that do not meet the customer's requirements and or specifications.
2. Over production. Producing more than what the customer requires, or producing before the customer needs it, or producing faster than the customer's consumption.
3. Over processing. Adding processing activities to a production process that it is not necessary and the customer is not willing to pay, for. That is having someone remove excess material on a molded part, when it was supposed to be perfect going out of the molding machine.
4. Transportation waste. Moving materials (during the production process) or products around unnecessarily.
5. Motion waste. People moving around during the production process unnecessarily.
6. Inventory waste. Keeping more inventory than strictly necessary to keep the process flowing continuously.
7. Waiting time. Time that an operation stops production flow waiting for an input (material, machine, people, order).
8. Talent waste. Waste for not using people's talent to improve the process.

In summary, Lean Manufacturing uses the principles and waste definitions described above to identify and eliminate waste, looking to achieve the shortest Lead-time, and the lowest cost possible. Additionally, the Lean path also uses a set of different tools and manufacturing approaches to support achieving the Lean objectives. The most popular tools used in the Lean path are: Value Stream Map, Takt time and Standard Work, VA/NVA analysis, and Lead time analysis. On the other hand, looking to achieve a Just-in-Time manufacturing scenario, Lean uses the following production approaches: Pull system, Cellular manufacturing, Poka-Yokes, SMED, TPM, Kanban, Mix-model production and Heijunka. And to capture people's talent, the Kaizen Newspaper, and Kaizen events are considered the most used and effective.

6.1.2 *Six Sigma*

Six Sigma is a continuous improvement methodology that focuses primarily in the identification and reduction of processes' variation. As a structured methodology Six Sigma was first introduced and implemented in Motorola during the 1990s (Pande et al. 2000).

The objective of Six Sigma is to achieve a defective part rate of not more than 3.4 defective parts per million parts processed, and detected after the production process. As shown in Fig. 6.1, the Six Sigma process is divided in five steps: Define-Measure-Analyze-Improve and Control (DMAIC) and encompasses a variety of problem-solving and statistical tools applied mainly to quality-related problems. Also Fig. 6.1 presents a summary of the extensive set of tools available for each phase of the DMAIC methodology. Due to these characteristics, there is a need for using Handbooks that contain a short explanation on how to use each tool when using this methodology. A good example of a useful manual was published by Rath and Strong (2000).

6.1.3 *Lean-Six Sigma*

The merging of these tested and efficient strategies and methodologies was proposed by George (2002). By developing a quality culture and improving the operational metrics to satisfy and exceed the customer requirements, George's model focuses on creating robust processes. Known as Lean-Six Sigma approach, this path takes advantage of the structured DMAIC road map from Six Sigma and the tools from Lean approach creating a large but more robust set of tools for each phase of the methodology (George et al. 2005). From the problem definition to the control phase, Lean-sigma combines the use of Lean and Six Sigma elements to create harmonious synergies for improving the quality and in consequence the production deliverables. Also, as a main characteristic of a project-oriented approach, the model emphasizes the completion of every phase before initiating the next, and uses a tollgate review to ensure that all Lean and Six Sigma components have been implemented and completed.

By merging the tools and approaches from both Lean and Six Sigma it is possible to find faster and more reliable solutions and to structure well validated improvement programs. Nowadays, Lean Six Sigma is defined as a powerful and proven method for improving business efficiency and effectiveness. Where the *value* is defined as the amount. The customer is willing to pay for the right products and services, at the right time, at the right price, at the right quality. Lean Six Sigma key principles are defined as:

- Focus on the customer.
- Identify and understand how the work gets done (the *value stream*).
- Manage, improve and smooth the process flow.

Six Sigma Improvement Process					
Objectives	•Confirm Team Goals and Validate Improvement Opportunity. •Define Current State	•Define Current State • Collect and display Baseline Data • Identify and define critical Requirements • Determine Process Capability	•Identify Potential Root Causes. • Explore cause/effect Relationships • Investigate potential root causes • Clarify Problem Statement. • Narrow Potential KPIV' s • KPOV' s • Validate Root cause • Generate Potential Solutions.	•Actively Validate Root causes. • Determine Optimal Solution • Document Future Process • Estimate Financial Benefit • Develop Implementation Plan.	•Develop Control Plan. • Monitor Performance. • Develop Communication Plan • Identify and Develop Replication and Standardization Opportunities. • Develop Project Closeout Plan.
Phases	DEFINE	MEASURE	ANALYZE	IMPROVE	CONTROL
Potential Tools	• Process Mapping • SIPOC • Affinity Diagram • Stakeholder' s Analysis • Rolled Throughput Yield (RTY) • Voice of Customer • Critical to Quality Tree (CTQ) • Project Charter • Process Thought Map	• Process Mapping • SIPOC • Brainstorming • Nominal Group Technique • Affinity Diagram • Pareto Charts • Cause and Effect Diagram • Cause and Effect Matrix • Check Sheets • Run Charts • Control Charts • Gage R&R • Kappa • Process Capability	• Graphical Tools and Techniques • Sampling Strategy • Probability • Hypothesis Testing • Simple Regression • Components of Variation • Multivari Chart • FMEA • Process Mapping • Multiple Subjective Evaluation • Kappa / ICC	• Multiple Regression • One-way ANOVA • Two-way ANOVA • Full Factorial Experiments • Fractional Factorial Experiments • Financial Basics • Process Mapping • FMEA • Process Capability • Graphical Tools and Techniques	• CUSUM Control Charts • EWMA Control Charts • Process Control Plan • Implementation Plan • Communication Plan • Poka-Yoke • 5S • Kaizen

Fig. 6.1 Roadmap of DMAIC phases and the potential tools that can be used for achieving the objectives in each phase (Estrada 2003)

- Remove Non-Value-Added steps and waste.
- Manage by fact and reduce variation.
- Involve and equip the people in the process.
- Undertake improvement activity in a systematic way.

To achieve and keep the Lean-Six Sigma principles (as mentioned before) is recommended to use the framework DMAIC as follow:

- **Define**: Clearly define the project's goals and/or the problem that needs to be solved. Coordinating and making clear to everyone involved, their corresponding role.
- **Measure**: by measuring the actual conditions or variables in this stage it is necessary to clarify things by seeing how the work gets done and how well.
- **Analyze**: during this stage it is necessary to find among the possible causes, the root cause of the problem or condition to focus on.
- **Improve**: once the root cause has been identified, obtain the possible solutions that address the root cause. Afterwards, select the best solution and test it out.
- **Control**: finally, it is necessary to make sure that the goals are obtained and develop a control plan to hold the achieved condition or gains.

6.1.4 *Lean-Sigma*

Lean-Sigma methodology goes back to the basic engineering principle of solving a problem not based on the annualized savings, but based on eliminating a waste or to overcome an obstacle to the continuous manufacturing flow.

The Lean-Sigma objective may be understood as solving a problem in the shortest time possible by creating synergy between the quick-fix lean approach, and the deep-statistical-analysis Six Sigma approach. Solving a problem will not always bring a process' performance to a six-sigma level, but instead, the Lean-Sigma methodology uses the incremental Lean approach to take the process quality to the desired level through several kaizen, or improvement events.

In general and for this chapter, Lean-Sigma is defined as a problem-solving oriented methodology, rather than a project-oriented methodology as Lean-Six Sigma is. Each step of the methodology is flexible and open to use any available tool from the engineering, Lean, or Statistical arenas. Lean-Sigma methodology is based on five rapid improvement steps:

1. Identify and measure the problem. How big is it?
2. Root cause analysis. What is the root cause of the problem?
3. Develop a solution. Identify the alternative solution that best solves the problem.
4. Verify the solution. Make sure that the problem is eliminated by the proposed solution.
5. Control Plan. Make a quick and effective plan so that the previous situation does not come back.

6.2 Case Study

The versatility of Lean- Sigma allows to be used for a wide range of applications and solutions in the production systems; such applications can go from the design stages to the optimization of the production process. Following is a case study that serves as an example of how to apply the concept of Lean-Sigma. This case study explores the possibility of using the Lean-Sigma approach as a problem-solving technique, rather than as a project-based methodology. The fundamental thought behind this study is to show how the Lean-Sigma synergy is released if the improvement is achieved by using the power of a deep statistical analysis (from Six Sigma) and the speed and simplicity of Lean, in other words, achieving a fast and effective improvement. For this purpose, a combination of Lean and Six Sigma tools is used to identify, to measure and analyze the root cause of variation, to develop a solution and to make a control plan. Showing and proving that with this Lean-Sigma approach, the whole solution process is completed in a short period of time.

6.2.1 Problem Description

The manufacturing facility used for this study, is dedicated to the production of plastic components using an injection molding process. The Machine number four (for reference), has been experiencing difficulties complying with the delivery schedule with its highest selling product. This product is a metal-polymer component, a plastic extrusion over-molded on a metal part. For this, a multiple cavity extrusion mold is use to produce five pieces at the same time. Initial reports show that several weeks back, a ~99 % percentage of the product was exhibiting flash (excess of material) on one side of the component. As a containment action, an operator was placed right before the quality inspection station. His main task was to use a special flash-removal tool, and make sure the product got corrected before final inspection and sending to the customer. After some time, the flash problem got worse to a point where a group of four operators were not able to keep up with the production rate, generating a process bottleneck and increasing the orders' backlog.

6.2.2 Methodology

The red signals coming from machine four reaches the top management of the factory who decides to face the problem as follows. A continuous improvement team was assembled to address the situation. The management over-emphasized that this team had to find an effective solution. The goal was simple, identify the root cause of the problem, correct it, and bring the process back to control as soon as possible. So, the dead line for implementing a permanent and efficient solution was set in 1 week. Finally, the top management specified that this problem-solving process should go through the manufacturing process to make lean the total lead-time, meaning that the waste in waiting time should be removed.

Consequently, all other evident waste in the process had to be eliminated: defects, over-processing, WIP and unnecessary transportation. Once the team was assembled, the decision was to use a problem-solving methodology based on the Lean-Sigma approach, in other words, use the power of a deep Six Sigma statistical analysis at the speed of Lean. The team agreed to follow a five-step methodology. Identify and measure the problem, conduct a root cause analysis, develop solution, verify solution, and control as shown in Fig. 6.2.

Based on in the DMAIC framework the team started to measure the problem by using a process mapping and a capability study (see Fig. 6.2). Using the information acquired a brain storming exercise was conducted to find the potential causes. Digging deeper in the analysis of such causes, the five whys helped in finding the root cause. Once the root cause was defined a design of experiments (DOE) was used for figure out the solution. After, a statistical validation was conducted; the initial conditions were compared to the implemented solution.

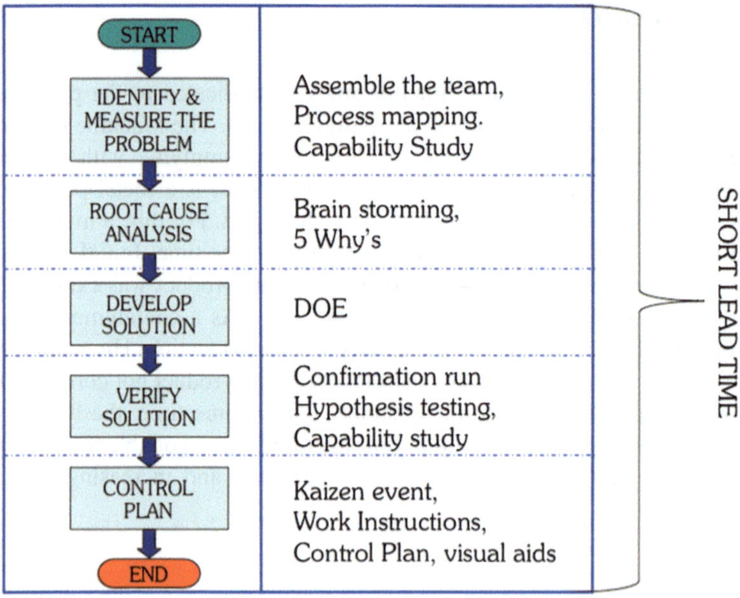

Fig. 6.2 Flow diagram of the lean-sigma steps and the tools used in this case study (Estrada 2003)

Finally, several control plans and kaizen events were implemented to keep the solution and to identify any change in the process.

6.2.3 Initial Data Analysis

Following the lean-sigma approach, the team immediately proceeded to collect the information about the flash issue in machine four Running in the initial conditions. Considering that each cavity in the mold may have a different variability, five samples (one from each cavity in the mold) were taken from the extrusion molding process for 20 random runs. The flash was measured (for reference see Table 6.1) and recorded. As shown Fig. 6.3, flash readings were ranging from 0.5 to 1.3 mm in an erratic pattern, having an average of 0.96 mm. From the design specifications for this product, the limits of the flash were established as −0.1 for the Lower limit (LSL) and 0.4 mm for the Upper limit (USL).

Initial analysis showed that the process was not in a state of statistical control (Fig. 6.4). Beside the mentioned mean, the Xbar-R chart show 40 % of the sample means out of the control limits, and unexpected changes in the working parameters. In the same Fig. 6.4, the ranges are of 0.5 mm or less which is close to the specified for this design. This and the capability study indicate that is possible to

Table 6.1 Collected flash measurements (in mm) from the initial conditions

Sample	Cavity				
	1	2	3	4	5
1	0.825	1.0492	1.0132	0.6952	1.1754
2	1.1774	0.8611	1.0991	0.6972	1.2393
3	0.9791	0.9721	1.0236	0.5411	1.1714
4	1.0773	0.8252	1.1462	0.5912	1.1632
5	1.0212	0.6773	1.0873	0.5711	1.2435
6	1.0733	0.8251	1.0032	0.6805	1.3095
7	1.3095	0.9231	1.1676	0.5895	1.0512
8	1.0833	0.9276	1.0914	0.6108	1.2733
9	1.1473	0.8652	0.9912	0.5833	0.9271
10	1.1113	0.9053	1.1754	0.6933	1.2495
11	1.0072	0.8972	0.8855	0.7731	1.1393
12	1.0993	1.0893	0.9951	0.6674	1.0292
13	1.0873	0.5973	1.0953	0.7471	1.2893
14	1.0012	0.9053	0.9311	0.5193	1.1253
15	1.1373	0.8733	1.0666	0.6373	1.1053
16	1.2094	0.8552	1.0694	0.7253	1.0873
17	1.0212	0.9534	1.1754	0.6505	1.2592
18	1.1053	0.8734	0.7831	0.5791	1.1512
19	1.0472	0.5595	1.0694	0.6943	1.0795
20	1.1073	0.8734	0.8994	0.7251	1.1393

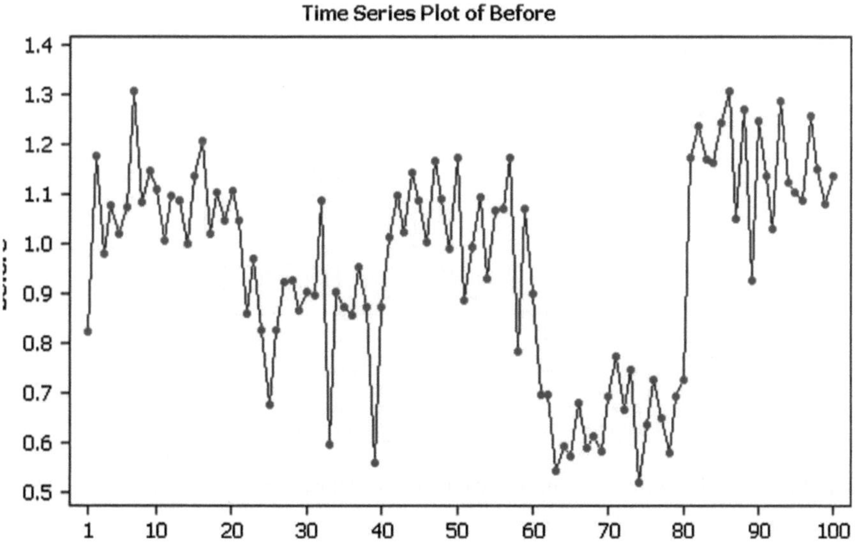

Fig. 6.3 Run chart for the flash measurements under the initial conditions

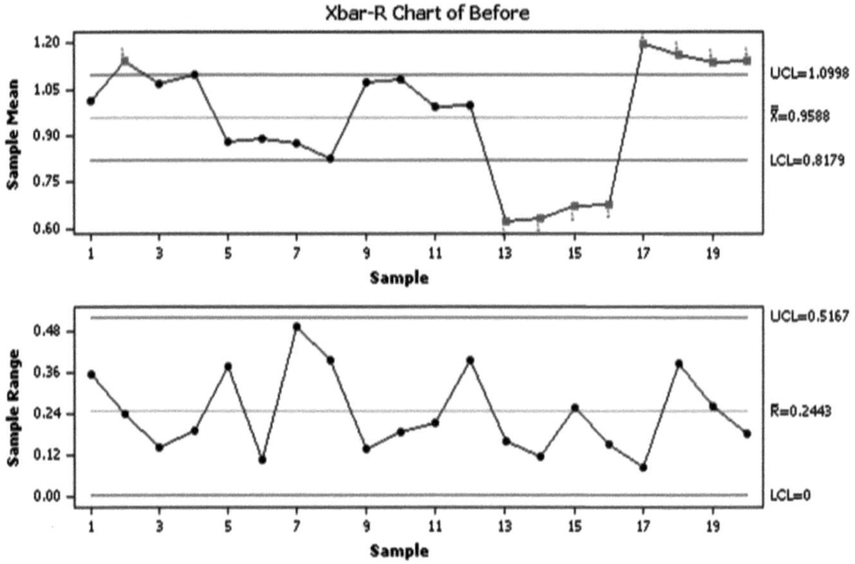

Fig. 6.4 Xbar-R chart for the flash measurements under initial conditions

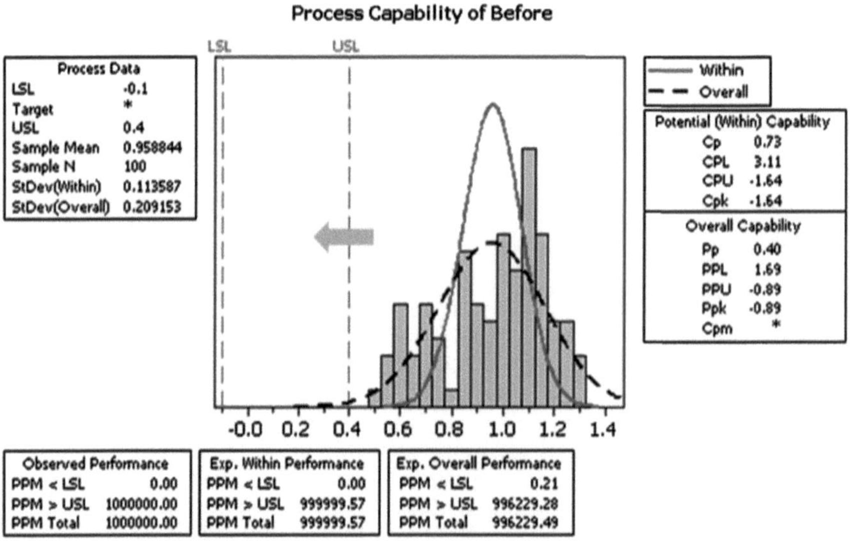

Fig. 6.5 Capability analysis for data of the initial conditions in machine four

put back in control this machine. The non-controlled conditions are better visualized in the results of the initial capability study (see Fig. 6.5). Figure 6.5 illustrates the performance shift (see arrow), a negative capability index (Cpk) was calculated (-1.64). By plotting in the capability- study graph the LSL and USL, it is easy to observe that such shift in machine 4 will result in over 99 % of the parts above the upper specification limit, with an expected overall process performance of 996,229 defective parts per million, as seen in data from Fig. 6.4.

6.2.4 Root Cause Analysis

Having the shown data and its analysis, as a second step the team conducted a brain storming session. For making more effective this session, the solving team included other key players from the production area: the production supervisor, machine operators, and facilities and set-up technicians. The goal for this session was to identify the factors that could be affecting the response variable (flash), and the potential causes for this problem. During the session, after considering a large list of ideas generated and after applying the 5 whys technique the list was filtered. The potential causes were reduced to five variables: temperature of the barrel, injection pressure, injection speed, injection cutting position 5 (the section where the flash was observed) and physical condition of the die. Lastly, because four of the variables were suitable for controlling and to run a DOE, the discussion was focused in the molding-die's cavities, considered by the key players (from the production area), as an impacting factor in the flash (response variable). Due to the uneven wear out of the cavities, they were seen as a critical factor to consider during the solution developing step.

6.2.5 Developing a Solution

The team finally decided to conduct an experiment using the four controllable factors related to the operation of the machine, and to consider the molding-die's cavities as one noise factor. Such noise was calculated by considering the variance among the five cavities within the die. Table 6.2 shows the factors and levels used for this study during the DOE. Since the process operates under the influence of a noise factor, the team decided to use a Robust Parameter Design approach. An L8 orthogonal array is chosen for the controllable factors, and one noise factor with 5 levels is used for the outer array.

Table 6.3 shows the experimental array and the response variable results obtained (in mm). Using the results in Table 6.3 it is possible to calculate the Signal to Noise ratio values for each level of the controllable factors (Table 6.4). It is possible to rank them by effect, in this case considering the injection pressure as the factor with the highest ratio and effect. Also and in order to have a Lean (fast) decision process this value are used to select the factors' levels (see bolt and

Table 6.2 Controllable factors and levels used for this study

Controllable factors	Level 1	Level 2	Noise factor
Temperature	255 °C	275 °C	Uneven wear out of die cavities (5)
Injection speed	20 mm/s	40 mm/s	
Injection pressure	30 bar	50 bar	
Injection cut position (Position 5)	25 cm	31 cm	

shadow values in Table 6.4) as follow: low level for the temperature and high level for the injection pressure, speed and position **(5).**

The selected values for the controllable factors are shown in Table 6.5, such values are then used to calculate and predict the system response. The signal to noise ratio is predicted as 12.34, the mean value for the flash is expected to be ~0.115 that is close to the target value for the flash (zero), see Table 6.6.

6.2.6 Verify the Solution

To verify that the selected levels solves the problem, the settings and values shown in Table 6.5 were used in machine 4 and a verification run was conducted. Again, 20 random runs were recorded using the 5 samples molded in each cavity of the mold. Table 6.7 shows the results in measuring the flash of each sample. The Run chart shown in Fig. 6.6 show less variation among the flash values. Here the readings range from 0.0 to 0.4 mm, values that are between the limits established for the flash (−0.1 for LSL and 0.4 mm for USL).

A noticeable process shift is observed in the Xbar-R Charts in Fig. 6.7. The right section shows the charts for the values after using the selected values for the controllable factors. These control charts shows that the process is stable and predictable, therefore it is in a state of statistical control. Around 40 % of the variation reduction is observed after use the selected levels (Fig. 6.7). As result, the flash mean moved from 0.96 to 0.14 mm. The UCL and LCL shift from 0.81 and 1.1 to 0.23 and 0.04 mm respectively. As shown in the range chart (see Fig. 6.7), all values are within the limits for this design.

The capability study shows that the proposed solution put the process back in control (see Fig. 6.8) a PPL value of 1.29 and a PPU value of 1.19 and a Cpk value is calculated in 1.19. Data also shows that is expected to have a process overall performance of 218 defective parts per million. From Table 6.6 is observed that the flash values are between the limit. However, from the capability study is possible to predict that after using the proposed levels, most of the 218 ppm defective parts (162 ppm) will be below the LSL and only 56 ppm are predicted to be above the USL.

Figure 6.9 shows a graphical representation of a comparison of the capability studies before and after using the selected levels. The Cpk moves from −1.640 to

Table 6.3 Experimental design array and the response variable results obtained (in mm)

	Inner array				Noise factor							
					Cavity							
Run	Temperature	Pressure	Position 5	Speed	1	2	3	4	5	S/N	S	Mean
1	255	30	25	20	1.1393	1.3956	1.0152	0	1.0933	5.3782	0.5384	0.9287
2	255	30	31	40	1.1874	−3.14	−3.14	0.849	−3.1397	−7.1607	2.2805	−1.4765
3	255	50	25	40	0.9992	1.2815	0.9471	1.0532	1.3195	15.42	0.1694	11,201
4	255	50	31	20	0.7669	0.7889	0.845	0.9231	1.2415	14.2776	0.1933	0.9131
5	275	30	25	40	1.1313	0	0	0.9271	1.0793	4.7651	0.5778	0,6275
6	275	30	31	20	0.9912	0	0.9311	0.9811	1.0592	7.0254	0.4454	0,7925
7	275	50	25	20	1.2214	3.3259	1.0512	0.9571	1.1353	-0.0361	1.0042	1.5382
8	275	50	31	40	1.0232	1.0472	0.9551	0.9031	1.3095	16.0808	0.1570	1.0476

Table 6.4 Signal/noise ratio results for the controllable variables

Level	Temperature	Injection pressure	Injection position 5	Injection speed
1	6.979	2.502	6.382	6.661
2	6.959	11.436	7.556	7.276
Rank	4	1	2	3

Table 6.5 Selected levels for the controllable factors

Temperature	Injection pressure	Injection position 5	Injection speed
255 °C	50 bar	31 cm	40 mm/s

Table 6.6 Predicted values

S/N Ratio	Mean	StDev	Ln(StDev)
12.3400	0.115733	0.729374	−1.42070

Table 6.7 Collected data after using the selected levels (in mm)

RUN	Cavity				
	1	2	3	4	5
1	0.1871	0.2168	0.1912	0.1957	0.1752
2	0.2469	0.0573	0.1669	0.1795	0.1992
3	0	0.1784	0.1291	0.0389	0.1563
4	0.1754	0.1625	0.1551	0.0459	0.1397
5	0.1393	0	0.2578	0.2321	0.0678
6	0.145	0.1914	0.0856	0.0754	0.1451
7	0.1618	0.1519	0.0771	0.1725	0.0456
8	0.1871	0.1893	0.0954	0.2215	0.0753
9	0.1972	0.125	0.1321	0.0845	0.1433
10	0	0	0.3872	0	0.1572
11	0.2372	0.1912	0.0975	0.1896	0.0945
12	0.1118	0.2791	0.1019	0.1605	0.1951
13	0.1015	0.1215	0.1372	0.0754	0.0895
14	0.1973	0.0984	0.0895	0.1145	0.1257
15	0.1669	0.1993	0.1532	0.0895	0.1955
16	0.1791	0.2472	0.1545	0.1974	0.1612
17	0.0967	0.1915	0.1442	0.1835	0.1837
18	0.1735	0.1845	0.1826	0.1374	0.1011
19	0.1661	0.0124	0.0911	0.0945	0.1571
20	0.1393	0.0348	0.1028	0.0468	0.1896

1.190. The standard deviation calculated for both conditions show a shifting from 0.209 to 0.067 (see Figs. 6.5 and 6.8 respectively) showing a variability reduction that in Fig. 6.9 is represented graphically. Figure 6.9 also shows the change for the

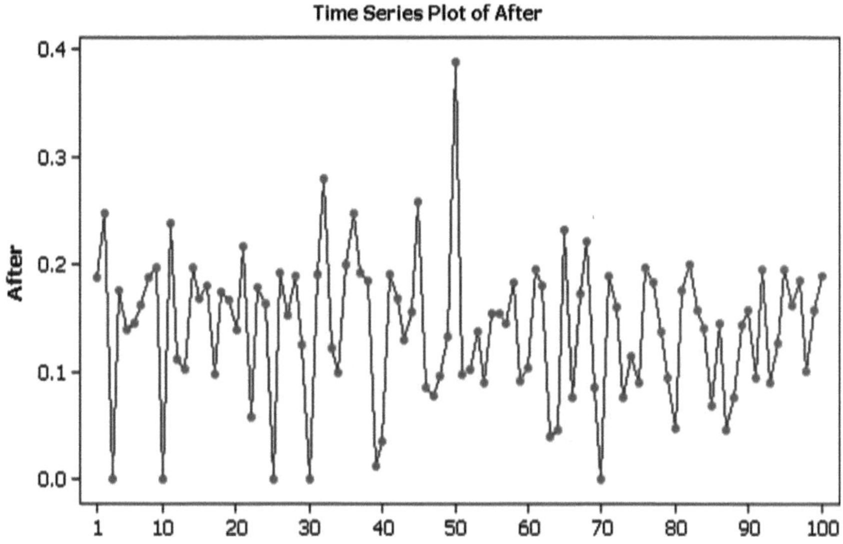

Fig. 6.6 Run chart for the flash reading after using the selected levels

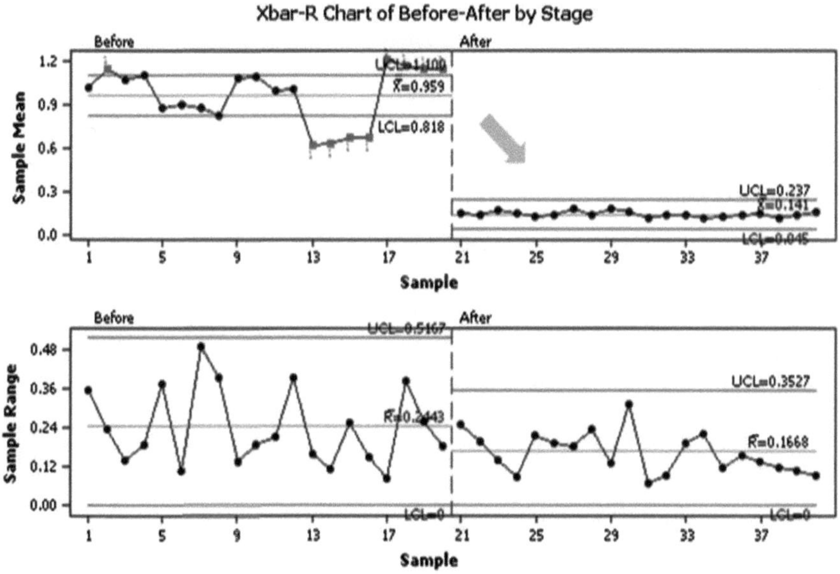

Fig. 6.7 Comparison Xbar-R charts for the flash measurements under **a** Initial conditions (*left charts*) and **b** After using the selected levels (*right charts*)

defective parts from 996,229 ppm expected for the initial conditions to 218 ppm after using the selected levels. The major change and cause for this shifting was due to the mean modification.

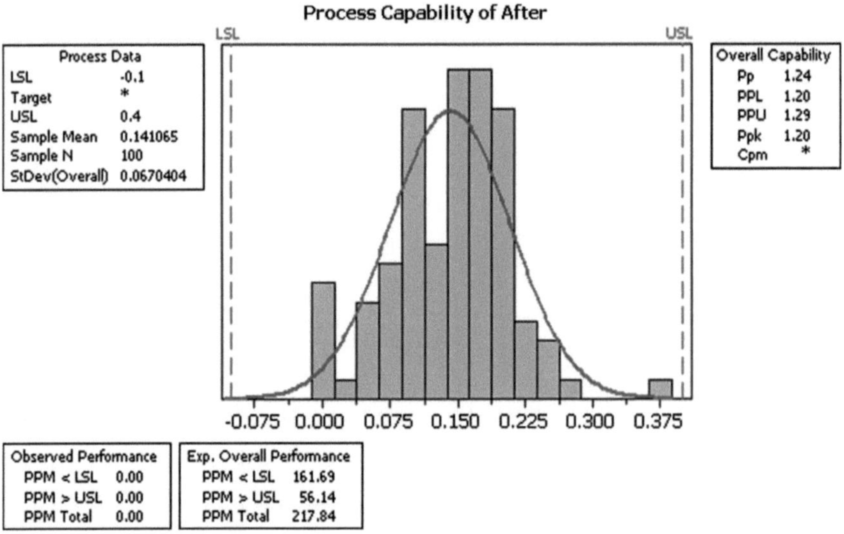

Fig. 6.8 Capability analysis for machine four after using the selected levels

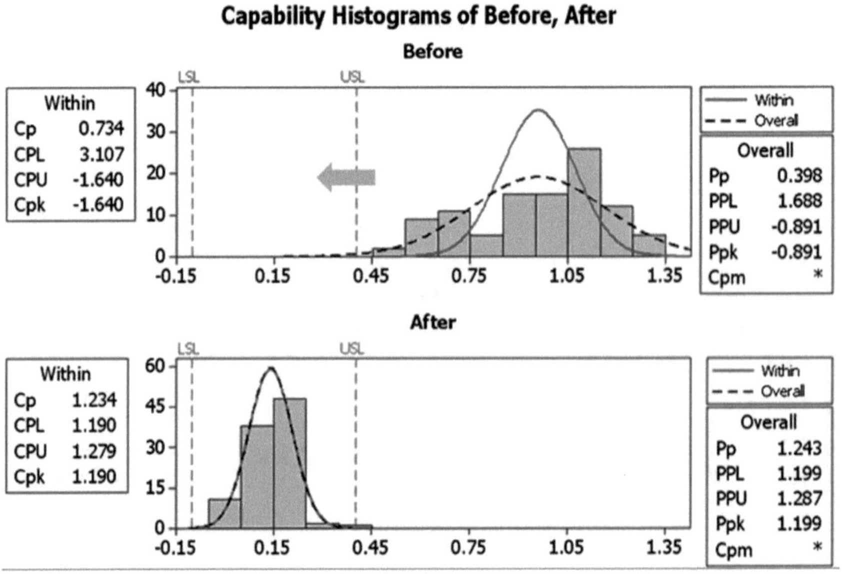

Fig. 6.9 Capability histograms for the flash measurements comparing the: **a** Initial conditions (*up-right bell*) and **b** After using the selected levels (*left-bottom bell*)

As part of the verification step and to be sure that the mean value of the response variable (flash) was indeed reduced, a two-sample-t hypothesis testing was conducted comparing data obtained from the process and in particular from

(a)

Two-Sample T-Test and CI: Cavity1 Before, Cavity1 After

```
Two-sample T for Cavity1 Before vs Cavity1 After

                N    Mean   StDev  SE Mean
Cavity1 Before  20  1.0814  0.0985    0.022
Cavity1 After   20  0.1505  0.0644    0.014


Difference = mu (Cavity1 Before) - mu (Cavity1 After)
Estimate for difference:  0.9309
95% lower bound for difference:  0.8863
T-Test of difference = 0 (vs >): T-Value = 35.38  P-Value = 0.000  DF = 32
```

(b)

Two-Sample T-Test and CI: Before, After

```
Two-sample T for Before vs After

          N    Mean   StDev  SE Mean
Before  100   0.959   0.209    0.021
After   100  0.1411  0.0670   0.0067


Difference = mu (Before) - mu (After)
Estimate for difference:  0.8178
95% lower bound for difference:  0.7814
T-Test of difference = 0 (vs >): T-Value = 37.23  P-Value = 0.000  DF = 119
```

Fig. 6.10 Two-sample-T test: **a** Comparing data before and after using the selected levels, for mold's cavity No. 1 data, and **b** Comparing data before and after using the selected levels for all the cavities in the mold

each of the mold cavities. Figure 6.10a shows the two-sample-t test for cavity number one and Fig. 6.10b shows the two-sample-t test for the process (all the cavities as a group). These analyses give statistical validation to the means' shifts for each cavity output and in general for the process outcome.

6.2.7 Control Plan

For the last step of the proposed methodology, a one-day Kaizen event was held at the production floor, to explain the new operating conditions, the importance of avoiding tampering the process, and updating working instructions and visual aids. Production-related personnel participated on verifying the new results, and on the preparation of updated work instructions and visual aids. Additionally, a Kaizen newspaper was posted on the production area for communicating further improvement ideas.

6.3 Conclusions

Initial data shows that the process initially had an overall performance of 996,229 ppm, which is equivalent to a sigma level of 0.1. Almost 100 % of the pieces had to be reworked. The process was not stable, and non-predictable, therefore it was considered not to be in a state of statistical control. Additionally, data shows evidence of tampering activities. That is, trying to bring the process to acceptable performance levels, excessive adjusting activity took place, which conducted to a worse performance level.

After applying the proposed Lean-Sigma methodology, the data shows the process is stable, predictable, and therefore considered to be in a state of statistical control. An overall performance of 218 ppm, which is equivalent to a sigma level of 5.0, is achieved. Additionally, other parallel objectives were accomplished, such as the relocation of the four people doing the rework into another productive function, reduction of the transportation fees due to "urgent" shipments, elimination of overtime worked, and some others.

The whole process was completed during a period of 7 days trying to keep up with management's requirement to maintain the improvement event's lead-time as short as possible. In summary, it may be concluded that using the Lean-Sigma methodology oriented as a problem-solving technique, rather than as a project, helps to achieve the quality goals and maintain the lead-time short.

References

Deming, W. E. (1982). *Out of the Crisis* (First MIT Press Edition, 2000, ©1982 and 1986). The W. Edwards Deming Institute, ISBN: 0-262-54115-7.

Estrada, F. J. (2003). *Lean-Sigma research notes*. Juarez

George, M. L. (2002). *Lean six sigma: Combining six sigma quality with lean speed*. New York: McGraw-Hill.

George, M. L., Rowlands, D., Price, M., & Maxey, J. (2005). *The lean six sigma pocket tool book*. Ma-Graw-Hill, ISBN 0-07-144119-0.

Hopp, W., Spearman, M. (2000). *Factory physics* (Vol. 110, pp. 164–168, 2nd ed.). McGraw-Hill, ISBN 0256247951.

Ohno, T. (1988). *Toyota production system—beyond large-scale production* (pp. 25–28). Productivity Press, ISBN 0915299143.

Pande, P. S., Neuman, R. P., & Cavanach, R. R. (2000). *The six sigma way*. New York: McGraw-Hill.

Rath & Strong. (2000). *Six sigma pocket guide*. Rath & Strong management consultants, Sixth Printing 2000. New York: McGraw-Hill

Schonberger, R. J. (1986). *World class manufacturing: The lessons of simplicity applied* (1st ed.). The Free Press, ISBN 0-02-929270-0.

Womack, J. P., & Jones, D. T. (1996). *Lean thinking* (1st ed.). Simon & Schuster, ISBN 0-684-81035-2.

Womack, J., Jones, D. T., & Roos, D. (1990). *The machine that changed the World*, (First Harper Perennial edition 1991). ISBN 0-06-097417-6.

Chapter 7
Automatic Product Quality Inspection Using Computer Vision Systems

Osslan Osiris Vergara-Villegas, Vianey Guadalupe Cruz-Sánchez, Humberto de Jesús Ochoa-Domínguez, Manuel de Jesús Nandayapa-Alfaro and Ángel Flores-Abad

Abstract The ability to visually detect the quality of a product is one of the most important issues for the manufacturing industry because of the demand made by the consumers is increasing. This process is typically carried out by human experts; unfortunately experts frequently make mistakes because this process could be tedious and tiring even for the most trained operators. A lot of solutions have been proposed to solve this problem, such as the use of lean manufacturing and computer vision systems. This chapter, presents a detailed explanation about the stages involved to create a system to automatically verify the quality of an object using computer vision and digital image processing techniques. First, a revision of state of art researches is presented. This work also focuses on a discussion of the issues involved with computer vision applications. Afterwards, a detailed explanation about the design of two study cases to inspect fabric and apple defects, and its correspondent results are presented. Finally, a point of view about the trends in automatic quality inspection systems using computer vision is offered.

Keywords Quality control · Computer vision · Automation

7.1 Introduction

Visual inspection is the result of a processing carried out by a part of the brain of the luminous information that arrives to the eyes, and is one of the main data sources of the real world (Sannen and Van Brussel 2012). The information

O. O. Vergara-Villegas (✉) · V. G. Cruz-Sánchez · H. de Jesús Ochoa-Domínguez · M. de Jesús Nandayapa-Alfaro · Á. Flores-Abad
Universidad Autónoma de Ciudad Juárez, Instituto de Ingeniería y Tecnología, Ciudad Juárez, Chihuahua, Mexico
e-mail: overgara@uacj.mx

J. L. García-Alcaraz et al. (eds.), *Lean Manufacturing in the Developing World*,
DOI: 10.1007/978-3-319-04951-9_7,

perceived with the sense of sight is processed in distinct ways based on the specific characteristics needed for the future tasks to execute. As a result of an image analysis process, the representation of an object is obtained. Immediately a decision is taken to define what to do with the visual information, which typically implies the recognition of the object(s) detected inside a scene and the reactions realized by a body part.

Every day, human beings recognize objects inside a particular scene observed by means of the vision system. This process is done unconsciously (with a minimum effort) even with a lack of the complete knowledge or description of the object to be recognized. Recognition is a term used to describe the ability of human beings to identify the objects around based on previous knowledge (Lee et al. 2010).

The task of visual inspection to recognize objects and to evaluate its quality constitute one of the most important processes in several industries such as the manufacturing industry and alimentary industry, in which, due to the customer demands, it is mandatory to assure the quality of a product (Satorres et al. 2012; Razmjooy et al. 2012; Peng et al. 2008; Sun et al. 2010). The issue of inspecting objects in order to detect defects such color, scratches, cracks or checking surfaces for a proper finish is related to visual quality inspection.

Typically, the quality inspection is made by human experts. However, the experts frequently make mistakes, because the process could be tedious and tiring even for well-trained operators. The problem increases usually due to, the workdays of the inspectors are very long (more than 6 h).

This leads to several industries to look for alternatives to avoid the mistakes made by human inspectors. One of the alternatives adopted by many industries, to remain competitive, is the promoting of lean manufacturing, in which the practices can work synergistically to create a streamlined, high quality system that produces finished products at the pace of customer demand with little or no waste (Sullivan et al. 2002; Abdulmalek and Rajgopal 2007). Unfortunately, the existent evidence suggests that several organizational factors may enable or inhibit the implementation of lean practices among manufacturing plants (Shah and Ward 2003).

Another alternative is to provide to a computer with the ability of inspect and recognize objects automatically. The use of a computer together with other mechanisms, such as cameras, sensors, the knowledge provided by the human expert, and complex algorithms, allows having a capable tool for automatic inspection of the product quality. Then, automation becomes a necessary task for inspection and recognition of objects in order to guarantee the quality of a product.

In this chapter, the issue of the automatic inspection of the quality of an object using computer vision (CV) is addressed. The rest of the chapter is organized as follows.

In Sect. 7.2, a revision of several current designs which use computer vision to inspect the quality of an object is presented. A brief explanation of the issues involved to create a computer vision system using digital image processing techniques is shown in Sect. 7.3. In Sect. 7.4, two study cases are discussed, the first case addressing the problem of fabric color defects detection and the second case addressing the problem of golden apples defect detection. A point of view about the trends in the design of computer vision systems is presented in Sect. 7.5. Finally, Sect. 7.6, presents the conclusions obtained with this work.

7.2 Literature Review

In the literature there exist several researches which have addressed the problem of automatic inspection of an object quality using a CV system. Machine vision has been a widely used technology in the industry for the past three decades. It has been an excellent tool for many industrial inspection tasks such as: brake disks, printed circuit board (PCB), float glass, electric contacts, tiles, chickens, fruits, vegetables, fabric, gears, chip alignment, led, to name a few.

The work proposed by (Lerones et al. 2005) defines a solution for the automatic raw foundry brake disk dimensional characterization and visual inspection for the automotive industry. To solve the problem, three CV techniques were used: (a) a calibrated 3D structured-light, for dimensional characterization and inspection, (b) a 3D uncalibrated structured-light, for local fault detection, and (c) a common 2D-vision technique for further local fault recognition. The industrial results show that the described system is appropriate for brake disk dimensional characterization as well as for the detection of hard masses, featheredges, pores, hole jump obstruction, ventilation slot obstruction and veining, providing more efficiency on the production line and bettering working conditions for human operators.

A region-oriented segmentation algorithm for detecting the most common peel defects of citrus fruits is shown in (Blasco et al. 2007). The histogram of an input image is computed to obtain the peak in the lower values corresponding to the background, since it was a uniform black color that contrasted against the orange color of the fruits. Then, the main stage was performed to detect a region of interest consisting of the sound peel, the steam and the defects. This is made by a region growing algorithm followed by a region merging. The algorithm is robust against different varieties and species of citrus fruit and does not need previous manual training or adjustments to adapt the system to work with different batches of fruit or changes in the lighting conditions.

In the chapter presented in (Peng et al. 2008), an online defects inspection method of float glass based on machine vision is presented. The method inspects defects through detecting the change of image gray levels caused by the difference in optic character between glass and defects. Initially, the noise is reduced by means of image filtration based on gradient direction. To remove the back ground stripes, a downward segmentation is implemented. The possible defect core and its distortion are segmented with fixed threshold method and the OTSU algorithm with gray range restrict. Finally, fake defects are eliminated through the method based on defect texture detection. The success of this inspection method provides a reference for defects detection on other materials, such as armor plate.

A CV for the automation of online inspection to differentiate freshly slaughtered wholesome chickens from systemically diseased chickens is presented in

(Yang et al. 2009). The system consisted of a camera used with an imaging spectrograph and controlled by a computer to obtain line-scan images quickly on a chicken processing line of a commercial poultry plant. The system scanned chicken carcasses to locate the region of interest (ROI) of each chicken to extract useful spectra from the ROI as inputs to the differentiation method, and to determine the condition of each carcass as being wholesome or systemically diseased. The high accuracy obtained with the evaluation results showed that the machine vision system can be applied successfully to automatic online inspection for chicken processing. Table 7.1, shows a summary of 11 CV systems contrasting its main features and showing the accuracy recognition rates.

7.3 Computer Vision and Digital Image Processing

CV is concerned with modeling and replicating the human vision using computer software and hardware. It is also the main theory for building artificial systems to extract information from images. The main challenge is to combine knowledge in computer science, electrical engineering, mathematics, physiology, biology, and cognitive science in order to understand and simulate the operation of the human vision system. Specifically, CV is a combination of concepts, techniques and ideas from Digital Image Processing (DIP), Pattern Recognition (PR), Artificial Intelligence (AI) and Computer Graphics (CG).

CV describes the automatic deduction of the structure and the properties of a (possible dynamic) three-dimensional world from either a single or multiple two-dimensional images of the world. The computer vision great trick is to extract descriptions of the world from pictures or sequences of pictures (Nalwa 1993).

As a consequence, computer vision systems need the digital image processing techniques to enhance the quality of the acquired images for future use or interpretation. Then, DIP is concerned with taking one array of pixels as input and producing another array of pixels as output, which, in some way, represents an improvement of the original array. DIP is the science of modifying digital images by means of a computer (Forsyth and Ponce 2012).

Frequently, computer vision and digital image processing are erroneously used as a same term. The main difference is in the goals, not in methods. For example, if the goal is to enhance an image for later use it, then this may be called digital image processing, on the other hand, if the goal is to emulate human vision like object recognition, defect detection or automatic driving, then it is closer to computer vision.

There are no clear-cut boundaries in the continuum from DIP at one end to CV at the other. However, one useful paradigm is to consider three types of computerized processes: low-level, mid-level and high-level. Low-level processes involve primitive operations for image preprocessing such as denoising and deblurring, this level is characterized by the fact that both inputs and outputs are images. Mid-level processes involve tasks such as segmentation and description of

Table 7.1 A summary of 11 computer vision systems to detect defects in objects

Author (year)	Country	Object	Technique	Defect	Accuracy (%)
Lerones et al. (2005)	Spain	Brake disk	3D structured light	Circularity, defective grinding, hard masses, feather edges	80
Blasco et al. (2007)	Spain	Citrus	Region oriented segmentation	Peel	95
Peng et al. (2008)	China	Float glass	Texture reverse subtraction	Bubbles, lards, optical distortion	98
Yang et al. (2009)	USA	Chickens	Thresholding	Wholesome, diseased	89
Sun et al. (2010)	China	Electric contacts	Blob analysis and PSO	Deckle edge, back cracks, side cracks, eccentricity	96.7
Mery et al. (2010)	Chile	Corn tortillas	SVM	Size, hedonic scale, production level	96
Gadelmawla (2011)	Saudi Arabia	Spur gears	Tolerances	Outer diameter, diameter pitch, circular pitch	99
Perng et al. (2011)	China	LED	Thresholding	Missing components, incorrect orientations	95
Razmjooy et al. (2012)	Iran	Potato	SVM and ANN	Size, color	96.86
Možina et al. (2013)	Slovenia	Pharmaceutical tablets	PCA	Debossing shallow	84
Lin and Fang (2013)	China	Tile	Corner convergence and clustering	Alignment	90.5

individual objects. Finally, the high-level involves making sense of an ensemble of recognized objects and performing the cognitive functions normally associated with vision (Gonzalez and Woods 2009).

As afore mentioned, two broad categories are defined; methods whose inputs and outputs are images and methods whose inputs may be images, but whose outputs are attributes extracted from those images. According to (Gonzalez and Woods 2009) the fundamental stages comprising a CV system for digital image processing are: (a) image acquisition, (b) preprocessing, (c) segmentation, (d) feature extraction (representation and description), (e) Recognition (interpretation and classification), and (f) knowledge base.

In Fig. 7.1 a graphical representation of each stage and the mode of interaction of each with the other are shown. Additionally, in the following subsections a brief explanation of each stage is presented.

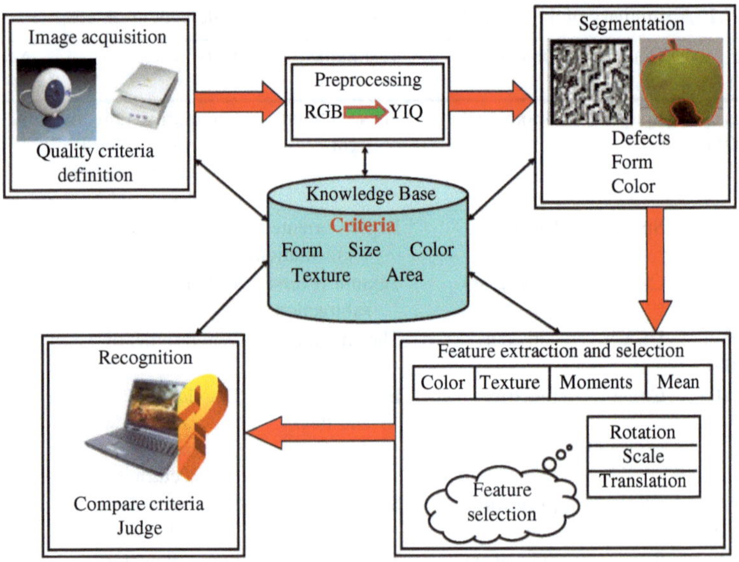

Fig. 7.1 Fundamental steps of a computer vision for digital image processing

7.3.1 Image Acquisition

Before any image processing can start, an image must be captured by a camera and converted to a manageable entity. Thus, in order to acquire a digital image, an image sensor and the ability to digitize the signal produced by that sensor are needed (Wandell et al. 2002). The sensor can be a television camera, a line scan camera, video, scanner, etc. If the output of the sensor is not digital, then an analog to digital converter is necessary to digitize the image. The digital image is obtained as a result of sampling and quantization of an analog image or created already in digital form.

Typically, a digital image is represented as a bi-dimensional matrix of real numbers. The convention $f(x,y)$ is used to refer an image with size $M \times N$, where x denote the row number, and y the column number. The value of the bi-dimensional function $f(x, y)$ at any pixel of coordinates $(x0, y0)$, denoted by $f(x0, y0)$, is called the intensity or gray level of the image at that pixel.

7.3.2 Preprocessing

After a digital image has been acquired, several preprocessing methods can be applied in order to enhance the data of the image prior to the computational processing. In this stage the image is processed and converted into a suitable form for further analysis (Choi et al. 2011). Most of the computer vision applications require taking care in designing the processing stage in order to achieve acceptable results. Preprocessing operations are also called filtration.

Examples of such operations include smoothing, exposure correction and color balancing, noise reduction (denoising), increasing sharpness, image deblurring, image plane separation, normalization, etc. The image obtained after this stage is the input to the segmentation step.

7.3.3 Segmentation

Autonomous segmentation is one of the most difficult stages in DIP. Segmentation is the process of partitioning a digital image into multiple segments (sets of pixels, also known as superpixels). The goal is to simplify and to change the representation of an image into something that is more meaningful and easier to analyze (Padma and Sukanesh 2013). A coarse segmentation process delays the outcome of a satisfactory solution in a DIP problem. On the other hand, a weak segmentation process in most of the cases will lead to errors.

As a result of the segmentation process, the raw pixel data that constitute the boundaries among the regions or information about which pixel belongs to which region is obtained. The image segmentation is the operation that marks the transition between low-level to mid-level image processing. Among the most commonly used segmentation methods are: thresholding, contour detection, region based, morphological watersheds and region growing.

7.3.4 Feature Extraction

Once an image has been segmented, the resulting individual regions can be described. Feature extraction, also called image representation and description, is the operation performed to extract and highlight features with some quantitative information which is essential to distinguish one class of objects from another. It is a critical step in most computer vision solutions because it marks the transition from pictorial to non pictorial data representation.

In order to store the characteristics extracted from an object an $n \times 1$ array called feature vector is built. The feature vector is a compact representation of an image and its content can be symbolic, numerical or both. The main challenge in this step is that the features extracted must be invariant to changes in rotation, scale, translation and contrast. Obtaining the invariants ensures that the computer vision system will be able to recognize objects even when they appear with different contrast, size, position and angle inside the image (Mullen et al. 2013).

The important features extracted include points, straight lines, regions with similar properties, color, textures, shapes, and a combination. The boundary descriptors such as statistical moments or Fourier descriptors are often used to tackle the problem of invariance.

7.3.5 Recognition

This stage constitutes the high-level of image processing. Recognition is the process to assign a label to an object, based on the information provided by its descriptors. Moreover, it involves assigning meaning to a set of recognized objects (Xiao and Wang 2013). The recognition algorithms analyze the numerical properties of various image features and classify data into categories. This stage is composed of two phases: training and testing.

All the classification algorithms are based on the assumption that the image in question depicts one or more features and that each of these features belongs to one of several distinct and exclusive classes. The classes may be stated a priori by an analyst specifying the number of desired categories (supervised classification) or automatically clustered (unsupervised classification) into sets of prototypes.

Quite often, the algorithms used for recognition include artificial neural networks (ANN), support vector machines (SVM), distance and similarity measures (k-Nearest Neighbors, Bayesian, Euclidean, Manhattan), and matching.

7.3.6 Knowledge Base

The a priori knowledge about a specific image processing problem is coded in the form of a knowledge database. The database may be as simple as detailing regions of an image where the information of interest is known to be located, thus limiting the search that has to be conducted for seeking information; or can be quite complex such as inter related list of all major possible defects in a material inspection problem (Gonzalez and Woods 2009). The knowledge base guides the operation of each processing module and also controls the interaction among all the modules.

7.4 Study Cases

As described in Sect. 7.2, several computer vision systems have been created in different industries. Even, when almost all the systems are build based on the steps explained in Sect. 7.3, not always all the steps are used. Sometimes, other more specific steps are included. The purpose of the current section is to present two study cases, one is to offer the specific details on how to build a computer vision system using digital image processing techniques to inspect the color quality fabric, and the other to inspect defects in apples.

7.4.1 Automatic Inspection of Fabric Quality

In the textile industry, defects detection is of crucial importance in the quality control of manufacturing (Bissi et al. 2013). The common inspection process is made by human experts, but since this task is dull and repetitive many errors are frequently committed. It has been estimated that the price of fabrics is reduced from 45 to 65 % due to the presence of defects (Kumar 2008).

The fabric is affected by yarn quality, loom defects and stamping process. The textile industry has the necessity to verify the quality of a stamping fabric process. The following subsections explain the design and implementation of a computer vision system to inspect color degradation in fabric.

7.4.1.1 Definition of the Quality Criteria

In the textile industry a common error that causes defects in the stamping process of a fabric roll is that some of the colors being used goes finishing. For a human inspector, this color degradation can be detected until the color difference between the original and the degraded fabric is very noticeable. For this example, the quality of a fabric is defined by the conservation of colors at the stamping process.

The input to the system is a criterion about the quantity of color variations accepted to consider a fabric without defects (good quality). At this stage a range between 1 and 10 % of color degradation is accepted for a good quality fabric, while out of that range the fabric is considered of bad quality.

7.4.1.2 Acquisition of Fabric Images

The stamping process was simulated using artificial fabric images, due to real stamping process imaging could not be achieved. Three databases of images were selected, each one containing what is called a texton that is the main element that compose a texture. The public databases used were: "Artificial Color Texture", "Forrest textures", and "Texture Gallery". A subset of ten textons was selected from each database.

7.4.1.3 Preprocessing of Fabric Images

To automatically simulate the creation of a fabric roll, each texton was repeated one after another 60 times. This operation was made randomly with rotations of 90

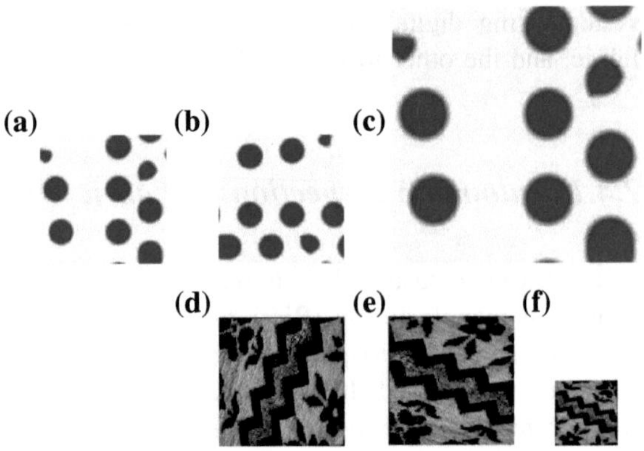

Fig. 7.2 Examples of textons. a Original texton: 125_exhaustive_p1, b 125_exhaustive_p1 rotated 90° clockwise, c 125_exhaustive_p1 doubled, d Original texton: S_S_cloth2, e S_S_cloth2 rotated counterclockwise, and f S_S_cloth2 rotated clockwise and halved

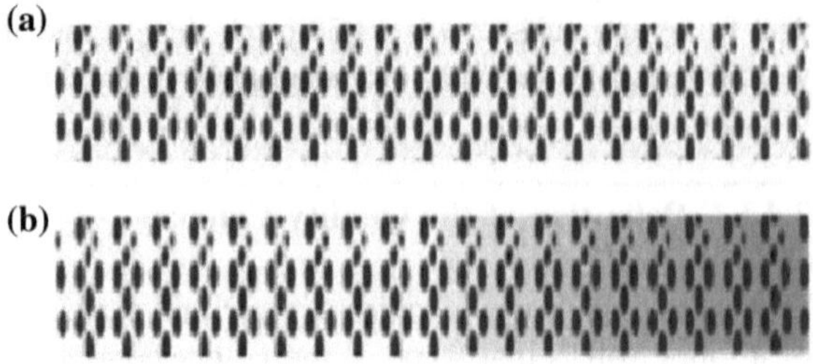

Fig. 7.3 Examples of fabric rolls. a Good quality fabric, and b Bad quality fabric

and 180° clockwise, scaling of doubled and halved of the original size, scaled and rotated together, and with addition of 10 and 40 % of salt and pepper noise with zero mean and a variance of 400. A random degradation in red, green and blue (RGB) colors of 1, 2, 5, 10, 20, 40 and 50 % was applied.

At the end of this process 1400 images were created, 700 images of good quality fabric and 700 images of bad quality fabric. In Fig. 7.2, several textons with different changes are depicted. In Fig. 7.3 an example of good quality fabric and bad quality fabric is shown.

7.4.1.4 Feature Extraction from Fabric Image Rolls

At this stage, a feature vector to store the statistical features associated with each fabric roll was computed. The process of the texture feature extraction was carried out in three RGB (Red, Green, Blue) planes and in the intensity plane of the HSI (Hue, Saturation, Intensity) model. In summary, the features extracted to handle image invariance were: 10 normal moments, 10 central moments, 7 Hu moments, 6 Sidharta Maitra moments, mean, variance and standard deviation in red, green and blue, and mean for the intensity of the HSI model. At the end of this step, a feature vector with 43 characteristics was obtained for each fabric roll.

7.4.1.5 Feature Selection

This stage consists on the selection of the more discriminate features to distinguish one fabric from another. In order to perform this stage, the so called testor theory was used. The algorithm used was the well known BT algorithm to obtain typical testors.

The BT algorithm defines a testor for two classes T_0 and T_1 as: "A set $t = (i1,\ldots, is)$ of columns of the table *T* (and the respective features $xi1,\ldots, xis$), is called testor for $(T_0, T_1) = T$, if after eliminating from *T* all of the columns except those belonging to *t*, does not exist any row in T_0 equal to T_1".

A testor is called irreducible (typical) if upon eliminating any columns it stops being testor for (T_0, T_1), where *s* is called the length of the testor. The BT algorithm for typical testors is based in the idea of generate boolean vectors of $\alpha = (0,\ldots.., 0, 1)$ until arrive to $\alpha = (1,1,\ldots.,1, 1)$. For each case, it is verified if the set of columns that correspond to the coordinates of the *n*-tuple generated is a typical testor.

At this point, it is needed to have the learning matrix which contains the descriptions of objects in terms of a set of features; the matrix of differences, which it is obtained from the learning matrix comparing the values of object features pertaining to different classes; and the basic matrix, that is formed exclusively by basic rows (incomparable rows).

After computing the BT algorithm, a measurement of the importance of a single feature was calculated. After the typical testors were obtained, an irreducible combination of fewer features was obtained, each feature was very important to assure the difference from one class to another (bad vs good quality). Therefore, if a feature appears in many testors, it cannot be eliminated because it helps to preserve the class separation.

After this phase, the feature vector was reduced by a 76.7 %. Only 10 features were selected from the original set of 43 features to be used in the final recognition step. The features selected were: 2 central moments, 3 Maitra moments, red and blue color mean, color green variance, illumination mean and blue standard deviation.

7.4.1.6 Recognition of Fabric Images

In the recognition step, the voting algorithm (Alvot), which is based on partial precedence or partial analogies, was used. The premise of the algorithm is the following: if an object may looks like another, but not totally, and that the parts that look alike can offer information about possible regularities. Thus, a final decision is taken. The model of voting algorithms is described in six steps: (a) obtain the system of support sets, (b) computation of the similarity function, (c) evaluation by row for a given support set, (d) evaluation by class for a given support set, (e) evaluation by class for all the system of support sets, and (f) obtaining the solution rule.

After the computation of all the stages involved in voting algorithm, the discrepancies among the object inspected and its correspondent model were determined and a decision about the membership to good or bad quality class of the inspected object is emitted.

7.4.1.7 Experimentation and Results

Five different types of tests were carried out: *Case* (*a*) *Validation of the system learning ability:* Consists on validating if the system performs the training in a correct way. In these test, the system does not fail when an image of the training set is used as input. *Case* (*b*) *Rotation changes:* The goal is to verify if the system can handle rotation effects at 90 and 180° with regard to the original fabric images. *Case* (*c*) *Scale changes:* The goal is to verify if the system can handle scaling effects of halved and doubled the size of the original fabric images. *Case* (*d*) *Rotation and scale changes:* This test was made to verify the case b and c together. *Case* (*e*) *Noise insertion:* The objective is to evaluate if the system can handle images contaminated with salt and pepper noise, with a probability of 10 % for good quality and 40 % for bad quality.

The results obtained in each case are depicted in Table 7.2.

As can be seen in Table 7.2, the techniques utilized were effective and an average recognition of 71.2 % was obtained. The main contribution is that the system can handle invariance to rotation, scale and noise in a separated and in a combined way.

7.4.2 Automatic Inspection of Apple Quality

The creation of computer vision applications for agricultural industry has increased considerably in recent years (Cubero et al. 2011). The agricultural industry has the necessity to verify the quality of fruits and vegetables during the storage process. Typical target applications of inspection systems include grading and quality estimation from external parameters or internal features. Much of the sorting and grading processes are still not automatic. Manual inspection of fruit is tedious and can cause eye fatigue. Frequently, the inspection process is subject to errors due to different judgment emitted by different persons.

This study case, addresses the problem of design and implement a computer vision system for the automatic inspection of the quality fruit, particularly the golden apples. Apples are very susceptible to damage and the presence of bruises on the apple skin affects not only the appearance of the apple, which is an important indicator of quality, but also accelerates its deterioration (Santos and Rodrigues 2012). Two main problems were detected for the implementation of a computer vision system for apple grading: (1) how to acquire the whole surface

Table 7.2 Results obtained for fabric roll color inspection

Case	# images	% of success	% of failure
a	200	100	0
b	360	69.7	30.3
c	360	64.15	35.85
d	240	72.31	27.69
e	240	50.2	49.8
Total	1400	556.36	143.64
Average		71.2	28.8

image of an apple by cameras at an on-line speed, and (2) how to quickly identify the stem, calyx and the presence of defects (Li et al. 2002).

The solution presented for apple classification tackles both problems. Additionally, the system not use only the typical numerical information obtained from a computer vision system but also a symbolic knowledge obtained from a human expert is added to enhance the ability of the system to classify the apples. The approach presented is called a neuro-symbolic hybrid system (NSHS).

7.4.2.1 Definition of the Quality Criteria

A hybrid system offers the possibility to integrate two or more knowledge representations of a particular domain in a single system. One of the main goals is to obtain the complementary knowledge that allows improving the efficiency of the global system. A specific example of a hybrid system is the so called NSHS, which is mainly based on the symbolic representation of an object obtained from a human expert in the form of rules and a computer vision system to obtain the numerical information.

The quality criteria to evaluate a golden apple were obtained from a human expert in apple classification. For the case of golden apples, a category is assigned depending on the value of the external attributes. There exist four categories: category extra, category I, category II, and category III. In this example, only the category extra is evaluated, by that an apple can belong to one of two classes: good or bad quality. In Fig. 7.4, an example of good and bad quality apples are depicted. Additionally, in Table 7.3 a resume of the external attributes of a golden apple with its associated variable name, as well as the value and type are shown.

7.4.2.2 Image Acquisition

For the problem of category extra golden apple inspection, 148 images were obtained by means of a digital camera. The golden apples were acquired from crop fields at Puebla city in Mexico. For the complete set of images an operation of rotating 90 and 180° clockwise, doubled and halved the scaling and noise addition, similar to the study case of fabric was performed. At the end of this process the set

Fig. 7.4 Examples of category extra golden apples. **a** Good quality apple, and **b** Bad quality apples

Table 7.3 Quality criteria determined to measure the quality of an apple

Attribute	Acronym	Type	Value
Lengthened defect	LD	Range	0–6
Spotted defect	SD	Range	0–2.5
Various defects	VD	Range	0–5
Stem	S	Binary	True/false
Red color	RC	Range	0–255
Green color	GC	Range	0–255
Blue color	BC	Range	0–255

of 148 apples were divided into two categories: bad (74) and good (74) quality. In Fig. 7.5, an example of different apples after changing in rotation and scale are depicted.

7.4.2.3 Image Preprocessing

The stage consists of the image conversion from RGB color model to YIQ (Luminance, Inphase, Quadrature) color model. The main reason to perform such conversion is to facilitate the image feature extraction. The YIQ model was computed to separate the color from the luminance component because of the ability of the human visual system to tolerate more changes in reflectance than to changes in shade or saturation. The main advantage of the model is that reflectance (Y) and the color information (I and Q) can be processed separately. Reflectance is proportional to the amount of light perceived by human eye.

7.4.2.4 Apple Feature Extraction

The characteristics of each image were extracted based on the information defined by the human expert in the form of rules and by means of image processing in the

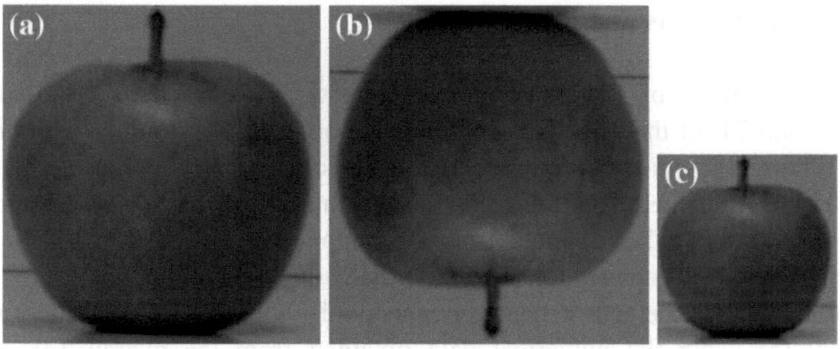

Fig. 7.5 Golden apples of category extra. **a** Original apple image, **b** Apple rotated 180° clockwise, and **c** Apple halved the original size

form of numerical data. These two types of knowledge information were combined in order to obtain the overall representation of an apple.

The number of rules defined by the experts were four, an example of a single rule is the following: *"IF an apple has the corresponding color, and has the stem, and has lengthened defects that do not exceed 2 cm, and has several defects that do not exceed 1* cm^2*, and has spotted defects that do not exceed ¼* cm^2*, THEN the apple belongs to category extra with good quality"*.

In order to obtain the numerical values, the same process of feature extraction for the problem of fabric was performed.

At the end of this stage, the rules were compiled with a knowledge based artificial neural network (KBANN), in order to obtain a representation of the information that further can be combined with the numerical results extracted from the computer vision system. The combination was made using the method called Neusim, which is based on the cascade correlation Fahlman algorithm.

7.4.2.5 Apple Classification

The output of the apple feature extraction stage was a joined representation of the symbolic and numerical knowledge. In order to further classify a golden apple, a refinement of that knowledge must be made. That refinement is performed by running again the Neusim method, but now not to join knowledge but use it as a classifier.

The main advantage of use Neusim algorithm is that one can see the number of hidden units added during the learning process, this is very useful to monitor the complete process of incremental learning. The output of this stage is the decision about the quality of an apple in one of two classes, bad or good.

7.4.2.6 Experiments and Results

From the total set of 128 golden apple images, 74 were used for the stage of training and 74 for the stage of recognition. In order to carry out the experiments, three different approaches were selected: (a) a connectionist approach which uses only the data obtained from the computer vision system, (b) a symbolic approach which uses only the data obtained from the compiled rules, and (c) a NSHS, which is a combination of connectionist and symbolic approach.

For the case of the tests using the connectionist approach, three scenarios were defined: (a) using the numerical data obtained from the overall 148 images (100 %), (b) using the data obtained only from 111 images (75 %), and (c) using only data from 74 images (50 %). Three rules were used to obtain the results referring to the test case using the symbolic approach. The first rule called 7 involves the following seven attributes: LD, SD, VD, S, RC, GC, and BC. The second rule called R5 considers the following five attributes: RC, GC, BC, S, and LD. Finally, the third rule called R4 includes the following four attributes: LD, SD, VD, and S. For the case of NSHS approach, a combination of connectionist and symbolic was made. Three rules R7, R5, and R4 were combined with 100, 75 and 50 % of the total number of examples. The overall results obtained are shown in Table 7.4.

One of the typical problem causing failures in computer vision systems is the lack of the complete description of an object. This can be observed by verifying the results obtained in Table 7.4 with symbolic and connectionist approaches. This disadvantage can be withdrawn by using a method to complete the information with the data defined by the knowledge of a human expert. The systems which allow this combination types are called NSHS, as can be seen with results shown in Table 7.4; these systems are very efficient for complementing the necessary knowledge for an automatic object inspection by means of a computer vision system.

For example, in the pure symbolic approach, the rule R4 was not enough to classify the apples correctly, but when it is integrated with the group of numeric examples (100, 75, 50 %), a substantial improvement is obtained, because the knowledge that does not contain the rule, is complemented with the numerical base of examples.

7.5 The Future of Computer Vision Systems

Even when the last few years have shown many computer vision systems which have been proven to be very efficient for the particular task for which they were created, there are still many challenges to be solved. The researchers divide these challenges into two categories, issues referring to hardware design techniques and issues concerned to software algorithms (Andreopoulos and Tsotsos 2013).

Table 7.4 The results obtained for the three approaches proposed for tests

Approach	Compiled rules	% of examples used	Accuracy (%)
Connectionist	–	100	95.14
	–	75	91.21
	–	50	90.54
Symbolic	R7	–	93
	R5	–	90.12
	R4	–	14.19
NSHS	R7	100	96.62
	R7	75	95.27
	R7	50	90.54
	R5	100	95.27
	R5	75	95.94
	R5	50	96.62
	R4	100	91.22
	R4	75	93.24
	R4	50	94.59

7.5.1 Issues Concerning Hardware

This challenge concerns the design and build of specific hardware to solve the typical problems of computer vision, due to that there is a significant gap in terms of the input size and the computational resources needed to reliably process those inputs. The solution needs to take into account variables such as real time acquisition huge storage space, distributed information, depth information, parallel execution, and portability. Following several of the hardware trends are explained.

Designing of sensors such Microsoft Kinect: With the invention of the low-cost Microsoft Kinect sensor, high-resolution depth and visual RGB sensing has become available for widespread uses. The complementary nature of the depth and visual information provided by the Kinect sensor opens up new opportunities to solve fundamental problems in computer vision and to design and build other powerful sensors based on this technology. The main topics that will be solved include preprocessing, object tracking and recognition, human activity analysis, hand gesture analysis, and indoors 3-D mapping.

Distributed smart cameras: This implies the design and implementation of real time distributed embedded systems that perform computer vision using multiple cameras. This approach has emerged thanks to a confluence of simultaneous advances in disciplines such: computer vision, image sensors, embedded computing, and sensor networks. Processing images in a network of distributed smart cameras introduces several complications such as the tracking process. The distributed smart cameras will represent key components for future embedded computer vision systems and that smart cameras will become an enabling technology for many new applications.

Multicore processors: The recent emergence of multi-core processors enables a new trend in the usage of computers. Computer vision applications, typically requires heavy computation and lots of bandwidth, which makes difficult to run the applications in real time. The use of multi-core processors can potentially serve the needs of such workloads. In addition, more advanced algorithms can be developed utilizing the new computation paradigm. The main advantage offered will be the execution of parallel algorithms, for example by taking multiple cameras inputs of a scene, extracting useful features, and performing statistical inference almost at the same time. In resume, the fact of parallelizing the workload will speed up the performance of the computer vision systems.

Graphics processing units (*GPU*): Computer vision tasks are computationally intensive and repetitive, and they often exceed the capabilities of the CPU, leaving little time for higher level tasks. However, many computer vision operations map efficiently onto the modern GPU, whose programmability allow a wide variety of computer vision algorithms to be implemented. The GPU provides a streaming, data parallel arithmetic architecture which carries out a similar set of calculations on an array of image data. The single instruction multiple data (SIMD) capability of the GPU makes it suitable for running tasks, which often involve similar calculations operating on an entire image.

Mobile devices: Since mobile devices are gaining increasingly powerful processors, great quality cameras and high speed network access, they will be suitable to implement computer vision algorithms. The ubiquitous platform offered by a mobile device can be used for a wide range of applications such as: smart surveillance, assembly line troubleshooting aid, controlling specific devices with gestures or screen touches, and entertainment. Additionally, the sensors often included in mobile devices such as: GPS, gyroscopes and accelerometers can contribute to enhance the computer vision tasks. The main disadvantage is the life time battery.

7.5.2 Issues Concerning Software

One of the main problems to solve is to find a way to generalize the knowledge as well as the human vision system, enhance the way for the representation of objects, and the ability for learning and inferencing the models used. Following several of the software trends are explained.

Multispectral and hyper spectral image analysis: The data of the images is obtained at specific frequencies across the electromagnetic spectrum. The spectral imaging allows the extraction of additional information which human eye fails to capture with its receptors for red, green and blue. After the representation in the electromagnetic spectrum, the images are combined to form a three-dimensional hyper spectral data cube for processing and analysis. The main difference between multispectral and hyper spectral images is the number of bands obtained. The main areas for the application of this type of imaging are agriculture, mineralogy, physics, and surveillance.

RGB-D images for representation and recognition: This type of images are formed with the classical red, green and blue information and adding information about the scene depth, which is obtained by a technique called structured light implemented with an infra red laser emitter. This type of images is obtained using a kinect sensor and the data containing visual and geometric information of the scene. This image offers the advantage to obtain the depth information using only one device instead of using the classical pair of images. The features obtained with RGB-D images will be very useful to improve the process of shape estimation, 3D mapping and localization, path planning, navigation, pose recognition and people tracking.

Real time display of obscured objects: This kind of algorithms could help car drivers and airplane pilots to see through fog, and submarines explore under the sea. The algorithms could provide safety features for future intelligent transportation systems. This means that computer vision software modules can distinguish specific objects more accurately from the rest of a scene, even in complete darkness.

Deep and feature learning: One of the main challenges for computer vision and machine learning is the problem of learning representations of the perceptual world. The learning and perceptual methods allows automatically learn a good representation of the input unlabeled data, offering a time reduction against typical learning algorithms which spend a lot of time to obtain the input feature representation. Since these algorithms mostly learn from unlabeled data, they have the potential to learn from vastly increased amounts of data (since unlabeled data is cheap), and therefore also achieving a vastly improved performance. The representation of multilevel hierarchies obtained are useful not only for low level features such as edge or blob detectors, but also are useful for high level concepts such as face recognition.

Mid-level patch discovery: The technique allows discovering a set of discriminative patches that can serve as a fully unsupervised mid-level visual representation. The desired patches need to satisfy two requirements (1) to be representative, they need to occur frequently enough in the visual world, and (2) to be discriminative, that is, they need to be different enough from the rest of the visual world. The patches could correspond to parts, objects, visual phrases, etc. but they are not restricted to be any one of them. The patches are simple to compute, and offers very good discriminability, broad coverage, better purity, and improved performance compared to visual world features. These approaches can be used mainly on scene classification, beating bag-of-words, spatial pyramids, object bank, and scene deformable-parts models.

Augmented reality: Allows enhancing the senses of a user by manipulating virtual objects superimposed on top of the real world scenes. In other words, AR bridges the gap between the real and the virtual in a seamless way. Specifically, because of the improved ease-of-use of augmented reality interfaces, these systems may serve as new platforms to gather data, as imagers may be pointed by users to survey and annotate objects of interest to be stored in different kind of systems such as educational learning, games, training skills and geographical information systems (GIS).

Gesture control: Human–computer interfaces (HCI) needs to evolve from mouse-keyboard based interaction, multi-touch screens and other exotic approaches such as using special gloves or other devices, until reach the use of gestures to translate human actions into control applications. There were and still are a series of attempts to produce computer control scripts via gesture-based interfaces, and the research literature is abundant in chapters on this subject. The main challenge is that the computer-based control via gestures has to be robust and in real-time. Any lag in the result can lead to users abandoning it.

7.6 Conclusions

In this chapter the design and implementation of two computer vision systems to verify the quality of fabric and apples was presented. The main challenges to build the computer vision systems were explained exhaustively. A brief explanation and contrasting of several computer vision systems already presented in the literature were discussed. Additionally, a highlighting was made about its ability or accuracy to detect defects of an object. Finally, a brief explanation about the future of computer vision systems for quality inspection was presented.

Despite the numerous studies developed in the computer vision area, there is still not a standardized method which could be proposed for the assessment of the quality of different types of objects. The particular characteristics of an object require a computer vision system to be customized; this implies an exhaustive research process and not only the purchase of expensive equipment. Besides, in order to obtain a better performance of the system, the acquisition of the knowledge from the human experts and the techniques to represent it in terms of numerical information is mandatory.

References

Abdulmalek, F., & Rajgopal, J. (2007). Analyzing the benefits of lean manufacturing and value stream mapping via simulation: A process sector case study. *International Journal of Production Economics, 107*(1), 223–236.

Andreopoulos, A., & Tsotsos, J. (2013). 50 years of object recognition: Directions forward. *Computer Vision and Image Understanding, 117*(8), 827–891.

Bissi, L., Baruffa, G., Placidi, P., Ricci, E., Scorzoni, A., & Valigi, P. (2013). Automated defect detection in uniform and structured fabrics using Gabor filters and PCA. *Journal of Visual Communication and Image Representation, 24*(7), 838–845.

Blasco, J., Aleixos, N., & Moltó, E. (2007). Computer vision detection of peel defects in citrus by means of a region oriented segmentation algorithm. *Journal of Food Engineering, 81*(3), 535–543.

Choi, J., Ro, Y., & Plataniotis, J. (2011). A comparative study of preprocessing mismatch effects in color image based face recognition. *Pattern Recognition, 44*(2), 412–430.

Cubero, S., Aleixos, N., Moltó, E., Gómez-Sanchis, J., & Blasco, J. (2011). Advances in machine vision applications for automatic inspection and quality evaluation of fruits and vegetables. *Food and Bioprocess Technology, 4*(4), 487–504.

Forsyth, D., & Ponce, J. (2012). *Computer vision: A modern approach* (2nd ed.). New Jersey, USA: Pearson.

Gadelmawla, E. (2011). Computer vision algorithms for measurement and inspection of spur gears. *Measurement, 44*(9), 1669–1678.

Gonzalez, R., & Woods, R. (2009). *Digital image processing* (3rd ed.). New Jersey, USA: Prentice Hall.

Kumar, A. (2008). Computer-vision-based fabric defect detection: A survey. *IEEE Transactions on Industrial Electronics, 55*(1), 348–363.

Lee, S., Kim, K., Kim, J., Kim, M., & Yoo, H. (2010). Familiarity based unified visual attention model for fast and robust object recognition. *Pattern Recognition, 43*(3), 116–1128.

Lerones, P., Fernández, J., García-Bermejo, J., & Zalama, E. (2005). Total quality control for automotive raw foundry brake disks. *International Journal of Advanced Manufacturing and Technology, 27*(3–4), 359–371.

Li, Q., Wang, M., & Gu, W. (2002). Computer vision based system for apple surface defect detection. *Computers and Electronics in Agriculture, 36*(2–3), 215–236.

Lin, K., & Fang, J. (2013). Applications of computer vision on tile alignment inspection. *Automation in construction, 35*(November), 562–567.

Mery, D., Chanona-Pérez, J., Soto, A., Aguilera, J., Cipriano, A., & Veléz-Rivera, N., et al., (2010). Quality classification of corn tortillas using computer vision, *Journal of food engineering*, 101(4): 357–364.

Možina, M., Tomaževic, D., Pernuš, F., & Likar, B. (2013). Automated visual inspection of imprint quality of pharmaceuticals tablets. *Machine Vision and Applications, 24*(1), 63–73.

Mullen, R., Monekosso, D., & Remagnino, P. (2013). Ant algorithms for image feature extraction. *Experts Systems with Applications, 40*(11), 4315–4332.

Nalwa, V. (1993). *A guided tour of computer vision* (1st ed.). Boston, Massachusetts, USA: Addison-Wesley Longman Publishing Co.

Padma, A., & Sukanesh, R. (2013). Wavelet statistical texture features-based segmentation and classification of brain computed tomography images. *IET Image Processing, 7*(1), 25–32.

Peng, X., Chen, Y., Yu, W., Zhou, Z., & Sun, G. (2008). An online defects inspection method for float glass fabrication based on machine vision. *International Journal of Advanced Manufacturing Technology, 39*(11–12), 1180–1189.

Perng, D., Liu, H., & Chang, C. (2011). Automated SMD LED inspection using machine vision. *The International Journal of Advanced Manufacturing Technology, 57*(9–12), 1065–1077.

Razmjooy, N., Somayeh, B., & Soleymani, F. (2012). A real-time mathematical computer method for potato inspection using machine vision. *Computers and Mathematics with Applications, 63*(1), 268–279.

Sannen, S., & Van Brussel, H. (2012). A multilevel information fusion approach for visual quality inspection. *Information Fusion, 13*(1), 48–59.

Santos, J., & Rodrigues, F. (2012). Applications of computer vision techniques in the agriculture and food industry: A review. *European Food Research and Technology, 5*(6), 989–1000.

Satorres, S., Gómez, J., Gámez, J., & Sánchez, A. (2012). A machine vision system for defect characterization on transparent parts with non-plane surfaces. *Machine Vision and Applications, 23*(1), 1–13.

Shah, R., & Ward, P. (2003). Lean manufacturing: context, practice bundles, and performance. *Journal of Operations Management, 21*(2), 129–149.

Sullivan, W., McDonald, T., & Van Aken, E. (2002). Equipment replacement decisions and lean manufacturing. *Robotics and Computer Integrated Manufacturing, 18*(3–4), 255–265.

Sun, T., Tseng, C., & Chen, M. (2010). Electric contacts inspection using machine vision. *Image and Vision Computing, 28*(6), 890–901.

Wandell, B., Gamal, A., & Girod, B. (2002). Common principles of image acquisition systems and biological vision. *Proceedings of the IEEE, 90*(1), 5–17.

Xiao, B., & Wang, G. (2013). Generic radial orthogonal moment invariants for invariant image recognition. *Journal of Visual Communication and Image Representation, 24*(7), 1002–1008.

Yang, C., Chao, K., & Kim, M. (2009). Machine vision system for online inspection of freshly slaughtered chickens. *Sensors and Instruments for Food Quality and Safety, 3*(1), 70–80.

Chapter 8
Critical Success Factors for Kaizen Implementation

Denisse Rivera-Mojica and Lizeth Rivera-Mojica

Abstract Currently the business organizations are forced to stay in the competitive global market, for this reason, it is necessary to adapt to various changes. Kaizen is one the most important methodologies used in the companies, with the goal of reducing the times of processes and increases the economic benefits among others. However, it is unknown what the Critical Success Factors for the implementation, by what this chapter presents an investigation in which is carried out a review of literature which identifies the Critical Success Factors that contribute to the success of the implementation of Kaizen, subsequently presented the results of a survey that consists of 37 activities and 14 benefits, which was applied to a personnel of 258 with responsibilities in continuous improvement programs in the company, which should be answered on a Likert scale. The questionnaire was validated by using the index of Cronbach's alpha and applied a factorial analysis using the method of principal components and for the purpose of obtaining a better interpretation of the factors, there was a rotation by the Varimax method of the factors extracted to facilitate the interpretation of the same. Finally, we performed a Confirmatory Factor Analysis, with the purpose of validating the relationships between variables and factors, which is gender a Structural Equation Model.

Keywords Kaizen · Project improvement · SEM · Factor analysis

8.1 Introduction

Today business organizations are forced to stay in the competitive global market, for this reason, it is necessary to adapt to various changes and aspects: economic, political, social, technological and commercial, why must develop the capacity to

D. Rivera-Mojica (✉) · L. Rivera-Mojica
Department of Industrial Processes, Technological University Paso del Norte, Pez Lucio No. 10526, Col. Puerto Anapra 32107, Ciudad Juárez, Chih, México
e-mail: denisserivera.utpn@gmail.com

J. L. García-Alcaraz et al. (eds.), *Lean Manufacturing in the Developing World*,
DOI: 10.1007/978-3-319-04951-9_8,

confront changes and react appropriately to environmental factors and in this way to proactively anticipate their needs and conditions (Cantu 2006).

Lean Strategies play a determinant role since the companies can have a competitive advantage that allows them to stay ahead of their competitors. Some tools used in Lean manufacturing optimization of operations are: Poka Yoke, JIT, Kanban and Kaizen among others.

In this context, there is the philosophy "Kaizen" as the key to competitive success, Japanese word, in simple terms means improvement. The concept of Kaizen has been given various definitions during development, but all contain the essence, expressed in different words. The meaning of Kaizen comes from two Japanese characters: "Kai" means change and "Zen", which means to improve (Savolainen 1999; Newitt 1996). This term implies a culture of constant change to evolve toward best practices (Imai 1996), that is, what is commonly known as continuous improvement or principle of continuous improvement (Lillrank 1995). In summary, it is important to indicate that after analyzing the literature that Kaizen has been a term that is still in evolution, which has resulted, different meanings depending on time and the organizational context in which it was submitted (Tozawa and Bodek 2002).

It has been documented in the literature several benefits when you apply the Kaizen philosophy. Some of these benefits are: reduction of resources used, time reduction of processes, systematization of the measurement of labor, allows a better orientation to the organization toward the customer, contributes to a better vision of the organization and can reach to encourage the participation, communication and teamwork between employees and managers (Dale et al. 1997; Manos 2006).

According to Farris et al. (2009) in recent years, the term Kaizen has gained increasing importance in organizations because it has popularized the use of its more practical side of Kaizen events. However in spite of its potential and popularity, there are few studies on the factors that actually influence for an effective implementation. Different studies have identified the difficulty that in many cases companies have to implement and sustain their improvements, due to the work culture of each organization (Prajogo and Sohal 2004).

Based on the above, as an example, a study in Mexico and Spain on the sustainability of Kaizen, shows that companies do not give continuity to its use, this method is abandoned in a large percentage in this country, because of the resistance to organizational change, the lack of implementation and monitoring of actions (Jaca 2010). Figure 8.1 shows the results for both regions, which discloses the percentage of companies that have abandoned their breeding programs, where Toluca -Lerma represents the industrial sector in Mexico and CAV companies Navarra is located in Spain. Figure 8.1 shows the level of neglect of the Kaizen methodology in Mexico is 10 %, indicating a dropout rate relatively high.

According to the contributions of existing articles and publications found in a literature review, we found different elements that contribute to the successful implementation of Kaizen and Table 8.1 shows the coincidence of the authors with respect to each particular activity. One can see that the element that matches had by the authors was the commitment and motivation of the team. Other factors found in the literature and had few references were: customer approach (Romero et al. 2009),

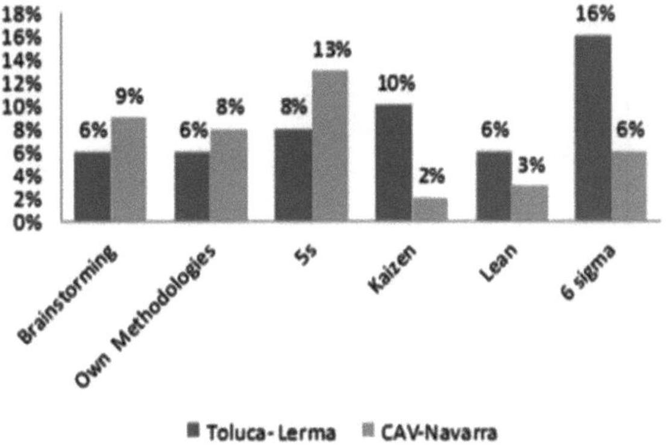

Fig. 8.1 Dropout rate improves systems by type and region (on the company). *Adapted from* Jaca et al. (2010)

development of quality management systems, form a learning organization approach to critical processes, establishing long-term goals (Kaye and Anderson 1999), making operating practices (Landa 2009), development of structures to stop failures, consistent focus improvement activities (Upton 1996), resistance to change (Melnyk et al. 1998), implementation of 5s, recognition of the need for change (Bateman 2005), application of methodologies to understand customer voice, effective internal processes (Dale et al. 1997), and safety at work (Jørgensen and Boer 2004).

The major critical success factors that have been determined over the applications and research of Kaizen are illustrated in Table 8.1, the star(*) shows the relation between the authors and the elements. According to the above, it is clear that despite the popularity and growth has been Kaizen philosophy, unfortunately most implementations of Kaizen in Mexican companies are not successful. There have been several studies on the benefits of Kaizen, yet little is known about the reasons for continuous improvement implementations fail, should be done to ensure the success of Kaizen in the manufacturing industry in the region. In this regard, we take into account as a subject of study Juarez maquiladora industry and research purpose to determine those key factors contributing to the success of the changes proposed and implemented the Kaizen in business organizations. The research methods used in the search for critical success factors in the implementation of Kaizen were the literature review and survey. Subsequently, for information analysis was done using the SPSS statistical package to determine different key success factors and variables that compose it, which is known as Exploratory Factor Analysis, then performed a confirmatory factor analysis. Finally the results of the study were used to make recommendations for organizations that implement Kaizen philosophy and lay the foundation for future research.

8.2 Benefits of Implementation of Kaizen

We found several benefits about implementing Kaizen in the literature, for example, Manos (2006) categorizes as qualitative and quantitative benefits. The quantitative benefits are those in which the results are measurable, these include; economic

Table 8.1 Coincidence of the authors with respect to each particular activity

Elements	Authors Total																		
	A	B	C	D	E	F	G	H	I	J	K	L	M	N	O	P	Q	R	S	
Commitment and motivation of the team			*		*	*	*							*	*	*	*	*	*	10
Support from top management	*					*			*			*				*	*	*		7
Assigned resources				*		*	*	*	*				*		*					7
Leadership			*					*				*				*		*	*	6
Developing a culture of continuous improvement			*				*					*				*	*			5
Set goals for improvement programs				*			*	*								*		*		5
Using an appropriate methodology					*				*	*							*			4
Standardization and process measurement			*	*				*								*				4
Organization of support equipment	*		*									*								3
Presence of a facilitator to support the program	*					*				*										3
Interdepartmental communication													*					*	*	3
Conflict between improvement approach and the existing culture		*														*	*			3
Employee attitude								*									*			2

(continued)

Table 8.1 (continued)

Elements	Authors Total																			
	A	B	C	D	E	F	G	H	I	J	K	L	M	N	O	P	Q	R	S	
Interdepartmental cooperation														*				*		2
Follow the PDCA cycle			*											*						2
Training								*										*		2
Heterogeneity of the improvement teams								*									*			2
Evaluation system				*											*					2
Skills and experience		*																		2
Establish policies, objectives and structure									*											2
Clarify common goals and ideas of the direction	*	*																		2

A Upton (1996), *B* Bessant and Caffyn (1997), *C* Dale et al. (1997), *D* Melnyk et al.(1998), *E* Kaye and Anderson (1999), *F* Rapp and Eklund (2002), *G* Jørgensen and Boer (2004), *H* Farris (2003), *I* Cooney and Sohal (2004), *J* Prajogo and Sohal (2004), *K* Bradley and Willett (2004), *L* Bateman (2005), *M* Bhuiyan et al. (2007), *N* Bisgaard (2007), *O* Readman (2007), *P* Suarez and Miguel (2009), *Q* Landa (2009), *R* Romero et al. (2009), *S* Tapias et al. (2010)

benefit, saving time contributing to the economic benefit, reduced material handling distance, less staff required, reducing waiting time or cycle time, reduced process steps and reducing inventory. The qualitative benefits are more difficult to measure because the results focus to the human factor, within which is the staff motivation and self-esteem. On the other hand Lefcovich (2007) mentions that some benefits of Kaizen are: Decrease the number of accidents, reduced inventories, goods in process and finished form of fomentation process-oriented thinking, greater emphasis on the planning stage, reduction in failure of the equipment and tools, reduced setup times of machines, increased levels of customer satisfaction and consumers, increased levels of inventory turnover, significant drop in the levels of failures and errors, improved self-esteem and motivation, high increases in terms of productivity, significant reduction in costs, improved product designs, lower levels of waste and waste, reductions in design cycles and operational improvement flows cash, lower customer turnover and employees, more and better financial balance, improvement in the attitude and competence of managers and staff to implement continuous changes, ability to compete in global markets and finally acquired a better ability to accommodate continuously to sudden market changes.

8.3 Methodology

The methodology that was used in this research project is based on the data compilation made in six phases that are explained in the following paragraphs.

8.3.1 Phase 1: Identification of the Critical Success Factors of Kaizen and it Benefits

The initial phase was the selection of the elements to implement Kaizen through a review of literature. Variables were obtained 35 and 14 by different authors cited benefits, which were dimensioned in 7 factors and 3 benefits.

8.3.2 Phase 2: Questionnaire Construction and Validation

To research a questionnaire was developed based on a review of literature and research by experts from different maquiladoras in Juarez, validation was content. Was performed a total of 37 questions related to critical factors obtained, 14 related to the benefit of companies to implement Kaizen and made five demographic questions.

The questionnaire should be answered on a Likert scale, which comprised between 1 and 5, in response to the global question "extent to which they have implemented the following activities" (Cox et al. 2006; Cua et al. 2001; Devaraj et al. 2004; Flynn and Sakakibara 1995; Kaya 2006; Long and Shields 2005; Ooi et al. 2007; Schroeder et al. 2002; Zacharatos et al. 2005).

The final part of the questionnaire contains questions about the participant's demographics and the company where he works as the business sector, to which it belongs, as the respondent, gender, seniority among others. Table 8.2 has the activities and benefits as well as its abbreviation.

For validation of the questionnaire was used Cronbach's alpha index (Hair et al. 1995), is an index used to measure the reliability of the type of internal consistency of a scale, to assess the extent to which the items of an instrument are correlated. According to Cortina (2003) the minimum acceptable value for Cronbach's alpha is 0.70; below this value the internal consistency of the scale used is low. Moreover, the maximum expected value is 0.90, above this value is considered as redundancy or duplication. Several items are exactly the same element measuring a construct; therefore, redundant items must be removed.

8.3.3 Phase 3: Application of the Questionnaire

The sample was taken to apply the questionnaire consists of 258 persons (directors, managers or bosses) and are involved in the operation of the Kaizen philosophy, in areas of quality, continuous improvement, Lean, 5S, among others. The survey was conducted in manufacturing industries in Juarez with various commercial sectors; data collection was performed by sending the survey through e-mail, personally or by telephone.

8.3.4 Phase 4: Descriptive Analysis of the information

After application of the survey and data capture for analysis of information used SPSS 17. For determine the feasibility of factor analysis examined the correlation matrix, which will allow us an idea of potential success of factor analysis is to check whether a number of correlations above 0.3 (Nunnally and Bernstein 2005) and whose p-values should be less than 0.05. Later KMO index was obtained to measure the correlation between variables was applied Barlett sphericity. The KMO measure of sampling adequacy (Kaiser–Meyer–Olkin) tests whether the partial correlations among variables are small enough. Compares the magnitude of the observed correlation coefficients with the magnitude of the correlation coefficients. The KMO statistic varies between 0 and 1. Kaiser, Meyer and Olkin advise that if $KMO \geq 0.75$ the idea of factor analysis is good, if $0.75 > KMO \geq 0.5$ the idea is acceptable and if $KMO < 0.5$ is unacceptable (Pardo 2002) commonalities were subsequently analyzed for each of the activities.

Table 8.2 Activities and abbreviations

Activity	Abbreviation
Operators are trained perform the job in various positions	OpRecCap1
The company managers receive training in teamwork and problem solving	ManRecFor2
Groups are organized to make suggestions for improvement of product/ process or solve problems: quality circles, suggestion schemes in groups	OrgGrSug3
Commitment and support are providing from management	ExisApoCom4
There is managerial leadership and departmental in the implementation of Kaizen	ExisLidGer5
There is an internal or external facilitator to help coordinate effectively the improvement program	ExisFacInExt6
The management provides the necessary resources for programs (economic, physical space, time)	GeProvRec7
The organization has a culture that encourages change	OrgaCulCam8
Programs exist within the enterprise security	ExisSegEmp9
Effective internal processes are used	UtiliProIntEfi10
There is communication between projects leaders and senior management, managers and their staff	ExisComGerA11
The management level of the company communicates to their workers if they are conducting a good job or not	NivGerEmCoEmp12
Operators are advised on strategic and/or economic information from enterprise	ComOperInfEst13
Meetings are held and workers can express their ideas or opinions	ReaReuExpId14
The company is considerate a learning organization	OrgAprend15
There is a motivation and engages for each person in the company	MotPer16
There is skill and experience by the improvement team members	ExisHabInt17
There is developed culture of continuous improvement	DesaCulMej18
Participants in Kaizen teams include representatives of different disciplines	ParRepDis19
Participants will consider all hierarchical levels	ParDifNivGer20
They are recognized by team members achievements and efforts of the event	RecMieEs21
There are restrictions to implement changes proposed improvement	ExisResCam22
There are restrictions to continue with the implementation of changes to improve	ExisResConImp23
There is support from persons affected in changed work areas	ApoyAfecReal24
There negativity to make changes for improvement	ExisNegRealCa25
They set goals and work plans in the short and long term and is aligned with a vision	FijObjPlanCL26
Assessment systems are used regularly to measure the performance of activities	SIsEva27
There are standards, indicators, guidelines or procedures derived for process measurement	ExisEstaIndPa28
Formats are used control records necessary for the performance of activities	ForRegNece29
There is implemented 5s methodology as a strategy of Kaizen	Desarr5s30
Participants demonstrate a positive attitude to make changes	ParActPos31
The points that are pending are documented and there is a charge which follows up	PunPenDoc32

(continued)

Table 8.2 (continued)

Activity	Abbreviation
It continuously measures progress towards goals	MidConPro33
We develop a framework for determining errors	DesEstFall34
It follows the PDCA cycle (Plan-do-Check-Act) as strategy of Kaizen	PDCA35
It takes into account the opinion of the companies' customers of the to make changes at work	OpinClieReaMod36
Methodologies are applied to understand the customer's voice	AplicMetVoCli37
Reducing the number of accidents at the company	RedAcc38
The company responds to customer needs	CoResNecClin39
The successful implementation of Kaizen increases customer satisfaction	ImplExiaumSat40
Increases staff motivation	AumMot41
By implementing Kaizen improves the attitude and competence of managers and staff	AumAct42
Increased the staff self-esteem	AumAut3
Increase in the rate of introduction of new products	AumVelInt44
The company meets the deadlines and pledges	FechEntrCan45
By implementing Kaizen reduces the percentage of defective products	KaizRedPorPro46
Improvement of product quality	MejCalPro47
Decreases manufacturing unit costs	DisCosUniMan48
Reduced inventory levels	RedNivInv49
The company reduced the time between order receipt and customer delivery as close to zero	CoReRec50
By implementing Kaizen generate increases in productivity	IncMatPro51
Business sector	SecEm
Number of employees	NumEmpl
Position of interviewee	PuestEnt
Sex	Sex
Years of experience in the position	Anti

8.3.5 Phase 5: Exploratory and Confirmatory Factor Analysis

We performed a factor analysis, the factor extraction method used was the principal component, the goal is to explain most of the total variability of all variables, with the least possible number of common factors as criteria for factors to determine the number extract was used auto criterion values greater than one (Hair et al. 1995). In order to obtain a better understanding of the factors was performed by the method Varimax rotation of the extracted factors to facilitate interpretation thereof.

Finally we performed a confirmatory factor analysis, in order to validate the relationship between variables and factors.

Table 8.3 Validation of the instrument and scale construction

Ítem	Cronbach's Alpha if the item is deleted (iteration)								
	1	2	3	4	5	6	7	8	9
OpRecCap	0.877	0.880	0.885	0.885	0.889	0.890	0.890	0.890	0.890
ManRecFor	0.876	0.880	0.884	0.885	0.889	0.889	0.890	0.890	0.890
OrgGrSug	0.877	0.880	0.885	0.886	0.889	0.890	0.890	0.890	0.890
ExisApoCom	0.874	0.877	0.882	0.883	0.886	0.887	0.887	0.887	0.887
ExisLidGer	0.880	0.883	0.888	0.889	0.892	0.893	0.894	0.894*	
ExisFacInExt	0.881	0.884	0.889	0.890	0.893*				
GeProvRec	0.878	0.882	0.886	0.887	0.890	0.891	0.892	0.891	0.892
OrgaCulCam	0.880	0.883	0.887	0.888	0.891	0.892	0.893	0.893	0.893
ExisSegEmp	0.872	0.876	0.880	0.881	0.885	0.885	0.886	0.885	0.885
UtiliProIntEfi	0.872	0.876	0.881	0.881	0.885	0.885	0.886	0.886	0.886
ExisComGerA	0.880	0.883	0.888	0.889	0.892	0.893	0.893	0.893	0.893
NivGerEmCoEmp	0.876	0.879	0.884	0.885	0.889	0.890	0.890	0.890	0.890
ComOperInfEst	0.878	0.882	0.886	0.887	0.891	0.891	0.892	0.892	0.892
ReaReuExpld	0.877	0.880	0.885	0.886	0.889	0.890	0.891	0.891	0.891
OrgAprend	0.878	0.881	0.886	0.887	0.891	0.891	0.892	0.892	0.892
MotPer	0.878	0.881	0.886	0.887	0.891	0.891	0.892	0.892	0.892
ExisHabInt	0.881	0.884	0.889	0.890	0.893	0.894*			
DesaCulMej	0.873	0.876	0.881	0.882	0.886	0.886	0.887	0.887	0.886
ParRepDis	0.876	0.880	0.885	0.886	0.889	0.890	0.890	0.890	0.890
ParDifNivGer	0.879	0.882	0.887	0.888	0.891	0.892	0.893	0.892	0.892
ParEntTec	0.877	0.880	0.885	0.886	0.889	0.890	0.891	0.890	0.890
ExisResCam	0.880	0.883	0.888	0.889	0.893	0.893	0.894*		
ExisResConImp	0.883	0.886	0.891*						
ApoyAfecReal	0.878	0.881	0.886	0.887	0.890	0.891	0.892	0.892	0.892
ExisNegRealCa	0.883	0.886*							
FijObjPlanCL	0.879	0.882	0.887	0.888	0.891	0.892	0.893	0.893	0.893
SIsEva	0.878	0.882	0.886	0.887	0.891	0.891	0.892	0.892	0.892
ExisEstaIndPa	0.873	0.876	0.881	0.882	0.885	0.886	0.887	0.886	0.886
ForRegNece	0.876	0.879	0.884	0.885	0.888	0.889	0.890	0.889	0.889
Desarr5s	0.877	0.880	0.884	0.885	0.888	0.889	0.889	0.889	0.889
RecMiemEs	0.877	0.880	0.885	0.886	0.889	0.890	0.891	0.890	0.890
PunPenDoc	0.884*								
MidConPro	0.882	0.885	0.890	0.892*					
DesEstFalla	0.874	0.877	0.882	0.883	0.886	0.887	0.887	0.887	0.887
PDCA	0.878	0.881	0.886	0.887	0.890	0.891	0.892	0.892	0.892
OpinClieReaMod	0.879	0.882	0.887	0.888	0.891	0.892	0.893	0.892	0.893
AplicMetVoCli	0.878	0.881	0.885	0.886	0.890	0.891	0.891	0.891	0.891
Alfa de Cronbach Total	0.881	0.884	0.886	0.891	0.892	0.893	0.894	0.894	0.894

8.3.6 Phase 6: Structural Equation Model

It generated a structural equation model, for which we used the AMOS 18 software. According to Ruiz et al. (2010) structural equation models are a family of multivariate statistical models that estimate the effect and the relationships between multiple variables. In the structural equation model, the ovals represent latent variables and rectangles are the activities, which were observed or measured.

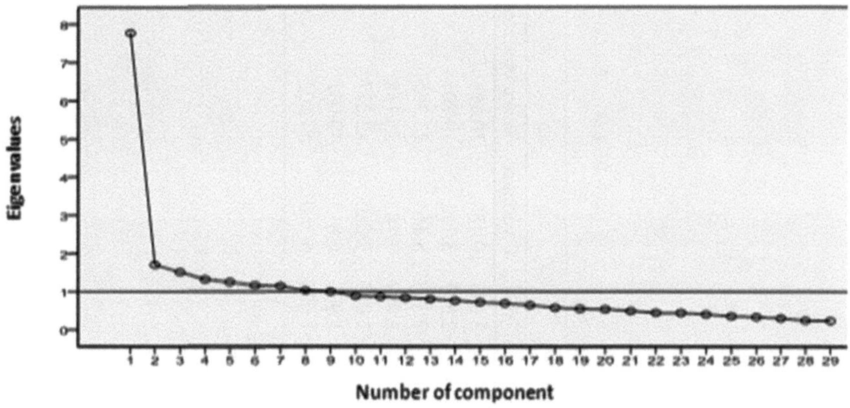

Fig. 8.2 Sedimentation graph

To measure the efficiency of structural equation model was used the value of chi-square, the degrees of freedom of the model and the relationship between the two parameters. In order to obtain a sufficiently explanatory model, we used the goodness of fit index. We also analyzed the comparative fit index in order to analyze the improvement of one model to another, also sought to maintain an acceptable measure of the root mean square residual.

Finally there was the adequacy of the sample size and ensured that the amendments were valid with Hoelter critical index.

8.4 Results

8.4.1 Composition of the Sample

In the data collection period yielded a total of 258 questionnaires of which 230 were taken as valid, which came from 40 companies located in Ciudad Juarez, Chihuahua, Mexico. The positions where 92 respondents belonged to engineers, 64 engineering technicians, 43 managers, 27 supervisors, 26 operators and 6 administrative staff. Respondents were of both sexes where 66.3 % were male and 33.7 % female.

8.4.2 Validation Questionnaire Scale Construction and Identification of Items

Was obtained Cronbach's alpha index to the 37 initial items and found that removing any of these are maintained or improved the internal consistency of the instrument, so that the final list of items to be analyzed is obtained iteratively, by

Table 8.4 Total variance explained of activities

Component	Initial eigenvalues	Sums of squared loadings extraction	Saturations sum of the square of the rotation						
	Total	% variance		Total	% variance	% accumulated	Total	% variance	% accumulated
1	7.775	26.810	26.810	7.775	26.810	26.810	3.042	10.488	10.488
2	1.705	5.880	32.690	1.705	5.880	32.690	2.527	8.712	19.201
3	1.519	5.238	37.928	1.519	5.238	37.928	2.368	8.164	27.365
4	1.323	4.561	42.489	1.323	4.561	42.489	2.257	7.783	35.148
5	1.256	4.330	46.820	1.256	4.330	46.820	2.053	7.080	42.227
6	1.169	4.030	50.849	1.169	4.030	50.849	1.976	6.815	49.042
7	1.146	3.953	54.802	1.146	3.953	54.802	1.670	5.760	54.802

Table 8.5 Rotated component matrix of activities

Factor number	Ítem	Factor load	Component description
1	Operators are trained to perform the work in various positions	0.711	Training and education: This component refers to the training plans to ensure that staff qualified enough to tackle organizational challenges
	Groups are organized to make suggestions for improvement of product/process or solve problems: quality circles, suggestion schemes in groups	0.683	
	The management level of the company receives training in teamwork and problem solving	0.659	
	There is support and commitment from management	0.657	
2	There is communication between projects leaders and senior management, managers and their staff	0.683	Communication process: It refers to the exchange of ideas, information and opinions among workers
	The management level of the company communicates to their workers if they are doing a good job or not	0.58	
3	Formats are used and/or control records necessary for the performance of activities	0.714	Documentation and evaluation: Dedication and degree of care that you have to record everything about continuous improvement activities and regular evaluation of the corresponding system
	There are standards, indicators, guidelines or procedures for measuring derivative processes	0.619	
	Assessment systems are used regularly to measure the performance of activities	0.55	
4	Operators are informed about information is a strategic and/or economic policy of the company	0.701	Workers integration and award: This component relates to worker participation in aspects related with continuous improvement activities
	Meetings are held where workers can express their ideas or opinions	0.659	
	They are recognized by team members achievements and efforts of the event	0.572	
5	The organization has a culture that encourages change	0.69	Culture for change and improvement: This component relates to the values and ways of thinking that characterize the behavior of staff at all levels of the company
	Participants show a positive attitude to make changes for improvement	0.518	
6	The management provides the necessary resources for breeding programs (economic, physical space, time)	0.76	Management commitment: This component is related to the level of commitment of the senior management and department heads to achieve the planned objectives regarding Kaizen, such as providing the necessary resources and motivate staff of the company
	It motivates and engages each person in the company	0.656	

(continued)

Table 8.5 (continued)

Factor number	Ítem	Factor load	Component description
7	It takes into account the opinions of the customers of the company to make changes in the work	0.796	Customer focus: Identifying of customer needs and dedicated attention
	Methodologies are applied to understand the customer's voice	0.661	

eliminating items on the questionnaire, this improved their overall internal consistency. In so doing we have obtained the results shown in Table 8.3.

In Table 8.3 shows the values of coefficient alpha if each item is removed in different iterations. In each iteration has set the highest value with an asterisk (*) corresponding to the item to be removed in the next iteration, having eliminated variables, PunPenDoc, ExisNegRealCa, ExisResConImp, MidConPro, ExisFacl-nExt, ExisHabInt, ExisResCam, ExisLidGer, the final list of items contains only 29 variables, that are the basis for subsequent analyzes.

8.4.3 Feasibility of Exploratory Factor Analysis of the Activities

Was obtained KMO (Kaiser–Meyer–Olkin) value of 0.874 and Bartlett's test of sphericity gave Chi-square value of about 2121,723, with 406 degrees of freedom representing a significance of 0.000 and the value determining the correlation matrix was 0.000. Thus we investigated the feasibility of factor analysis and proceeded to apply it.

8.4.3.1 Exploratory Factor Analysis of the Activities

In exploratory factor analysis considered only those factors with an eigenvalue greater than unity in Fig. 8.2 shows the scree plot.

According to the results obtained will be discussed the first 7 components, Table 8.4 shows the value of each of the eigenvalues, the variance explained for each and the cumulative variance.

Table 8.4 shows that the number of principal components is seven, which explain a total of 54.80 % of the total variance of the 29 items analyzed. With the above information, the items identified conformed to each component, for which use was made of the rotated component matrix, same as shown in Table 8.5.

Table 8.5 shows that four activities comprising the first component, two in the second component, three in the third component, three activities in the fourth, two activities in the fifth, two activities in the sixth, and finally two seventh component.

Table 8.6 Total variance explained

Component	Initial eigenvalues	Sums of squared loadings extraction	Saturations sum of the square of the rotation						
	Total	% variance	% accumulated	Total	% variance	% accumulated	Total	% variance	% accumulated
1	5.992	42.800	42.800	5.992	42.800	42.800	3.039	21.709	21.709
2	1.473	10.519	53.319	1.473	10.519	53.319	2.884	20.599	42.308
3	1.020	7.288	60.607	1.020	7.288	60.607	2.562	18.299	60.607

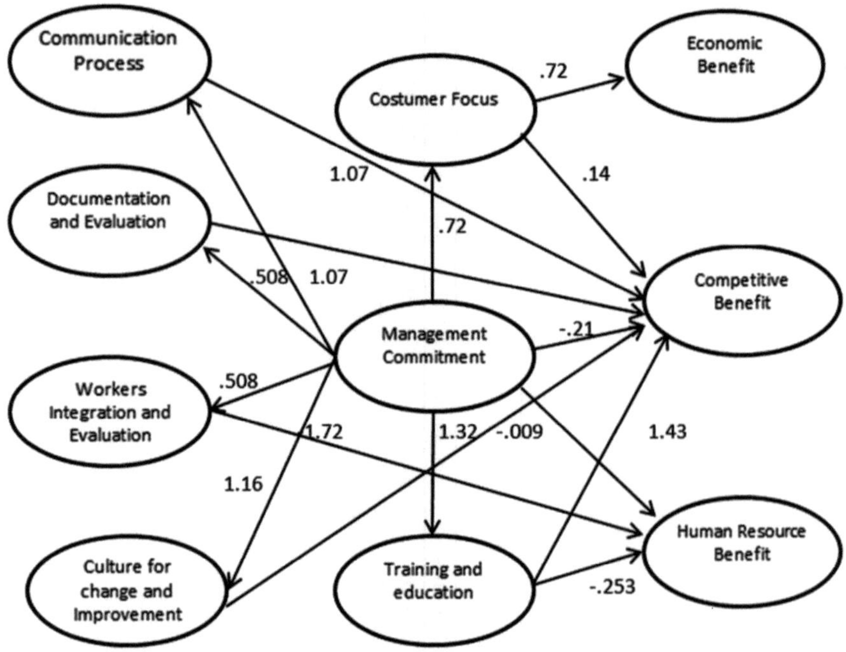

Fig. 8.3 Structural equation model

8.4.4 Feasibility of Exploratory Factor Analysis of Benefits

To identify the benefits of implementing Kaizen yielded a Cronbach's alpha of 0.896, 0.896 KMO index, Chi-square approximately 1532.96 with 91 degrees of freedom that represents a significance of 0.000.

8.4.5 Exploratory Factor Analysis of Benefits

According to the results obtained are analyzed first 3 components, Table 8.6 shows the value of each of the eigenvalues, the variance explained for each and the cumulative variance.

Table 8.6 shows that the number of principal components is three, which explain a total 60.60 % of the total variance of the fourteen analyzed items. With this information, we identified the items that made up each component, for which use was made of the rotated component matrix.

Initial model was generated based on the facts found in the AFE, which is illustrated in Fig. 8.3.

Table 8.7 Estimated values of parameters and initial model relations

Variables	Relatión	Variables	ETM	ES	RC	P
Customer focus	<—	Management commitment	0.722	0.161	4.499	***
Workers integration and award	<—	Management commitment	1.021	0.206	4.967	***
Documentation and Evaluation	<—	Management commitment	0.508	0.128	3.975	***
Culture for change and improvement	<—	Management commitment	1.161	0.226	5.141	***
Training and education	<—	Management commitment	1.372	0.251	5.459	***
Communication process	<—	Management commitment	1.074	0.204	5.272	***
Competitive benefit	<—	Customer focus	0.914	0.189	4.847	***
Economic benefit	<—	Customer focus	0.925	0.154	6.015	***
Economic benefit	<—	Management commitment	0.145	0.152	0.948	0.343
Competitive benefit	<—	Management commitment	−0.214	0.242	−0.885	0.376
Competitive benefit	<—	Training and education	−0.009	0.058	−0.153	0.878
Competitive benefit	<—	Communication process	−0.061	0.09	−0.679	0.497
Competitive benefit	<—	Documentation and Evaluation	0.182	0.097	1.874	0.061
Benefit human resource	<—	Training and education	−0.253	0.129	−1.961	0.05
Competitive benefit	<—	Culture for change and improvement	0.032	0.072	0.447	0.655
Benefit human resource	<—	Workers integration and award	−0.172	0.123	−1.401	0.161
Benefit human resource	<—	Management commitment	1.432	0.387	3.701	***

The asterisks *** Indicate that P is significant at the 0.001 level

The initial model had a CMIN value of 818–448 degrees of freedom and their relationship was 1.826. The goodness of fit index (GFI) was 0.839 and the comparative index (CFI) of 0.859.

The square error (RMSEA) was 0.057 and the minimum sample required was 164.En Table 8.7 shows the estimated values of the initial structural equation model, illustrates the dependent and independent variables and the direction of the relationship where ETM indicates unstandardized estimates, ES is the standard error of the estimated parameter, RC is the critical ratio estimator and P is the significance level.

It can be observed that not all values of RC are greater than 1.96; therefore, not all parameters are different from zero.

We analyzed the indices of modification of the model it was observed that the error of AplicMetVoCli (e18) and OpinClieReaMod (e17) auto correlated, so they added a relationship of covariance between these errors, thus generating a new model and their indexes have considerably improved.

Again were analyzed indices needed and made the necessary adjustments to reach the final model shown in Fig. 8.4.

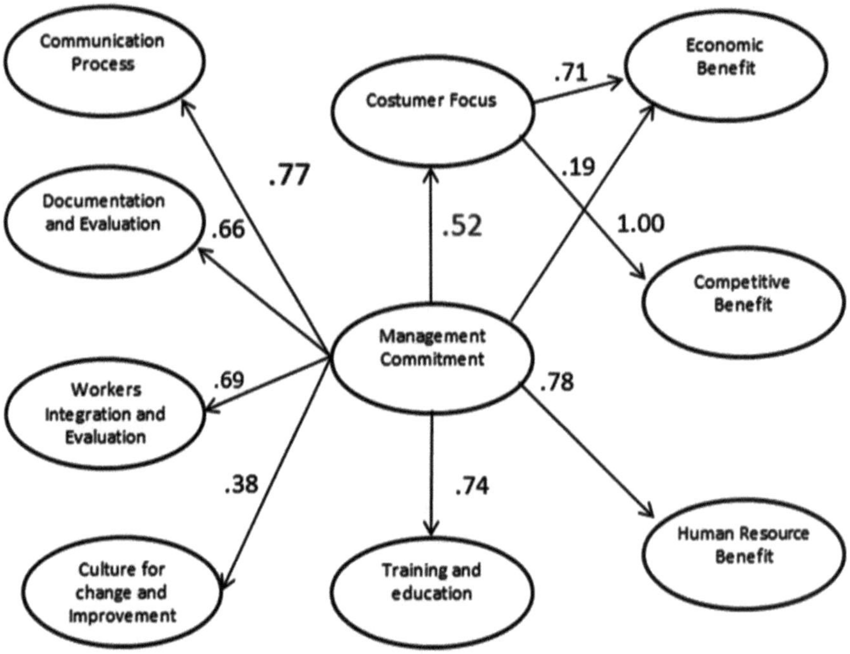

Fig. 8.4 Final structural equation model

The final model has a value of CMIN 770 with 453 degrees of freedom their relationship was 1,701. The index of goodness of fit (GFI) was 0.845 and the comparative index (CFI) of 0.879, the square of the error (RMSEA) was 0.051 and the minimum sample required was 176.

In the Table 8.8 shows the estimated values of the final structural equation model.

It can be seen that all the values of RC are greater than 1.96, therefore ETM all the parameters are different from zero and therefore are significant.

On the basis of the structural equation model end where the parameters are displayed and standardized its significance. Rositas (2009), proposes the following criteria in standardized coefficients: coefficients less than or equal to 0.10 the impact is imperceptible, or despicable; 0.11–0.15 barely perceptible; 0.16–0.19 considerable; 0.20–0.29 of significant impact, of 0.30–0.50 strong impact and finally coefficients greater than 0.50 is considered a very strong impact. From the detailed observation of the model, new relationships are with degrees of very strong impact because they have coefficients above 0.50, a significant connection which is the impact the management commitment in the economic benefit.

In addition to final model can be seen that all coefficients were with correct sign, this is greater than zero. On one side of each coefficient is annotated to that level is statistically significant, it is shown by an asterisk when it is a significant relationship to 0.05, with two asterisks significant relationships to 0.001.

Table 8.8 Estimated values of parameters and relationships in the final model

Variables	Relatión	Variables	ETM	ES	RC	P
Customer focus	<—	Management commitment	0.588	0.132	4.466	***
Competitive benefit	<—	Customer focus	0.853	0.135	6.325	***
Economic benefit	<—	Customer focus	0.893	0.137	6.509	***
Workers integration and award	<—	Management commitment	0.985	0.2	4.933	***
Documentation and evaluation	<—	Management commitment	0.527	0.133	3.948	***
Culture for change and improvement	<—	Management commitment	1.222	0.237	5.149	***
Benefit human resource	<—	Management commitment	0.624	0.141	4.424	***
Economic benefit	<—	Management commitment	0.269	0.107	2.524	0.012
Training and education	<—	Management commitment	1.294	0.24	5.394	***
Communication process	<—	Management commitment	1.11	0.213	5.22	***

The asterisks *** Indicate that P is significant at the 0.001 level

8.5 Conclusions

On the basis of the analysis exploratory and confirmatory factor, applying the information and the results presented above, it is concluded that the main success factors for the implementation of Kaizen are the management commitment, focus on customer, training, communication process, integration of human resource, organizational culture, documentation and evaluation. The confirmatory factor analysis is valid the information obtained in the factorial analysis was accurate, however it was found that several items were correlated and therefore this changed its content. In the final model were all the interrelations with correct sign and the vast majority was important and statistically significant. In the structural equation model end can be seen that the only independent factor is the management commitment, because that is the only latent variable that is not explained by other, achieving a management commitment is obtained:

An approach to the Customer, take into account the voice of the customer and applied methodologies for understand, also, for each unit of variance in the management commitment, focus on client does so record at 0.588 units, according to the non-standardized coefficients.

An organizational culture where count with an organization that fosters the exchange and the participants show a positive attitude to make the changes. For each unit of variance in the management commitment, the organizational culture makes it 1.22 units, according to the non-standardized coefficients.

An integration of human resource, where the worker is involved in the activities of the company. For each unit of variance in the management commitment, the integration of human resource makes it 0.985 units, according to the non-standardized coefficients.

A communication process where workers exchange ideas, opinions and relevant information about the company. For each unit of variance in the management commitment, the process of communication makes it 1.11 units, according to the non-standardized coefficients.

Assessment and documentation where you register everything relating to the activities of continuous improvement and for the regular evaluation of the corresponding system. For each unit of variance in the management commitment, the documentation and evaluation factor makes it 0.527 units, according to the non-standardized coefficients.

The training where the staff is sufficiently skilled to tackle the organizational challenges. For each unit of variance in the management commitment, the factor of training makes 1.294 units, according to the non-standardized coefficients.

Likewise, in the final model can be seen the relationship of the variables observed, as for example:

When the company responds to the needs of the customer, increasing customer satisfaction. For each unit of variance of the first, the second makes it in 0.362.

Finally we can conclude that the management commitment and the approach to the customer are the most important factors to ensure the benefits offered by the Kaizen.

A practical use of this type of research, as suggested Kanji (1998), can serve as a basis for benchmarking companies, with implementations of Kaizen or continuous improvement projects. In addition it is suggested that complement the present results, both with a focus on structural equations as with other comparative approaches. Also, it is recommended that you extend and support these methods of analysis to the deployments of Kaizen in other sectors, such as medical and educational, as it was mentioned previously, this research sheds the results only of the manufacturing industry of Ciudad Juarez, Chihuahua, Mexico.

References

Bateman, N. (2005). Sustainability: The elusive element of proccess improvement. *International Journal of Operations & Production Management, 25*(3), 261–276.

Bessant, J., & Caffyn, S. (1997). High-Involvement innovation through continuos improvement. *International Journal of Technology Management, 14*(1), 7–28.

Bhuiyan, N., Baghel, A., & Wilson, J. (2007). a sustainable continuous improvement methodology at an aerospace company. *International Journal of Productivity and Performance Management, 55*(8), 671–687.

Bisgaard, S. (2007). Quality management ann Juran's legacy. *Quality and Reliability Engineering International, 23*, 665–677.

Bradley, J. R., Willett, J. (2004). Cornell students participate in Lord Corporation's Kaizen projects. *Interfaces, 34*(6), 451–459.

Cantú, H. (2006). *Desarrollo de una Cultura de Calidad (Tercera edición ed.)*. México: McGraw-Hill.

Cooney, R., & Sohal, A. (2004). Teamwork and total quality management: A durable partnership. *Total Quality Management & Business Excellence, 15*(8), 1131.

Cox, A., Zagelmeyer, S., & Marchington, M. (2006). Embedding employee involvement and participation at work. *Human Resource Management Journal, 16*(3), 250–267.

Cua, A., Mckone, K., & Schroedr, R. (2001). Relationships between relationships between. *Journal of Operations Management, 19*(6), 675–694.

Dale, B., Boaden, R., Wilcox, M., & McQuarter, R. (1997). Sustaining total quality management: What are the key issues? *The TQM Magazine, 9*(2), 372–380.

Devaraj, S., Hollingworth, D., & Schroedew, R. (2004). Generic manufacturing strategies and plant performance. *Journal of Operations Management, 22*(3), 313–333.

Farris, J. (2003). A standard frame work for sustaining Kaizen events. *Master's Thesis, Department of Industrial and Manufacturing*. Wichita, KS.

Farris, J., Van, A., & Doolen, T. (2009). Critical success factors for human resource outcomes in Kaizen. *International Journal of Production Economics, 20*(3), 42–65.

Flynn, B. B., & Sakakibara, S. (1995). Relationship between JIT and TQM: Practices and performance. *Academy of Management Journal, 38*(5), 1325.

García, E., Gil, J., & Rodriquez, G. (2000). *Análisis Factorial. Cuadernos de Estadística*. Mexico: La muralla.

Hair, J. F., Anderson, R. E., Tatham, R. L., & Black, W. C. (1995). *Multivariate data analysis*. New Jersey: Prentice Hall.

Imai, M. (1996). *Kaizen-Clave de La Ventaja Competitiva*. México: Editorial CECSA.

Jaca, C., Mateo, R., Tanco, M., Viles, E., & Santos, J. (2010). Sostenibilidad de los sistemas de mejora continua en la industria: Encuesta en la CAV y Navarra. *Intangible Capital, 6*(1), 51–77.

Jørgensen, F., & Boer, H. (2004). Development of a team-based framework for conducting self-assessment of continuous improvement. *Journal of Manufacturing Technology Management, 15*(4), 343–349.

Kanji, G. (1998). *Measurement of business excellence*. UK: Sheffield University School of Computing and Management Sciences.

Kaya, N. (2006). The impact of human resource management practices and corporate entrepreneurship on firm performance: Evidence from Turkish firms. *International Journal of Human Resource Management, 17*(12), 2074–2090.

Kaye, M., & Anderson, R. (1999). Continuous improvement: the ten essential criteria. *The International Journal of Quality & Reliability Management, 16*(5), 485.

Kwok, W., & Sharp, D. (1998). A review of construct measurement issues in behavioral accounting research. *Journal of Accounting Literature, 17*(1), 37–74.

Landa, A. (2009). Factores de éxito y permanencia en eventos Kaizen. *Sinnco*, 1–20.

Lefcovich, M. (2007). Ventajas y beneficios del Kaizen. Retrieved Septiembre 10, 2011, from http://www.tuobra.unam.mx/publicadas/040816180352.html.

Lillrank, P. (1995). The transfer of management innovations from Japan. *Organization Studies, 16*(6), 971–989.

Long, R. J., & Shields, J. L. (2005). Best practice or best fit? High involvement management and base pay practices in Canadian and Australian firms. *Asia Pacific Journal of Human Resources, 43*(1), 52–75.

Manos, A. (2006). *Lean Kaizen: A simplified approach to process improvement* (pp. 47–49). Milwaukee: ASQ Quality Press.

Melnyk, S., Calantone, R., Montabon, F., & Smith, R. (1998). Short-term action in pursuit of long-term improvements: Introducing Kaizen events. *Production and Inventory Management Journal, 39*(4), 69–76.

Newitt, D. (1996). Beyond BPR & TQM—Managing through processes: Is Kaizen enough? *Industrial Engineering Conference Proceeding, 1*, 100–110.

Nunnally, J., Bernstein. (2005). Psychometric Theory, third edition. New York: McGraw-Hill.

Ooi, K. B., Arumugam, V., Safa, M. S., & Bakar, N. A. (2007). HRM and TQM: Association with job involvement. *Personnel Review, 36*(6), 939–962.

Pardo, M. (2002). *Guía para el análisis de datos*. Madrid: McGraw-Hill.

Perry, F. (2004). Rewired for Success. *Air Transport World, 41*(9), 38–39.

Prajogo, D., & Sohal, A. (2004). The sustainability and evolution of quality improvement Programmes- an Australian case study. *Total Quality Management, 15*(2), 205–220.

Rapp, C., & Eklund, J. (2002). Sustainable development of improvement activities—The long-term operation of a suggestion scheme in a Swedish company. *Total Quality Management, 13*(7), 945–969.

Readman, J. (2007). What challenges lie ahead for improvement programmes in the UK? Lessons from the CINet Continuous Improvement Survey 2003. *International Journal of Technology Management, 37*(3), 290–305.

Rockart, J. (1979). Chief executives define their own data needs. *Harvard Business Review, 57*(2), 81–92.

Romero, R., Noriega, S., Escobar, C., & Ávila, D. (2009). Factores Críticos De Éxito: Una Estrategia De Competitividad. *CULCYT, 6*(31), 5–14.

Rositas, M. (2009). Factores críticos de éxito en la gestión de calidad total en la industria manufacturera. *Ciencia UANL, 7*(2), 181–193.

Ruiz, M., Pardo, A., & Martin, F. (2010). *Modelo de ecuaciones. Papeles del Psicólogo, 31*(1), 34–45.

Savolainen, T. (1999). Cycles of Continuous Improvement. Realizing Competitive Advantage Through Quality. *International Journal of Operations & Production Management, 19*(11), 1203–1222.

Schroeder, R. G., Bates, K. A., & Junttila, M. A. (2002). A resource-based view of manufacturing strategy and the relationship to manufacturing performance. *Strategic Management Journal, 23*(3), 105.

Suárez, B., & Miguel, J. (2009). "En la búsqueda de un Espacio de Sostenibilidad: un estudio empírico de la aplicación de la Mejora Continua de Procesos en Ayuntamientos Españoles. *INNOVAR Journal of Administrative and Social Sciences, 19*(35), 47–64.

Tapias, A., Yeison, A., Correa, R., & Hernan, J. (2010). Kaizen: Un caso de estudio. *Scientia et Technica, 16*(45), 59–64.

Tozawa, C., & Bodek, N. (2002). *Kaizen Rápido y Fácil*. Madrid: TGP Hoshin.

Upton, D. (1996). Mechanism for building and sustaining operations improvement. *European Management Journal, 14*(3), 1996.

Zacharatos, A., Barling, J., & Iverson, R. D. (2005). High-performance work systems and occupational safety. *Journal of Applied Psychology, 90*(1), 77–93.

Chapter 9
Critical Success Factors Related to the Implementation of TPM in Ciudad Juarez Industry

José Torres

Abstract In the search for more effective and efficient use of machines and equipment in the industry generates planning staff training. It is therefore essential that managers are aware of all that is at stake in maintaining an excellent system. Industrial or service level, costs, productivity, quality, safety, customer satisfaction and meeting deadlines depend largely on the proper functioning of the equipment and the benefits of it. The remarkable importance of Total Productive Maintenance (TPM) in its acronym in English, waste disposal gives it a special place in both the Kaizen system as Just in Time System. Still a multitude of small and medium enterprises have failed to take due account of the great importance for improving economic performance implementation of systems to improve maintenance of equipment. This paper presents the results of detecting what the Critical Success Factors Related to the Implementation of Total Productive Maintenance in Ciudad Juarez.

Keywords TPM · Industrial maintenance · Factor analysis

9.1 Introduction and Concepts

1. TPM aims to create an enterprise system that maximizes the efficiency of production systems (improving the overall efficiency of the operation (Nakajima 1988).
2. TPM creates a system to prevent the presence of any losses in the production line and focuses on the final product. This includes systems to achieve the goals

J. Torres (✉)
Department of Industrial Processes, Technological University of Juarez City,
Avenida Universidad Tecnológica No. 3051, Lote Bravo II C.P. 32695,
Ciudad Juárez, Chihuahua, México
e-mail: jose_torres@utcj.edu.mx

J. L. García-Alcaraz et al. (eds.), *Lean Manufacturing in the Developing World*,
DOI: 10.1007/978-3-319-04951-9_9,

of "zero accidents, zero defects and zero breakdowns" throughout the life cycle of the production system (Nakajima 1984).
3. TPM is applied in all sectors, including production, development, and administrative departments (Nakajima 1989).
4. TPM is based on the participation of all members of the companies, which act in alignment (Nakajima 1988).
5. TPM can eliminate losses through improvement activities carried out in small teams of workers (Takahashi and Osada 1990).

Total Productive Maintenance is actually a set of activities in a process involving several people from different departments with the same goal, which is to improve operating efficiency.

TPM is actually an evolution of Total Quality Manufacturing, derived from the concepts of quality that Dr. W. Edwards Deming's so positively influenced Japanese Industry (Nakajima 1988).

The TPM was practiced for the first time on the ground Nippodenso, an automotive electrical parts manufacturing in Japan in the late 1960s. Seiichi Nakajima a senior official of the Japan Institute of Plant Maintenance (JIPM), is credited with having defined the concepts of TPM and its implementation in hundreds of plants in Japan (Nakajima 1988).

The times and needs changed in 1960 established new concepts, Productive Maintenance was the new trend that determined a more professional, so highest responsibilities were assigned to people related to the maintenance and considerations were made about the reliability and equipment design and plant. It was a profound change and generated the term "Plant Engineering" instead of "Maintenance", the tasks included a higher level of knowledge of the reliability of each component of machines and facilities in general (Nakajima 1988).

9.2 Benefits of TPM Implementation

The TPM can differentiate an organization in relation to its competition because of the impact on reducing costs, improving response times, reliability of supply, the knowledge that people have and the quality of final products and services.

9.2.1 Benefits with Respect to the Organization

- Improved quality of work environment.
- Better control of operations.
- Increased employee morale.
- Creating a culture of responsibility, discipline and respect for the rules.
- Lifelong learning.

- Creating an environment where participation, collaboration and creativity a reality.
- Effective communication networks.

9.2.2 Benefits with Regard to Safety

- Improved environmental conditions.
- Culture of preventing adverse health events.
- Increased ability to identify potential problems and corrective actions search.
- Understanding why certain rules, rather than how.
- Prevention and elimination of potential causes of accidents.
- Radically eliminates contamination and pollution sources.

9.2.3 Benefits with Regard to Productivity

- Eliminates loss affecting plant productivity.
- Improved reliability and availability of equipment.
- Reducing maintenance costs.
- Improving the quality of the final product.
- Lower financing costs for spare parts.
- Improved Technology Company.
- Increased responsiveness to market movements.
- Create competitive capabilities from the factory.

Once a good program of Total Productive Maintenance (TPM) takes place, the benefits start flowing to the entire organization. It is time that all people begin to support the system. Participants usually feel encouraged and confident to share their ideas in the new attitude of "willingness to listen" to the entire team.

To create the right atmosphere, we must always meet the most basic requirements:

- Total commitment from top management.
- Proper dissemination plan and its results.
- Authentic delegation of the responsibility to decide and mutual respect at all levels.

9.3 Fundamental Processes TPM (Pillars)

The scope **"Production TPM"** contains five interdependent goals represent minimum requirements for the TPM program development by Nakajima (1989), for what some authors consider only steps 7 and 8 in the implementation of TPM, defined as interdependent pillars numbered 1–5 (Ahuja and Khamba 2008; Bamber et al. 1999; Chan et al. 2005; Cigolini and Turk 1997; Gupta et al. 2006; Ireland and Dale 2001; Nasurdin et al. 2005; Shamsuddin et al. 2005; Sharma et al. 2006), in likewise, there are authors who consider the steps 9–11 of the new vision extended the reach-Company. Wide TPM, also as additional pillars to the five previous listing them as pillars of 6–9 (Ahuja and Khamba 2008; Cigolini and Turco 1997; Gurinder 2006; Shamsuddin et al. 2005).

Fundamental processes have been called by the JIPM as "pillars". These pillars serve as support for the construction of a production system ordered. Are implanted following a disciplined methodology, powerful and effective. The pillars considered by the JIPM as necessary for the development of TPM in an organization are:

9.3.1 Focused Improvement or Kobetsu Kaisen, Kokuo (1996)

These activities are developed with the involvement of the different areas involved in the production process in order to maximize Overall Equipment Effectiveness, processes and plants, all this through an organized work and cross- functional teams that employ specific methodology and focus their attention on eliminating existing losses in industrial plants.

It is about developing the continuous improvement process similar to that in the process of applying Total Quality Control procedures and maintenance techniques. If an organization has similar improvement activities simply can incorporate into their process, Kaizen or enhancements, new tools developed in the TPM environment. No need to modify your current improvement process.

TPM techniques ostensibly help eliminate equipment failures. The procedure followed for focused improvement actions follow steps known Deming Cycle or PDCA (Plan-Do-Check-Act).

Developing **Kobetsu Kaizen** activities are performed through the steps shown in Fig. 9.1.

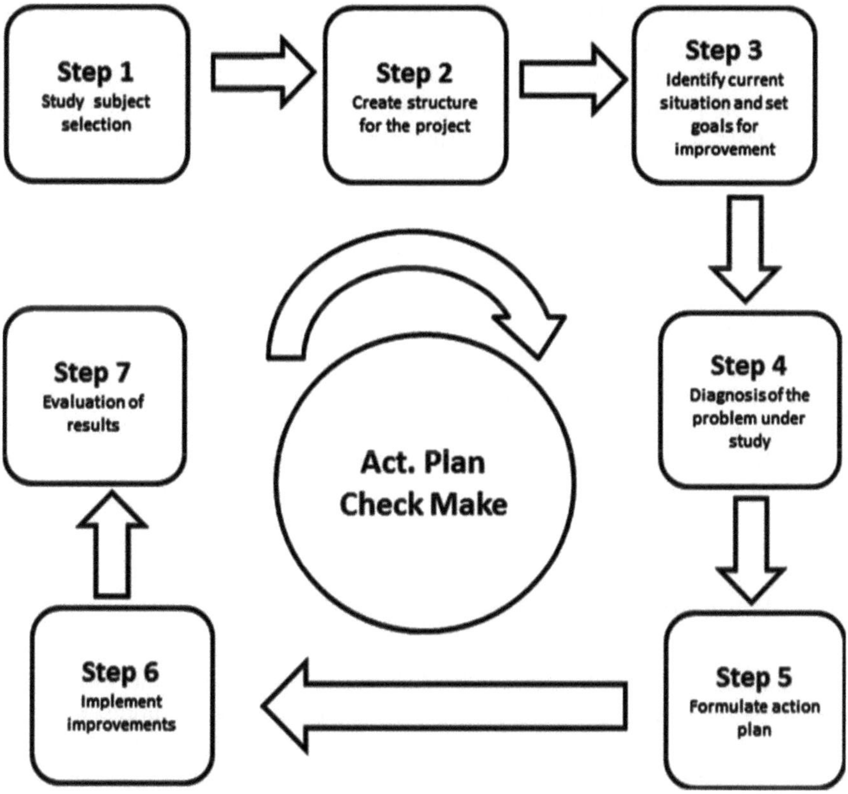

Fig. 9.1 Deming or PDCA cycle

9.3.2 Autonomous Maintenance or Jishu Hozen

One of the TPM system activities is the production staff participation in maintenance activities. This process is one of the highest impacts in improving productivity. Its purpose is to engage the operator in the care of equipment through a high degree of training and professional development, respect of operating conditions, maintenance of work areas free of contamination, dirt and disorder.

The autonomous maintenance is based on the knowledge that the operator has to master the equipment conditions, that is, mechanisms, operational aspects, care and conservation, management, troubleshooting, etc. With this knowledge operators can understand the importance of the preservation of the working conditions, the need for preventive inspections, participate in problem analysis and performing light maintenance work in a first stage, and then assimilate maintenance actions more complex.

The Autonomous Maintenance consists of a set of activities performed daily by all workers operating equipment, including inspection, lubrication, cleaning, minor procedures, tools and spare change, studying possible improvements, analyzing and solving computer problems and actions that lead to maintain equipment in top operating condition. These activities should be performed following standards previously prepared with the collaboration of the workers themselves. Operators must be trained and must have the skills to master the system operating.

9.3.3 Planned Maintenance or Progressive

The objective of the planned maintenance is to eliminate computer problems through improvement actions, prevention and prediction. For the proper management of maintenance activities is necessary to have information bases, obtaining knowledge from data, programmability of resources, maintenance management technology and power of motivation and coordination of the team responsible for these activities.

9.3.4 Maintenance of Quality or Hinshitsu Hozen

Such maintenance aims to improve product quality by reducing the variation by controlling the conditions of the components and equipment conditions that have direct impact on the quality characteristics of the product. Often understood in the industrial environment problems occur when teams fail and stop, however, there may be faults that do not stop the operation of the equipment but losses occur due to changes in the quality characteristics of the final product. Maintaining quality is a kind of preventive maintenance care oriented conditions resulting product.

9.3.4.1 Quality Maintenance is Not

- Apply quality control techniques to maintenance.
- Apply ISO system to the maintenance function.
- Use techniques of statistical quality control maintenance.
- Apply continuous improvement actions to the maintenance function.

9.3.4.2 Maintaining Quality Is

- Perform maintenance actions team oriented care that this does not lead to quality defects.
- Preventing quality defects certifying that the equipment meets the conditions for "zero defects" and that these are within the technical standards.

- Observe the changes in the characteristics of the equipment to prevent defects and take action ahead of the potential abnormal situation.
- Perform computer engineering studies to identify the items of equipment that have a high impact on the quality characteristics of the final product, making control of these elements of the machine and intervene these elements.
- Preventing quality defects certifying that the equipment meets the conditions for "zero defects" and that these are within the technical standards.
- Observe the changes in the characteristics of the equipment to prevent defects and take action ahead of the potential abnormal situation.
- Perform computer engineering studies to identify the items of equipment that have a high impact on the quality characteristics of the final product, making control of these elements of the machine and intervene these elements.

9.3.5 Prevention Maintenance

Improvement activities are those that are made during the design, construction and commissioning of the equipment in order to reduce maintenance costs for their exploitation. A company that seeks to acquire new equipment can make use of the history of the behavior of the machinery that has, in order to identify potential improvements in the design and drastically reduce the causes of failures from the same time negotiating a new team. The preventive maintenance techniques are based on reliability theory; this requires having good databases frequency of breakdowns and repairs.

9.3.6 Maintenance Administrative Areas

This kind of activity does not involve the production team. Departments such as planning, development and administration produced no direct value as output but facilitate and provide the support necessary for the production process run efficiently, with lower costs, opportunity and requested the highest quality. Your support is usually offered through a productive process of information.

9.3.7 Training and Skills Development Operation

The skills have to do with the correct way to interpret and act according to the conditions for the functioning of processes. Knowledge is acquired through experience and reflection in daily work for a while. The TPM requires personnel who have developed skills to perform the following activities:

- Ability to identify and troubleshoot equipment.
- Understand the operation of the equipment.
- Understanding the relationship between the mechanisms of the equipment and the quality characteristics of the product.
- Able to analyze and solve operational problems and process operations.
- Ability to retain knowledge and teach other students.
- Ability to work and cooperate with areas related to industrial processes.

9.4 The 5S's, a Philosophy Essential

Based on Japanese words that begin with an "S", this philosophy focuses on effective work place organization and standardized work processes. 5S's simplifies the work environment, reduces waste and non-value added activities, while increasing safety and quality efficiency.

- *Seiri* (order or arrangement), the first "S" refers to the work area to eliminate everything that is not necessary. An effective way to identify these elements have to be eliminated is called "red tagging". Indeed a red card (ejection) is placed on each item that is considered not necessary for operation. Next, these items are taken to a temporary storage area. Later, when it was confirmed that they were unnecessary, they are divided into two classes, those that are usable for another operation and will be discarded useless. The Sorting is an excellent way to free floor space and eliminate such things as broken tools, obsolete jigs and fixtures, scrap and excess raw material. This step also helps eliminate the mentality of "Just in Case".
- *Seiton* (Everything in Its Place) is the second "S" and focuses on saving systems efficient and effective.

  (a) What do I need to do my job?
  (b) Where do I need to have?
  (c) How many pieces do I need?

Some strategies for this process of "everything in His place" are: painting floors clearly defining work areas and locations, silhouettes tables and modular shelving and/or cabinets instead have things like a trash can, one broom, mop, bucket, etc. Imagine how much time is wasted looking for a broom is in place! The broom should have a place where everyone who needs it, the halle. "A place for everything and everything in its place".

- *Seiso* (shine!) Once you've eliminated the amount of clutter and even garbage, and we do need relocated, comes a super-clean the area. When this is accomplished the first time, will have to maintain a daily cleaning to maintain good appearance and comfort of this improvement. Workers will develop pride in clean and tidy so you have your work area. This cleaning step really develops a sense of ownership among workers. At the same time begin to show obvious problems that were previously hidden by clutter and dirt. So realize oil leaks, air, coolant, parts of excessive vibration or temperature, contamination risks, fatigued parts, bent, broken, misalignment, etc. These elements, when not addressed, can lead to equipment failure and loss of production, factors affecting company profits.

- *Seiketsu* (Standardize) In implementing the 5S's, we should concentrate on standardizing best practices in our work area. Let employees participate in the development of these standards or standards. These rules are very valuable sources of information in regard to their work, but often they are not taken into account. Consider what McDonald's, Pizza Hut, UPS, U.S. Army would be without effective work standards or standards.
- *Shitsuke* (Hold) the "S" is the most difficult to reach and implement. Human nature is to resist change and not a few organizations have found themselves in a dirty cluttered shop a few months following their attempt to implement the "5S's". There is a tendency to return to the tranquility of the "Status Quo" and the "traditional" way of doing things. The Sustain focuses on defining a new "status quo" and a new set of rules or standards in the organization of the work area.

Once fully implemented, the 5S process boosts morale, create positive impressions on customers and increase organizational efficiency. Not only do workers feel better about where they work, but that the effect of continuous improvement generates less waste, better quality products and faster revolving, any of which, makes our organization more remunerative and competitive in the market.

9.4.1 Some Cases Studies have at the Moment Are

Ford, Eastman Kodak, Dana Corporation, Allen Bradley, Harley Davidson, which are just some of the companies that have successfully implemented TPM. All report increased productivity with TPM. Kodak reported a $5 million investment resulted in an increase of $16 million in profits that could be tracked and contributed directly to the implementation of a TPM program (Swamidass 2000).

Appliance manufacturer reported a time to die changes in a press forming was several hours to 20 min. This is the same as having two or three million additional dollars machines available for use on a daily basis without having to buy or rent them (Nakajima 1986).

Texas Instruments reported increased production figures up to 80 % in some areas. Almost all of the above companies reported 50 % or greater reduction in time, reducing spare parts inventory, and increased on-time deliveries. The need for partial or complete outsourcing of a product line has been significantly reduced in many cases (Takahashi 1990). Total Productive Maintenance is currently one of the key systems to achieve overall efficiency, based on which it is feasible to achieve full competitiveness. The current trend to increasingly improve competitiveness means raising in unison and to a maximum extent as efficiency, time and cost of production and involves the firm in Total Productive Maintenance in conjunction with Total Quality Management.

Among Kaizen systems, there is a special place Total Productive Maintenance which in turn makes another system variable holding Kaizen practice that the system is "Just in Time".

In recent years, the staff is more involved in implementation activities and the achievement of Total Productive Maintenance program. The company Coclisa SA DE CV and many other companies implement the TPM based on the experiences of other companies without knowing exactly which activities should be performed or ignoring the critical success factors that would ensure the success of Nippon Enterprises located in Juarez, some of them are Harnesses Juarez, Appliances and Harnesses, Leads Juarez Technology, Epson, Nichirin, Diversified Electrical Products, Toshiba-Electromex.

The implementation of a continuous improvement system, as TPM will bring economic benefits to the organization operating since through the pursuit of efficiency and continuous improvement of its processes, the organization build a culture of constant change and innovation. This will help to gain prestige in the industry providing increasingly better for the customer.

Unknown until the factors of success, to implement TPM in process manufacturing maquiladora industry. Most manufacturing companies in Ciudad Juarez are of foreign capital, the Mexican employee in these companies has a picture of attitudes at work, the barrier to communication managers and operational staff are one of the main points the difficulty of implementing a Total Quality culture (Nogueira 2010).

9.5 TPM Costs and Disadvantages

List of costs and disadvantages of implementing TPM

- Equipment failure, producing unexpected loss of time.
- Tuning and adjustments of machines that produce loss of time when starting a new operation or another stage of her. For example, early in morning, changing workplace, changing an array or make an adjustment.
- Marches empty, waiting and minor arrests during the operation normal time losses occur either by operation detectors, full mailboxes, obstructions on roads.
- Reduced operating speed, which causes loss of time by not obtained the design speed of the process.
- May. Defects in the process, producing loss of time by having to redo parts of it or repair defective parts or complete activities unfinished.
- Loss of time characteristic of the implementation of a new process, trial run, probation.

9.6 TPM Implementation Process

The existence of case studies that examine different approaches of each company to TPM implementation methodologies in successful deployments is wide (Ireland and Dale 2001), finding authors that show improvements TPM activities and advise procedures implantation (Blanchard 1997; Kaizen 1997; Patterson et al. 1996; Suzuki 1992). From the literature review it can be concluded that the model referenced in most articles is that developed by Seiichi Nakajima, TPM and published initially as tenkai by the Japanese Institute of Plant Maintenance (JIPM) in 1982 and later published in English (Nakajima 1989), this publication is introduced to the principles of TPM in the context of a program designed for a Japanese company manufacturing and assembly of medium size, indicating that the implementation should be carried out in three stages: preparation, implementation and stabilization. Although its development can be done in many ways and can be facilitated in many cases by consultants (e.g. JIPM) (Andreassen et al. 2004).

The TPM is usually implemented in four phases, which can be broken down into twelve-step:

1. Preparation
2. Introduction
3. Implantation
4. Consolidation.

9.6.1 The Twelve Steps of Implementation of TPM

It is very important the commitment of senior management which must be committed not only involved but also the commitment must be disclosed at all levels indicating the intentions and expectations with respect to the method.

It is recommended that the decision to implement the TPM formally disclosed by documents circulating on the company for the knowledge of all employees Senior management must be aware and sure to meet the following points and get success in implementation of TPM:

Step 1: Statement of senior management to introduce TPM

It is very important the commitment of senior management which not only must be committed but also involved and this commitment must be disclosed to all levels indicating the intentions and expectations with respect to the method.

Step 2: Campaign introductory training

The goal of TPM is to restructure the business culture through improvement of both human resources and equipment and facilities. Based on this, it must develop a program of introductory education at all levels.

Step 3: Create a promotional structure TPM

This step establishes an implementation coordination committee (preferably heads of departments) which in turn appoints their teams in each area.

Step 4: Establish policies and objectives for the TPM

Promotion of TPM as part of a policy and objective administration, clarifying their integration in the medium and long term with the company policies and the introduction of its goal in the company's business objective.

Step 5: Create the master plan for the development of TPM

It should establish a pilot scheme for the support from the preparation for the introduction of TPM to its final implementation to enable verification of the progress made, set parameters and compare with current development patterns changing, if necessary.

Step 6: Commencement of Implementation

You must have completed the introductory education process to TPM to all employees before the start of the implementation.

Step 7: "Kobetsu-Kaisen" to obtain the efficiency of equipment and facilities

"Kobetsu–Kaisen" is the detailed survey of the needs to improve a team, carried out by a multidisciplinary team of engineers, line managers, maintenance personnel and operators.

Step 8: Establishment of "Jishu-Hozen" (autonomous maintenance)

The "Jishu–Hozen"is a development method that allows the same operator to control his own team. The "Jishu–Hozen" is developed in seven steps, going from one to another after completing the previous one with the support and evaluation of managers.

Step 9: Effectiveness of equipment for production engineering (operation and maintenance)

Implementation of the methodology in the pilot team, normalizing and transforming routine, everything that was provided in the previous step.

Step 10: Establishment of the system to obtain the overall efficiency in the areas of administration

Support for increasing production efficiency in both the office and the teams.

Step 11: Establishment of the system, looking for ideal conditions to promote safety, health and working atmosphere

Analysis and implementation of "recommendations secured".

Step 12: Full implementation of TPM and increase of the respective levels

In this step becomes an extension of the TPM to the other computers on the ground, defining new goals and challenges and takes a consultancy for the implementation of adjustments.

9.7 Problems Implementing TPM

It is important to understand up front that Total Productive Maintenance is the most difficult of all the "lean tools" to implement in companies for two reasons:

A TPM implementation requires the greatest amount of culture change (as compared to implementing other lean tools) from different groups of people within the organization almost simultaneously.

1. Lack of support by senior management for not fully understands the purpose of the TPM and the effort required
2. Trainers and consultants used no experience
3. Organizational structure for implementing the strategic needs inadequate to support the TPM teams and activities
4. Unclear Objectives
5. Do not publish and display the results the measurable relevant and benefits achieved in financial terms
6. Key constituents to implement the TPM use or possess power sources hinder the practice correct abutment
7. Minimum focus to redesign of equipment, using the existing information
8. Allow work with margins of error due to inefficient institutionalized reveals being trapped by sloth organizational
9. Poor and discouraging environment area work in the absence of implementing 5S
10. Apathy and disability senior management implementing work practices safe area work
11. Declare victory too soon, before change penetrate the culture of the company (between 5 and 10 years), this generates regression to less leverage persuaded to go back.

9.8 Case of Study: Problem Definition

9.8.1 Problem Definition and Research Objective

The main problems encountered in industrial plants due to lack of organization that exists in many of the activities that constitute the process in the maquiladora industry in Ciudad Juarez, Mexico, the goals and objectives usually posed at the beginning of each year, have defaults, the application has methodologies to improve processes necessary and identify success factors of these methodologies.

In recent years, the staff is more involved in implementation activities and the achievement of Total Productive Maintenance program. The company Coclisa SA DE CV and many other companies implement the TPM based on the experiences of other companies without knowing exactly which activities should be performed or ignoring the critical success factors that would ensure the success of Nippon Enterprises located in Juarez, some of them are Harnesses Juarez, Appliances and Harnesses, Leads Juarez Technology, Epson, Nichirin, Diversified Electrical Products, Toshiba-Electromex.

The implementation of a continuous improvement system, as TPM will bring economic benefits to the organization operating since through the pursuit of efficiency and continuous improvement of its processes, the organization build a culture of constant change and innovation. This help you get prestige in the industry providing increasingly better for the customer.

At the time of which are unknown success factors when implementing TPM in process manufacturing maquiladora industry. Most manufacturing companies in Ciudad Juarez are of foreign capital, the Mexican employee in these companies has a picture of attitudes at work, the barrier to communication managers and operational staff are one of the main points the difficulty of implementing a Total Quality culture (Nogueira 2010).

The objective are determine the key success factors of Kaizen, based on an empirical analysis of surveys conducted in Ciudad Juarez, Chihuahua, Mexico.

The objective specifics are:

- What kind of activities does each person involved directly and indirectly in the process, and how they relate to the success of TPM?
- Identify the Critical Success Factors of TPM
- Determine importance of Critical Success Factors of TPM
- Determining the relationship of Critical Success Factors of TPM with the results.

9.8.2 Research Questions

The research questions that arise are:

1. What are the Critical Success Factors in TPM?
2. How to determine the importance level of Critical Success Factors of TPM?
3. What is the relationship between the Critical Success Factors of TPM and the results?

9.8.3 Hypotheses and Variables

The desired test hypotheses are:

1. Critical success factors in implementing TPM are management commitment, integration of operators in the performance of TPM and tidiness of the workplace.

2. The level of importance of the key success factors of TPM is determined by factor analysis and is explained by the value of the eigenvalue of the factor.
3. The relationship to the TPM success factors and their relationship to the results, is determined by structural equation.

9.8.4 Methodology

The methodology of the research has been carried out from the literature, which has a scope of the study to be performed, for which classification adopted Danhke (1989), who divided the types of research: exploratory, descriptive, correlational and explanation. This classification is very important because the type of study depends on the research strategy (Gomez 2006).

Was chosen research method was correlational and explanatory as they determine the critical success factors related to TPM Implementation through a survey of data through an assessment instrument designed with the variables of study, once data collected will be used an analysis technique, guiding the development of the methodology in the following materials and steps.

The method is divided into two sections, relating to the materials used in the method applied.

9.8.5 Materials

The materials that are occupied in the development of the research were a questionnaire and software. The questionnaire was the basic document for research information. The questionnaire was developed with a total of 22 questions divided into nine sections.

9.8.5.1 The Survey

The first section of questions is a total of 17, and is aimed at knowing how to have the Company to implement TPM in the Company. In the second section of questions is a total of 6 and were designed in the form of an administrative system management regarding accepted improvement in this case was the Administrative department management regarding the TPM.

9.8.5.2 Questionnaire Design

The instrument where it got the information of the research was a questionnaire.

The questionnaire was the basic document for research information. The questionnaire was developed with a total of 22 questions divided into two sections.

First Section of the Questionnaire

The first section of questions is a total of 16, and is aimed at knowing how to have the Company to implement TPM in the Company.

Section of the questionnaire questions

- What is the way to implement TPM in your company?

1. Do you think that the maintenance staff training is adequate?
2. Did tracks the progress of the maintenance program and its evaluation?
3. When goals are not achieved? TPM are left without explanation?
4. Do you think that there is ignorance on the part of operators in the handling of equipment and machinery in charge?
5. Is there commitment to superior immediate supervisors and maintenance personnel with the functionality of the machines?
6. Is there leadership in implementing TPM programs by senior management?
7. Is there leadership in implementing TPM programs by the responsible production and engineering?
8. Is there leadership in implementing TPM programs by the maintainers?
9. Are there differences in interest in the availability of the machines for maintenance? (Production and maintenance).
10. Does the operator know the maintenance schedule for teams operating ?
11. Critical systems are known in which a machine can fail?
12. Maintenance programs are based on the useful life of systems and component parts of the machine?
13. Are there differences between the lifespans of parts and components provided by suppliers and those reported by the company in the daily performance of the equipment?
14. Briefings are conducted by the team responsible for maintaining?
15. Investments are carried novel tools for easy maintenance?
16. When investing in equipment or machinery do you consider the maintenance as purchasing decision criteria?

As management is handled with respect to TPM

1. Do all department heads within our company accept their responsibility to the TPM?
2. Does the company management have personal leadership in the execution of TPM programs?
3. Do you carry out working meetings between the maintenance department and production?
4. Are business managers promoting employee participation in maintenance and upkeep of equipment?
5. Does the company management creates and communicates a vision focused on quality and maintenance?

6. Does the management of the company is personally involved in maintenance projects?

The questionnaire was applied when what is being measured is quantitative. The questions should be at the level of knowledge of the people who will implement the questionnaire. It should use natural language, presenting direct questions. There should be extensive in terms of number of questions.

9.8.6 The Software

SPSS is a statistical program used in the social sciences and market research firms. At present, the symbol is used to designate both the statistical and the company that produces it. It includes a number of utilities.

9.8.7 Questionnaire Validation

Process validation of a questionnaire is designed once the final draft, an once defined the information, formulated the questions, defined the number of them that we will include in the questionnaire and ordered the questions, carry out corresponding conducting pilot testing and evaluation of the psychometric properties of the scale. For Cronbach's alphas, there is a general formula, according to the provisions Hernández et al. (2003), the formula is as follows.

$$\propto = \frac{\overline{N}p}{1 + \overline{p}(N - 1)} \tag{9.1}$$

where:

$\overline{N}$ = number of questions *and*
p = average correlations

Cronbach's alphas obtained vary, but all are greater than 0.65, which is the minimum set by Cronbach for an instrument to be reliable.

He has managed to build a good questionnaire when it helps you get the information necessary in relation to the purposes of the research, when taken into account the needs and reactions of the subject without face, and also creates a favorable environment. Another characteristic of a good questionnaire is: keep the interest of the respondent to obtain genuine and meaningful content as the subject and the research problem. The qualities that the questionnaire as a research tool has (Cordova 2008):

- It is consistent with the problem and research objective.
- Get information that cannot be achieved by other means.

- Fits the resources available for research.
- It has an attractive appearance.
- No writing or spelling errors.
- It is as short as possible.
- Contains appropriate sections to the case of concern.
- Each section is clearly identifiable.
- The presentation section clarifies the scope and the reasons of the study.
- The instructions are clear and complete.
- It has a compelling cooperation request.

The study or research method used is quantitative correlation, which is aimed at evaluating the relationship that exists between two or more concepts, categories or variables (quantified relationships). Is, presumably related measure each variable and then also measured and analyzed the correlation. Such correlations are expressed hypotheses tested.

9.8.8 Sample

The sample was selected in the list of the Maquiladora Companies who are discharged from the Association of Maquiladoras, AC Ciudad Juarez Chihuahua (AMAC). With a population of 782.

The sample size was estimated considering two essential requirements the reliability and validity. As the reliability of a measurement instrument grade, in which repeated application to the same subject or object produces equal results. And the validity broadly refers to the degree to which an instrument actually measures the variable being measured (Hernández Sampieri 1998).

The size of the sample followed the following steps

$$n_{\infty} = \frac{Z^2_{\infty/2}\sigma^2}{e^2} \tag{9.2}$$

where:

$Z_{\infty/2}$ = Z corresponding to the chosen confidence level

σ^2 = Population Variance

e = Maximum error

2. Check marks

$N > n_{\infty}\ (n_{\infty} - 1)$

Respondents were chosen for the list of AMAC, (association of maquiladoras of Juarez City, Chihuahua, Mexico) with a quote by phone, scheduling a visit day and time to complete the questionnaire.

9.8.9 Application of the Questionnaire

The application of the questionnaire was contacting by telephone the person in the company to schedule an appointment day and time that he could assist me. The application of the questionnaire was personally, was handed a pair of leaves to answer the questions. Sometimes we were received and treated to the first appointment, sometimes on the second date, but when the person we rescheduled third appointment, we decided that the person was busy and had no time to assist us.

No questionnaire was sent by email, because the questionnaire is a document that does not take more than 5 min to answer it. The implementation of the survey took a period of 6 months to contact the person of the Company, by contacting an appointment, and filling the survey.

9.8.10 Information Capture

Once applied the questionnaires in Annex A, the responses obtained were transferred to a matrix, according to Table 9.1 Coding of answers, the answers were for questions.

To capture the information, we used the scale indicated in Table 9.1, which takes values from 1 (never), 3 (sporadically), 5 (often), 7 (Very often or almost always), 9 (always) 2, 4, 6, 8, (intermediate values).

The database the Table 9.2 was the result of the capture of each of the responses of each survey (interview), in the program Statistical Package for the Social Sciences (SPSS). In the database, the rows are the number of people interviewed, with a total of 203 and the columns are the number of questions, with a total of 16.

The variables were coded automatically by the program, where VAR00001 coding is equivalent to the first question VAR00002, equivalent to the second question so on.

9.8.11 Capturing Data and Database Design

They design a table in the SPSS software for each question in the questionnaire is inserted into column, and for each of the people involved in the questionnaire is generated in a line, and the intersection is placed according to the Coding of According to Likert scaling method.

Table 9.1 Coding of according to Likert scaling method

1	3	5	7	9	2, 4, 6, 8
Never	Sporadically	Frequently	Very often or almost always	Always	Intermediate values

Table 9.2 Capturing data and database design

	001	002	003	004	005	006	007	008	009	010	011	012	013	014	015	016	017
1	5	7	9	7	7	7	3	9	9	1	1	3	1	3	1	3	7
2	9	5	5	5	7	5	7	1	9	9	9	1	3	9	1	7	7
3	5	7	7	7	5	3	5	3	5	5	3	3	5	5	5	5	3
4	9	9	9	9	9	5	5	1	1	1	1	1	9	3	1	3	1
5	9	5	5	9	1	5	1	9	9	1	1	9	9	9	9	9	9
6	5		5	5	7	5	5	3	3	9	9	3	1	9	9	3	9
7	9	9	9	9	9	9	9	9	9	9	5	9	9	9	0	0	0
8	9	9	9	9	9	9	9	9	9	9	9	5	9	9	0	0	0
9	9	8	8	8	9	9	9	8	9	8	8	7	7	5	7	8	9
10	9	9	9	9	9	9	8	9	9	8	5	7	7	5	7	8	8
11	1	3	2	3	2	5	2	2	2	2	1	2	5	3	5	3	1
12	7	8	8	5	8	8	8	3	5	3	3	7	9	5	3	2	1
13	5	5	5	3	3	1	3	5	5	5	5	9	3	9	3	3	5
14	9	9	9	9	7	9	5	5	5	9	9	3	9	3	5	5	9
15	5	7	5	3	1	3	1	3	3	3	1	3	9	3	9	9	9
16	9	9	9	9	9	8	9	8	7	8	9	3	9	9	9	9	5
17	5	7	5	6	7	5	5	5	7	7	5	1	7	7	5	7	1
18	9	7	7	9	9	9	3	9	3	7	9	5	1	9	9	9	7
19	9	9	7	9	3	9	3	7	9	5	1	9	9	9	7	7	5
20	9	9	9	9	9	9	7	5	9	9	9	5	5	9	5	5	9
21	9	9	7	9	9	9	5	1	9	7	3	7	7	9	7	7	7
22	9	8	8	8	8	8	5	9	9	9	9	9	5	1	8	8	8
23	5	5	5	7	5	7	5	5	5	5	7	5	5	5	5	5	5

9.8.12 Information Analysis

The information analysis is carried out in stages, which are outlined below:

Stage One. Identifying and Creating Instrument Attributes

During this stage was emphasized in the literature review to locate research related to the status of a case study or problem by determining the key attributes which were considered to choose a college and make a rational validity, which allowed developing a preliminary questionnaire containing 30 questions to which was applied to a population of 68 students from three higher education institutions located in Juarez, Chihuahua, Mexico, which allowed for a response validation. Nevertheless allowed a space for students to manifest other attributes respondents personally consider for choosing the university and were not considered in the initial questionnaire. With new attributes for a total of 13 that were identified to be removed was completed five of the attributes that were considered initial way as they had no relevance to the respondents also that the variation was minimal, thus obtaining a total of 38 attributes, with which he built a final questionnaire, obtaining content validation.

The final questionnaire with 38 questions divided into five dimensions, infrastructure with five questions, economic questions seven, eleven questions academic, administrative nine questions, and finally, social and performance with six questions. The questionnaire should be answered on a Likert scale, which includes values between one and five where the number one notes that this attribute was not considered outstanding for the university during college when selecting oppositely number five represented a of paramount importance for the choice of college.

Second Stage. Application of the Questionnaire

The development of the second stage focused on contact managers of higher education institutions in Ciudad Juarez, Chihuahua, Mexico. Of the institutions that applied the questionnaire to request permission for the questionnaire. Following this process, three Institutions of Higher Education agreed that their students were surveyed in contravention four Higher Education Institutions flatly refused the application by protecting its refusal questionnaire in the high levels of insecurity in the area.

The sampling was designed in a simple random, always based on the balance between the number of students that the institution had, so does the number of runs in its possession and the semester for which the student was enrolled and same as the basis of sampling, was surveyed requirement that students take at least an average of three courses per semester, this ensured that students were part-time, regardless of the semester you were enrolled, however, focused more to students in the first semester of their careers as these had been recently this process.

Stage Three. Capture of Information and Instrument Validation

During this stage the development of the information discussed in the software called Statistical Package for Social Sciences (SPSS), version number 18. intended to measure the internal consistency index for this purpose questionnaire was used Cronbach Alpha Index long before any analysis of the questions and validating the result compared with that generated by performing the partition of the sample, resulting in a again Cronbach's alpha index. Is important to emphasize that some of the attributes eliminated, since the reliability of the instrument is increased considerably. To construct validation was obtained correlation coefficient girders five dimensions that composed the questionnaire, seeking that these were significant at 95 % confidence.

Fourth Stage. Descriptive Analysis of Information

During the development of this stage was a descriptive analysis of the information. We obtained the median, and mode as measures of central tendency, obtaining that data obtained though and they were expressed numerically, were represented on an ordinal scale and were subjective. The highest median values indicate that these attributes had been prominent for students when selecting a higher education institution and conversely, low values denoting minor. In relation to fashion, the values obtained in the indicated attributes and value consensus more were answered by respondents.

At the same time, as a measure of dispersion is considered the first and third quartile of each question, and the difference between them, which is called interquartile range (RI) and represents 50 % of the data and includes the median, which is represented by the second quartile. The high values in the interquartile range indicate that there was a consensus among respondents regarding the level of importance that had this attribute, while low values represent poor dispersion and therefore greater consensus among respondents regarding the importance level.

Fifth Stage. Exploratory Factor Analysis (EFA)

During the development of this phase was determined the feasibility of exploratory factor analysis, we performed a detailed analysis of the correlation matrix and visualized most of the correlations between the attributes were greater than 0.3 also analyzed the diagonal of the matrix anti—image correlation matrix with the intention of observing the adequacy of the sample. Also, was obtained KMO (Kaiser–Meyer–Olkin), which was applied to the Bartlett sphericity test to measure the adequacy of the sample and analyzed the communities of each of the questions or attributes to validate your contribution, setting 0.5 as the cutoff.

In order to be able to determine the critical factors that students consider when selecting a University Institution conducted a factor analysis by the principal components method using the correlation matrix for the extraction of the components were taken as significant and all those factors with a value greater than or equal to unity in their Eigen values, conditioning them to 100 interactions search for convergence of a result. Furthermore, in order to obtain a better understanding of the critical factors is performed by the method Varimax rotation.

9.9 Results

9.9.1 Sample Description

The companies that participated in the survey are appended in the table below (Table 9.3); respondents were operational staff, technical staff, and warehouse staff and personnel quality inspectors.

9.9.2 Analysis of Results

The results obtained according to the survey to determine the critical factors in the process of implementation of TPM were using the software Statistical Package for

Table 9.3 Companies surveyed

ANSELL PERRY DE MEXICO, S.A. DE C.V.	CADIMEX, S.A. DE C.V.	CUMMINS JUAREZ S.A. DE C.V.
ANSELL EDMONT DE MEXICO, S.A. DE C.V.	CONVERTORS DE MEXICO, S.A. DE C.V.	DATAMARK DE MEXICO S.A. DE C.V.
COMPANIA ARMADORA, S.DE R.L DE C.V.	CAPACITORES COMPONENTES DE MEXICO, S. DE R.L. DE C.V.	DELMEX DE JUAREZ, S. DE R.L. DE C.V.
TOSHIBA ELECTROMEX	CONTROLES DE PRESION DE CIUDAD JUAREZ, S.A. DE C.V.	DELPHI DE MEXICO S.A. DE C.V.
COCLISA DE MEXICO S.A DE C.V.	CONTROLES DE TEMPERATURA, S.A. DE C.V.	DULCES BLUEBERRY S.A. DE C.V.
ROBERT BOSCH	CORDIS DE MEXICO, S.A. DE C.V.	EAGLE OTTAWA S.A DE C.V.
EDM, S. DE R.L. DE C.V.	EDUMEX, S.A. DE C.V.	EES, S.A. DE C.V.
ELECTRO COMPONENTES DE MEXICO S.A. DE C.V.	ELECTRONICA DALE DE MEXICO, S.A. DE C.V.	EPIC TECHNOLOGIES DE JUAREZ, S. DE R.L. DE C.V.
TED DE MEXICO, S.A. DE C.V.	TERMOCONTROLES DE JUAREZ, S.A. DE C.V.	VENUSA DE MEXICO S.A. DE C.V.

Table 9.4 Reliability of the questionnaire

Cronbach's alpha	Variable number
0.868	17

Social Sciences (SPSS), this is a statistical computer program widely used in the social sciences and business market research.

Getting the Cronbach Alpha which is a parameter that measures the reliability of the questions that are prosecuted for what we want, with a score of 0.868, based on a total of 17 questions in a sample of 203 respondents, according to Table 9.4 The minimum for a reliable instrument is $\infty = 0.65$.

In Table 9.5 presents the evidence of KMO (Measure of Kaiser–Meyer–Olkin) and Bartlett, (and test of sphericity Bartlett's) and study of the matrix and anti-image correlation matrix and its significance are satisfactory for that factor analysis can provide good results, the correlation matrix was 0.001, It conforms to the identity matrix.

Table 9.6 contains the variables initially commonalities (Initial) and communalities, reproduced by the factor solution (extraction). The communality of a variable is the proportion of its variance that can be explained by the factor model obtained.

In our model the variable Do managers and senior management are examples of housekeeping in the work area? It's the worst explained, the model is able to reproduce only 0.34 % of the original variability. To reach this factor solution, we used a method of extraction called principal components.

Table 9.5 Validation of the questionnaire

Statistical variables total				
	A	B	C	D
What company emphasizes putting all tools and accessories instead?	93.75	528.12	0.524	0.859
What company are work areas clean and orderly?	94.17	532.61	0.546	0.859
Does your company remains lim-employed at all times and in all departments?	94.12	532.216	0.564	0.859
Do employees receive the workout to perform multiple tasks properly?	94.14	519.324	0.656	0.855
Do employees are responsible for maintaining the machines start operating?	94.55	524.989	0.449	0.863
Are employees between-Swim in this plant so they can meet others if needed?	94.43	521.097	0.554	0.858
Do you make an effort so that all operators provide opinions and ideas before making a decision about maintenance?	95.77	526.807	0.421	0.864
Do you get a log of work and maintenance of each of the machines?	93.78	531.375	0.437	0.863
Does programming man-entertainments performed in consensus with the production department?	94.33	527.923	0.447	0.863
Is the maintenance schedule of machines is carried out with the view of the operators?	96.93	526.089	0.488	0.861
Does the operator is taken into account in the drafting of the reports in logbook?	96.63	520.125	0.48	0.861
Does the operator can apply maintenance directly to the respective department when there is a fault?	94.76	517.063	0.475	0.862
Can the operator to stop a production line if it detects that problem exist on your computer maintenance?	94.61	528.269	0.413	0.864
Do managers and senior management are examples of housekeeping in the work area?	94.06	533.296	.478	.861
Do caregivers provide appropriate maintenance manuals of the machines?	93.98	521.934	0.543	0.858
Do providers provide technical assistance related to equipment maintenance?	94.13	524.667	0.486	0.861
Does the company signed some sort of agreement with suppliers of machinery associated with maintaining it?	95.01	525.085	0.491	0.861

where
A scale is average if the variable is eliminated
B is the scale of variance if the variable is eliminated
C Correlation corrected variables of total
D Cronbach's alpha if the variable was eliminated

In the Table 9.7 Rotate Matrix Components represents the four levels or critical factors reflected by staff that are involved in an industrial process in the city industrial maquiladora Juarez, Mexico.

Table 9.6 Communality of variables

Variables or item	Extraction
Do managers and senior management are examples of housekeeping in the work area?	0.34
Do employees are responsible for the initial maintenance of machines operating?	0.37
Can the operator to stop a production line if it detects that there are maintenance problems on your computer?	0.40
Do you get a log of work and maintenance of each of the machines?	0.44
Are employees trained in this plant so they can meet others if needed?	0.48
Does the operator can apply directly to the respective department maintenance when there is a fault?	0.53
Do employees receive training to adequately perform multiple tasks?	0.57
Does the scheduling of maintenance is done in agreement with the production department?	0.60
Do you make an effort so that all operators provide opinions and ideas before making a decision regarding maintenance?	0.63
What company emphasizes putting all tools and accessories instead?	0.68
Is the maintenance schedule of machines is done with the view of the operators?	0.70
Does the operator is taken into account in the drafting of the reports in logbook?	0.71
Does your business is kept clean at all times and in all departments?	0.72
Does the company signed some sort of agreement with suppliers of machinery related to the maintenance of it?	0.72
Do caregivers provide appropriate maintenance manuals of the machines?	0.75
Do providers provide technical assistance related to equipment maintenance?	0.80
What company are work areas clean and orderly?	0.81

9.10 Concluding Remarks

In evaluating the Critical Success Factors Related to the Implementation of TPM Maquiladora Industry in Ciudad Juarez, the following results were obtained.

The analysis of the data indicates that the factors considered in the 17 questions 203 sample, where the cases are valid 201, with 99 % and 2 are excluded, with 1 %, information that indicates the Table 9.3.

The result for the index Cronbach's alpha was 0.868 which shows that it is not necessary to perform the removal of any items.

Therefore it is important to consider these aspects in the implementation of TPM in an industrial process in CD. Juarez.

The community of a variable is the proportion of its variance that can be explained by the factor model obtained. In our model the variable Do managers and senior management are examples of housekeeping in the work area? It's the worst explained, the model is able to reproduce only 0.34 % of the original variability. To reach this factor solution was used an extraction method called principal components, this information can be seen in Table 9.4.

Table 9.7 Rotate matrix components

Rotate matrix components		
Variables	Components	
What company are work areas clean and orderly?	0.882	These activities are key to any company or workplace, representing by the 5 S
Does your company remains lim-employed at all times and in all departments?	0.809	
What company emphasizes putting all tools and accessories instead?	0.785	
Do employees receive the workout to perform multiple tasks properly?	0.546	
Do providers provide technical assistance related to equipment maintenance?	0.874	The second key reflects our service provider, i.e. customer service
Do caregivers provide appropriate maintenance manuals of the machines?	0.829	
Does the company signed some sort of agreement with suppliers of machinery associated with maintaining it?	0.811	
Does the operator is taken into account in the drafting of the reports in logbook?	0.815	The third key is to involve operational staff in decision-making, when you want to improve a business process
Is the maintenance schedule of machines is done with the view of the operators?	0.802	
Do you make an effort so that all operators provide opinions and ideas before making a decision regarding maintenance?	0.763	
Does the scheduling of maintenance is done in agreement with the production department?	0,719	The fourth key to improve a process reflects communication between departments
Does the operator can apply directly to the respective department maintenance when there is a fault?	0.671	
Can the operator to stop a production line if it detects that there are maintenance problems on your computer?	0.561	
Do you get a log of work and maintenance of each of the machines?	0.553	

In the end it is 4 levels of results with a group of activities, are represented by the 5 S, the second shows us the service provider, i.e. customer service, the third key is to involve staff in making operational decisions and finally the fourth is reflected in the communication between the department. Therefore the benefits of TPM are vertical down method, which starts from the pinnacle of the hierarchy and reaches the operator (Torres 2005).

References

Ahuja, I. P. S., & Khamba, J. S. (2008). Strategies and success factors for overcoming challenges in TPM implementation in Indian manufacturing industry. *Journal of Quality in Maintenance Engineering, 14*(2), 123–147.

Andreassen, M., Gertsen, F., Christiansen, T. B., & Michelsen, A. U. (2004). Status and trends in the development of Total Productive Maintenance (TPM)—a review of international articles. *Proceedings from CINet 2004*. Sydney, Australia.

Bamber, C. J., Sharp, J. M., & Hides, M. T. (1999). Factors affecting successful implementation of total productive maintenance: A UK manufacturing case study perspective. *Journal of Quality in Maintenance Engineering, 5*(3), 162–181.

Blanchard, B. S. (1997). An enhanced approach for implementing total productive maintenance in the manufacturing environment. *Journal of Quality in Maintenance Engineering, 3*(2), 69–80.

Chan, F. T. S., et al. (2005). Implementation of total productive maintenance: A case study. *International Journal of Production Economics, 95*(1), 71–94.

Cigolini, R., & Turco, T. (1997). Total productive maintenance practices: A survey in Italy. *Journal of Quality in Maintenance Engineering, 3*(4), 259–272.

Cordova, F. G. (2008). *Recomendaciones metodológicas para el diseño de cuestionario*. Sonora, México: Editorial Limusa SA de CV.

Danhke, G. L. (1989). *Investigación y Comunicación. En: Metodología de la Investigación*. México DF: McGraw Hill.

Gómez, M. M. (2006). *Introduction to the methodology of scientific research*. Cordova, Argentina: Editorial Witches.

Gupta, S., Tewari, P. C., & Sharma, A. K. (2006). TPM concept and implementation approach. Retrieved January 18, 2013, from http://www.maintenanceworld.com/articles/sorabh/research_paper.pdf

Gurinder, S. B. (2006). Keeping the wheels turning [total productive maintenance]. *Manufacturing Engineer, 85*(1), 32–35.

Hernández, R. (1998). *Metodología de la Investigación* Mc. Graw Hill- Interamericana de México, S.A. de C.V. México, DF.

Hernández, R., Fernández, C., & Baptista, L.P. (2003). *Metodología de la Investigación*, Mc Graw Hill, Tercera Edicción México D.F.

Ireland, F., & Dale, B. G. (2001). A study of total productive maintenance implementation. *Journal of Quality in Maintenance Engineering, 7*(3), 183–192.

Kaizen, K. (1997). *Focused equipment improvement for TPM teams* (1st ed.). London, England: Productivity Press.

Nakajima, S. (1984). *Introduction to TPM*. London, England: Productivity Press.

Nakajima, S. (1986). TPM a challenge to the improvement of productivity by small. *Maintenance Management International, 6*, 73–83.

Nakajima, S. (1988). *Total productive maintenance*. New York, NY, USA, Cambridge: Productivity Press.

Nakajima, S. (1989). *TPM Development Program Productivity*. Inzynieria Utrzymania Ruchu 6.

Nasurdin, A. M., Jantan, M., Wong, W. P., & Ramayah, T. (2005). Influence of employee involvement in total productive maintenance practices on job characteristics. *International Journal of Business, 7*(3), 287–300.

Nogueira, L. S. (2010). La Calidad y la Cultura en Empresas Maquiladoras. *CULCYT, 7*(38–39): 16–29.

Patterson, J. W., Fredendall, L. D., Kennedy, W. J., & Mc.Gee, A. (1996). Adapting total productive maintenance to Asten, Inc. *Production and inventory management journal, 37*(4): 32–36.

Shamsuddin, A., Masjuki, H. H., & Zahari, T. (2005). TPM can go beyond maintenance: excerpt from a case implementation. *Journal of Quality in Maintenance Engineering, 11*(1), 19–42.

Sharma, R. K., Kumar, D., & Kumar, P. (2006). Manufacturing excellence through TPM implementation: a practical analysis. *Industrial Management & Data Systems, 106*(2), 256–280.

Suzuki, T. (1992). *New directions for TPM*. New York, NY, USA: Productivity Press.

Swamidass, P. M. (2000). *Encyclopedia of production and manufacturing managament*. Norwell, Massachusetts, USA: Kluwer Academic Plublishers.

Takahashi, Y. d. (1990). *TPM*. NY, North America: Kraus Organization.

Takahashi, K., & Osada, T. (1990). *TPM-Productive Maintenance*. Asian Productivity Organisation, Hong Kong.

Torres, L. (2005). *Mantenimiento: Su Implementación y la Introducción de Mejoras en la Producción*. Argentina: Ediciones Universitaria.

Chapter 10
Critical Success Factors for the Implementation of JIT

Lizeth Rivera-Mojica and Denisse Gabriela Rivera-Mojica

Abstract The Just In Time manufacturing system (JIT) has been one of the most investigated topics in the operations management area because of its success in the Japanese industry and has been developed intensively for more than three decades. Various benefits have been reported due JIT implementation, for example, reducing inventory improving efficiency of operations and a faster response to the client, among others. Therefore, successful implementation of JIT is vital for many companies. The main objective of this research is to identify the Critical Success Factors of JIT implementation and built and integral model which evaluated the relationship between these factors and performance indicators, which was applied a questionnaire in a sample composed of managers, supervisors and technicians within the manufacturing sector of Ciudad Juarez, Chihuahua. The research results show that there is a significant relationship between JIT and performance indicators and they are interrelated with other factor such as management commitment, supplier strategy, equipment layout, and quality management. Future research opportunities were also identified.

Keywords JIT · Critical success factors · Integral model

10.1 Introduction

Globalization has created a new outlook for the manufacturing industry which is characterized by competition, frequent product introductions and rapid changes in product demand (Koren 2010). Companies must make strategic changes in manufacturing system consistent with the requirements of its environment

L. Rivera-Mojica (✉) · D. G. Rivera-Mojica
Technological University of Ciudad Juárez, Av. Universidad Tecnológica No. 3051, Lote Bravo II 32695, Ciudad Juárez, CHIH, Mexico
e-mail: litzyrivmoj@hotmail.com

J. L. García-Alcaraz et al. (eds.), *Lean Manufacturing in the Developing World*,
DOI: 10.1007/978-3-319-04951-9_10,

(Sandanayake et al. 2008) and reconfigure the supply chain (Koren 2010), providing high quality and reduce delivery times.

One way to achieve a competitive advantage in manufacturing is to leverage the excellent production, inventory control systems and ensure cost leadership position (Matsui 2007). Just in Time (JIT), Advanced Manufacturing Technologies and Total Quality Management, among others, are some of the tools that should be used as part of the strategic manufacturing system settings to improve efficiency and customer responsiveness (Yasin et al. 2003).

Consistent with Mackelprang and Nair (2010) JIT has remained popular in practice and it is still widely used in businesses around the world. There have been many authors like Ahmad (2003), Yasin et al. (2003), Fullerton et al. (2003), Matsui (2007) and Ferreira Mota (2008), Maiga and Jacobs (2009), Mackelprang and Nair (2010) and others who have dedicated their time and in-depth analysis.

10.2 JIT Definition

JIT definitions have been developed since the strict sense of the production just in time to reach the concept of a general management philosophy to satisfy customers and gain a competitive advantage in the market (Chang and Lee 1996).

To clarify in a better way the meaning of JIT, a series of definitions proposed by various authors is listed:

- Ohno (1982), who is a pioneer of JIT, defined JIT as having the right part at the right time and amount.
- McWatters and Fullerton (2002) propose that JIT manufacturing is a philosophy that emphasizes excellence through continuous improvement in productivity and quality in all phases of the industrial cycle.
- Wakchaure et al. (2006) defined as a JIT manufacturing philosophy that aims to minimize raw materials, work in process, and finished goods inventory helping to expose more serious deficiencies in the production cycle.
- Singh and Garg (2011) propose that the JIT manufacturing system is based on a philosophy of eliminating waste where the central idea is to expose problems and utilize the full capacity of each worker to obtain the maximum benefit.

10.3 JIT Benefits

Research has shown that successful implementation of JIT has the potential to increase organizational effectiveness and efficiency. To identify the benefits of JIT 17 articles published between 1995 and 2011 were analyzed and the results are shown in Table 10.1.

Table 10.1 Benefits from JIT

Author	a	b	c	d	e	f	g	h	i	j	k	l	m	Total
Improved performance and delivery time	•	•	•			•	•			•	•	•	•	9
Reduced inventory (WIP, raw materials, finished product)	•	•	•		•		•		•		•	•		8
Increase productivity and equipment utilization		•	•			•	•		•			•	•	7
Improved customer service	•	•	•		•							•		5
Reducing unit cost of manufacturing	•						•			•		•	•	5
Flexibility to change volume and product mix				•				•		•	•		•	5
Reduced cycle time								•			•	•		3
Improved customer satisfaction and employee efficiency					•					•				2
Remove waste material and rework		•					•							2
Promotes cooperative relations with suppliers					•							•		2
Maximize use of space facilities						•	•							2
Increasing financial performance				•								•		2
Improving operational efficiency									•					1
Standardized work methods							•							1
Promotes cooperative relations with the client							•							1
Increase in the quality of incoming material							•							1
Increase accuracy in demand forecasting							•							1
Improved performance and delivery time													•	1

Authors: *a* Lawrence and Hottenstein (1995); *b* Gunasekaran et al. (1998); *c* Dean and Snell (1996); *d* Mckone et al. (2001); *d* Petersen (2002); *e* Shah and Ward (2003); *f* Aghazadeh (2003); *g* Yasin et al. (2003); *h* Swink et al. (2005); *i* Avitatthur and Swamidass (2007); *j* Matsui (2007); *k* Dal Pont et al. (2008); *l* Caifeng (2009); *m* Teeravaraprug et al. (2011)

As you can appreciate the increase in quality of conformance of the product was the benefit most concordance was among the authors, followed by improved performance and delivery time, reduced inventory levels and increase productivity and utilization equipment.

According to Ahmad et al. (2003) JIT minimizes the use of expensive buffers (such as WIP) and eliminates scrap at every stage of the production process. Therefore, the unit cost of manufacturing is reduced. The reduction in the size of the buffers also provides warnings of quality problems in the production process then the root cause of the problems can be identified and resolved.

Reduce preparation time allows machinery to run a mixed model production where a small number of different products can be manufactured each day. In addition, a pull system linked to customer and suppliers enables a production being agile which improves delivery reliability.

10.4 Critical Success Factors (SCF)

Romero et al. (2009) define the SCF as variables to be taken into account before and during the execution of a project, as they provide valuable information to achieve the goals and objectives of the company and emphasizes the importance of analyzing and determining the factors that are key to the initiation and development of a project for which it need a comprehensive literature review.

A wealth of literature has emerged as part of the efforts of academic researchers who have tried to determine the CSFs for the successful implementation of JIT. From a methodological perspective, research on JIT during the 1980s lacked reliable and valid measures (Walleigh 1986; Voss and Robinson 1987; Wildeman 1988; Willis and Suter 1989). These limitations led to the development of rigorous methods for defining and measuring the central constructs underlying JIT (Mackelprang and Nair 2010). Dimensions proposed by some researchers are presented in Table 10.2.

Subsequent investigators have used the JIT dimensions identified by these authors (Forza 1996; Sim and Curatola (1999); Fullerton and McWatters 2001; McWatters and Fullerton 2002; Ahmad et al. 2003; Fullerton et al. 2003; Narasimhan et al. 2006). The dimensions and elements reported by Ramarapu et al. (1995) provided the basis to identify the dimensions and JIT elements used to guide this research.

Table 10.3 shows the concordance between the elements related to production factor in which 22 articles mentioning more than three techniques a literature review from 1992 to 2011.

Table 10.4 shows the concordance between elements factor related to participation of suppliers in which 14 items of a review of related literature from 1992 to 2011.

Table 10.2 Authors and dimensions on JIT implementation

Authors	Dimensions
Golhar and Stamm (1991)	Waste removal
	Employee participation in decision making
	Participation by suppliers
	Total quality control
Davy et al. (1992)	Operational structure and control
	Production scheduling
	Implementation of quality
Mehra and Inman (1992)	Management commitment
	JIT production strategy
	Provider strategy
	JIT education strategy
Sakakibara et al. (1993)	Human resource management
	Simplifying the physical flow
	Supplier management
Ramarapu et al. (1995)	Management commitment
	Control and quality improvement
	Participation of employees
	Participation by suppliers
	Elimination of waste and production strategy

Table 10.5 shows the concordance between the elements related to the quality factor in which 14 articles mentioning more than three techniques a literature review from 1992 to 2008 are related.

Table 10.6 shows the concordance between elements related factor productive force participation twelve articles in which a review of related literature from 1992 to 2011.

Table 10.7 shows the concordance between the elements related to management commitment factor in which items of a review of related literature from 1992 to 2011.

10.5 Software Used in Industry to Apply Technical

The information is captured and analyzed using SPSS 18® software. (Statistical Package for the Social Science). Statistical analysis includes the correlation of the items and the critical factors to reduce the number of variables and then factor analysis for grouping constructs is performed. SPSS software was used for its high diffusion and it provides the researcher with a wide range of methods and analysis tools.

Table 10.3 Elements linked to JIT practices

Element	a	b	c	d	e	f	g	h	i	j	k	l	m	n	o	p	q	r	s	t	u	v	Total
Reduced time set up	•	•	•	•	•	•		•	•	•	•	•	•	•	•	•			•			•	17
Layout	•	•	•	•	•	•		•		•	•	•	•	•	•			•	•		•		16
Uniform work load			•		•				•		•		•	•			•		•		•	•	10
Kanban		•					•	•		•		•	•						•	•	•	•	10
MRP adapted to JIT							•					•							•				3
Compliance program production		•																	•	•			3
Poka yoke				•		•				•													3
5 s									•	•													2
Visual control								•															1

Authors: *a* Mehra and Inman (1992); *b* Sakakibara et al. (1993); *c* Zhu and Meredith (1995); *d* Lawrence and Hottenstein (1995); *e* Ramarapu et al. (1995); *f* Forza (1996); *g* Chang and Lee (1996); *h* Zayko et al. (1997); *i* Hanckok and Zayko (1998); *j* Callen et al. (2000); *k* White and Prybutok (2001); *l* Fullerton and Mcwatters (2001); *m* Ahmad et al. 2003; *n* Yasin et al. (2003); *o* Li et al. (2005); *p* Swink. et al. (2005); *q* Narasimhan et al. (2006); *r* Ward and Zhou (2006); *s* Matsui (2007); *t* Dal Pon et al. (2008); *u* Sandanayake et al. (2008); *v* Inman et al. (2011)

Table 10.4 Elements linked to provider factor

Element	a	b	c	d	e	f	g	h	i	j	k	l	m	n	Total
Frequent supplier delivery	•	•	•	•		•		•	•	•	•			•	10
Reduction in the number of suppliers			•	•	•	•	•			•					6
Long term contracts with suppliers			•	•	•	•						•			5
Supplier certification	•			•		•					•				4
Location of suppliers				•		•								•	3

Authors: *a* Mehra and Inman (1992); *b* Zhu and Meredith (1995); *c* Lawrence and Hottenstein (1995); *d* Gelinas (1999); *e* Prybutok and White (2001); *f* Fullerton and Mcwatters (2001); *g* Yasin et al. (2003); *h* Li et al. (2005); *i* Narasimhan et al. (2006); *j* Ward and Zhou (2006); *k* Matsui (2007); *l* Dal Pon et al. (2008); *m* Sandanayake et al. (2008); *n* Inman et al. (2011)

Table 10.5 Elements linked to quality factor

	a	b	c	d	E	f	g	h	i	j	k	l	m	n	Total
Quality circles	•		•		•	•		•	•	•			•	•	9
Preventive maintenance	•	•				•	•	•	•			•		•	8
Total quality management (TQM)				•				•	•				•	•	5
Statistic process control (SPC)				•			•					•			3

Authors: *a* Mehra and Inman (1992); *b* Zhu and Meredith (1995); *c* Lawrence and Hottenstein (1995); *d* Forza (1996); *e* Chang and Lee (1996); *f* Zayko et al. (1997); *g* Hankock and Zayko (1998); *h* Jusko (1999); *i* Gelinas (1999); *j* Prybutok and White (2001); *k* Fullerton and Mcwatters (2001); *l* Yasin et al. (2003); *m* Li et al. (2005); *n* Sandanayake et al. (2008)

Table 10.6 Elements linked to productive force participation factor

Practica JIT	a	b	c	d	e	f	g	h	i	j	k	l	Total
Multifunctional employees	•	•		•		•	•			•	•	•	8
Training	•		•		•				•	•	•		6
Workers participation in decision making		•	•					•					3
Rewards program		•								•			2
Recruitment and selection		•								•			2

Authors: *a* Mehra and Inman (1992); *b* Lawrence and Hottenstein (1995); *c* Chang and Lee (1996); *d* Callen et al. (2000); *e* Gelinas (1999); *f* Prybutok and White (2001); *g* Fullerton and Mcwatters (2001); *h* Oral et al. (2003); *i* Yasin et al. (2003); *j* Matsui (2007); *k* Sandanayake et al. (2008); *l* Inman et al. (2011)

To analyze the relationship between different variables and be able to confirm by some metric, the Software is used AMOS 18. This software was chosen because the difficulty level is relatively low; the readout is very easy and can graph the structural equation model.

Table 10.7 Elements linked to management commitment factor

Practica JIT	a	b	c	d	e	f	g	h	Total
Management presence at the plant	•	•	•		•	•	•		6
Senior management commitment in implementing JIT	•	•		•	•			•	5
Promoting teamwork		•					•		2
Interdepartmental coordination			•				•		2

Authors: *a* Mehra and Inman (1992); *b* Lawrence and Hottenstein (1995); *c* Chang and Lee (1996); *d* Callen et al. (2000); *e* Gelinas (1999); *f* Oral et al. (2003); *g* Matsui (2007); *h* Inman et al. (2011)

10.6 Case Study

10.6.1 Problem Description and Objective

Some researchers suggest that the unsatisfactory results of JIT are associated with ineffective and incomplete implementations (Clode 1993; Milgrom and Roberts 1995). Prybutok and White (2001) argue that the benefits will not be fully until all critical elements of JIT are integrated.

The problem identified in this study is that there is uncertainty about what the critical success factors and variables that make these factors to ensure the successful implementation of JIT manufacturing industry bouquet of Juarez are.

The problem identified in this study is that there is uncertainty about what are the critical success factors and variables to ensure the successful implementation of JIT manufacturing industry of Juarez.

10.6.2 Methodology

Figure 10.1 shows a flowchart of the steps in the research method.

Pilot questionnaire was developed based on the literature review. The questionnaire was subjected to the evaluation of two research professors who have extensive knowledge in the area of operations management, a quality manager and a production manager of the maquiladora industry, which determined the ability of the questionnaire to assess all dimensions is intended to measure.

The experts were asked: (i) if it had included the correct items (ii) if the questions were easy to understand (iii) if needed include some other item. The necessary changes were made according to the comments and suggestions from the experts. In the final version of the questionnaire three types of information is mainly collected:

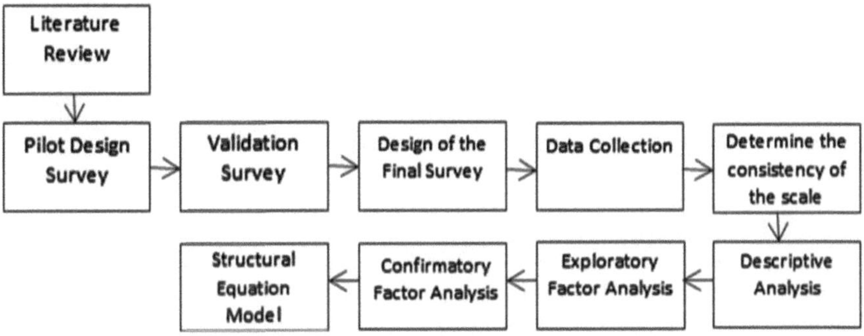

Fig. 10.1 Flowchart for methodology

1. Information to allow evaluation of the extent to which the plant uses the JIT techniques.
2. Information that could evaluate the benefits of JIT obtained in the company.
3. Plant characteristics and the person who answered the questionnaire.

The measuring instrument includes 47 items divided into five dimensions: Commitment management, participation of productive force, production techniques, and disposal of waste, supplier Involvement, quality management.

Also includes eight performance indicators: unit manufacturing cost (Ahmad et al. 2003; White and Prybutok 2001; Matsui 2007), reduced inventory levels (White and Prybutok 2001; Mackelprang and Nair 2010), quality of product conformance (Lawrence and Hottenstein 1995; Ahmad et. al. 2003; Matsui 2007; Mackelprang and Nair 2010), delivery time (White and Prybutok 2001; Ahmad et. al. 2003; Matsui 2007, Mackelprang and Nair 2010); flexibility in introducing new products (Matsui 2007) and efficient use of machinery and equipment (Fullerton and McWatters 2001).

The consistency of the questionnaire was confirmed through Cronbach's alpha. It is considered to have good internal consistency when the alpha value is greater than 0.7 (Nunnally 1970). A Likert scale of five points was used like scoring system where 1 indicates unimplemented and 5 completely implemented. Activities and benefits are illustrated in Table 10.8 such abbreviation used throughout this work of research.

A thorough search of case studies published between 1992 and 2011 based on the factors reported by Ramarapu et al. (1995) was performed. According to (Malhotra 2004) a sample of four times the number of items, in this case 47 items, resulting 188 the sample size. Although there was a sample defined in this study, we sought to apply maximum possible surveys. In total 300 questionnaires were distributed.

The questionnaire was applied to a sample in companies belonging to the sector of the manufacturing industry in Ciudad Juárez, Chihuahua, Mexico. Convenient sampling methods based on personal contacts were used. The questionnaire was

Table 10.8 Activities and abbreviations

Abbrevation	Activity
DifDep	Interdepartmental coordination and communication
SupFom	Promoting teamwork by supervisors
GerIng	Constant interaction of managers, engineers and operators
ComEj	Support and commitment from senior management in the implementation of JIT
OrgCul	Implementing a culture that encourages change
SisAuto	Implementation of systems that allow the operator self-inspection machines
OpRes	Responsibility of operators of machinery and equipment maintenance
MejHab	Improve the skill and knowledge of workers
CapCruz	Training employees to multitask
RotEst	Rotation operators between workstations
OpCon	Recruitment of operators for their ability to solve problems and teamwork
CirCal	Training teams to solve production problems
ProgrRec	Rewarding workers to learn new skills
OpResProb	Implementing a Suggestion System
OpResIns	Responsibility of operators to inspect their own work
OpAutDet	Operators have authority to stop production
DisSetUp	Emphasis on reducing setup times of equipment
OrgCelMan	Organization in manufacturing cells
MaqPeq	Using small, flexible and movable equipment
SisKan	Use of a system for controlling kanban production
RedNum	Implementation of measures to reduce the number of processes
DispPYK	Implementation of Poka-Yoke devices
DisPro	Nearest distribution processes and machinery
JitMrp	Using both JIT and MRP or MRP II for control and production planning
CumDiar	Compliance daily production schedule
Ballinea	Take steps to reduce fluctuations in workload by balancing lines and stable cycle rates
ProgProd	Production leveling
DisMezSim	Design the product mix to be similar to the mixture of the forecast demand
ProdManFluj	The product is manufactured in a continuous flow within the value chain
ProEst	Standardization of processes
ProvIntSis	Integration of suppliers to the company through a system of pull
EntrDiar	Daily deliveries of most suppliers are received
ContrLarg	Long-term contracts with suppliers
ProvCert	Supplier certification
RedNum	Reducing the number of suppliers
SPC	Using statistical monitoring to control and reduce process variation (SPC)
TQM	Implementation principles and tools of total quality management (TQM)
TPM	Implementation of a program of total productive maintenance (TPM)
Benefits of Implementing JIT	
RedCost	Reducing costs of unit manufacturing costs
RedNiv	Reducing inventory levels of raw material

(continued)

Table 10.8 (continued)

Abbrevation	Activity
DisWip	Decreased inventory levels of work in process (WIP)
DisInv	Reduced inventory levels of finished product
DisTiem	Reduced lead time (lead time)
IncEfUt	Increased efficiency and utilization of machinery and equipment
MejCal	Improvement in product quality
AumVel	Increase in the rate of introduction of new products

applied for managers, engineers, technicians and supervisors within the organization that was considered had sufficient knowledge of the operations to complete it.

Information was captured and analyzed using SPSS 18 (Statistical Product and Service Solutions). Statistical analysis included correlation of critical items to reduce the number of variables and then factor analysis for grouping constructs factor.

In the phase of exploratory factor analysis was determined that observable variables loaded on latent variables. The stage of exploratory factor analysis is not essential but is highly recommended (Lévy and Varela 2003).

To determine the feasibility of factor analysis variables were correlated (Malhotra 2004). The Bartlett test of sphericity was used to verify if the factor model is appropriate (Malhotra 2004). Kaiser Meyer-Olkin (KMO) index was obtained to compare the magnitudes of the observed correlation coefficients with magnitudes of partial correlation coefficients, setting a higher value of 0.80 (Lévy and Varela 2003).

A factor analysis was performed by the method of analysis of principal components to determine the minimum number of factors that explain the greatest variance of the data for use in subsequent multivariate analysis.

Varimax rotation method in order to minimize the number of variables with large loads on a factor was used, which improved the playability factor.

At the stage of confirmatory factor analysis (CFA) AMOS 18 software was used in order to reach an optimal model. To validate the relationships between variables and factors, the value of the parameters obtained and the critical ratio of each estimate was analyzed.

To measure the efficiency of the AFC the minimum value of the chi-square, $\chi 2$ (CMIN), the degrees of freedom (DF) of the model and the ratio of these two parameters (CMIN/DF) was used (Byrne 2006). To have a good enough explanatory model, the goodness of fit index (GFI) which is a measure of efficiency and is recommended to have values above 0.9 was used (Tanaka and Huba 1985).

Several models were generated, they were improved in function iteratively modification indexes. The comparative fit index (CFI) was analyzed for improvement between one model and another, accepting the changes if the difference in CFI is greater than 0.01 (McDonald and Marsh 1990) and sought to maintain a measure of the Index root mean square error of approximation (RMSEA), which should have a value between (0.05 and 0.08) (Lévy and Varela 2003).

In order to observe the adequacy of the sample size in each model and ensure that the changes are valid and restricting their size is not violated, the critical index N of Hoelter analyzed at a confidence level of 95 % (Bolleng and Liang 1988).

The model is studied based on the following criteria: first conducted an analysis of fit indices (CMIN, CMIN/DF, CFI, GFI and Critic Hoelter N). After identified if all the estimators of the model were in order to remove the relationship between latent variables were not significant. In addition, modification indices were analyzed to visualize the impact of a new relationship between two variables in measuring the chi-square and allow the re-specification of the model. Finally, the model that best fit was chosen among several competing models.

10.6.3 Results

All questionnaires were subjected to a step of manually reviewed to determine if they had been fully completed or were answered in a correct way, 41 questionnaires were discarded in that it did not contain reliable information. Leaving a sample of 205 valid questionnaires, representing 68.33 % of the total questionnaires sent.

In order to determine the reliability of the instrument as a measure of internal consistency Cronbach's alpha coefficient was used. Following the recommendations of authors like Cortina (1993), Kamata et al. (2003), Lévy and Varela (2003) and Streiner (2003) coefficient was calculated for each dimension. The evaluation of the Cronbach alpha coefficient for each resulting scale is shown in Table 10.9. As you can see every scale has a value greater than 0.70 (Nunnally 1978).

10.6.3.1 Exploratory Factor Analysis of the Activities

At first instance the correlation matrix was analyzed and we observed a substantial number of correlations greater than 0.30 (Nunnally and Bernstein 2005), most of them are highly significant with a "p" equal or close to zero, which shows that it is feasible to make factor analysis.

Assessing the adequacy of the factor model to the data, the anti-image matrix was analyzed, in which most of the off-diagonal elements were small and the diagonal elements were large. In addition, the percentage of larger absolute residuals of 0.05 was 22 % so it is considered that the factor model is appropriate.

A value of the determinant of the correlation matrix equals 1.58E-011 was obtained indicating that the variables are linearly related and that the correlations are very high.

The KMO equals 0.917 which is considered very good and indicates that it is appropriate to use factor analysis. The principal components method was used to

Table 10.9 Assessment cronbach alpha coefficient

Dimension	Cronbach alpha (α)
Factor linked to management commitment	0.871
Factor linked to productive force participation	0.847
Factor linked to JIT practices	0.913
Factor linked to strategy suppliers	0.794
Factor linked to quality management	0.848
Performance factor	0.893

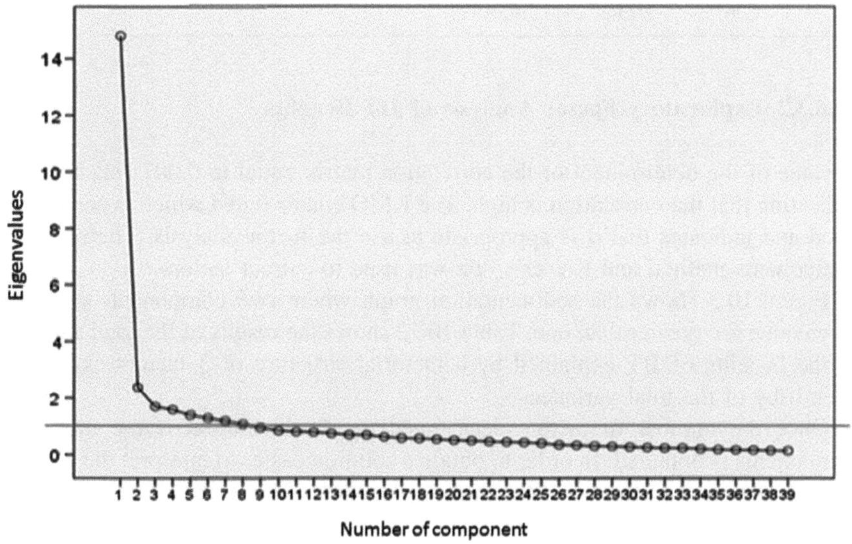

Fig. 10.2 Sedimentation chart activities

extract the factors. Figure 10.2 shows a graph of sedimentation, where the number of components can be appreciated with a eigenvalue greater than one.

In terms of total variance explained there was no big difference between a structure of 7 components and a structure of 8 components. Based on these results it was decided to remove a component in order to simplify the problem and have a simpler model. Table 10.10 shows the results of the total variance explained by a factorial structure of 7 factors explaining overall by 62.394 % of the total variance.

Once the number of factors was determined, the final solution is obtained, which it is the matrix of components. Items with a value less than 0.50 (Lévy and Varela 2003) were deleted. In order to obtain a solution easier to interpret the array of components is rotated by the varimax method. The results are shown in Table 10.11.

Table 10.10 Total variance explained with 7 components

Axis	Initial eigenvalues			Sum of squared loadings rotation		
	Total	% variance	% acumulate	Total	% variance	% acumulate
1	14.81	37.97	37.97	4.956	12.709	12.709
2	2.374	6.087	44.057	3.996	10.247	22.956
3	1.7	4.36	48.417	3.874	9.933	32.889
4	1.592	4.082	52.499	3.249	8.331	41.22
5	1.395	3.576	56.075	3.197	8.197	49.417
6	1.283	3.29	59.366	2.964	7.601	57.018
7	1.181	3.029	62.394	2.097	5.376	62.394
8	1.067	2.736	65.131			

10.6.3.2 Exploratory Factor Analysis of JIT Benefits

A value of the determinant of the correlation matrix equal to 0.007 was obtained indicating that the correlation is high. The KMO equals 0.894 which is considered good and indicates that it is appropriate to use the factor analysis. The principal components method and Kaiser's rule was used to extract factors.

Figure 10.3 shows the sedimentation graph where two components with one eigenvalue are greater than one. Table 10.12 shows the results of the total variance of the benefits of JIT explained by a factorial structure of 3 factors explaining overall by of the total variance.

Once the number of factors determined the final solution is the matrix of components is obtained. In order to obtain a solution easier to interpret the array of components is rotated by the varimax method. The results are shown in Table 10.13.

10.6.3.3 Confirmatory Factor Analysis

According to Thompson (2004) should not only confirm the fit of a theoretical model but should compare the fit indices of several alternative models to select the best. Table 10.14 summarizes the fit indices for alternative models.

As shown in Table 10.14 model number 5 fits acceptably (GFI 0.839, CFI 0.906, RMSEA 0.073) and so significantly higher (significant change in the value of chi-square) to alternative models. It can also be noted that changes between models are justified because the difference in CFI between each model is greater than 0.01. In Fig. 10.4 the factor model with five factors (model 5) is shown.

An initial model based on the factor derived from the AFC was generated. The model suggested that the implementation of the management dimensions of suppliers, quality management practices JIT and distribution of the plant results in a performance improvement in inventory levels and operational performance and management commitment is related to each of the dimensions of the activities. Figure 10.5 shows the proposed model.

Table 10.11 Rotated component matrix of activities

JIT practices		Management commitment		Quality control		Supplier strategy		Layout		Team work		Authority and responsibility	
Item	Load	Item	Load	Item	Load	Item	Load	Item	Load	Item	Load	Item	Load
ProdMan	0.71	GerIng	0.751	TQM	0.73	ProvInt	0.73	OrgCel	0.69	OpCon	0.77	OpRes	0.78
CumDiar	0.64	DifDep	0.636	SPC	0.73	EntrDiar	0.73	MaqPeq	0.63	Progr	0.66	OpAut	0.66
DisMez	0.63	ComEj	0.622	DispPYK	0.64	ProvCert	0.59	DisPro	0.57	CirCal	0.65		
ProgProd	0.62	SupFom	0.602	TPM	0.60	RedNum	0.56						
SisKan	0.58	OrgCul	0.571	SisAuto	0.55	ContrLarg	0.52						
Ballinea	0.57	MejHab	0.557										
ProEst	0.54												
JitMrp	0.53												

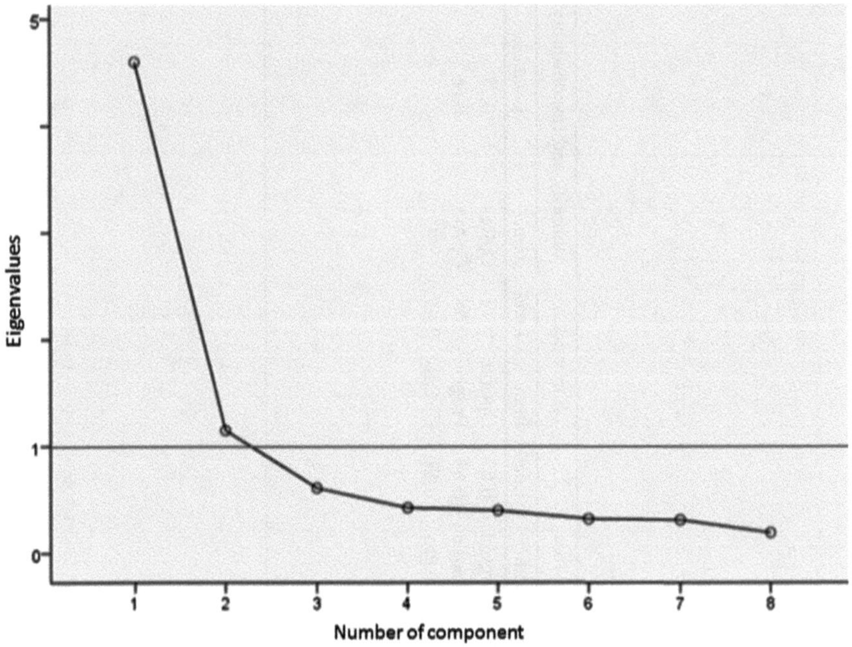

Fig. 10.3 Sedimentation graph of JIT benefits

Table 10.12 Total variance explained of the benefits of JIT

Axis	Initial eigenvalues			Sum of squared loadings rotation		
	Total	% variance	% acumulate	Total	% variance	% acumulate
1	4.596	57.446	57.446	3.249	40.608	40.608
2	1.148	14.355	71.801	2.495	31.193	71.801
3	0.609	7.607	79.408			

In Table 10.15 the values of the estimated parameters are not standardized. Model shown in Fig. 10.5 shows the dependent and independent variables and the direction of the relationship, SE is the standard error of the estimated parameters.

Based on the results shown in Table 10.15 the hypotheses were verified:

H1 The relationship between management commitment and supplier strategy is positive and significant at the 0.001 level with an estimated 0.897 and 7.28 t_value so you have evidence to reject H_{01}.

H2 The relationship between management commitment and quality management is positive and significant at the 0.001 level with an estimated 0.923 and 7.671 t_value of so you have evidence to reject H_{02}.

Table 10.13 The rotated matrix components JIT benefits

Axis	Component	Load	Description
1	The successful implementation of JIT results in an increase in the rate of introduction of new products	0.839	Operational performance
	Due to the implementation of JIT reduces lead time	0.818	
	Because implementing JIT product quality is improved	0.759	
	The successful implementation of JIT increases the efficiency and utilization of machinery and equipment	0.735	
	The successful implementation of JIT results in a reduction in unit costs of manufacturing costs	0.708	
2	The successful implementation of JIT resulting in reduced inventory levels of finished product	0.863	Reduction inventory levels
	Due to the successful implementation of JIT inventory levels of raw materials are reduced	0.853	
	Due to the implementation of JIT inventory levels decrease in work in process (WIP)	0.823	

Table 10.14 Indices of adjustment of confirmatory factorial models

Model	CMIN	DF	CMIN/DF	GFI	CFI	RMSEA	HOELTER	
							0.05	0.01
Initial	943.720	443	2.130	0.742	0.851	0.074	108	113
2	829.770	390	2.126	0.792	0.865	0.074	108	113
3	708.170	335	2.114	0.808	0.879	0.074	110	115
4	549.590	260	2.114	0.824	0.896	0.074	111	118
5	462.447	220	2.102	0.839	0.906	0.073	113	120

H3 The relationship between JIT practices and management commitment is positive and significant at the 0.001 level with an estimated 0.809 and 8.169 t_value of so you have evidence to reject H_{03}.

H4 The relationship between management commitment and distribution of the plant is positive and significant at the 0.001 level with an estimated 0.751 and 6.132 t_value so you have evidence to reject H_{04}.

H5 The relationship between suppliers and inventory strategy is positive and significant at the 0.05 level with an estimated 0.274 and 2.29 t_value so you have evidence to reject H_{05}.

H6 The relationship between strategy and operational performance of suppliers is positive and significant at the 0.05 level with an estimated 0.275 and t value 2.192 so you have evidence to reject tH_{06}.

H7 The relationship between quality management and inventory is positive but not significant with an estimated 0.097 and 1,146 t_value so there is insufficient evidence to reject H_{07}.

H8 The relationship between quality management and operational performance is positive but not significant with an estimated 0.055 and 0.613 t_value of so you do not have sufficient evidence to reject H_{08}.

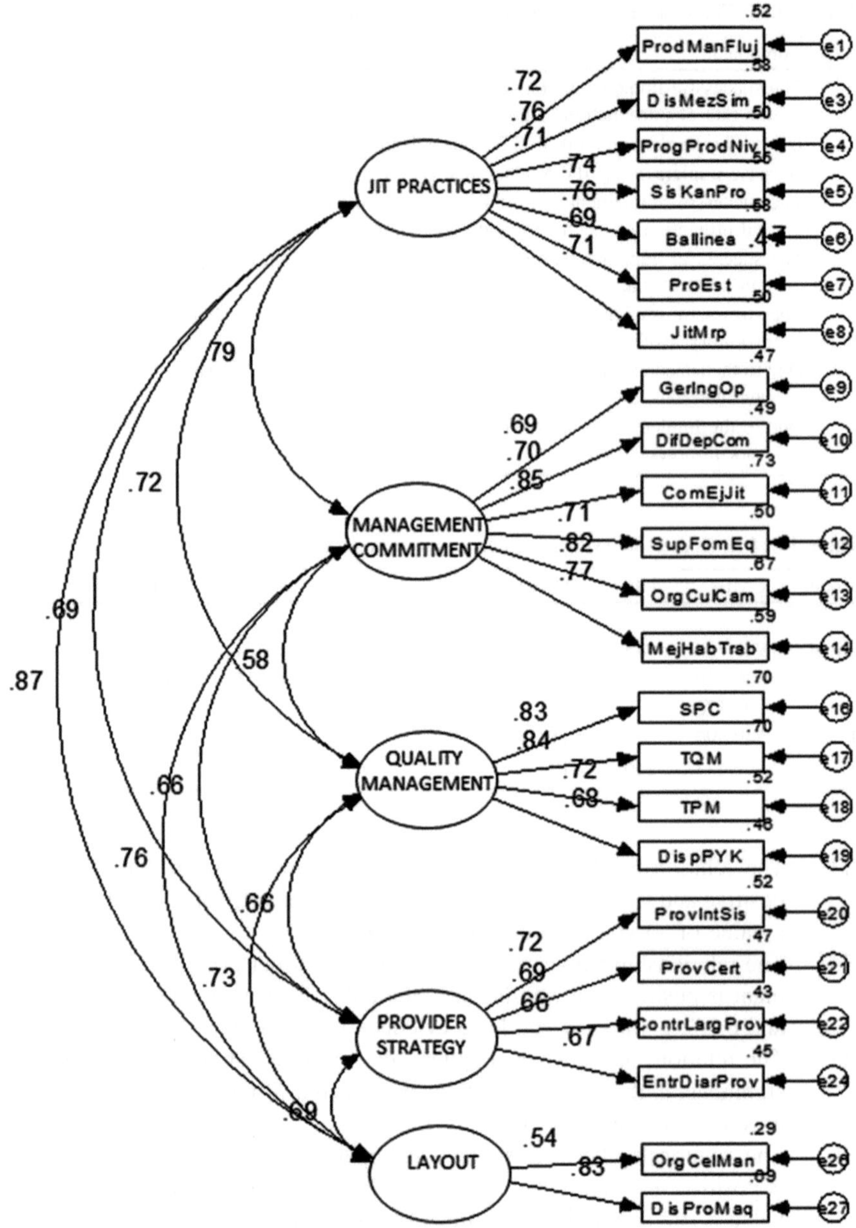

Fig. 10.4 Five factors confirmatory model

H9 The relationship between JIT practices and operational performance is positive and significant at the 0.001 level with an estimated 1,354 and 5,359 by t_value have enough evidence to reject H_{09}.

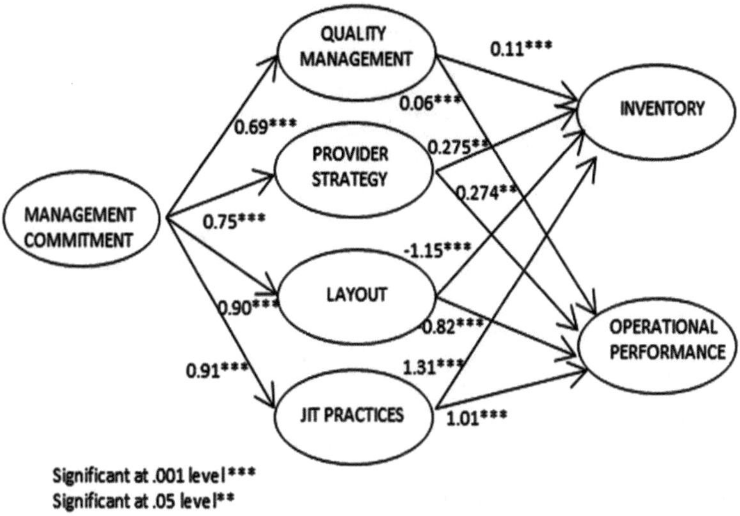

Fig. 10.5 Initial structural equation model

Table 10.15 Estimated parameter values of the initial model

			Estimate	S.E.	t-value	P
Provider_Strategy	<—	Management_Commitment	0.879	0.121	7.281	***
JIT_Practices	<—	Management_Commitment	0.809	0.099	8.169	***
Quality_Management	<—	Management_Commitment	0.923	0.120	7.671	***
Layout	<—	Management_Commitment	0.751	0.122	6.132	***
Inventory	<—	JIT_Practices	1.779	0.303	5.865	***
Operationa_Performance	<—	JIT_Practices	1.354	0.253	5.359	***
Inventory	<—	Quality_Management	0.097	0.085	1.146	0.252
Operationa_Performance	<—	Quality_Management	0.055	0.089	0.613	0.540
Operationa_Performance	<—	Layout	−1.178	0.294	−4.003	***
Inventory	<—	Provider_Strategy	0.275	0.120	2.294	0.022
Operationa_Performance	<—	Provider_Strategy	0.274	0.125	2.192	0.028
Inventory	<—	Layout	−1.673	0.388	−4.316	***

H10 The relationship between JIT practices and inventory is positive and significant at the 0.001 level with an estimated 1,779 and 5,865 t_value so it has enough evidence to reject H_{010}.

H11 The relationship between layout and operational performance is negative and significant at the 0.001 level with an estimated 1178 of 4,003 t_value and so do not have enough evidence to reject H_{011}.

H12 The relationship between plant distribution and inventories is negative and significant at the 0.001 level with an estimated 1673 to 4316 and t_value so there is insufficient evidence to reject H_{012}.

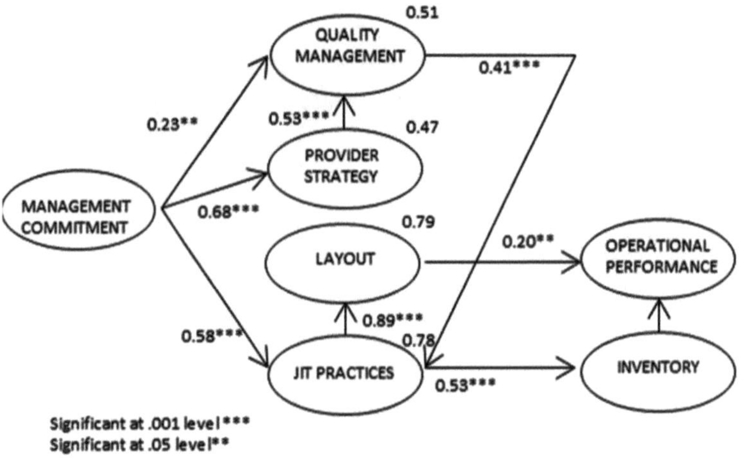

Fig. 10.6 Final structural equation model

The negative relationship between layout and operational performance and inventory is contrary to the results found by some researchers that concluded the distribution of the plant is one of the important factors in the performance contributions (Sakakibara et al. 1993; Swink et al. 2005; Narasimhan et al. 2006; Matsui 2007; Mackelprang and Nair 2010). These results combined with the values in the model modification indices led to rethink the model removing some of the relationships and adding some indirect relationships.

In Fig. 10.6 the final structural equation model with standardized parameters and their significance is shown. Rositas (2009) proposes the following criteria for grading each verbal impact (standardized coefficients): less than or equal to 0.10 absolute coefficients imperceptible or negligible impact (D); from 0.11 to 0.15 barely noticeable (P); from 0.16 to 0.19 (C) considerable; from 0.20 to 0.29 significant impact (I); from 0.30 to 0.50 strong impact (F); and finally coefficients greater than 0.50 are considered very strong impact (MF).

Each component receives an effect that has a multiple correlation coefficient, which indicates the percentage of the variability in the component that is explained by the variability of the set of components, setting a minimum account value of 0.15 (Rositas 2009).

In the structural equation model shown in Fig. 10.6. can be seen that six relationships have very strong impact (greater than 0.50), the relationship of quality management practices on JIT has a strong impact (0.41) and finally, the distribution ratio of the plant operating performance (0.20) and the relationship of management commitment in quality management (0.23) present significant impacts. In addition, all relationships were statistically approved at a level of 0.05 or 0.001. Table 10.16 shows the values of the parameters and estimates non standardized end structural equation model.

Table 10.16 Values of estimated parameters of the final model

			Estimate	S.E.	t_value	P
Provider_Strategy	<—	Management_Commitment	0.688	0.109	6.298	***
Quality_Management	<—	Management_Commitment	0.297	0.135	2.201	0.028
Quality_Management	<—	Provider_Strategy	0.676	0.157	4.314	***
JIT_Practices	<—	Management_Commitment	0.495	0.077	6.457	***
JIT_Practices	<—	Quality_Management	0.271	0.048	5.64	***
Inventary	<—	JIT_Practices	0.713	0.108	6.612	***
Layout	<—	JIT_Practices	0.921	0.141	6.553	***
Operational_Performance	<—	Layout	0.252	0.102	2.475	0.013
Operational_Performance	<—	Inventory	0.584	0.081	7.219	***

Table 10.17 Indices of adjustment structural equation model

Modelo	CMIN	DF	CMIN/DF	GFI	CFI	RMSEA	HOELTER	
							0.05	0.01
Initial	917.004	422	2.173	0.775	0.865	0.076	105	110
Final	747.068	422	1.770	0.816	0.911	0.061	129	135

Table 10.17 shows the fit indices of structural equation of initial model and final alternative model.

As seen in Table 10.17 the final model fits acceptably (GFI 0.816, CFI 0.911, RMSEA 0.061) and so significantly (significant change in the value of chi-square) the initial model.

It can also be seen that the changes between models are justified because the difference in CFI between each model is greater than 0.01 (Bentler 1990). Critical Hoelter index suggests that the required sample size for a confidence level of 95 % is 129. Therefore, as N = 205 can ensure that the changes are valid.

10.7 Conclusions

The resulting SCF for the implementation of JIT are: Management Commitment, Plant Distribution, Quality Management, Supplier Strategy and JIT practices. Based on the final structural equation model we see that the JIT practices relate to other areas such as quality management, layout and management commitment.

The JIT production system contributes to improve performance in inventory levels. Moreover, the distribution of the plant has a significant impact on operational performance. The JIT production system impacts the operational performance indirectly through other areas such as layout and performance on inventory levels.

The JIT production system influences in some areas while others support the JIT production system. Companies are recommended to use these synergies to enhance their competitiveness in the market. The results of this study also showed that successful implementation of JIT requires strong management commitment.

A direction for future research would be to study the process of implementation and how the JIT practices and infrastructure practices can be implemented to achieve a superior competitive performance within the plant. In addition, other factors may be incorporated into the model as the manufacturing strategy and linked to the client JIT.

Results of the impact of JIT in SCF performance indicators deserves to be considered as part of the manufacturing strategy in order to improve competitiveness.

References

Aghazadeh, S. (2003). JIT inventory and competition in the global environment: A comparative study of American and Japanese values in auto industry. *Cross Cultural Management, 10*(4), 29–42.

Ahmad, S., Schroeder, R., & Sinha, K. (2003). The role of infrastructure practices in the effectiveness of JIT practices: Implications for plant competitiveness. *Journal of Engineering and Technology Management, 20*, 161–191.

Avittathur, B., & Swamidass, P. (2007). Matching plant flexibility and supplier flexibility: Lessons from small suppliers of U.S. manufacturing plants in India. *Journal of Operations Management, 25*(3), 717–735.

Bentler, P. M. (1990). Comparative Fit Indexes in structural Models. *Psychological Bulletin*, 238–246.

Bollen, K., & Liang, J. (1988). Some properties of Hoelter's CN. *Sociological Methods and Research, 16*, 492–503.

Byrne, B. M.(2006), Structural Equation Modeling with Eqs.: Basic Concepts, Applications and Programming (Multivariate Applications). *SAGE Publications*.

Cai-feng, L. (2009). Research on a fast delivery production system: Just-in-time production system, *Canadian Social Science*, *5*(3), 121–126.

Callen, J. L., Fader, C., & Krinsky, L. (2000). Just in time: A cross-sectional plant analysis. *International Journal of Production Economics, 63*(3), 277–301.

Chang, D., & Lee, S. (1996). The impact of critical success factors of JIT implementation on organization performance. *Production Planning and Control, 7*(5), 329–338.

Clode, D. M. (1993). A survey of UK manufacturing control over the past 10 years. *Production and Inventory Management Journal*, *2*, 53–56.

Cortina, J. (1993). What is coefficient alpha? an examination of theory and applications. *Journal of Applied Psychology, 78*, 98–104.

Dal Pont, G., Furlan, A., & Vinelli, A. (2008). Interrelationships among lean bundles and their effects on operational performance. *Operations Management Research, 1*, 150–158.

Davy, J. A., White, R. E., Merritt, N. J., & Gritzmacher, K. (1992). A derivation of the underlying constructs of just-in-time management systems. *Academy of Management Journal, 35*, 653–670.

Dean, J., & Snell, S. (1996). The strategic use of integrated manufacturing: an empirical examination. *Strategic Management Journal, 17*, 459–480.

Forza, C. (1996). Achieving superior operating performance from integrated pipeline management: an empirical study. *International Journal of Physical Distribution and logistics Management*, *26*(9), 36–63.

Fullerton, R., & McWatters, C. (2001). The production performance benefits from JIT implementation. *Journal of Operations Management, 19*, 81–96.

Fullerton, R., McWatters, C., & Fawson, C. (2003). An examination of the relationships between JIT and financial performance. *Journal of Operations Management, 21*, 383–404.

Gelinas, R. (1999). The just-in-time implementation project. *International Journal of Project Management, 17*(3), 171–179.

Golhar, D. Y., & Stamm, C. L. (1991). The just-in-time philosophy: A literature review. *International Journal of Production Research, 29*(4), 657–676.

Gunasekaran, A., Goyal, S. K., Martikainen, T., Yli–Olli, P. (1998). A conceptual framework for the implementation of zero inventory and Just In Time manufacturing concepts. *Human Factors and Ergonomics in Manufacturing, 8*(1), 63–78.

Hanckok, W., & Zayko, M. (1998). Lean production: Implementation problems. *IIE solutions 30*(6).

Inman, R., Sale, S., Green, K, Jr, & Whitten, D. (2011). Agile manufacturing: Relation to JIT, operational performance and firm performance. *Journal of Operations Management, 29*, 343–355.

Jacobs, F., & Maiga, A. (2009). JIT performance effects: A research note. *Advances in Accounting Incorporating Advances in International Accounting, 25*, 183–189.

Kamata, A., Turhan, A., Darandari, E. (2003). Estimating reliability for multidimensional composite scale scores. *Encuentro anual de la American Educational Research Association*, Chicago.

Koren, Y. (2010). *Globalization and manufacturing paradigms, in the global manufacturing revolution: Product-process-business integration and reconfigurable systems*. Hoboken: Wiley.

Lawrence, J., & Hottenstein, N. (1995). The relationship between JIT manufacturing and performance in Mexican plants affiliated with US. companies. *Journal of operation management, 13*, 3–18.

Lévy, J. y Varela, J. (2003). Análisis Multivariable para las Ciencias Sociales. PearsonEducación, Madrid

Li, S., Rao, S. S., Ragu-Nathan, T. S., Ragu-Nathan, B. (2005). Development and validation of a measurement instrument for studying supply chain management practices. *Journal of Operation Management, 23*(6), 618–641.

Mackelprang, W., & Nair, A. (2010). Relationship between just-in-time manufacturing practices and performance: A meta-analytic investigation. *Journal of Operations Management, 28*, 283–302.

McWatters, C., & Fullerton, R. (2002). The role of performance measures and incentive systems in relation to the degree of JIT implementation. *Accounting, Organizations and Society, 27*, 711–735.

Malhotra, N. (2004). *Investigación de Mercados-Un enfoque apropiado, cuarta edición* (pp. 562–570). México: Prentice Hall.

Matsui, Y. (2007). An empirical analysis of just in time production in Japanese manufacturing companies. *International Journal of Production Economics, 108*, 153–164.

McDonald, R., & Marsh, H. (1990). "Choosing a multivariate model: Noncentrality and goodness of fit. *Psychological Bulletin, 107*, 247–255.

McKone, K., Schroeder, R., & Cua, K. (2001). The impact of total productive maintenance practices on manufacturing performance. *Journal of Operations Management, 19*(1), 29–58.

Mehra, S., & Inman, R. A. (1992). Determining the critical elements of just-in-time implementation. *Decision Sciences, 23*, 160–174.

Milgrom, P., & Roberts, J. (1995). Complementarities and fit strategy, structure and organizational change in manufacturing. *Journal in accounting and Economics, 19*(2–3), 179–208.

Mota, M., & Ferreira, R. (2008). A study on Just In Time implementation in Portugal. *Brazilian Journal of Operation & Production Management, 5*(1), 5–22.

Narasimhan, R., Swink, M., & Kim, S. (2006). Disentangling leanness and agility: An empirical investigation. *Journal of Operations Management, 24*(5), 440–457.

Nunnally, J. (1978). *Introduction to psychological measurement*. Japan: McGraw Hill.

Nunnally, J., & Bernstein, H. (2005). *Teoría psicométrica*. Mexico: McGraw Hill Interamericana.

Ohno, T. (1982). How the Toyota production system was created. *Japanese Economic Studies, 10*(4), 83–101.

Oral, E., Mistikoglu, G., & Erdis, E. (2003). JIT in developing countries—a case study of the Turkish prefabrication sector. *Building and Environment, 38*, 853–860.

Petersen, P. (2002). The misplaced origin of just-in-time production methods, *Management Decision, 40*(1): 82–8.

Ramarapu, N. K., Mehra, S., & Frolick, M. N. (1995). A comparative analysis and review of JIT implementation research. *International Journal of Operations and Production Management, 15*(1), 38–49.

Romero, R., Noriega, S., Escobar, C., & Ávila, D. (2009). Factores Críticos De Éxito: Una Estrategia De Competitividad. *CULCYT, 6*(31), 5–14.

Rositas, J. (2009). Factores Críticos de Éxito en la Gestión de Calidad Total en la Industria Manufacturera Mexicana. *Ciencia UANL, 12*(2), 181–193.

Sakakibara, S., Flynn, B., & Schroeder, R. (1993). A framework and measurement instrument for just-in-time manufacturing. *Production and Operations Management, 2*(3), 177–194.

Sandanayake, Y., Oduoza, F., & Proverbs, D. (2008). A systematic modelling and simulation approach for JIT performance optimisation. *Robotics and Computer-Integrated Manufacturing, 24*, 735–743.

Shah, R., & Ward, P. (2003). Lean manufacturing: Context, practice bundles, and performance. *Journal of Operations Management, 21*, 129–149.

Sim, K. L., & Curatola, A. P. (1999). Time-based Competition. *International Journal of Quality and Realiability Management, 16*(7), 659–674.

Singh, S., & Garg, D. (2011). JIT System: Concepts, benefits and motivations in Indian Industries. *International Journal of Management & Business studies, 1*(1), 26–30.

Streiner, D. (2003). Being inconsistent about consistency: When coefficient alpha does and doesn't matter. *Journal of Personality Assessment, 80*, 217–222.

Swink, M., Narasimhan, R., & Kim, S. (2005). Manufacturing practices and strategy integration: Effects on cost efficiency, flexibility, and market-based performance. *Decision Sciences, 36*(3), 427–475.

Tanaka, J., & Huba, G. (1985). A fit index for covariance structure models under arbitrary GLS estimation. *British Journal of Mathematical and Statistical Psychology, 38*, 197–201.

Teeravaraprug, J., Ketlada, K., & Nuttapon, S. (2011). Relationship model and supporting activities of JIT, TQM and TPM. *Songklanakarin Journal of Science and Technology, 33*(1), 101–106.

Thompson, B. (2004). *Exploratory and confirmatory factor analysis: Understanding concepts and applications* (1st ed.). Washington: American Psychological Association.

Voss, C. A., & Robinson, S. J. (1987). Application of just-in-time manufacturing techniques in the United Kingdom. *International Journal of Operations & Production Management, 7*, 46–52.

Wakchaure, V., Venkatesh, M., Kallurkar, S. (2006). Review of JIT practices in Indian manufacturing industries. *2006 IEEE International Conference on Management of Innovation and Technology 2, 34*(2), 1099–1103.

Walleigh, R. (1986). What is your excuse for not using JIT? *Harvard Business Review*, 2–7.

Ward, P., & Zhou, H. (2006). Impact of information technology integration and lean/just-in-time practices on lead-time performance. *Decision Sciences, 37*(2), 177–203.

White, R., & Prybutok, V. (2001). The relationship between JIT practices and type of production system. *Omega, 29*, 113–124.

Wildemann, H. (1988). Just In Time production in West Germany. *International Journal of Production Research, 26*, 521–538.

Willis, T. H., & Suter, W. C. (1989). The five Ms of manufacturing: A JIT conversion life cycle. *Production and inventory Management, 30*, 53–57.

Yasin, M., Small, M., & Wafa, M. (2003). Organizational modifications to support JIT implementation in manufacturing and service operations. *Omega, 31*(3), 213–226.

Zayko, M. J., Broughman, D. J., & Hanckok, W. M. (1997). Lean Manufacturing yield word class improvement for small manufcaturer. *IEE solution*, April.

Zhiwei, Z., & Meredith, P. H. (1995). Defining critical elements in JIT implementation: a survey. *Industrial Management & Data Systems, 95*(8), 21–28.

Chapter 11
Supplier Selection in a Manufacturing Environment

Rodrigo Villanueva-Ponce, Jaime Romero-González and Giner Alor-Hernández

Abstract Suppliers evaluation and selection processes are key elements in Supply Chain Management, because the right selection reduces costs, improves quality and promotes long term relationships between companies increasing supply chain competitiveness. Supplier evaluation and selection processes have been studied extensively in the past years providing help to the researchers and decision makers. This chapter presents the multi-criteria decision making (MCDM) techniques, the most used attributes in supplier selection and the description of two MCDM techniques that are useful in the selection process.

Keywords Supply chain · Supplier selection process · Multicriteria supplier selection

11.1 Introduction

The current pressure of global competition is demanding faster, reliable, more flexible and inexpensive products, therefore companies have to continuously improve their systems and establish new procedures and methods to keep up with customer demands. Supply Chain Management (SCM) is one of the strategies used nowadays to address these demands.

Supply Chain Management (SCM) eases the integration of a list of suppliers that provide the raw material, manufacturers that fabricate the products and

R. Villanueva-Ponce (✉) · J. Romero-González
Institute of Engineering and Technology, Autonomous University of Ciudad Juarez, Av. Del Charro 450 Norte. Col. Partido Romero, Ciudad Juárez, CHIH, Mexico
e-mail: rodrigo.villanueva@delphi.com

G. Alor-Hernández
Division of Research and Postgraduate Studies, Instituto Tecnológico de Orizaba, Av. Oriente 9, 852. Col Emiliano Zapata C.P, 94320 Orizaba, Mexico

J. L. García-Alcaraz et al. (eds.), *Lean Manufacturing in the Developing World*,
DOI: 10.1007/978-3-319-04951-9_11,

distributors to move these goods to a final customer. All these three members are important, but it can be said that the supplier has the ability to direct where the whole Supply Chain (SC) will go. Therefore selecting the right supplier from this list of options will help avoid problems in the long run for the SC.

The supplier selection problem is presented as an important activity in the SC. The objective of supplier selection is to identify suppliers with the highest potential for meeting company's needs consistently that will provide benefits to the SC such as better quality, low costs and on-time deliveries for all products. Then the following question arises: What needs to be considered in order to select the right supplier? There are multiple attributes that can be considered when selecting a supplier other than just cost, quality and delivery time of the product. These attributes are selected within the decision making group according to the needs that need to be fulfilled. The attributes should be related to the importance given to the product to be fabricated. The way to analyze these group of characteristics is an important task and it requires a close attention. Therefore using a multi-criteria or multi-attribute decision making techniques, created to structure, scrutinize and evaluate the given attributes is required as a way to identify the best fit.

This chapter provides a list of useful concepts related to the SC, SCM, attributes and techniques used in the supplier selection problem. The chapter presents a list of multi-criteria and multi-attribute decision making techniques that can be applied to the supplier selection problem depending on the situation being evaluated and the characteristics presented. It also offers a theory description on two of the multi-criteria decision making techniques; Analytic Hierarchy Process (AHP) and Technique of Order Preference by Similarity to Ideal Solution (TOPSIS). To finalize the chapter an example of a supplier selection problem is presented and solved using TOPSIS.

11.2 Supply Chain Management

During the recent rapid growth of technology and economic globalization, contemporary industry has been driven to initiate a division of labor. Thus, companies are focused on developing and strengthen inner capabilities and outsourcing the activities that are not handled internally as a way to enhance their competitive advantage. Consumers are demanding new custom made products and the companies must offer diverse products to satisfy these demands. Therefore, enterprises are motivated to invest in strategies that will help address these highly changing needs and establish a competitive advantage among competitors.

SCM is one of the most important competitive strategies used by companies nowadays; when several enterprises establish their own SC. It is usually integrated by the network management of three main components; raw material supplier, manufacturing plant and distribution center.

SCM is the administration of an interrelated business system involved in the provision of product and/or services required by an end customer. SCM comprehends all movement and storage of raw material, work-in-process inventory, and finished goods from point of origin to point of consumption. SCM as the design, planning, execution, control, and monitoring of supply chain activities with the objective of creating net value. Building a competitive infrastructure worldwide, synchronizing supply with demand and measuring performance globally in each member of the SC.

One of the reasons why the SCM is so important relates to cost, raw material's price frequently represents almost 70 % of the final product's cost. Therefore, the purchasing department within the SC plays an important role in reducing overall components costs. In other words, the purchasing department of any company that intends to introduce components to the global market needs to give importance to the supplier's selection process (Ghodsypour and O'Brien 1998; Aksoy and Öztürk 2011).

11.3 Supply Chain

Blanchard (2010) has defined the SC as the sequence of events that cover the entire lifecycle of a product or service from beginning (raw material distribution) to end (final customer use). In other words, SC comprehends all of the functions involved and needed in the process to receiving processing and finally satisfying a request. These tasks include and are not limited to a new product development, marketing, operations, distribution, finance and customer service (Chopra and Meindl 2006). Thus, a SC consists of all the parties involved directly or indirectly in fulfilling a customer's request.

As shown in Fig. 11.1 the supply chain is formed by three main members:

- Supplier (raw material)
- Manufacturer (product)
- Distributor (to end user).

A SC can become extremely complex since integrates independent organizations that now work together to develop, control and manage a product that will be used by a final customer (Wu et al. 2012). The integration of independent companies lead for SC to become an important element in current global economy, some authors declare that SC is the new strategy for competence (Ngai et al. 2011).

11.4 The Supplier

Suppliers have a direct impact on quality, cost and delivery time of final products. Suppliers may dictate the path the complete SC will follow. This is important due to the fact that the cost of raw materials and component parts that conform their

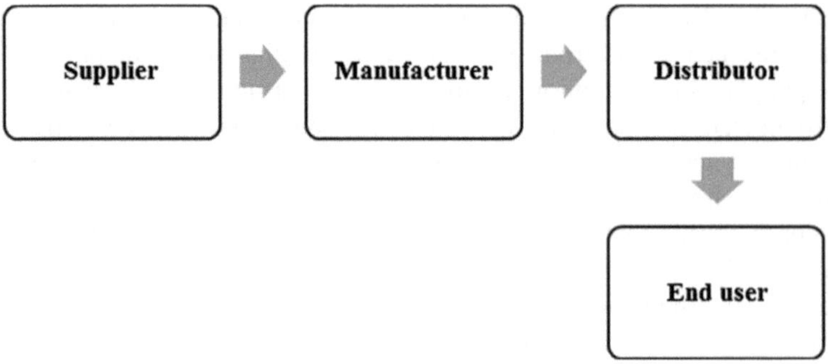

Fig. 11.1 Typical supply chain

final products represents the largest investment. Each activity performed at the supplier level will impact the whole SC. For instance, if for any particular reason the delivery time for the raw material gets extended, the manufacturing of the final product and the distribution of it may also be impacted. Hence, it is important that suppliers have a robust system in order to avoid problems and have a successful SC process.

One of the key strategies in contemporary supply management is to maintain long term relationship with suppliers, and use fewer but reliable suppliers (Ho et al. 2010). Therefore choosing the right supplier involves much more than scanning a series of attributes.

11.5 Supplier Selection

The suppliers are the first element in a SC and have a direct impact on quality, cost and delivery of new products. Therefore supplier's selection is a relevant process in the SCM. Depending on the companies' role and incorporation within the SC, the degree on integration with its supplier is established. The supplier's characteristics must match the SC integration process to achieve the integration between organizations.

Companies should use different strategies depending on the difficulty of obtaining supply materials and their economic impact. The greater the difficulty of obtaining the material the harder to achieve financial profit. Thus, companies should maintain good relationship with their suppliers as a strategy for future businesses. In this way the companies will be creating a vast list of suppliers to fulfill the needs. When a company and its suppliers are integrated based on norms and strategic objectives, they can achieve more successful incorporation as they can understand each other's organizational culture and upbringing.

Corporations should always have more than one option to rely on in the case of an eventuality. Having multiple supplier options to source the material will help the companies act quickly in the case of a disruption of the sourcing of the raw material due to an environmental catastrophe, failed commercial agreement or any other event.

Choosing the right supplier will help establish and maintain long term partnerships for future businesses, and improve the SC flow. Supplier selection consists of analyzing and measuring the performance of various options, ranking their characteristics and choosing the one that best fulfills the needs. This process is not an easy task, as various potential suppliers may have similar performance characteristics for different attributes. Therefore, companies must define a plan through the evaluation of suppliers to assess their internal organization and review the strengths, weaknesses and opportunities.

Most of the times, suppliers' strengths and weaknesses are varied leading to though decision making processes when selecting one option (Shaw et al. 2012).

11.6 Supplier Selection Process

Regardless of the technique, all follow the next methodology proposed by (De Boer et al. 2001):

(1) Supplier selection problem definition
(2) Identification of criteria
(3) Supplier selection by chosen technique
(4) Supplier selection.

De Boer et al. (2001) identifies the main techniques supporting supplier selection and selection as shown in Table 11.1.

In order to apply De Boer's methodology, there has to be a way to determine and define the main attributes to be integrated in a technique for the supplier selection process.

11.6.1 Identification of Criteria

Once a supplier selection problem has been defined, the next step is to define the main attributes to be integrated in the chosen technique. In contemporary supply chain management, the performance of potential suppliers is evaluated against multiple attributes rather than considering a single factor-cost (Ho et al. 2010).

The attributes can be categorized into two groups; qualitative and quantitative attributes.

Table 11.1 Methods for supporting supplier choice

Method	Author and year	Purpose in supplier selection.
TOPSIS	Hwang and Yoon (1981)	Suppliers are compared to best-in-class and worst-in-class performance
Distance from target	Hwang and Yoon (1981)	Suppliers are compared to a virtual 'ideal' supplier
Maximin	Chen and Hwang (1991)	Based on the idea that a 'chain' is as strong as its weakest 'link'
Linear assignment	Chen and Hwang (1991)	Required the purchaser to rank order suppliers only on each criterion separately
STEM	Vincke (1986)	Interactive procedure which does not require fixed criteria weights
Even swaps	Hammond et al. (1998)	Systematic procedure which aims at converting a comparison of suppliers on several criteria to a comparison on fewer and ultimately one criterion

Quantitative attributes in simple terms is everything that can be measured, such as cost, lead time, speed, size, capacity, etc., and all tangible attributes that physically describe a product.

Qualitative attributes are subjective, these can be used by assigning weight to each depending on its importance to the organization.

There is quite information in relation to the SC and the suppliers, based on this it can be identified the main attributes utilized in a supplier selection process. Figure 11.2 shows the results of a review of literature that comprehends published papers from 2001 to 2011 developed by Ho et al. (2010), Genovese and Lenny (2010), where quality, cost and delivery are the main attributes being evaluated in most cases.

Many studies on supplier selection defined numerous criteria, many empirical researches revealed that the relative importance of various criteria is similar. There is no doubt that cost, quality and delivery are the top characteristics to choose but there are other attributes also being considered such as Service, Capacity, Technology, etc. Depending on the industry role the need to look for different attributes arises for instance, a company where the production is constantly changing criteria such as capacity and flexibility are as important as the top three, mainly because the demands may change and the companies will reduce their requirements forcing he suppliers to adjust to abrupt changes in their production.

11.6.2 Supplier's Selection by Chosen Technique

Due to the importance of supplier's selection and selection, academics, government and researchers have developed techniques to minimize the painful selection and selection processes. There are three main categories; the economic techniques

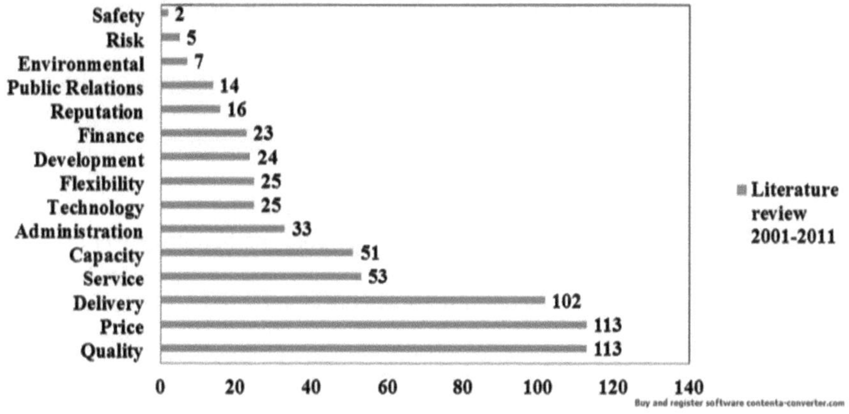

Fig. 11.2 Selection criteria in supplier selection, *Source* Ho et al. (2010), Genovese et al. (2010)

focused in economic attributes, the strategic techniques focused in strategic attributes of the company and the analytic techniques, that can integrate in the selection process the objective and subjective attributes from a supplier.

11.6.2.1 Economic Techniques

The economic techniques do represent the industrial practice, however these do not integrate the qualitative aspects of the business into the analysis, and therefore these techniques are extremely criticized due to the poor problem analysis. Yusuff et al. (2001) states, the decisions taken following the recommendation of the economic techniques tend to fail.

11.6.2.2 Strategic Techniques

These techniques mainly represent the objectives and mission of the company, however do not integrate the economic aspects of the selection. These are mainly used by the senior management but not entirely accepted by the rest of the company.

11.6.2.3 Analytic Techniques

These techniques are characterized by the integration of both qualitative and quantitative attributes. These techniques integrate all aspects of the problem being evaluated. Most of the researchers and academics do prefer the use of these techniques due to the multi-criteria and multi-attribute focus these have.

11.6.3 Multi-criteria Decision Making Techniques (MCDM)

Extensive multi-criteria decision making techniques have been proposed for supplier selection due to the contemporary supply chain management need to choosing the right supplier. These techniques involve both quantitative and qualitative attributes. According to Ho et al. (2010) have conducted a literature review about this important topic; they report an extensive analysis on the selection techniques.

The use of MCDM techniques provides a reliable methodology to rank alternatives when numerous objectives are defined. Regardless of the large number of available MCDM techniques none of these is considered the best technique for all types of problems. In other words, the techniques that fit better the identified problem or decision to make will be the ones applied.

These selection techniques can be applied in two ways:

- Individual application.
- Combined application.

11.6.3.1 Individual Selection Techniques

Some examples of those applied individually are shown in Table 11.2.

Data Envelopment Analysis (DEA) used mainly by Sarkis and Talluri (2002), Talluri and Narasimhan (2004), Garfamy (2006), Ross et al. (2006), Saen (2006, 2007), Seydel (2006), Talluri et al. (2006), Wu et al. (2007) among others.

Mathematical Programming (MP) is primarily used by Talluri and Narasimhan (2003, 2005), Hong et al. (2005), Narasimhan et al. (2006), Wadhwa and Ravindran (2007), Ng (2008) among others.

Analytic Hierarchy Process (AHP) and Case-Based Reasoning (CBR) used mainly by Chan and Chan (2004), Liu and Hai (2005), Chan and Kumar (2007), Hou and Su (2007), Choy et al. (2004a, 2005) and Chan (2003).

Analytic Network Process (ANP) and Fuzzy Logic (FL), has been used mainly by Sarkis and Talluri (2002), Bayazit (2006), Gencer and Gürpinar (2007), Chen et al. (2006), Sarkar and Mohapatra (2006) and Florez-Lopez (2007).

Simple Multi-attribute Rating Technique (SMART) There are reports of genetic algorithms and the technique mainly used by Barla (2003), Huang and Keska (2007) and Ding et al. (2005).

11.6.3.2 Integrated Selection Techniques

Some authors generally use integrated techniques in their supplier selection investigations because the combination brings better results. The integrated techniques are commonly used and the combination relies on the characteristics being evaluated and the information available. One of the most individual techniques used in combination with others is Analytic Hierarchy Process (AHP).

Table 11.2 Individual selection techniques

Technique	2009	2010	2011	2012	Total
Fuzzy logic	14	9	19	14	56
Analytic hierarchy process	3	1	1	4	9
Mixed integer programming	0	1	4	1	6
DEA	1	2	2	0	5
ANP	2	1	0	1	4
Multi-objective programming	1	1	1	0	3
Statistics	1	0	1	0	2
Neural networks	0	0	2	0	2
Nonlinear integer programming	1	0	1	0	2
Tests of hypotheses	2	0	0	0	2
Goal programming	0	2	0	0	2

Table 11.3 presents a typical combination of techniques that could be applied together. The rows show the first technique and the columns illustrate the second technique used in the combination, the last column shows the total number of combinations that had a first technique and the last row shows the total number of combinations that had a secondary technique. The intersection of a row and a column indicates the number of times these techniques were combined.

AHP is usually combined with Goal Programming and has been employed by Çebi and Bayraktar (2003), Wang et al. (2004, 2005), Perçin (2006), Kull and Talluri (2008) and Mendoza and Ventura (2010). Furthermore, AHP has been combined with DEA, CBR and Fuzzy Logic and used by Ramanathan (2007), Saen (2007), Sevkli et al. (2007), Choy et al. (2003, 2004b), Kahraman et al. (2003), Chan and Kumar (2007), and Ourkovic and Handfield (2006).

It is seen that Fuzzy logic is the approach that gets combined the most with another techniques; it is combined six times with ANP, eight times with QFD, seven times with TOPSIS, for a total of 40 different combinations. Regarding the secondary techniques most widely used, it can be seen that AHP presents a total of eleven combinations being used along with Fuzzy logic in eight articles, followed by multi-objective programming, goal programming and dynamic programming.

Another widely used technique is TOPSIS, which has been combined with fuzzy logic in seven different articles, and in one article is combined with data envelopment analysis, utility theory and Taguchi methods.

11.6.4 Analytic Hierarchy Process and Individual Technique

The Analytic Hierarchy Process (AHP) is a systematic multi-criteria decision technique widely used for decision problems and falls into the analytic type of selection techniques as it applies both qualitative and quantitative criteria. It has been proposed by Saaty (1980). Due to the great capacity to handle multi-criteria decision problems it can be used for supplier selection problems. This technique

Table 11.3 Integrated selection techniques

T 1/ T 2	AHP	GP	DEA	FL	MP	INP	MOP	ANP	TOPSIS	DP	DEMATEL	ELECTRE III	DEMPSTER-SHAFER	LP	SWOT	VIKOR	LAGRANGIANA	TOTAL
AHP	0	0	1	0	1	0	0	0	0	0	0	0	0	0	0	0	0	2
GP	0	0	0	0	0	0	0	1	0	0	0	0	0	0	0	0	0	1
DEA	0	0	0	0	1	0	0	0	1	0	0	0	0	0	0	0	0	3
FL	4	0	1	0	0	0	1	8	6	0	1	1	1	2	1	2	0	31
NN	0	0	0	1	0	0	0	0	0	0	0	0	0	0	0	0	0	1
MOP	0	0	0	0	1	0	0	1	0	0	0	0	0	0	0	0	0	2
ANP	0	0	0	1	0	1	0	0	0	0	0	0	0	0	0	0	0	4
BA	0	0	0	0	0	0	0	0	0	0	0	0	0	0	0	0	0	1
EST	0	1	0	0	0	0	0	0	0	1	0	0	0	0	0	0	0	2
LP	1	0	0	0	0	0	0	0	0	0	0	0	0	0	0	0	0	1
DP	0	0	0	0	0	0	0	1	0	0	0	0	0	0	0	0	0	1
SEM	0	0	0	1	0	0	0	0	0	0	0	0	0	0	0	0	0	1
Total	6	1	2	3	3	1	1	11	8	1	1	1	1	2	1	2	1	46

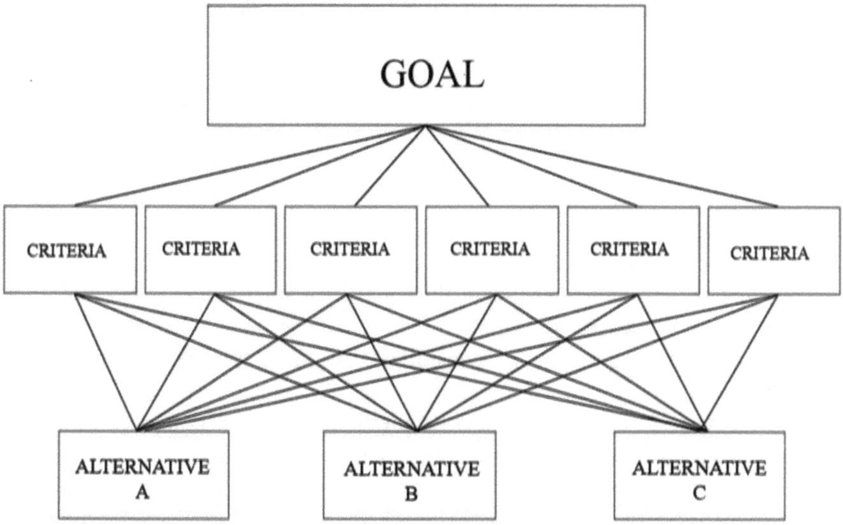

Fig. 11.3 Analytical hierarchy process technique (AHP)

performs a modeling of the problem that provides a formation of a hierarchy representative of the decision schema.

The AHP technique is a procedure designed to incorporate the managerial judgments about the importance of each of the attributes being considered during the decision taken process. It is a reliable tool to facilitate systematic and logical decision making processes, and determine the importance of a set of criteria. Figure 11.3 Yao (2010) shows the way the AHP technique is applied in a typical multicriteria decision process.

The AHP technique splits the decision problem into three simple phases:

(1) Problem decomposition—The decision problem is presented in the top level of a hierarchy that is broken down into different levels in which the final level is usually the alternatives to be selected.
(2) Pairwise Comparison Matrix (PCM)—Captures priorities from paired comparison judgments assigning a rating if importance of the elements of the decision with respect to each of their parent criteria and arranges these in a matrix. Pairwise comparison is an important step in AHP to be completed by the experts and in order to avoid having judgments not conscientiously only reasonable amounts of criteria is evaluated. The rating is established following the next scale:

    1 = preferred
    3 = slightly preferred
    5 = strongly preferred
    7 = very strongly preferred
    9 = extremely preferred

Intermediate values such as 2, 4, 6, 8 are permissible to help in the grading (Fig. 11.3).

(3) Normalized Matrix (NM)—Develop the normalized matrix dividing each number in a column of the matrix of pairwise comparison for the total sum of the column.
(4) Criteria priority vector—Calculated by the average of each normalized matrix's row. This average by row vector represents the priority of the alternative with respect to the criterion considered.
(5) Consistency—The consistency of the opinions used in the pairwise comparison matrix can be determined through the Consistency Ratio (CR). An acceptable CR should be less than 0.1. For those cases where CR > 0.1 the opinions and judgments are considered as taken arbitrarily and must be reconsidered.
(6) Priority Matrix (PM)—After all the previous steps are complete for all the established criteria, the results are summarized in a PM, the row represent the alternatives and the columns the criteria.
(7) Criteria Comparison Matrix (CCM)—Develop a matrix pairwise similarly to what was done for alternatives in steps 2, 3, 4.
(8) Global priority vector—Develop the global priority vector multiplying the criteria priority vector by the PM.This method assumes that the decision maker must establish the relative importance of each of the objectives and then define a structure of preferences among the alternatives identified. The end result of an alternative classification, indicates a general preference associated with each of them, allowing the identification of the best alternative.

11.6.5 Technique of Order Preference by Similarity to Ideal Solution and Individual Technique

One of the most widely used techniques for selection processes is TOPSIS, which was originally developed by Hwang and Yoon (1981) with further developments by Yoon (1987) and Hwang et al. (1993).

Table 11.4 illustrates the main areas of application that TOPSIS has had (Behzadian et al. 2012). The table shows the level of acceptance that TOPSIS has in administrative decisions in supply chain and manufacturing environments.

This technique is a multi-attribute method that measures the closest euclidean distance to an ideal positive solution and the longest euclidean distance to a negative ideal solution. It is based on the concept that the chosen alternative is the one selected from a set of options that will be evaluated comparing these to an ideal-positive and ideal-negative solutions. The alternatives are assumed as points in an N-dimensional space (N is the number of attributes that represent an alternative).

Table 11.4 Areas of application for TOPSIS

Areas	N	%
Supply Chain Management and Logistics	74	27.5
Design, Engineering and Manufacturing Systems	62	23
Business and Marketing Management	33	12.3
Health, Safety and Environment Management	28	10.4
Human Resources Management	24	8.9
Energy Management	14	5.2
Chemical Engineering	7	2.6
Water Resources Management	7	2.6
Other topics	20	7.4
Total	269	100

Source Behzadian et al. (2012)

TOPSIS is a useful technique in dealing with multi-criteria decision making problems in several areas. The TOPSIS technique can be expressed in a series of steps presented in coming subsections.

(1) Develop decision matrix
(2) Develop the standardized decision matrix
(3) Define the Positive Ideal Solution (PIS)
(4) Define the Negative Ideal Solution (NIS)
(5) Calculate distance to ideal solutions
(6) Calculate the relative proximity to ideal solutions
(7) Preference ordering.

11.6.5.1 Creating the Decision Matrix

TOPSIS evaluates the decision matrix as shown in Table 11.5. Where m Alternatives Si, for i = 1,…, m, are evaluated in terms of n attributes Cj, for j = 1,…, n. These may have different levels of importance wr for r = 1,…, n. Particularly, xij denotes the value of i-th alternative in terms of j-th attribute.

11.6.5.2 Standardizing the Attributes in the Decision Matrix

It is possible that in TOPSIS the different attributes being evaluated are expressed in different measurement scales. For example currency units for costs, meters for length, square feet for area, m/s for speed, etc.; therefore, these attributes need to be dimensionless values.

In TOPSIS every attribute is considered a vector in an m-dimensional space (m is the total number of alternatives). The Euclidean norm is applied to every

Table 11.5 Decision matrix for TOPSIS

	w_1	w_2		w_j		w_n
	C_1	C_2	...	C_j	...	C_n
S_1	X_{11}	X_{12}	...	X_{1j}	...	X_{1n}
S_2	X_{21}	X_{22}	...	X_{2j}	...	X_{2n}
...	...	...	...	...	...	...
S_m	X_{m1}	X_{m2}	...	X_{mj}	...	X_{mn}

attribute to remove the units, as shown in Eq. 11.1. Thus, an element nij of the decision matrix normalized N = [n ij] mxn is calculated as follows:

$$n_{ij} = \frac{x_{ij}}{\sqrt{\sum_{j=1}^{m} (x_{ij})^2}} \quad for j = 1, \ldots n; \; i = 1, \ldots m \tag{11.1}$$

11.6.5.3 Generating the Normalized and Weighted Decision Matrix

The weighted normalized value v̄ij of the weighted and normalized decision matrix $V = [\bar{v}_{ij}]_{mxn}$is calculated by Eq. 11.2.

$$\bar{v}_{ij} = w_j \otimes \bar{n}ij, \quad for\, j = 1, \ldots n; \; i = 1, \ldots m \tag{11.2}$$

where $\boldsymbol{w_j}$ represents the attribute weight and $\sum_{j=1}^{n} w_j = 1$. Determining the attribute weights is not an easy task these frequently vary from person to person, however there is enough information that helps on this process. The most commonly used techniques to achieve this step are Fuzzy Logic and AHP (Analytic Hierarchy Process).

11.6.5.4 Positive and Negative Ideal Solutions

The basic concept of TOPSIS is that each of the m alternatives may represent a point in a n-dimensional space. It also considers two hypothetical alternatives that are integrated with the best and worst ratings in the set of attributes.

The first hypothetical ideal-positive solution is represented by S+ and can be estimated by Eq. 11.3.

$$\bar{S}^{+} = \{\bar{v}_1^{+}, \ldots \bar{v}_n^{+}\} = \left\{ \left(\frac{Max}{i} \bar{v}_{ij} j \in J \right) \left(\frac{Min}{i} \bar{v}_{ij} j \in J' \right) \right\} \quad for\, i = 1, \ldots, m \tag{11.3}$$

The second alternative is called ideal-negative solution and is represented by S,—and can be estimated by Eq. 11.4.

$$\bar{S}^- = \{\bar{v}_1^-, \ldots \bar{v}-\} = \left\{ \left(\frac{Min}{i} \bar{v}_{ij} j \in J \right) \left(\frac{Max}{i} \bar{v}_{ij} j \in J' \right) \right\} \quad for\, i = 1, \ldots, m \tag{11.4}$$

where J is associated with attributes that must be maximized and J′ is associated with attributes that must be minimized.

11.6.5.5 Calculation of Distance to Ideal Alternatives

The separation of each given alternative (Si) to the ideal- positive solution $\bar{S}^+$ is given by Eq. 11.5, while the separation of each given alternative (Si) to the ideal-negative solution $\bar{S}^-$ is given by Eq. 11.6.

$$\bar{d}_i^+ = \sqrt{\sum_{i=1}^{n} \left(\bar{v}_{ij} - \bar{v}_j^+ \right)^2} \quad for\, i = 1, \ldots, m \tag{11.5}$$

$$\bar{d}_i^- = \sqrt{\sum_{i=1}^{n} \left(\bar{v}_{ij} - \bar{v}_j^- \right)^2} \quad for\, i = 1, \ldots., m \tag{11.6}$$

11.6.5.6 Calculation of the Relative Proximity to Ideal Solutions

The relative proximity between both ideal solutions and each alternative is given by $\bar{R}_i$ and is defined by Eq. (11.7).

$$\bar{R}_i = \frac{\bar{d}_i^+}{\bar{d}_i^+ + \bar{d}_i^-} \quad for\, i = 1, \ldots, m \tag{11.7}$$

11.6.5.7 Preference Ordering

The alternatives are ranked according to $\bar{R}_i$ in ascending order, selecting the alternative with the lowest value in $\bar{R}_i$.

11.6.6 Case of Study Using TOPSIS

An automotive company that handles aesthetic product is facing the need to select a packaging supplier that will be able to distribute the products to the final customer without any damage to the parts. The company has to select a supplier and has six options available. The company has defined five attributes to evaluate each supplier.

These attributes are the following:

- The cost of the packaging service (PC)
- The delivery time (DT)
- The quality of service (QS)
- Administration (A)
- Technology (T).

The final decision matrix is shown in Table 11.6. The table presents the six suppliers listed in each row. The columns show the attributes that describe each supplier previously defined by the company as these are the main points to evaluate.

The positive and negative ideal solutions are shown in Table 11.7. The data is obtained by following Eqs. 11.3 and 11.4. The last row (w) in the table represents the weight assigned to each attribute, in other words the importance given to each attribute. The determinant of the correlation matrix is 0.009, this value is small and relatively close to zero.

The attributes selected are measured in different units therefore a normalization process needs to be performed to eliminate the units and leave all values unit-less to be able to perform the calculations between these. The weighed normalized matrix obtained by Eq. 11.2 is shown in Table 11.8.

The separation of each given alternative to the ideal positive and negative solutions is given by Eqs. 11.5 and 11.6, the table showing the results is given by Table 11.9.

After performing the calculation of the relative proximity to ideal solutions per Eq. 11.7, the results obtained are presented in Table 11.10.

The best supplier to choose based on the TOPSIS methodology is Supplier S1 followed by S4, S6, S5 and S2. These results provide a guide for the company to select the right alternative that will help achieve the goals established. The different selection techniques may or may not agree with this result and the reason is that these techniques take in consideration different factors. The selection technique should be chosen based on the needs that the decision makers have established.

Table 11.6 Final decision matrix

Alternatives/attributes	PC	DT	QS	A	T
S_1	185	6.5	7.4	7.6	7.4
S_2	290	7.5	5.8	6	7.4
S_3	310	7.6	7.4	7	5.6
S_4	245	6.5	7	7.4	6.4
S_5	325	7.55	7	7.6	7.6
S_6	235	6.85	6.2	6.6	6.2

Table 11.7 Positive and negative ideal solutions

Alternatives/attributes	PC	DT	QS	A	T
S+	185	6.5	7.4	7.6	7.6
S−	325	7.6	5.8	6	5.6
Objective	Min	Min	Max	Max	Max
w	0.179	0.202	0.202	0.202	0.213

Table 11.8 Weighed normalized matrix

Alternatives/attributes	PC	DT	QS	A	T
S_1	0.050	0.067	0.079	0.079	0.079
S_2	0.079	0.077	0.062	0.062	0.079
S_3	0.084	0.078	0.079	0.072	0.060
S_4	0.066	0.067	0.075	0.077	0.069
S_5	0.088	0.078	0.075	0.079	0.082
S_6	0.064	0.071	0.066	0.068	0.067
S+	0.050	0.067	0.079	0.079	0.082
S−	0.088	0.078	0.062	0.062	0.060

Table 11.9 Distance to ideal solutions

Alternative/distance	d+	d−
S_1	0.0021	0.0501
S_2	0.0386	0.0215
S_3	0.0421	0.0204
S_4	0.0213	0.0323
S_5	0.0397	0.0300
S_6	0.0263	0.0274

Table 11.10 Relative proximity using TOPSIS

Alternative	$\bar{R}_i$	Preference ordering
S_1	0.04108	1
S_2	0.64168	5
S_3	0.67368	6
S_4	0.39679	2
S_5	0.56962	4
S_6	0.48980	3

References

Aksoy, A., & Öztürk, N. (2011). Supplier selection and performance evaluation in just-in-time production environments. *Expert Systems with Applications, 38*(5), 6351–6359.

Barla, S. B. (2003). A case study of supplier selection for lean supply by using a mathematical model. *Logistics Information Management, 16*(6), 451–459.

Bayazit, O. (2006). Use of analytic network process in vendor selection decisions. *Benchmarking: An International Journal, 13*(5), 566–579.

Behzadian, M., Khanmohammadi, O. S., Yazdani, M., & Ignatius, J. (2012). A state-of the-art survey of TOPSIS applications. *Expert Systems with Applications, 39*(17), 13051–13069.

Blanchard, D. (2010). *Supply Chain Management Best Practices* (2nd ed.). New Jersey: Wiley.

Çebi, F., & Bayraktar, D. (2003). An integrated approach for supplier selection. *Logistics Information Management, 16*(6), 395–400.

Chan, F. T. S. (2003). Interactive selection model for supplier selection process: an analytical hierarchy process approach. *International Journal of Production, 41*(15), 3549–3579.

Chan, F. T. S., & Chan, H. K. (2004). Development of the supplier selection model—A case study in the advanced technology industry. *Proceedings of the Institution of Mechanical Engineers Part B—Journal of Engineering Manufacture, 218*(12), 1807–1824.

Chan, F. T. S., & Kumar, N. (2007). Global supplier development considering risk factors using fuzzy extended AHP-based approach. *Omega, 35*(4), 417–431.

Chen, S. J., & Hwang, C. L. (1991). *Lecture notes in economics and mathematical systems: Fuzzy multiple attribute decision making*. Springer, Berlin.

Chen, T. C., Lin, C. T., & Huang, S. F. (2006). A fuzzy approach for supplier evaluation and selection in supply chain management. *International Journal of Production Economics, 102*(2), 289–301.

Chopra, S., & Meindl, P. (2006). *Supply Chain Management* (3rd ed.). Upper Saddle River: Pearson/Prentice Hall.

Choy, K. L., Lee, W. B., & Lo, V. (2003). Design of an intelligent supplier relationship management system: A hybrid case based neural network approach. *Expert Systems with Applications, 24*(2), 225–237.

Choy, K. L., Lee, W. B., & Lo, V. (2004a). An enterprise collaborative management system–A case study of supplier relationship management. *The Journal of Enterprise Information Management, 17*(3), 191–207.

Choy, K. L., Lee, W. B., Lau, H. C. W., Lu, D., & Lo, V. (2004b). Design of an intelligent supplier relationship management system for new product development. *International Journal of Computer Integrated Manufacturing, 17*(8), 692–715.

Choy, K. L., Lee, W. B., & Lo, V. (2005). A knowledge-based supplier intelligence retrieval system for outsource manufacturing. *Knowledge-Based Systems, 18*(1), 1–17.

De Boer, L., Labro, E., & Morlacchi, P. (2001). A review of methods supporting supplier selection. *European Journal of Purchasing and Supply Management, 7*(2), 75–89.

Ding, H., Benyoucef, L., & Xie, X. (2005). A simulation optimization methodology for supplier selection problem. *International Journal Computer Integrated Manufacturing, 18*(2–3), 210–224.

Florez-Lopez, R. (2007). Strategic supplier selection in the added-value perspective: A CI approach. *Information Sciences, 177*(5), 1169–1179.

Garfamy, R. M. (2006). A data envelopment analysis approach based on total cost of ownership for supplier selection. *Journal of Enterprise Information Management, 19*(6), 662–678.

Gencer, C., & Gürpinar, D. (2007). Analytic network process in supplier selection: A case study in an electronic firm. *Applied Mathematical Modeling, 31*(11), 2475–2486.

Genovese, A., Koh, L. S. C., Bruno, G., Bruno, P. (2010). Green supplier selection: A literature review and a critical perspective. *8th International Conference on Supply Chain Management and Information Systems (SCMIS), 2010*, pp. 1–6.

Ghodsypour, S. H., & O'Brien, C. (1998). A decision support system for supplier selection using an integrated analytic hierarchy process and linear programming. *International Journal of Production Economics, 56–57*(1), 199–212.

Hammond, J., Keeney, R., Raiffa, H. (1998). Even swaps: A rational method for making trade-offs. *Harvard Business Review, 76*(2), 137–152.

Ho, W., Xu, X., & Dey, P. K. (2010). Multi-criteria decision making approaches for supplier evaluation and selection: A literature review. *European Journal of Operational Research, 202*(1), 16–24.

Hong, G. H., Park, S. C., Jang, D. S., & Rho, H. M. (2005). An effective supplier selection method for constructing a competitive supply-relationship. *Expert Systems with Applications, 28*(4), 629–639.

Hou, J., & Su, D. (2007). EJB–MVC oriented supplier selection system for mass customization. *Journal of Manufacturing Technology Management, 18*(1), 54–71.

Huang, S. H., & Keska, H. (2007). Comprehensive and configurable metrics for supplier selection. *International Journal of Production Economics, 105*(2), 510–523.

Hwang, C. L., & Yoon, K. (1981). *Multiple attribute decision making: Methods and applications*. New York: Springer-Verlag.

Hwang, C. L., Lai, Y. J., & Liu, T. Y. (1993). A new approach for multiple objective decision making. *Computers and Operational Research, 20*, 889–899.

Kahraman, C., Cebeci, U., & Ulukan, Z. (2003). Multi-criteria supplier selection using fuzzy AHP. *Logistics Information Management, 16*(6), 382–394.

Kull, T. J., & Talluri, S. (2008). A supply-risk reduction model using integrated multicriterio decision making. *IEEE Transactions on Engineering Management, 55*(3), 409–419.

Liu, F. H. F., & Hai, H. L. (2005). The voting analytic hierarchy process method for selecting supplier. *International Journal of Production Economics, 97*(3), 308–317.

Mendoza, A., Ventura, J. A. (2010). A serial inventory system with supplier selection and order quantity allocation. *European Journal of Operational Research, 207*(3), 1304–1315.

Narasimhan, R., Talluri, S., & Mahapatra, S. K. (2006). Multiproduct, multicriteria model for supplier selection with product life-cycle considerations. *Decision Sciences, 37*(4), 577–603.

Ng, W. L. (2008). An efficient and simple model for multiple criteria supplier selection problem. *European Journal of Operational Research, 186*(3), 1059–1067.

Ngai, E. W. T., Chau, D. C. K., & Chan, T. L. A. (2011). Information technology, operational, and management competencies for supply chain agility: Findings from case studies. *Journal of Strategic Information Sytems., 20*(3), 232–249.

Ourkovic, S., & Handfield, R. (2006). Use of ISO 9000 and baldrige award criteria in evaluation of supplier quality. *International Journal of Purchasing and Materials Management, 32*(2), 2–11.

Perçin, S. (2006). An application of the integrated AHP–PGP model in supplier selection. *Measuring Business Excellence, 10*(4), 34–49.

Ramanathan, R. (2007). Supplier selection problem: Integrating DEA with the approaches of total cost of ownership and AHP. *Supply Chain Management: An International Journal, 12*(4), 258–261.

Ross, A., Buffa, F. P., & Carrington, D. (2006). Supplier evaluation in a dyadic relationship: An action research approach. *Journal of Business Logistics, 27*(2), 75–102.

Saaty, T. L. (1980). The analytic hierarchy process: Planning, priority setting, resource allocation, McGray-Hill.

Saen, R. F. (2006). A decision model for selecting technology suppliers in the presence of nondiscretionary factors. *Applied Mathematics and Computation, 181*(2), 75–102.

Saen, R. F. (2007). Supplier selection in the presence of both cardinal and ordinal data. *European Journal of Operational Research, 183*(2), 741–747.

Sarkar, A., & Mohapatra, P. K. J. (2006). Evaluation of supplier capability and performance: A method for supply base reduction. *Journal of Purchasing and Supply Management, 12*(3), 148–163.

Sarkis, J., & Talluri, S. (2002). A model for strategic supplier selection. *Journal of Supply Chain Management, 38*(1), 18–28.

Sevkli, M., Koh, S. C. L., Zaim, S., Demirbag, M., & Tatoglu, E. (2007). An application of data envelopment analytic hierarchy process for supplier selection: A case study of BEKO in Turkey. *International Journal of Production Research, 45*(9), 1973–2003.

Seydel, J. (2006). Data envelopment analysis for decision support. *Industrial Management and Data Systems, 106*(81), 81–95.

Shaw, K., Shankar, R., Yadan, S., & Thakur, L. S. (2012). Supplier selection using fuzzy AHP and fuzzy multi-objective linear programming for developing low carbon supply chain. *Expert Systems with Applications, 39*(9), 8182–8192.

Talluri, S., & Narasimhan, R. (2003). Vendor evaluation with performance variability: A max-min approach. *European Journal of Operational Research, 146*(3), 543–552.

Talluri, S., Narasimhan, R. (2004). A methodology for strategic sourcing. *European Journal of Operational Research, 154*(1), 236–250.

Talluri, S., & Narasimhan, R. (2005). A note on a methodology for supply base optimization. *IEEE Transactions on Engineering Management, 52*(1), 130–139.

Talluri, S., Narasimhan, R., & Nair, A. (2006). Vendor performance with supply risk: A chance-constrained DEA approach. *International Journal of Production Economics, 100*(2), 212–222.

Vincke, P. (1986). Multi-criteria decision aid. New York: Wiley

Wadhwa, V., & Ravindran, A. R. (2007). Vendor selection in outsourcing. *Computers & Operations Research, 34*(12), 725–3737.

Wang, G., Huang, S. H., & Dismukes, J. P. (2004). Product-driven supply chain selection using integrated multi-criteria decision-making methodology. *International Journal of Production Economics, 91*(1), 1–15.

Wang, G., Huang, S. H., & Dismukes, J. P. (2005). Manufacturing supply chain design and evaluation. *International Journal of Advanced Manufacturing Technology, 25*(1–2), 93–100.

Wu, C. H., Chen, C. W., & Hsieh, C. C. (2012). Competitive pricing decisions in a two-echelon supply chain with horizontal and vertical competition. *International Journal of Production Economics, 135*(1), 265–274.

Wu, T., Shunk, D., Blackhurst, J., & Apalla, R. (2007). AIDEA: A methodology for supplier evaluation and selection in a supplier-based manufacturing environment. *International Journal of Manufacturing Technology and Management, 11*(2), 174–192.

Yao, J., (2010). Research on evaluation indicator system for construction of demonstrative building based on analytic hierarchy process theory. *2nd International Conference on Industrial Mechatronics and Automation*, 418–421.

Yoon, K. (1987). A reconciliation among discrete compromise situations. *Journal of Operational Research Society, 38*, 277–286.

Yusuff, R. M., Yee, K. P., Hashimi, M. S. J. (2001). A preliminary study on the potential use of the analytical hierarchical process (AHP) to predict advanced manufacturing technology (AMT) implementation. *Robotics and Computer-Integrated Manufacturing, 17*(5), 421–427.

Chapter 12
Megaplanning: Strategic Planning, Results Oriented to Improve Organizational Performance

Case Study: Planning Logistics Distribution Systems for Small and Medium Businesses in Obregon City, Sonora, Mexico

Ernesto Alonso Lagarda-Leyva, Javier Portugal-Vásquez, Lilia Guadalupe Valenzuela-Preciado, Tania Mendoza-Gurrola and Roger Kaufman

Abstract This article discusses how a societal-referenced approach to strategic thinking and planning can add measurable value to logistic distribution systems using small and medium business in the City of Cajeme, Sonora, Mexico as the platform. By enlarging the value chain to include societal value to organizational value, improvements in effectiveness and efficiency may be demonstrated. The increase in competition in the international market from organizations has led to the investigation and establishment of strategies that increase their competitiveness in the market that they operate, this has aroused the interest of studying the logistics systems of the companies that help you increase the level of service to the client and improve its logistics efficiency. Through the development of this research, the goal is to design the technological solution to first level of needs of the logistic system of supply and distribution of Small and Medium Enterprises in the sector of Obregon City, which favor the increase of competitiveness. This study has great importance because currently there is no information about the logistic system of Small and Medium Enterprises in the sector service of Obregon City, which allows employers, educational and Government institutions, support programs tailored to the requirements of these organizations. Due to the enormous importance of small and medium-sized enterprises in Mexico, the benefits for the

E. A. Lagarda-Leyva (✉) · J. Portugal-Vásquez · L. G. Valenzuela-Preciado · T. Mendoza-Gurrola · R. Kaufman
Sonora Institute of Technology, 5 de Febrero 818 Sur Col. Centro, Obregón, SON, México
e-mail: ernesto.lagarda@itson.edu.mx

R. Kaufman
Florida State University, 1123 Lasswade Drive, Tallahassee, FL 32312, USA

J. L. García-Alcaraz et al. (eds.), *Lean Manufacturing in the Developing World*,
DOI: 10.1007/978-3-319-04951-9_12,

development of this project are several, among those ones can quote: Since the results of this research, you wish to design the technological solutions from the point of view of supply and distribution system, in Obregon City service sector of Small and Medium Enterprises, providing necessary information to support its competitiveness.

Keywords Strategic planning · Supply chains · Logistics · Performance

12.1 Introduction

12.1.1 Performance Evaluation in Organizations

Authors like Chiavenato (1998), Grados et al. (2002) and Rodríguez (2009), are agree that the performance is to assess how employees perform the activities that are responsible for, and by adding measurable value. The definitions of these authors regarding the performance are included below: Process to formally evaluate the work behavior and provide feedback in which adjustments can be made in the same (Grados et al. 2002). Consists of assessing the efficiency with which the occupant of a position you run in a given period of time (Rodríguez 2009) and, a system for assessing the performance of an individual in the position and its potential development (Chiavenato 1998). The universalization of performance appraisal systems began around 1980 as a remuneration policy tool to gradually transform professional development tools (Rodríguez 2009). In the early 19th century, Robert Owen structured a set of books and notebooks, where a supervisor made daily reports and comments on the performance of each employee. At the beginning of the 21st Century Winslow Taylor, pointed out that "while the industry had a clear concept of the quantity and quality of work you can expect from a machine, Taylor did not have a vision of verifiable limits worker efficiency" having an estimate of the performance that a worker could show in one operation, performing their best, it would provide a standard to estimate the efficiency and performance of other employees in the execution of the same operation (Grados et al. 2002). Moreover, appropriate performance evaluation is possible "to identify strengths and weaknesses of an organization, the quality of subordinates, and the level of compliance of administrative functions and the effectiveness and efficiency the performance of duties". Some of the objectives of performance evaluated in organizations are: "To provide data on the performance of employees over time, so that appropriate decisions can be made". Contribute to decision making related to training, counseling, payments and promotion of personnel and other matters. "The treatment of human resources can be seen as a competitive advantage of the company and whose productivity can be developed, depending on the system of administration, among others". An assumption of these approaches is that improvement in individual performance will add measurable value to the organization and to external clients.

12.2 Strategic Planning Methodologies

Strategic planning today provides a framework for defining, justifying, and then conducting the activities of organizations aim at improving results based on their performance, Kaufman (2011) further suggests that strategic thinking and planning will best assure organizational success and value-added to include measurable objectives for community and societal value-added. He calls this approach Megaplanning. Organizations are increasingly concerned to provide information to their employees on how to achieve the goals from a perspective associated with the same strategy that is reflected in its vision as part of the end sought, using the commitment in its mission in order to provide the criteria for defining the means to reach the mission and align tactics and resources for effectiveness and efficiency.

12.2.1 Balanced Scorecard Approach Finance Learning and Staff Development

The concept of strategy is integral to the Balanced Scorecard proposed to provide a framework to describe and communicate strategies in a consistent and clear manner while assuring that there is a balance of contributing variables, not just financial in its own. Since the introduction of the Balanced Scorecard (BSC) in 1992, the strategic map is designed as a logical and comprehensive structure to describe a strategy (Kaplan and Norton 2001). The proposal of Kaplan and Norton, basically evaluates the performance of organizations in the following four perspectives: (1) Financial Perspective normally associated with the profitability of the company, (2) Customer perspective, associated with the product being delivered, (3) Internal perspective associated with processes and finally, (4) Learning and Growth Perspective, associated with what the human resource achieved in terms of skills and competencies for example, see Fig. 12.1. Kaplan and Norton (2005) claim that in addition to providing information to senior management from different perspectives, the BSC focuses on the most critical measurements for the organization.

Some authors, Kaufman (2000), have suggested that in order to make a score card really balanced it should include indicators for external client and societal value added. This is currently missing from conventional score card applications.

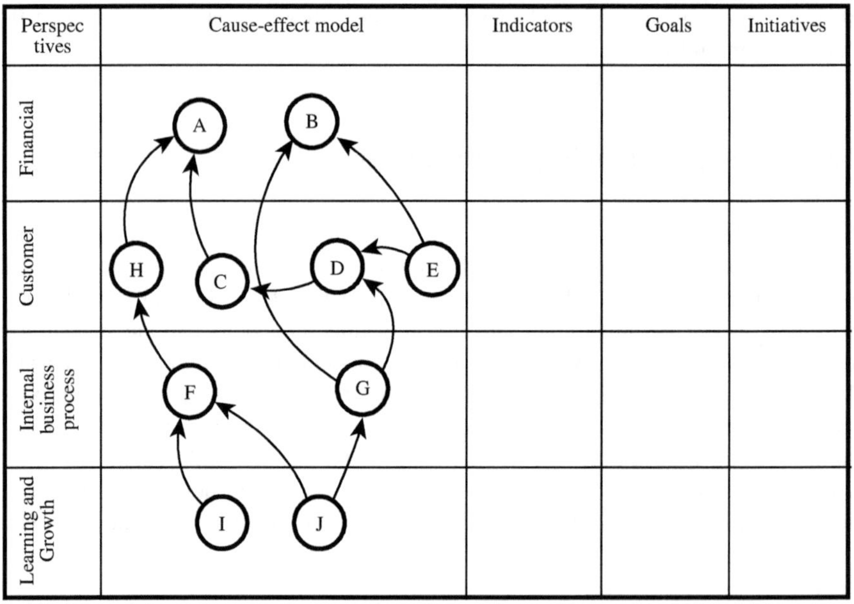

Fig. 12.1 Strategy map in terms of cause and effect. *Source* Personal communication

12.2.2 Hoshin Kanri Method: Focus on Processes

The Hoshin Kanri method is a technique that helps organizations to focus their efforts and analyze their activities and results. It is a systematic approach to identify, manage and resolve activities that require drastic changes or improvement. Hoshin Kanri is a tool for effective strategic planning and facilitates: Identify critical objectives, evaluate restrictions, establish performance measures, develop implementation plans and, review conducts boards periodically. The concept of strategic planning is simple: it is a management system that aligns the organization. It translates the vision and mission of an institution in an understandable arrangement of tactical objectives, which defined performance indicators and transforms them into a framework of project-based work, Socconini (2009).

12.2.3 SCOR Model: Focus on Processes in Supply Chains

Another model that can evaluate the performance of organizations is the Supply Chain Operations Reference Model (SCOR-model), in this sense Calderon and Lario (2005) indicate that the SCOR model is a tool to analyze, represent and set strings supply. This model provides a unique framework that links business to processes, technologies and management indicators in a unified structure to improve the efficiency of managing the supply chain and improve those activities related with the supply chain supporting communication among supply chain partners. The model has been able to provide a basis for improving the supply chain and local specific projects in global projects (Calderon and Lario 2005). The SCOR model, Fig. 12.2, allows a description of the business activities necessary to meet the customer demand.

Calderon and Lario (2005) indicate that the SCOR model covers all customer interactions (from entry order to payment of invoices), all physical transactions of materials (from the suppliers of the suppliers (i.e.) to the suppliers, customer to customer, customers, including equipment, supplies, spare parts, bulk product, software, etc.). And all market interactions (from aggregate demand to the fulfillment of each of the orders). This model contains sales and marketing, i.e. demand generation, does not contain product development, research and development; and contains some elements of after-sale customer service. The model covers, however implies human resources activities, training, systems, management and quality assurance among others (Calderon and Lario 2005). The council says that supply chain SCOR contains three levels of process detail, as shown in Fig. 12.3: the top level (types of processes), the configuration level (process categories), the level of elements processes (decomposition process), and finally, a fourth level of implementation. The SCOR model focuses on the first three levels and emphasizes the way in which each organization should conduct its business or design their systems and information flow. Therefore, each organization that implements improvements in their supply chain using SCOR model will need to extend the model, at least to level 4, using the processes, systems and practices of your organization (Calderon and Lario 2005). It does 'open the door' to measuring societal value added.

12.2.4 Megaplanning Organizational Elements Model Approach to Organizations Adding Measurable Value to External Clients and Society

The tend to be few organizations in their strategic map that contemplate social impact look like a measurable value in this regard, the contribution made by Kaufman and Guerra-Lopez (2013), is intended to measure in their Organizational Elements Model (OEM) developed in 2000, where required, to understand first what are Ends and Means, what and how. The Organizational Elements Model defines three levels of results and two types of processes, resources and media. Organizational Elements are: Societal results (Outcome): These are the results that add measurable value to society, community and customers outside the organization. Planning at this level is called MEGA. Organizational results (Output): Are the results an organization can do or deliver outside itself to an external customer and society. They are the contributions of the external results (MEGA). The

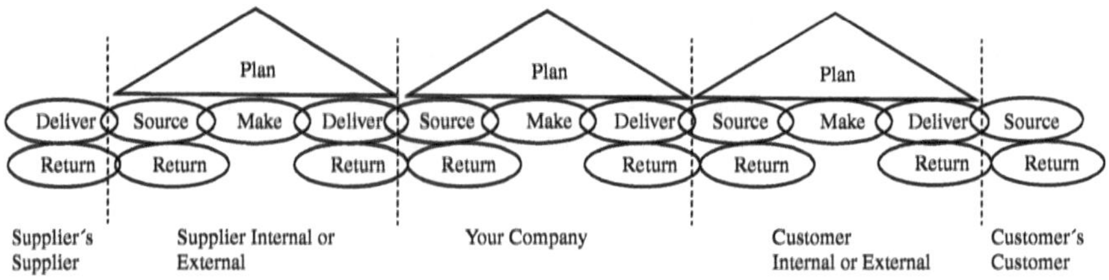

Fig. 12.2 Primary processes SCOR model management. *Source* (SCM Operations, 2011)

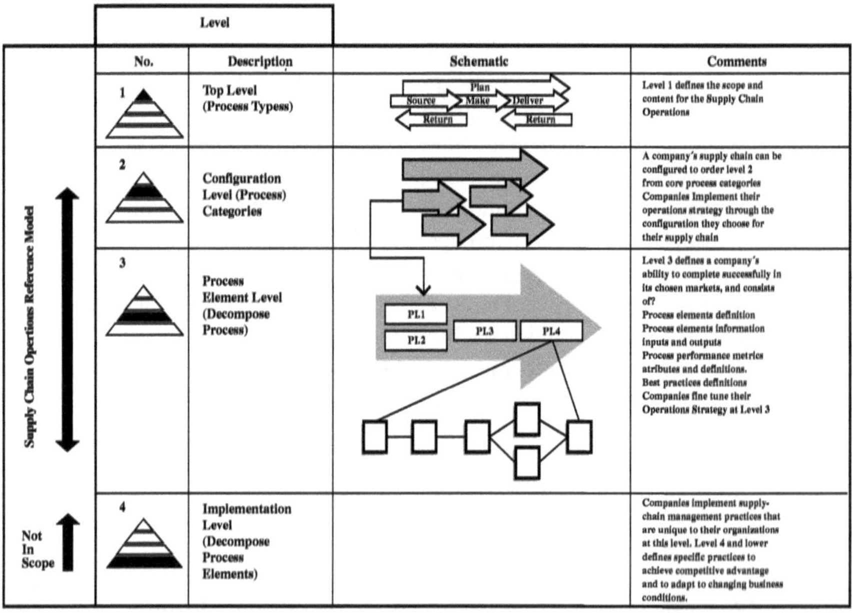

Fig. 12.3 SCOR model. *Source* (SCM Operations, 2011)

outputs can and should deliver MEGA consequences. At this level planning is called MACRO planning. Unfortunately, most of the time, the planning MACRO is called strategic planning. It is so called because it offers performance levels of more conventional strategic planning and therefore leaves the external values of the society in question or assumed. Individual and small group results (Products): These are the results that form the basis of the production of an organization; they have effects inside and outside the organization and affect payments to external clients and society. It is the primary focus of most organizational activities and resources applications. These are the building blocks of organizational contributions. Means, methods, procedures, and activities (Processes): They are the means, activities, interventions, programs and initiatives that an organization can do. Whatever is determined to use resources and provide results is a process.

The utility of a process depends on the relationship and the ability to deliver efficiently and effectively as useful results. Ingredients and resources (Inputs): Are the raw materials that an organization uses in its processes to deliver results (Kaufman 2011). None of the above organizational elements are more important than each others; they all have the same importance in organizations and likewise are interrelated with each other (Kaufman 2011). The goals must be identified and listed in the three levels of planning, MEGA, MACRO AND MICRO. (Kaufman 2011).

The processes, results, and contributions have a dynamic relationship with all that the organization does and the consequences of this. The results are results that must be met, while the processes are those possible techniques, procedures or methods that may be used to obtain desired results. The means are selected from the best based on achieved results, and are promoted because sooner or later performance or results will be the basic test that will be used to see if is a resource or a worth method (Kaufman 2000). Each level has specific implications and contributions so that the results have different levels of impact (Kaufman 2000). In order to develop useful purposes, there are three considerations: the ability to differentiate the means and results, relativity between the capacities of the measurement results and, organizational range covering elements MEGA macro and micro planning (Kaufman 2000). Also, while the objectives are related to the three types of purposes, these must be clearly linked to the five organizational elements.

The five organizational elements—Mega, Macro, Micro, Processes, and Inputs—must be defined with rigor and precision, and they must also relate one to the others so whatever you use, do, produce, and deliver will add value to our external clients and society. This primary (but not exclusive) attention to Mega—societal value added—is vital, critical, and missing from most other approaches to needs assessment and planning. The Organizational Elements must be linked and aligned if we are able to deliver Organizational improvement and success. Doing so enables you to ensure that everything you use (Inputs) and do (Processes) as well as individual results (Micro) and organizational contributions (Outputs) deliver useful societal results (Mega); all in the value chain, both well served.

The following table shows how the organizational elements are linked and aligned each other. On the other hand, it is possible to make organizations use adequately the methodologies proposed by different authors such as those in this project have been used, where it is important to have an awareness of the possibilities for change and improvement of competitiveness, the important thing is to from what, propose how, in this sense the Methodology Based Strategic Planning and System Dynamics with Scenarios (Lagarda-Leyva 2012) defines that one of the phase is the awareness where all employees recognize the importance to make a transformational change that will benefit everyone and the region itself in a project of cohesion, cooperation and consistency Table 12.1.

Table 12.1 How the OEM links to needs assessment and key stakeholders at each level (Kaufman and Guerra-López 2013)

Organizational elements	Examples	Needs assessment	Type of planning	Key stakeholders
Outcomes: societal results and consequences	Quality of life, health, sel-sufficiency, gainfully employed graduates	Mega	Strategic planning	Clients, client's clients, community, society
Outputs: organizational results	Profits, sales, patients discharged, graduates	Macro	Tactical planning	Organizational itself
Products: en-route results or building blocks note there may be multiple levels of products)	Competent employees, courses completed, assembled vehicles, medical procedures completed, accomplished/met standards	Micro	Operation planning	Individuals and groups of employees or performers
Processes: interventios, solutions, methods	Teaching: trainning, learning, manufacturing, selling, managing, marketing	Quasi	Action planning	Individual and groups of employees or performers
Inputs: resources	Funding, employees, equipment, regulations, standards	Quasi	Resource planning	Individual and groups of employees or performers

12.3 The Case Study: Planning Logistics Distribution Systems for Small and Medium Businesses in Ciudad Obregon, Sonora, Mexico

To the new conditions of competitiveness in today's world, proper management of supply chain and logistics, play a very important role for companies who export or produce for the local market, large or small (Secretaría de Economía 2011). According to the Council of Supply Chain Management Professionals (2010) logistics is defined as "the process of planning, implementation and control, efficiently and effectively, the flow and storage of goods and services, as well as related information, from the point of origin to the point of consumption in order to meet the requirements of customers". Since logistics directly affects the client-company relationship, it becomes a key factor in business competitiveness, to improve the levels of service to the customer, the quality of service, reliability, flexibility, response time and costs. According to Schwab (2013) the Mexico competitiveness index is located in 53th place; in regards to logistics performance around the world, in 2012 Mexico is ranked 47 of 155 countries with an evaluation, in: level of efficiency in the process of customs, clearance in customs, and the carried out by border agencies, quality of the infrastructure of transport and information technologies in the logistics field and, ease and ability to reach an affordable agreement in regards to international shipments (practice of foreign trade in terms of cost and feasibility in transport).

The Ministry of economy mentions that the efficiency of logistics is an element that is becoming more important in the field of global competitiveness. Other factors that determine the logistics performance are the quality of the infrastructure, the environment, business, reliability of the trading system and supply chain. According to an estimate made by the company AT Kearney (cited by the Ministry of economy 2008), in Mexico, the company's logistics costs represent an average of 12.6 % of sales; 40 % of this cost corresponds to the cost of transportation, while the remaining 60 % inventory, order processing, storage and transportation management planning provide it. The areas of warehousing, inventory and transportation management correspond to the costs associated with the distribution system.

A distribution system is part of the comprehensive logistics system and is defined, according to Kotler (2001) as the process of plan, implement and control the physical flow of materials, finished products and information from point of origin to point of consumption to meet the needs of customers, and for profits.

Competitiveness includes several factors, it is necessary to reduce transaction costs and improve the mobilization of goods times, ensuring that they are in the place and moment as well as shape and right cost; this in order to develop an economy with greater opportunities for attracting and retaining investment. According to the INEGI, in Mexico the 98 % of enterprises are small or medium-sized. Small business are the ones that are composed of a maximum of 50 employees in the industrial sector as well as in the service sector, and 30 for the trade sector; while in the median number of employees is between 51 and 250 for the sectors of Commerce and industry and from 31 to 100 for the service sector. According to Thompson (2007), "small business are an independent entity, created to be cost-effective that are not predominant in the industry to which it belongs, whose annual sales values does not exceed a certain limit, and the number of people that it does not exceed a certain threshold, and as a whole company, has aspirations, goals, achievements, material goods, technical and financial capabilities" "everything which allows you to engage in the production, processing and/or provision of services to meet certain needs and desires that exist in the society". According to the Multisectoral Investment Bank, medium-sized companies are that economic unity with the opportunity to develop their competitiveness on the basis of improving its organization and processes, as well as improve their business skills.

The following table shows some of the advantages and disadvantages of small and medium-sized enterprises that mention Díaz (2008) and Thompson (2007). In Table 12.2 you can see that the disadvantages of both types of company are due to economic reasons. Only to medium-sized it presents a more complex level than small one.

The President of the Association of executives in logistics, distribution and traffic, points out that in Mexico, the logistical cost of services of small and medium-sized enterprises (Small and Medium Enterprises) is one of the highest in the world, accounting for 14 % of revenues. In other parts of the world, such as Europe and the United States, the cost of logistics services reaches 11 % of the income of enterprises, which is explained by the level of efficiency that they manage in such activities. In Obregon City, small and medium enterprises in the service sector presents a series of logistical problems, so through this research aims to design technological solutions at the first level to the needs presented in the previous project called "Detection of needs in the logistic system of supply and distribution of small and medium-sized enterprises of Obregon City service sector". This is why the following problems are posed: what are the technological solutions of the logistical supply and distribution system prior to the needs of small and medium-sized companies in the sector service of Obregon City.

12.3.1 Objective

To design technological solutions to first level for the logistic system of supply and distribution of small and medium-sized companies in the sector service of Obregon City, to generate strategies that have an impact on competitiveness and enabling contribution to best practices of the logistics management of the companies.

Table 12.2 Advantages and disadvantages of small and medium enterprises

	Small	Medium-sized
Advantages	Economically more innovative, more apt to respond to the changing demands of the consumer, willing to create opportunities	They ensure the labour market through the decentralization of the labour force
	Ability to make alliances and partnerships	They have a major socio-economic effect, since it allows the concentration of income and productive capacity from a small number of companies to one higher number
	It acts as an entry point to the economy of new workers	Reduce social relationships closer personal terms between the employer and the employee
	It provides satisfaction and autonomy of work	They present greater technological adaptability to lower cost of infrastructure
Disadvantages	Less bargaining power	Lacks of adequate financing for the capital or work as a result of the difficulty of access to the financial market
	Less access to financing	Little attractive size for the financial sectors
	It implies a drastic change both in the economic way of life	Lack of level of qualification of the labour force occupied
	They usually close their doors more often than large companies, due to the lack of financial resources, technical capacity or sufficient customers	Difficulties to develop research plans

Source Personal communication

12.3.2 Procedure

After analyzing the—gaps in results–of the systems of supply and distribution of small and medium enterprises in the service sector, there were proposed the following steps to achieve the objective of the research, based on Porter (1994) and Ballou (2004), they are presented.

12.3.2.1 Analysis of Information Concerning Areas of Opportunity in the System of Supply and Distribution of Small and Medium Enterprises in the Sector Service

At this point, the results of the application of the draft instrument were used: detection of requirements of the logistic system of supply and distribution of the small and medium-sized companies in the sector service of Obregon City, Sonora.

The instrument was divided in two products of supply system and distribution with eight categories: selection of the supplier, development of the supplier, acquisition, storage, material handling, transportation, customer service and reverse logistics. The first three belong to the supply system and the remaining five to the distribution system. The results are presented in the Table 12.3.

12.3.2.2 Analysis of the Logistics System with Greater Number of Areas for Improvement of Small and Medium Enterprises in the Sector Service

Distribution system consists of five categories: storage, handling, materials, transportation, reverse logistics and customer service. While these categories, which the presented areas for improvement were materials, transportation and reverse logistics management, the system will always be composed of five categories before mentioned, so it is necessary to contemplate all the solution. The diagram of Ishikawa, showing in a relational manner the symptoms and the problem of logistics distribution system is shown in Fig. 12.4.

All of these symptoms lead to the problematic situation, which was defined as: not complying with the delivery of the product or service in accordance with the requirements of the client and, with an affordable cost. The analysis of relational map symptoms were adjusted so that they could be used as indicators and similarly, referencing Ballou (2004) by obtaining the following symptoms presented in the following Table 12.4. These indicators were to follow up on the actions that the Organization, in order to align the strategic structure of the organization in regard to the logistics distribution system.

12.3.2.3 Alignment of the Strategic Structure of the Organization in Regard to the Logistic Sector Distribution System Service

The Method of Hoshin Kanri, an effective tool that makes it easy to identify critical targets, evaluate restrictions, establish performance measurements, develop implementation plans and conduct to periodic review meetings used to carry out the alignment. The format presented below was used to carry out schematically the aspects above. The alignment of the indicators of the logistics distribution system with the Organization's strategy will help to improve the system of distribution of the Organization, without deviating from the same approach. Table 12.5 shows an example of Hoshin Kanri.

In the example above, you can see how we defined the guidelines of the Organization in relation to the categories of warehousing and customer service established for the distribution system, and with each of these respective indicators, as well as the strategy to be followed with the way in which is measured (indicator) and responsible for carrying it out. Key activities for the efficient implementation of these strategies are also listed.

Table 12.3 Areas for improvement by category of small and medium enterprises in the service sector

Category	Size of firms	Specification of improvement area
Material handling	Small	Actions for improvement of container and packaging
	Medium	Actions for improvement of container and packaging
Transport	Small	Costs related to the transportation of its products
		Planning of shipments of their products made on the basis of demand
		Percentage of goods damaged during transport
	Medium	Capacity utilization in its units of transportation (weight and volume)
		Costs associated with the transport of the products
		Planning of shipments of the products made on the basis of the demand of the customers
		Percentage of products damaged during the transportation process
		Control of products during transport (tracing, tracking, etc.)
Reverse logistics	Small	The number of products returned by the customer because of damage
		The number of products returned by the customer due to expiration
		The number of products returned by the customer because of warranty claim
		The reuse of products/materials that are returned
		The recycling of products/materials that are returned
	Medium	The number of products returned by the customer because of damage
		The number of products returned by the customer due to expiration
		The number of products returned by the customer because of warranty claim
		The reuse of products/materials that are returned
		The recycling of products/materials that are returned by the customer

Source Personal communication

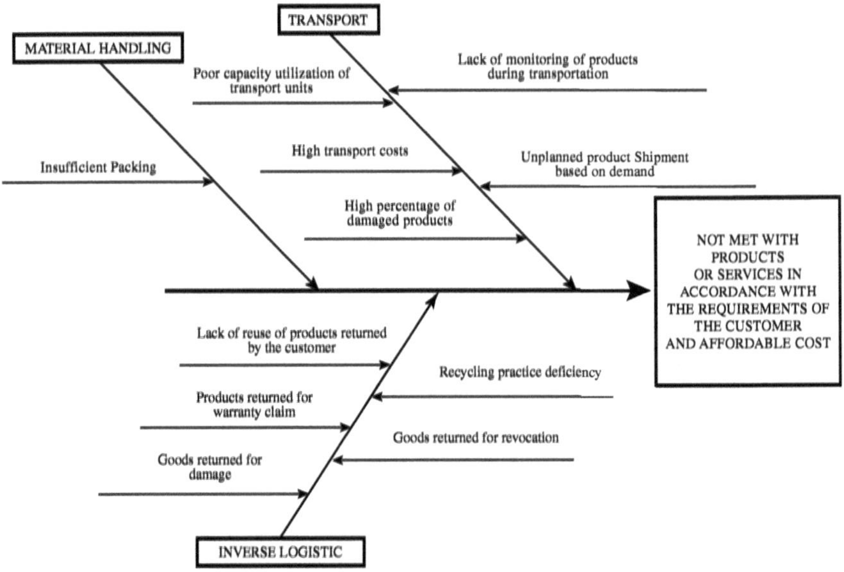

Fig. 12.4 Ishikawa diagram of distribution system. *Source* Personal communication

Table 12.4 Symptoms and indicators

Symptoms	Indicators
High percentage of damaged products	Percentage of damaged products
High transportation costs	Transportation cost
Improper use of the capacity of transportation units	Percentage of utilization of capacity of transportation units
Delivery of the product does not meet the requirements of the customer	Number of customers satisfied with the delivery of the product or service
Deficiency in the container and packaging products	Quantity of containers or defective packaging
Products returned for warranty claim	Number of products returned under warranty
Products returned for damages	Number of products returned by damage
Products returned for revocation	Number of products returned by expiration

Source Personal communication

12.3.2.4 Analysis of the Regulations Applicable to the Logistics System Under Study of Small and Medium Enterprises in the Sector Service

We analyzed the Mexican official standards applicable to each of the categories of logistics distribution system (storage, handling of materials, transportation, reverse logistics and customer service). Table 12.6 was the product obtained by this analysis.

Table 12.5 Example Hoshin Kanri

Management	Management planning	Performance				
Guidelines	Indicators	Strategies	Indicators	Responsible	Key activities	Leader
Warehousing	Number of damaged or expired products	Make a plan for proper stock rotation	Percentage of expired products	Company	1. Order Management 2. Inventory Management 3. Receiving and checking of goods	
Customer service	Percentage completion of deliveries	Surveys that measure customer satisfaction	Percentage of satisfied customers	Company	1. Identify customer needs 2. Apply questionnaires to assess customer satisfaction 3. Report the results	

Source Personal communication

You can see that for warehousing and material handling applies the same standard relating to safety in the warehouse and material handling. For transport, rules apply on carrying capacity and transit, in the category of service; the customer presents a standard related to hygiene and sanitation in the preparation of food offered in fixed establishments. On the other hand, in the category of reverse logistics there are laws regarding to the treatment and final disposal of waste.

12.3.2.5 Determination of the Design Requirements of the Logistical System Under Study of the Sector Service

Store requirements. Bowersox (2008) and with respect to the NOM-006-STPS-2000, certain requirements were to establish a store in the small and medium enterprises of Obregon City in the service sector. There are: halls, palletizing, ceilings, boxes or crates, shelves, selective rack double, pallets, demarcations, and signs.

Cooling Chambers. The products must be kept at a temperature of 7° C or less, with visible thermometer or devices for temperature that record and in good condition. Any loading, platform and shelf used to store them should be clean and 15 cm above the floor level, avoid contact with the ceiling and allow the flow of air between the products.

Freezing Chambers. They must be kept at a temperature of—18° C or lower, with visible thermometer or devices for recording temperature, in good conditions; and must allow the flow of air between the products.

Dry Store. The products must be kept at room temperature. They should be located in a closed area, dry, ventilated and clean. FIFO system should be applied in all food stores because they are cooling, freezing or drying; to ensure the organoleptic features of all products consumed. Storage of insecticides must be in a bounded place and separated from any area of handling or storage of food, and

Table 12.6 Regulations applicable to the logistics distribution system for the service sector

Category	Standard	Description
Warehouse	NOM-006-STPS-2000	Handling and storage of material-conditions and safety procedures
	NOM-093-SSA1-1994	Concerning stores in dry, cold room and freezing Chambers
Material handling	NOM-006-STPS-2000	Handling and storage of material-conditions and safety procedures
Transport	PROY-NOM-071-SCT-2-2000	Transport auto shipping from load–vehicle up to 4 tons of gross weight Vehicular-characteristics and technical specifications and safety ground-service
	Regulation of federal motor transport and ancillary services	Employed drivers must have the corresponding licenses
Customer service	NOM-093-SSA1-1994	Goods and services practices of hygiene and health in the preparation of food offered in fixed establishments
Reverse logistics	Law Federal metrology and standardization	"When the goods or services subject to compliance with certain Mexican official standard, does not meet the relevant specifications, the competent authority will immediately prohibit marketing, immobilizing products, even ir so far they are conditioned, reprocessed, repaired or replaced…"
	General law for the prevention and integral management of waste	Concerning the integrated waste management, as well as collection, transportation, treatment and final disposal

Source Catalogue of Mexican official standards of the Ministry of economy must have a strict control for distribution and use. They must be tagged or labeled in such a way to report on its toxicity and employment.

Requirements of Material Handling. To analyze for the category of material handling requirements, the management of equipment for materials and their specifications. Through Table 12.7, it is shown that there are several teams that can be used for the handling of materials. Referencing the standard NOM-006-STPS-2000 items that should be considered for safety and hygiene in the handling of the materials.

Transport Requirements. Based on the NOM-071-SCT-2-2000 it is concentrated the following requirements for the service sector transportation system: Maximum limits of weights and dimensions for transport units. Maximum weight of 4 tons. Maximum width of 2.40 m, this width maximum does not include mirrors of attachment and other attachments to the cargo securement. These accessories must not protrude more than 20 cm to each side of the vehicle. Maximum height of 2.70 m, and a maximum total length of 6.30 m. The compartment load must be separated from the driver's cab, quite closed and leak-proof, wear-resistant, waterproof and washable. Both conducting walls and containers in which the product is transported must be resistant, washable, non-toxic, absorbent non degradable material. Wood or plastic bags are not allowed. To deal with dry box units: must be double rolled, count with flip-up rear door, must have compartments and dry with crest. Drivers employed for the management of transportation units must have the appropriate driver license as established by the Federal trucking regulations.

Customer Service Requirements. Points out that the main requirements for the service to the client are those shown in Table 12.8.

On the other hand, an important technological tool in terms of service to the client is the CRM or Customer. Relationship Management, which is a powerful strategy to convince the customer and to be successful in the market because it is the only strategy that takes the customer as an active participant and it is not conventionally placed it on the pedestal, but involves it in the interactive process to stimulate their ability to express themselves and choose their own decisions Table 12.9.

Reverse logistic requirements. According to Bastos (2007) the basic requirements for a good program of reverse logistics development are the following: Mechanisms of return of excessive inventory. Customer returns. Meanwhile, Antun (2004) mentions ten strategic recommendations for the development and successful implementation of a reverse logistics, being only seven recommendations, which apply to the sector service: Develop a program that reduces the procurement and purchase of new materials, seeking to use recycled or reusable products. For reverse logistics best practices it is advisable to consult the Federal law on metrology and standardization, as well as the General Law for the prevention and integral management of waste.

Table 12.7 Material handling equipment

Equipment	Description	Features
Manual load	It is any operation of a freight transport by one or more operators, such as lifting, positioning, push, traction or displacement	Maximum manual load that the workers rise up is 50 kg; for minors is 35 kg, and for women is 20 kg. Women should not be this activity in pregnant and during the first 10 weeks post-partum
Forklift	Forklift trucks are mobile devices similar to cranes that perform work for transportation of unit loading. It allows that a person can lift and accurately place larger and heavier loads with little effort. It is purpose is to move, push, pull, and lift a load and then stack it or place it in a buffer shelf (in rows).	Access to very high levels Access to platforms placed in two rows Practical to narrow corridors It has the capacity for the handling of heavy loads. The most common are from 450 to 1800 kg and exceeding 4500 kg capacity They are less inexpensive tours of more than 90 m. Because its maximum speed is usually between 8 and 16 km/h
Trucks	It is widely used to load boxes or other things that can stack, while the almost universal wheel has greater maneuverability in small spaces	Useful for the handling of small orders They do not require fuel Low maintenance cost It has a capacity of 350 kg load
Hydraulic forklift	They are used for movements in workstation and sometimes in short distances	Their capacity are usually from 2500 to 4500 pounds
Trucks to supply orders	These trucks have a raised platform in front of the tower, from which the forklift and pallet can be handled	Their typical use are to take partial loads in narrow aisles and heights of up to 7.2 m, which allows significant savings in manpower and space

Source Personal communication

Table 12.8 Requirements of service to the customer

Customer service
Level of compliance of deliveries to customers
Delivery of requested product quantity
Delivery of the product in the requested time
The service index
Percentage of orders delivered on time, place and manner requested
Percentage of orders that fully satisfy
Percentage of returns from the customer for damages, errors and delivery deadline
Total satisfaction of the client
Quality of service
Continuous improvement of the process
Goods without transport damages
Safety of personnel involved in the delivery of the product

Source Personal communication

Table 12.9 Analysis CATWDA service sector

	Warehouse	Material handling	Transport	Customer service	Reverse logistics
C	Transport	Store, transport	Who requires the product (consumer)	User or consumer	The company store
A	Responsible for warehouse, loading and unloading	Responsible for material handling	Shipping manager, driver	Personnel of the company	Customer, staff of transport (driver, fleet manager)
T	1. Order management	1. Handle the load of a place at a specific point using the material in a proper way of handling equipment	1. Delivery planning	1. Identify the needs of customers	1. Procurement and purchasing
	2. Inventory management		2. Selection of the Middle	2. Customer service	2. Waste management
	3. Receipt of goods		3. Loading and unloading	3. Measure customer satisfaction	3. Return to origins
	4. Storage		4. Transfer of product	4. Report on the results	4. Administrative processes
	5. Audit		5. Product delivery		5. Recovery
	6. Expedition		6. Maintenance		
W	Keep the product value for disposal according to the customer requirements	Meet loads of efficiently handling according to material handling principles ensuring the properties and physical conditions of the product to discuss its value	Deliver the product/service fulfilling the requirements of the customer (quantity and time)	Comply with the requirements of the customer in relation to your product and service (quantity and service)	Return the materials and products from the point of consumption to the point of origin in order to retrieve value
D	Warehouseman	Warehouse manager	Shipping manager	Company	Company manager
A	Cost of services, failure of suppliers, environmental issues, regulations and regulatory policies	Climate, temperature, pests	Logistics infrastructure cost of fuel and tolls (economic conditions), transit, transportation, weather conditions and competition regulation laws	Competition, demand	Regulations, laws of traffic, policies, return

Source Personal communication

Fig. 12.5 Level 1. SCOR model. Scope of the system under study. *Source* Own elaboration

12.3.2.6 Design at the First Level of the Technological Solution for the Logistics System Under Study of the Service Sector

Once the requirements for each of the categories of logistics distribution system of small and medium-sized are established, companies in the sector service of Obregon City, a CATWDA analysis was developed to identify the relationships between the categories and then it was moved to make the first design of the technological solution of distribution-based logistics system 9.0 SCOR (Supply Chain Operations Reference Model) model. This took place in the first design of the logistical system under study of the sector service as the last step of the established method. The reference model (SCOR Model) version 9.0 of supply chain operations is used as an adapted key tool that is spreading only to level 3 to introduce and configure all elements of processes or categories that make up the logistics distribution system, since it provides a single framework that links indicators, key processes and best practices into a unified structure to improve the efficiency of the logistics system.

Level 1: *Definition of the process*. For process design, the scope of the system under study was defined in the first level of the SCOR model as the logistics distribution system for small and medium-sized enterprises in the sector service of Obregon City, which can be seen in Fig. 12.5.

Level 2: *Categories in the process*. At this setting, the categories that are part of the logistics distribution system which are warehouse, handling of materials, transportation, reverse logistics and customer service, are permanently imprinted, as shown in the following Fig. 12.6.

Level 3: *Elements of the process*. This third level details the key elements for each category comprising the system of distribution of Small and Medium Enterprises in the sector of service, defining the process, information about inputs and outputs, indicators and best practices for each category; the information collected was based on a bibliographic research and aided to elements defined in the CATWDA analysis presented above. Table 12.10 shows the indicators applicable to the logistics distribution system for Small and Medium Enterprises in the service sector, classified by category and establishing the respective equations for calculating each indicator.

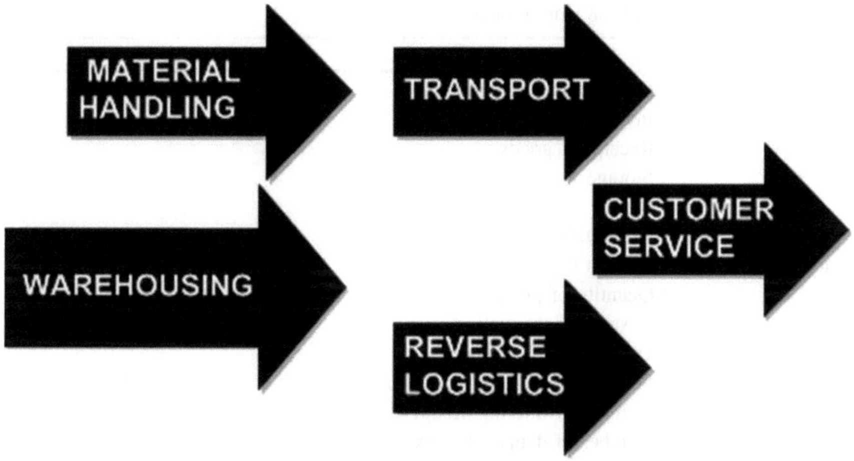

Fig. 12.6 Level 2. Categories of the process of the system under study. *Source* Personal communication

Table 12.10 Indicators for the categories of distribution system

Category	Indicator
Warehouse	Percentage of inventory turnovers
	Number of missing products
	Quantity of product damaged or expired
Material handling	Percentage of damaged products by material handling
	Delay of transferring material handling activities
Transport	Percentage of capacity utilization of the transport unit
	Transportation costs
	Percentage of damaged products during transport
	Percentage of products shipped to the customer
Customer service	Percentage of compliance with delivery on time
	Delivery time of the product
	Number of complaints received
	Safety of those using the delivered products
Reverse logistics	Amount of products returned for revocation
	Quantity of products returned by confusion of orders

Source Personal communication

Then the key elements for each process of the five categories that make up the distribution system were defined: warehouse, handling of materials, transport, the customer service and reverse logistics, based in Saints Soret (2006) and Sainz (2003).

Table 12.11 Elements of the store process

Warehouse	
Processes	Order management
	Inventory management
	Receipt of goods
	Storage
	Audit
	Expedition
Entry information	Product type
	Quantity of products
Output	Customer information
	Quantity of products
Indicators	Percentage of inventory turnovers
	Number of missing products
	Number of damaged or expired products
	Number of persons whit warehouse job (MEGA)
Best practices	See the store as an active element and as a channel for distribution
	Flexible design of warehouse (layout)
	Criteria for classification of materials
	Using warehouse management systems (Warehouse Management Systems) and MRP
	Outsourcing of storage features
	Using the technique of cross docking

Source Personal communication

Table 12.12 Elements of the process of material handling

Material handling	
Processes	1 Handle the load of a place at a specific point using the handling of materials in a proper way
Entry information	Type of products to move
	Amount of products to move
	Location of the products
Output	Quantity and type of products taken and moved
	Number of consolidated products
Indicators	Percentage of damaged products by material handling
	Delay of transferring material handling activities
	Number of persons without injuries (Mega)
Best practices	Inputs and outputs and exact positions of product registration

Source Personal communication

Whereas previously the indicators proposed in the five risk categorization that make up the distribution system: warehouse, handling of materials, transport, the customer service and reverse-logistics, some Mega indicators are included as a contribution of this purpose according to Kaufman and Guerra-López (2013).

Table 12.13 Elements of the transport process

Transport	
Processes	1. Delivery planning 2. Selection of the medium 3. Loading and unloading 4. Transfer of products 5. Product delivery 6. Maintenance
Entry information	Customer data Number of orders
Output	Delivery of the order (average delivery time, customer's requirements).
Indicators	Percentage of the capacity utilization of the transport unit Transportation costs Percentage of products shipped to customer Percentage of products damaged in transit Level control of shipments Reduce the CO2 effects (MEGA)
Best practices	Use of GPS devices Logistics outsourcing (3PL) Efficient planning of routes

Source Personal communication

Table 12.14 Elements of the customer service process

Customer service	
processes	1. Identify the needs of customers 2. Customer service 3. Measure customer satisfaction 4. Report on the results
Entry information	Requirements or customer complaint
Output	Claims for warranty or expiration Returns Customer dissatisfaction
Indicators	Percentage of time delivery compliance Delivery time of orders The number of claims received Self-sufficiency (Mega)
Best practices	Surveys that measure customer satisfaction Plans of action for customer satisfaction Procedures of customer loyalty Establishment of a customer call center

Source Personal communication

Table 12.15 Elements of the reverse logistics process

Reverse logistics	
Processes	1. Procurement and purchasing
	2. Waste management
	3. Return to origins
	4. Administrative processes
	5. Recovery
Entry information	Claims for damages, warranties or expiration
	Customer dissatisfaction
Output	Defective product
Indicators	Amount of products for revocation
	Number of products returned by not having anyone at the time of home delivery
	Number of products returned by confusion of orders
	Reducing pollutants (Mega)
Best practices	Register order or customer order requirements
	Prior notification of shipments

12.4 Conclusions

Every organization, whether or not it recognizes it, is a means to societal ends. As performance improvement expert Dale Brethower cautions us, *If you are not adding value to society*, you are probably subtracting value. Based on the above consideration requires different approaches to developing solutions in this regard to consider various methodologies provides important information to guide organizations. Derived from the first stage of development of the case have been considered basically four methodological approaches that allowed differences on each of the approaches and perspectives. In this way it was possible to consider in the analysis solution of indicators ranging from resources, processes, products and the impact it causes in society Tables 12.11, 12.12, 12.13, 12.14 and 12.15.

As a result of the application of the case study has derived from a statistical analysis that provided the results of that small and medium-sized enterprises of City Obregon, logistics distribution system service sector presented areas of improvements in their categories of transportation, reverse logistics and material handling, however, for the realization of the design of the solution were referred to the five categories: warehouse, handling of materials, transportation, reverse logistics and customer service. For the first level design was very helpful reference SCOR model 9.0, this allowed analyze, represent and set the distribution system and allowed to visualize the main processes of the categories studied comprehensively.

The proposed solution requires complying with regulations, requirements, best practices, use of technology, as well as controlling and measuring the performance of its activities for best results of the distribution system that increase the efficiency of companyies' operations and having an impact on competitiveness.

Finally, it is very important to include societal value to organizational value, improvements in effectiveness and efficiency through the value chain, this could be demonstrate using and implementing indicators in the different perspectives according to the vision and mission of organizations.

References

Antun, J. (2004). *Logistica inversa*. México: Universidad Nacional Autónoma de México.

Ballou, H. R. (2004). *Logística, Administración de la cadena de suministro*. México: Pearson Educación.

Bastos, B. A. (2007). *Distribución logística y comercial. La logística en la empresa*. España: Ideaspropias.

Bowersox, D. (2008). *Administración y logística, en la cadena de suministro*. México: McGrawHill.

Calderon, J. L., & Lario F. C. (2005). Análisis del modelo SCOR para la Gestión de la Cadena de Suministro. Informe publicado en el *IX Congreso de Ingeniería de Organización Gijón, España.*

Chiavenato, A. (1998). *Administracion de recursos humanos*. México: McGraw Hill.

Díaz, M. (2008). La Pequeña Empresa. Obtenido el 8 de abril de 2010. Retrieved from http://www.monografias.com/trabajos10/micro/micro.shtml

Grados, J., Beutelspacher, O., & Castro, M. A. (2002). *Calificación de Meritos. Evaluacion de competencias laborales*. México: Editorial Trillas.

Kaplan, R., & Norton, D. (2001). *La Organización focalizada en la estrategia. Cómo implementar el Balanced Scorecard*. España: Editorial Gestion 2000.

Kaufman, R., Forbes, R., & Parks, J. (2005) How to make the balanced score card really balanced. In M. Silberman & P. Philips (Eds.), *The 2005 ASTD Team & Organization Development Sourcebook*. Alexandria, VA: ASTD Press.

Kaplan, R., & Norton, D. (2005). El Balanced Scorecard: Mediciones que impulsan el desempeño. *Harvard Business Review*, *83*, 3–10.

Kaufman, R. (2000). *Mega planning. Practical tool for organizational succes*. Thousand Oaks: Sage Publications Inc.

Kaufman, R. (2011). *A manager's pocket guide to strategic thinking and planning*. Amherst, MA: HRD Press Inc.

Kaufman, R., & Guerra-Lopez, I. (2013). *Needs Assessment* (Primera ed.). Estados Unidos de América: ASTD.

Kotler, P., & Armstrong, G. (2001). *Marketing*. México: Pearson Educación.

Lagarda-Leyva, E. A. (2012). *Propuesta metodológica de planeación estratégica bajo un enfoque de modelación dinámica y por escenarios*. Sonora, México: Ciudad Obregón.

Porter, E. (1994). *Estrategia competitiva: Técnicas para el análisis de los sectores industriales y de la competencia*. México: Editorial Continental.

Rodríguez, D. (2009). *Logística para la exportación de productos agrícolas, frescos y procesados*. IICA: Costa Rica.

Sainz, J. (2003). *La distribución comercial: Opciones estratégicas*. ESIC: España.

Secretaría de Economía (2011). *Evaluación del desempeño logístico de las cadenas de suministro en México*. Presentación de Resultados. Retrieved from http://www.elogistica.economia.gob.mx/swb/work/models/elogistica/Resource/3/1/images/evolucion2011.pdf

Schwab, K. (2013). *The Global Competitiveness Report*, 2013–2014. Switzerland: WEF.

Socconini, L. (2009). *Lean Manufacturing, paso a paso*. México: Norma.
Soret, I. (2006). *Logística y marketing para la distribución comercial*. España: ESIC.
SCM Operations (2011). Retrieved from http://www.scm-operations.com/2011/01/scor-model-supply-chain-improvement.html
Thompson, I. (2007). *La pequeña Empresa*. Obtenido el 11 de febrero de 2010. Retrieved from http://www.promonegocios.net/empresa/pequena-empresa.html

Part III
Human Factors and Ergonomics in Lean Manufacturing

Chapter 13
Human Factors and Ergonomics for Lean Manufacturing Applications

Arnulfo Aurelio Naranjo-Flores and Ernesto Ramírez-Cárdenas

Abstract Ergonomic is a scientific discipline concerned with the understanding of interactions among humans and other elements of a system, and the profession that applies theory, principles, data and methods to design in order to optimize human well-being and overall system performance. One of the fields of application of this discipline is the prevention of occupational—risks in the workplace, preventing the presentation of musculoskeletal disorders (MSD) and thereby establishing the conditions for increasing productivity in organizations. This is precisely the reason that has led some companies to focus in continuous improvement process through the use of Lean Manufacturing which refers to the creation of value stream. However, it is necessary to have a methodology of intervention focused on the correct application of both concepts under the premise of achieving results without neglecting the human factor. In this context, the present work discloses the continuous improvement cycle adapted from various authors, explaining the steps and activities to be performed under a Lean Ergonomics perspective. This implies a new analysis approach in problem solving and/or elimination of or waste to generate value based on human factors in the pursuit of operational excellence without neglecting the health of workers. The chapter concludes by emphasizing the importance of considering the human above any method which will ensure success in organizations.

Keywords Ergonomics · Occupational health · Lean manufacturing

---

A. A. Naranjo-Flores (✉) · E. Ramírez-Cárdenas
Department of Industrial Engineering, Sonora Institute of Technology, 5 de Febrero 818, Col. Centro, Obregón, SON, Mexico
e-mail: arnulfo.naranjo@itson.edu.mx

J. L. García-Alcaraz et al. (eds.), *Lean Manufacturing in the Developing World*,
DOI: 10.1007/978-3-319-04951-9_13,

13.1 What Does Ergonomics Mean?

Ergonomics in recent years has generated interest from a number of specialists in several branches of science: Engineering, Medicine, Psychology, Sociology, Architecture, Design, among others. The scientific application of knowledge that it brings, has been revealed as an important element in reducing accidents and injuries, increasing productivity and quality of life of workers (Mondelo et al. 2000). Both aspects are achieved by reducing their exposure to risk, allowing healthy growth and development of the worker, in combination with increasing organizational efficiency. Besides impacting the welfare of the worker, Álvarez (2008), establish that ergonomics can and should generate profits directly into increasing productivity and competitiveness of enterprises essential to a global market.

According to Helander (1995), the main purpose of ergonomics is to design systems that are productive, safe, comfortable and enjoyable. It states that the design activities are to relate systems, jobs, products and environments to the skills and physical and mental limitations of individuals. Systems must be designed to enable people to use them intuitively without requiring special training or education. In that sense, Ramirez (1999) states that ergonomics results in rational conscience of the items, means and procedures involved in the work, with efficient and effective results that can be converted into increasing profitability, lower costs, greater motivation of staff, best quality, and excellent working environment.

The term ergonomics comes from the Greek Ergo (work) and Nomos (laws). It was first used in 1857 by Wojciech Jastrzebowski in his study "Ergonomics essays or science of work", based on the objective laws of science on the nature, in which, a model of labor human activity, was wanted to be built Karwowski and Marras (1999).

On the other hand, Farrer et al. (1994) define ergonomics as the multidisciplinary applied science that aims to the suitability of the products, systems and artificial environments to the characteristics, limitations, and needs of its users to optimize their effectiveness, safety and comfort. For Salvendy (1997) ergonomics is the scientific discipline concerned with the interaction between humans and artifacts and the design of systems where the man is involved. This is confirmed by Niebel and Freivalds (2001) as they mention that ergonomics is the design of the workplace, tools, equipment and the environment so as to adjust to the human operator. On the other hand, Kroemer et al. (2001) define it as the application of scientific principles, methods, and data from a variety of disciplines for the development of systems engineering in which the person plays a significant role.

Over the last 50 years, the term Ergonomics, has been used synonymously with Human Factors and Ergonomics (denoted HFE), and evolves a unique and independent discipline that focuses on the nature of human–artifact interactions, viewed from the unified perspective of the science, engineering, design, technology, and management of human-compatible systems, including a variety of natural and artificial products, processes, and living environments (Karwowski 2005).

HFE describes the interaction between the operator and the demands of the task being performed, and both are concerned with reducing unnecessary stress in these interactions. Ergonomics, however, has been traditionally focused on how work affects people. This approach includes studies of, physiological responses to physically demanding work, environmental stressors such as heat, noise, and illumination; complex psychomotor assembly and visual-monitoring tasks. It has been emphasized on methods to reduce fatigue by designing tasks so that they can adjust within people's work capacities. In contrast, the field of human factors, as practiced in the United States, has tradicionally been more interested in the human–machine interface, or human engineering. It has focused on people's behavior as they interact with equipment and their environment, as well as on size and strength of capabilities relative to product and equipment design. The emphasis of human factors is often on designs that reduce the potential for human error (Chengalur et al. 2004).

The most accepted definition of HFE is provided by the International Ergonomics Association (IEA), which defines it as scientific discipline concerned with the understanding of interactions among humans and other elements of a system, and the profession that Applies theory, principles, data and methods to design in order to optimize human well-being and overall system performance (IEA 2013). The Human Factors and Ergonomics Society has adopted this definition of ergonomics.

13.1.1 Goal of Ergonomics

The objective of ergonomics is to improve "quality of life" user interaction with a machine—tool, and at the same time it can be used in a domestic area, in both cases this objective focuses on reducing the risk of error, and therefore the increase in consumer welfare, reduction of injuries and occupational diseases, lowers costs for disabilities of workers, improving the quality of life and productivity both individually and collectively (Mondelo et al. 2000).

According Chapanis (1996) Ergonomics objectives are:

1. Basic operational objectives
   a. Reduce errors
   b. Increase safety
   c. Improve system performance
2. Objectives related to reliability, maintenance, availability and integrated logistics support
   a. Increase reliability
   b. Improve maintenance
   c. Reduce staffing requirements
   d. Reduce training requirements

3. Objectives affecting users and operators
   a. Improve work environment
   b. Reduce fatigue and physical stress
   c. Increasing human comfort
   d. Reduce boredom and monotony
   e. Increase usability
   f. Increasing user acceptance
   g. Increase aesthetic appearance
4. Other objectives
   a. Reduce wasting time and equipment
   b. Increase production economy.

Ergonomics has two goals: safety and productivity. The main purpose of ergonomics is to design systems so that they could be productive, safe and also comfortable and enjoyable. Helander (1995) establishes that the purpose of the design of activities is to relate systems, works, products and environments to the skills and physical and mental limitations of individuals. Systems must be designed to enable people to use them intuitively without requiring special training or education.

Agreeing with the productivity aspect mentioned by the author above, Ramirez (1999) defines it as the result of rational union of the elements, means and procedures involved in the work, all these with efficient and effective results that create translated into a higher profitability, lower costs, higher motivation, better quality and excellent working environment, and overall, productivity is defined as the input–output positive relationship in which ergonomics participates improving that relationship. Likewise Oborne (1992), states that ergonomics takes and integrates data to optimize safety, efficiency and reliability of the performance of the operator, to make the task easier and to increase the sense of comfort.

Therefore it can be concluded that the objective pursued by ergonomics is to identify risk factors, and to propose viable solutions, trying to be feasible financially for the company and operationally effective to the staff.

13.1.2 Scope of Ergonomics

The works of ergonomics is firstly determine the capabilities of the operator and then intend to build a working system that will support these capabilities (Oborne 1992). For preventive ergonomics priorities are: man physical, mental and social characteristics and then after identify these priorities design the work area, in those organizations in which it has not been designed a preventive ergonomics plan and this could be the cause of some problems, there is the alternative of applying corrective ergonomics which consists in the redesign of the work area, taking into

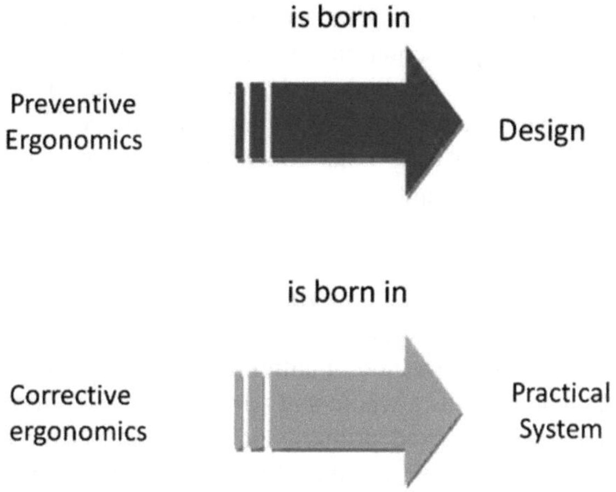

Fig. 13.1 Difference between preventive and corrective ergonomics

account as preventive ergonomics human characteristics, but only if the system has been designed and implemented, as shown in Fig. 13.1.

In opinion of Mondelo et al. (2000) there are three approaches in the scope of ergonomics. A first approach to ergonomics would place it in the position of the study of human beings in their work environment, enabling thinking of ergonomics as a technique to be applied in the project conceptualization phase (design ergonomics or preventive) or a redesign technique for the improvement and optimization (corrective ergonomics). A second view of ergonomics pick up the idea that, in fact, it should be an eminently prescriptive discipline, which should provide to the project managers the limits of the participation of users and thus adapt to artificial achievements human limitations, and finally, a third approach a bit more ambitious than the previous ones, understands this science as an interdisciplinary field of study that discusses the problems of what to plan and how to articulate the sequence of possible user interactions with the product, with services, or even with other users.

13.1.3 Disciplines Related to Ergonomics

The importance of ergonomics is generated from being a catalyst in the emergence of new disciplines and the use of isolated disciplines that when entering the field of ergonomics, are transformed to have new applications or as well contribute to the rapid growth of others.

As it is known ergonomics development is based on the knowledge of other sciences and disciplines. To mention some of them are industrial psychology, which provides a first starting point to ergonomics, the knowledge of man, based on the analysis of personality and in terms of their intrinsic components:

temperament, character, conduct, motivations and integration of oneself, for rational implementation activities (Ramirez 1999). Medicine that aims to promote and maintain the highest degree of physical, mental and social workers in all occupations; prevent any damage to their health caused by working conditions, protect against the risks arising from the presence of harmful agents to health, placing and maintenance of worker to a suitable employment according to physiological and psychological aptitude, to sum up, adapt the work to man and each man to his work (Chapanis 1996). The functional anatomy intertwines with ergonomics through morphologic, biochemical and mental condition and the analysis of the organs and systems of the human organism to which there is employed the method of anthropometric research, aiming to describe and measure human body as a whole and various parts (Ramirez 1999). *Timing and movement technique* referred primarily to increased performance by measuring and minimizing the time it takes to perform various operations (movements) Niebel and Freivalds (2001) Industrial Security comprising a series of legal norms, socioeconomics, technical, hygienic, and organizational that ensures the integrity of the individual in particular, and the facilities in general, in the vast panorama of the workforce. Ergonomics relies on the contributions of industrial safety complements and helps to develop them (Ramirez 1999). Sociology serves ergonomics investigating the problem of the adaptation of work handling some variables, such as age, level of education, income, housing, family environment, transportation and routes (Chapanis 1996). Anthropometry science is also important since it studies the proportions and measurements from different parts of the human body, such as arms, length, weight, height of men, height, ratio of the length of the legs and trunk, taking into account the diversity of individual measures against the average environment, analyzing, the operation of various levers muscle, and investigates the forces that can be applied depending on the position of different muscle groups (Mondelo 2000).

Diverse areas use ergonomics as a basic component of their knowledge or their working procedures. Development from basic sciences to several disciplines applied to ergonomics, and the use of knowledge for ergonomics in specific areas, is shown by Kroemer et al. (2001) in Fig. 13.2.

13.2 Musculoskeletal Disorders (MSDs)

One important field of application of ergonomics is the prevention of occupational hazards, which have developed different ergonomic assessment methods to prevent and detect occupational hazards, especially those related to prevention methods and identification of Musculoskeletal disorders (MSDs) that are injuries or disorders of the muscles, nerves, tendons, joints, cartilage and upper and lower limbs, and injuries that occur in the neck, lower back, aggravated by exertion or prolonged exposure to physical factors such as repetition, force, vibration or awkward posture (Llaneza 2009).

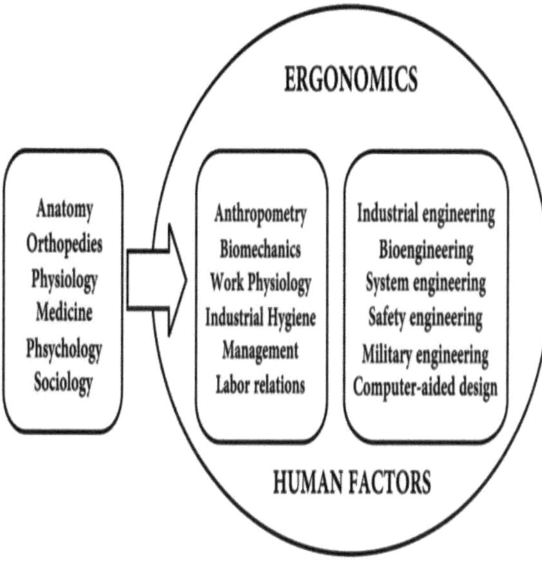

Fig. 13.2 Origins, developments, and applications of ergonomics. (Kroemer et al. 2001)

Symptoms related to the occurrence of musculoskeletal disorders include muscle pain and/or joint pain, tingling, loss of strength and decreased sensitivity. At the onset of disorders consists in three stages: (1) Onset of pain and fatigue during working hours, getting better beyond, during evenings and weekends. (2) Onset of symptoms to the start of the workday, without disappearing at night, disrupting sleep and decreasing the working capacity. (3) Persistence of symptoms during rest, making difficult task execution, even the most trivial. The predominant symptoms are pain associated with inflammation, loss of strength and functional disability decreased or anatomical area affected Secretary of Occupational Health (2008).

According to Occupational Health and Safety Council of Ontario (2007a), the (MSDs) occur in the workplace when demands exceed the capabilities of the person to do the job. There is a strong relationship between exposure to physical factors or risks in the workplace and the development of an MSD. There is also evidence that certain work organization factors are associated with an increased risk of (MSDs). These factors include: (1) perceptions of high job demands and workload, (2) monotonous work tasks (3) perceptions of low job control, (4) lack of clarity about the work, expectations or importance (5) low job satisfaction, and perceived low social support.

The International Labor Organization (ILO) in its report "Preventing Occupational Diseases" states that in 2012 2.34 million accident and illnesses occurred, from this number, 321,000 (14 %) were fatal and about 2,020,000 (86 %) as fatal diseases. The report states that the technological, social and organizational changes recorded in the workplace as a result of rapid globalization we live involve to carries some risks and new challenge, registering an increase of new types of

Fig. 13.3 Carpal tunnel syndrome

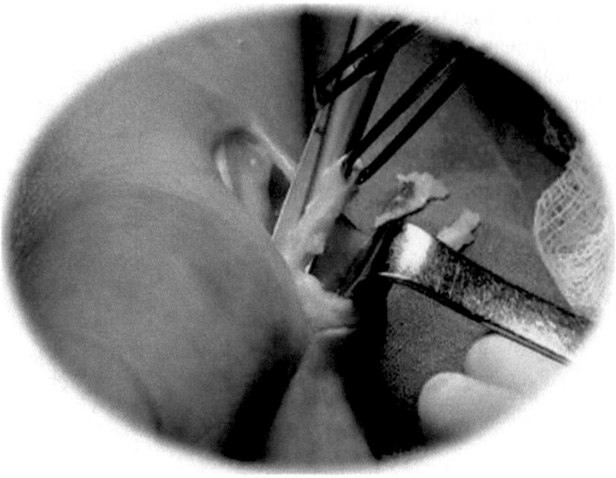

diseases such as (MSDs) or mental disorders, without applying prevention, protection and control measures.

MSDs are the leading cause of absenteeism with 49.9 % of all absences of more than 3 days and permanent disability with 60 % European Commission (2005). In the Republic of Korea, MSDs increased dramatically from 1,634 cases in 2001 to 5,502 in 2010. In Argentina there were 2,013 cases of occupational diseases, being the most common MSDs in 2010.

In 2005, the ILO states that the MSD called carpal tunnel syndrome, represents the 59 % of all occupational diseases registered by the European Occupational Diseases Statistics (European Agency for Safety and Health at Work 2010). MSDs have lately received more attention, mainly those ones that are related to hands and wrists. This kind of MSDs compresses and traps the median nerve as it passes through the wrist and the hand in the carpal tunnel (Fig. 13.3). The median nerve provides the sense of touch in the thumb, index, middle and ring fingers.

Table 13.1 lists those conditions that are most often associated with overuse disorders. Of course, the list is neither complete nor exclusive: New occupational activities occur all the time, such as use of a computer mouse beginning in the late 1900s, and several activities may be part of the same job-for example, use of the mouse and keyboard.

According to Kroemer et al. (2001), the occurrence of health problems related to MSD can be compared to a mountain (Fig. 13.4). These are stages that occur mainly according to the exposure time of the worker. These stages or manifestations ranging from an initial stage represented by a daily accumulation of common cases, such as tugs, fatigue, anxiety and discomfort during or after a long day of work. Going through a next higher level, is represented by occasional movements or posture problems, beyond mere tiredness, regularly accompanied by small pains that go away after a night's rest, to the top of the mountain where the injuries and disorders requiring special medical treatment, such as surgery, and fortunately a

Table 13.1 Example of musculoskeletal disorders. (U.S. Department of Labor, Ergonomics the Study of Work 2000)

Body parts affected	Symptoms	Possible causes	Workers affected	Disease name
Thumbs	Pain at the base of the thumbs	Twisting and gripping	Butchers, house-keepers, seam-stresses, cutlers	De Quervain's disease
Fingers	Difficulty moving finger; snapping and jerking movements	Repeatedly using the index fingers	Meatpackers, poultry workers, carpenters, electronic assemblers	Trigger finger
Shoulders	Pain, stiffness	Working with the hands above the head	Power press operations, welders, painters, assembly line workers	Rolator culf tendinitis
Hands, Wrists	Pain, swelling	Repetitive or forceful hand and whist motions	Core making, poultry processing, meatpacking	Tenosynovitis
Fingers, Hands	Numbness, tingling, ashen skin; loss of Keeling and control	Exposure to vibration	Chain saw, pneumatic hammer and gasoline-powered tool operators	Raynaud's syndrome (white finger)
Fingers, Wrists	Tingling, numbness, severe pain; loss of strength, sensation in the thumbs, index; or middle or half of the ring fingers	Repetitive and forceful manual tasks without time to recover	Most and poultry and garment workers, upholsterers, assemblers, VDT operators, cashiers	Carpal turned syndrome
Back	Low back pain, shooting pain or numbness in the upper legs	Whole body vibration	Truck and bus drivers, tractor and subway; warehouse workers; grocery cashiers; baggage handlers	Back disability

Fig. 13.4 Mountain of cumulative trauma partly obscured by fog. (Kroemer et al. 2001)

very small number of cases of disabilities that despite having been treated by complete medical options, these could not be resolved or relieved.

Currently there are a significant number of tools that are used by ergonomists as well as trained and experienced to use them for analysis, evaluation and prevention of MSD in the workplace. To mention a few: NIOSH Lifting Equation (1991. an approach to calculating a 'maximum' permissible load for lifting different circumstances. There exists in the original 1993 version and a newer 'revised version' ***Rapid Upper Limb Assessment (RULA).*** Assessment tool Provides a 'score' for upper limb Demands by McAtamney and Corlett (McTamney and Corlett 1994). ***Rapid Entire Body Assessment (REBA)***, similar to RULA but with a whole body focus (Hignett and McAtamney 2000). For more information on other methods see Neumann (2006) "Inventory of Tools for Ergonomic Evaluation".

Table 13.2 summarizes some methods that helps evaluate the MSD risk that can be used depending on the task performed by the worker as repetition and duration of the activity, grip strength and lifting, postures, vibrations, among others Occupational Health and Safety Council of Ontario 2007b).

We need a new paradigm focused on preventing occupational diseases and injuries not only at work. The recognition, prevention and treatment of occupational diseases and the improvement of recording and reporting systems, are priorities. Ergonomics is seen as a value-added function, helps industry to identify and remove barriers between people, production and quality. The goal of increasing production and quality with fewer injuries is obtainable if ergonomics is properly introduced as a way of doing businesses.

Table 13.2 MSD risk assessment methods summary table

Assessment tool	Repetition/ duration	Force: gripping/ pinching	Force: lift/ lower/ carry	Force: push/ pull	Posture	Vibration	Contact stress/ Impact	Neck/ shoulder	Hand/ wrist/arm	Back/ trunk/hip	Leg/knee/ ancle
Checklist methods (multiple hazards considered)											
MSD hazard risk assessment checklist	X	X	X	X	X	X	X	X	X	X	X
Washington State Checklist (Caution/Hazard Zone)	X	X	X	X	X	X	X	X	X	X	X
Manual material handing (lighting, lowering, pushing, pulling, carrying)											
ACGIH: lifting TLV	X		X		X			X		X	
NIOSH Lifting Equation	X		X		X			X		X	
Snock Tables	X		X	X	X			X		X	X
MAC (UK)	X		X		X			X		X	
Mital et al. tables	X		X	X	X			X		X	X
Upper Limb											
ACGIH: HAL	X	X	X	X					X		
RULA	X				X			X	X	X	
Strain Index	X	X	X	X	X				X		
CTD Risk Index (CTD-RAM)	X	X	X	X				X	X		
LUBA OCRA	X	X	X	X	X	X	X	X	X		
Combined methods (not checklist)											
QEC	X		X	X	X	X		X	X	X	X
REBA	X		X	X	X			X	X	X	X
ManTRA	X	X	X	X	X	X		X	X	X	X
OWAS	X	X	X	X	X			X		X	X

13.3 Lean Manufacturing

The Toyota Production System is a comprehensive system of production and management emerged in the Japanese automotive company of the same name Toyota Motor; Womack and Daniel (2003) describe a largely traditional benchmarking system that they called Lean Manufacturing ("Lean Manufacturing") and whose concept is based on a path to specify value, organize actions in the best sequence, conduct these activities without interruption whenever someone requests them and make them ever more efficient (Socconini 2009) states that it is a continuous and systematic process for identification and elimination of waste or excesses; defining excesses as all those activities that do not add value to the process, nevertheless it adds cost and labor. Moreover, this term is the tireless and continuous effort to create more effective innovative and efficient companies where there are administered processes or tasks that create value and make use of necessary resources (Amendola 2006).

Lean manufacturing seeks to eliminate any activity that does not add value to the process, commonly called waste such as: Overproduction refers to excess production; Excess of inventory, is any material or product or process that exceeds what is needed to meet customer demand; Defects ("scrap"), is the loss of resources used, when a faulty item or service machine investing time and labor is occurred, Transportation, consists of all shipments of materials that do not directly support the production system; Unnecessary processes are a result of the workshop needs, quality of manufacturing or planning bad deliveries; Unnecessary movements, refers to the after-side of people from one point to another in their workplace without providing value to the product or not contributing to the benefit of the client; Waiting, is the time lost when an operator wait for the machine to finish the work or when the operator waits for materials (Socconini 2009).

For Meyers (2006), Success in the elimination or reduction of waste is based on following a process of continuous improvement which basically requires the participation of all employees at various levels of the organization (Meyers 2006) to pursue goals aimed at reducing costs, improve processes, increase customer satisfaction and maintain profit margin, giving companies tools to survive in a global market that demands higher quality, faster delivery at lower prices and in the quantity required (Leyva 2005).

13.3.1 Lean Manufacturing Methodology to Prevent Musculoskeletal Disorders (MSDs)

In the prevention of MSDs, it is necessary to implement several steps considered under the methodological basis of continuous improvement cycle adapted from Edward Deming (PDCA) in conjunction with guidelines and procedures proposed by authors like Womack and Daniel (2003) Shingo (1990), Cuatrecasas (2005).

Fig. 13.5 Methodology for lean ergonomics implementation

This approach can be explained in Fig. 13.5. In the core of the figure, are established the steps related to:

(1) Plan
Planning the analysis of the situation of the company, where it is necessary to identify the problem situation, the causes of these and the actions that must be considered for the solution of the problem,
(2) Do
Perform the actions specified during planning,
(3) Verify
Check the results of the implementation of the solution of the problem, if it is not obtained, the planning stage will be taken, but if the results are satisfactory, the last step will be continued.

(4) Act
Proceed to achieve the continuous improvement cycle.

In the following sections every stage of the proposed methodology for Lean Ergonomics implementation will be explained.

13.3.1.1 Analysis of the Current Situation

This step starts with the identification of the participant responsible for carrying out the process of defining the problem and how activities must be carried out for its solution. It is necessary to find a facilitator who will be responsible for carrying out the organization of the lean manufacturing principles in HFE, specifically in the analysis, evaluation and prevention of MSDs. This person may be part of the business, or be an external agent. For the election this participant it has to be considered issues such as: Having extensive knowledge and experience in the application of Lean philosophy, ergonomics and human factors and their tools; demonstrate talent to change things around, i.e. a person creative, dynamic and challenging.

The improvement team to work with the agent may be constituted as follows: Sponsor, leader, facilitator (certified Black Belt and HFE), members and participants, recorder, and spokesperson Timekeeper.

Cuatrecasas (2005) states the development of a value stream map of the current state as the subsequent step to follow in this methodology. This requires information about the whole business process such as duration of an activity, the production capacity of the machine, the information flow among other characteristics. It is important that the team identifies all those activities that add value to the process or final service from the ones that do not, to have a continuous process flow and thus achieve customer requirements fullfillment. However, it is here when, according to the ideal form of the process, it should not be overlooked those risk factors that can lead ergonomic with time in MSD.

It is important to identify the wastes as mentioned above: Overproduction, Delays, Defects, Movements, Process, Inventory and Transportation, as well as, risk factors for MSDs like repetitiveness of the task, application of force to develop an activity, posture and body position, push and pull tasks, often lifting heavy objects, difficult grip hand tools, vibration body segments (primarily in hand—wrist), and mechanical stress concentrations, that must be evaluated during the improving process.

13.3.1.2 Define the Problem

Considering the prevalence of MSDs as the object of study, the problem is defined in terms of reviewing records of incidents, accidents, occupational diseases, frequency of visits to the medical and disability services and interviews with workers to identify stressful tasks or activities that contribute to the development of MSDs. It is also important that the problem would be drafted in terms of direct and indirect costs of injuries and work diseases.

13.3.1.3 Analyze Facts and Data to Isolate the Root Cause

With the gathered information from the previous step, responsible participants must analyze the data and find what cause can be originating the problematic situation. MSDs hazards can be caused by a number of different factors, so it is important to consider the different possible causes of any health problem and the ergonomic and safety. Several techniques are widely known to achieve this purpose: Technical Brainstorming diagram, cause—effect (Diagram Ishikawa, 5 questions Techniques, among others. It is important not to forget also the analysis of the information to look for patterns of trends regarding the root cause, such as: Pareto chart, trend chart, frequency histogram, scatter plot or correlation and analysis regression, Eight Disciplines Problem Solving, Failure Mode and Effects Analysis called FMEA. FMEA is carried in order to appreciate the relationship between a feature (effect) and factors (causes) that affect it.

13.3.1.4 Establish Actions to Eliminate the Root Cause

Once the root cause of the problem is identified, the following step is to proceed establishing actions that eliminate or reduce this cause. Lean Manufacturing tools are useful solving this problem in a systematic and documented way such: Eight Disciplines Problem Solving, Failure Mode and Effects Analysis (FMEA). These allow documenting the solution process and generating long-term solutions, which is promoted by the human factors and ergonomics in favor of the health of workers. The tools used in the analysis phase to identify causes have to be used again to compare process performance improved from baseline.

13.3.1.5 Communicating the Master Plan

Once the action plan through Eight Disciplines Problem Solving tools or Failure Mode and Effects Analysis (FMEA) has been established, the agent of change and the responsible participant for each activity should design the training and education of all staff channel to create a Lean Ergonomics environment, within the company. That is to educate staff and make it aware of the importance of implementing these actions and the benefit of personal health and productivity of the company and thereby create a culture of problem solving.

When the staff is well informed and convinced of the actions to be carried out for Lean Ergonomics transformation, the formal launch can be made, and when the application of this philosophy throughout the company can be realized and effective communication with to the staff done, the goals and objectives can be achieved at the end of this introduction. This allows to implementing the actions and tools set.

This gives way to implementing the actions and tools set.

13.3.1.6 Run the Actions Set

The change agent, collaborator and the applying representative of each action must organize working groups, which will be responsible for implementing each analysis?? Ergonomic??tool.

Once working groups are established, the specialized training of each tool for each team, takes place. The way how this should be is up to each Lean tool and ergonomic risk factors have to be eliminated It is important that improvements are implemented in a gradual and organized manner, involving the team work of all the people in the company to make changes without large capital investment and in a constant way. To achieve this, it is recommended to develop Kaizen event to implement improvement actions and assess their impact on immediate or short term. For more information we recommend Imai (2012).

13.3.1.7 Verify the Results

All staff with established results should compare them before and after the application, in order to have a vision of the improvement that was obtained. It is not sufficient that there has been a progress, the change agent, during the development of this process, established a goal; based on an end result, the company determines whether it did or not, for this verification it could be used a graphical solution—impact, checklists or plan control. Campaigns are recommended to promote and widespread the good results, organize site visits to observe the—making the Gemba—Broadcast Campaigning and set-up meetings.

13.3.1.8 Continuous Improvement

To ensure that procedures, practices and prevention activities related to MSDs risk factors can constantly run through the Lean Manufacturing philosophy, a culture of prevention within the organization must be promoted. In addition, maintaining the triumphs and set new challenges in the prevention of MSD as suggested by this methodology, it is important that the actions directed to protect the health of workers can be replicated in all areas of the organization on an ongoing basis; so it become a habit in the personnel and in the organization.

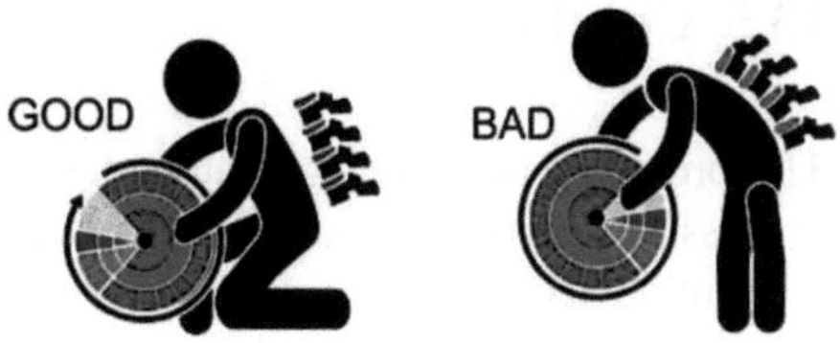

Fig. 13.6 Lean with human approach

13.4 Lean Ergonomics: Human as a Key Player in the Continuos Improvement

In this part Kester (2013), states that Lean Manufacturing (LM) has replaced the mass production of nearly half of all U.S. manufacturers. LM has helped entrepreneurs to improve operational efficiency and to retain their jobs. Unfortunately, many of them have seen incrementing their compensation costs for their workers' after implementing Lean concepts. The author refers to an automotive manufacturer that experienced an increase of 100 percent of cases of musculoskeletal disorders (MSDs), receiving a requirement of the Division of Occupational Safety and Health of California's Division of Occupational Safety and Health of California, for "insufficient attention to ergonomics" after implementing Lean during a an assembly line modification.

Unfortunately, Lean processes can make very repetitive work, which cause havoc on employees because of stressful postures due to the use of high strength during all day long, meanwhile the critical break time for employees is eliminated. In this sense the derivative financial savings gained from productivity and quality improvements are used to pay the high cost of claims for compensations for workers (Fig. 13.6).

The way of the interaction of the human factor with the application of lean manufacturing, where, although the method and its proper application plays a key role in the success of the improvement, at no time should be considered beyond to the person especially when it comes to prevent musculoskeletal injuries. The integration of ergonomics in the lean manufacturing process allows to identifying possible risk factors and production flow design, build stations and working methods, which reduce or eliminate the risk to workers. Since that lean manufacturing and ergonomics share the objectives of eliminating waste and adding value, in ergonomics are natural integration points in most of the set-two processes. Ergonomics is simply another tool that can be embedded into lean processes to make them more successful.

13.5 Conclusions

The proposed methodology integrates the human factors and ergonomics in the process of improving Lean Manufacturing, which seeks operational excellence without neglecting health worker. As mentioned above HFE is seen as a function of value added industries to help identify and remove the barriers between people, production and quality. The goal of increasing production and quality with fewer injuries is obtainable if HFE is properly introduced as a way of doing businesses as well as more Lean successful processes. It is because of all above that through this document there is presented a proposal regarding a new approach to the implementation of improvements, considering the human factor as a core part of the mentioned method and its implementation.

References

Álvarez, E. (2008). *MC Salud Laboral. No.7*. pág. 12. Polytechnic University of Catalonia. Spain.

Amendola, L. (2006). *Balanced systems of indicators in asset management* (2nd ed.). Brazil: World Congress on Maintenance.

Chapanis, A. (1996). *Human factors in systems engineering*. USA: Wiley.

Chengalur, S., Rodgers, S., & Bernard, T. (2004). *Kodak's ergonomics design for people at work*. USA: Wiley.

Cuatrecasas, L. (2005). *Methodology for the implementation of lean management in an industrial enterprise and mid-size independent*. Spain: Instituto Lean.

European Agency for Safety and Health at Work. (2010). *Annex to Report: Work—related musculoskeletal disorders—facts and figures* (Luxembourg, Office for Official Publications of the European Communities). Retrieved January 28, 2013, from https://osha.europa.eu/en/resources/tero09009enc-resources/europe.pdf.

Farrer, F., Minaya, G., Child, J., & Ruiz, M. (1994). *Ergonomics manual*. Madrid: Mapfre Editorial.

Helander, M. (1995). *A guide to the ergonomics of manufacturing*. USA: Taylor and Francis Inc.

Hignett, S., & McAtamney, L. (2000). Rapid entire body assessment. *Applied Ergonomics, 31*, 201–205.

IEA—International Ergonomics Association. (2013). *Definition of ergonomics*. Retrieved July 15, 2013, from http://www.iea.cc.

Imai M. (2012). *How to implement kaizen in the workplace (GEMBA)*. McGrawHill. México.

Karwowski, W., & Marras, W. S. (1999). *The Occupational Ergonomics Handbook*. Boca Raton, FL: CRC Press.

Karwowski, W. (2005). Ergonomics and human factors: The paradigms for science. *Engineering, Design, Technology, and Management of Human—Compatible Systems, Ergonomics, 48*(5), 436–463.

Kester, J. (2013). A lean look at ergonomic. Healthier continuous improvement processes can limit musculoskeletal disorders. Industria Engineer. *Engineering and Management Solutions at Work, 45*, 3.

Kroemer, K., Kroemer, H., & Kroemer, K. E. (2001). *Ergonomics, how to design for ease and efficiency* (2nd ed.) (pp. 1, 7, 387, 391, 392) . New Jersey: Prentice Hall Inc.

Llaneza, Á. J. (2009). *Handbook for specialist training*. Spain: Lex Nova S. A. de CV.

Leyva, I. (2005). *Lean applied in automotive production line*. Ph. D. Thesis. Technological Institute of Sonora, Mexico.

McTamney, L., & Corlett, E. N. (1994). *RULA - a Rapid Upper Limb Assessment tool, Contemporary Ergonomics*. London: Taylor & Francis.
Meyers, F. E. (2006). *Time and motion study for agile manufacturing* (2nd ed.). Mexico: Edit. Pearson education.
Mondelo, P., Gregori, E., & Barrau, P. (2000). *Ergonomics Fundamentals* (p. 7). Barcelona: UPC Editions Universal Mutual.
Neumann (2006). Inventory of tools for ergonomic evaluation. Berlin: National Institute for Working Life.
Niebel, B., & Freivalds, A. (2001). *Industrial engineering. Methods, Labour Standards and Design*. Mexico: Alfaomega editors.
Oborne, D. J. (1992). *Ergonomics in action* (2nd ed.). Mexico: Editorial Threshings.
Occupational Health and Safety Council of Ontario. (2007a). *Musculoskeletal disorders prevention series. PART 2,* Resource Manual for the MSD Prevention Guideline for Ontario. Canada.
Occupational Health and Safety Council of Ontario (2007b). *Occupational Health and Safety Council of Ontario's MSD Prevention Series. Part 2:* Resource Manual for the MSD Prevention Guideline for Ontario. Canada. (WSIB Form Number : 5158A) .
Ramirez, C. C. (1999). *Ergonomics and productivity*. Mexico: Limusa México.
Salvendy, G. (1997). *Handbook of human factors and ergonomics*. USA: A Wiley Interscience Publication.
Secretary of Occupational Health (2008). *Manual of musculoskeletal disorders*. Workers Committee of Castilla and León. Valladolid. Spain.
Shingo, S. (1990). *Toyota production system from the point of view of engineering*. USA: Productivity Press.
Socconini, L. (2009). *Lean Manufacting*. Mexico: Norma Editions S.A. de CV.
Womack, J. P., & Daniel, T. (2003). *Lean thinking: How to use lean thinking to eliminate waste and create value in the company* (p. 65). USA: Management 2000.

[illegible]

Chapter 14
Low Back Pain Risk Factors: An Epidemiologic Review

Lilia Roselia Prado-León

Abstract In the area of human-factors engineering (ergonomics), various ergonomic factors associated with Low Back Pain (LBP) are mentioned constantly. From this perspective the problem is conceptualized as a musculoskeletal disorder (MSD) of mainly occupational origin, among whose principal causes are repeated trauma ocurring over a long period of time. So this concept excludes injuries caused by acute or instantaneous trauma. In the United States, the cost of occupational injuries and illnesses exceeds $170 billion annually (Occupational Safety & Health Administration 2004). Mexico lacks of precise data regarding the impact of these problems, but according to data from the family medicine units of Mexican Social Security Institute (IMSS) for the year 2010, back pain ranked seventh among reasons for consulting a doctor nationwide. This literature review aims to identify occupational and non-occupational ergonomic factors and to develop a comprehensive model for the etiology of LBP. Thirty-two epidemiologic articles, from 1992 on, provided evidence regarding the relationship between LBP and risk factors—individual, psychological, socioeconomic and occupational—considered in this chapter.

Keywords Low back pain · Occupational safety · Musculoskeletal disorder

14.1 Introduction

Low back pain (LBP) is a huge common health problem (Dionne et al. 2006; Rapoport et al. 2004) and causes an enormous economic burden on individuals, families, communities, industry and governments (Thelin et al. 2008).

L. R. Prado-León (✉)
Ergonomic Research Center, University of Guadalajara Art, Architecture and Design Center, Independencia 5075, Huentitán El Bajo, C.P. 41300 Guadalajara, JAL, Mexico
e-mail: ailil_p@yahoo.com.mx

J. L. García-Alcaraz et al. (eds.), *Lean Manufacturing in the Developing World*,
DOI: 10.1007/978-3-319-04951-9_14, © Springer International Publishing Switzerland 2014

In the United States, the cost of occupational injuries and illnesses is more than $170 billion annually (Occupational Safety & Health Administration 2004). Most people who experience activity-limiting LBP go on to have recurrent episodes. Estimates of recurrence after 1 year range from 24 to 80 % (Hoy et al. 2010).

Mexico lacks of precise data regarding the impact of these problems, but according to figures from the Mexican Social Security Institute (MSSI) for the year 2010, back pain was seventh among reasons for consulting a doctor at family medicine units nationwide.

The current literature review includes thirty-three articles and aims to identify occupational and non-occupational ergonomic factors in, and to develop a comprehensive model for, the etiology of LBP.

14.2 Literature Review

The kind of studies carried out to address the problem of LBP and its associated factors may be grouped according to the following rubrics: biomechanical, physiological, psychophysical and epidemiologic.

The present review concentrates on the epidemiologic approach.

14.2.1 Epidemiologic Studies

This method concerns identification of the incidence, distribution and potential controls for damage to the population. Although risk factor studies cannot be linked with causes of LBP, they can give us the known variables that may influence the incidence or prevalence of the problem. This type of study can in turn be subdivided into three major categories according to the methodology used: (1) cross-sectional or prevalence, (2) case-control, and (3) longitudinal or cohort.

14.2.1.1 Cross-Sectional or Prevalence Studies

In this kind of study, all subjects are chosen at the same time, from a defined source population, and questions are aimed at identifying and quantifying exposures. There exists, therefore, at the start of the study, an indeterminate number of cases or controls, since these depend upon the prevalence of the illness and the prevalence of exposure within the sampling scheme being used. Comparisons may be made considering risk factors among people who do or do not have the condition under study. However, given that cases are included by prevalence or incidence, temporal relationships of cause and effect cannot be easily established because exposures and results are being evaluated at the same time. Cross-sectional studies have greater difficulty in selection bias because they include, for example, people working in a determined type of job, and only for the duration of the study (Bombardier et al. 1994).

14.2.1.2 Case-Control

Beginning with a group of people suffering from the disease under study (controls), exposure is identified and quantified. The number of subjects to include in each group is predetermined ahead of time by means of a sample size calculation.

14.2.1.3 Longitudinal or Cohort

The typical study includes a group of "healthy" people from a defined population (construction workers, for example) which is longitudinally followed for a period of time (10 years, for example) in order to record events of interest (i.e. on-the-job reports). Baseline information for suspected risk factors is gathered before the illness occurs. This "cohort" of people is generally followed (prospective studies) with relation to time, for the purpose of recording events of interest (exposure).

Cohort studies are expensive since on the one hand they require a much longer time period to complete, and on the other they need a very large subject sample in order to record sufficient events of interest (Rothman and Greenland 1998).

14.2.2 Literature Inclusion Criteria

Searches were performed in databases compiling biomedical and ergonomic literature, using the key words "low back pain", "risk factors", "back injury", "back pain", "musculoskeletal disorder risk", "case-control study", and "cohort study". Titles and abstracts of identified papers were read. Potentially useful papers (on the basis of title and abstract) were read in full.

14.3 Results

The following is a brief summary of the articles reviewed: divided, as mentioned above, into three types of epidemiological studies.

14.3.1 Cross-Sectional or Prevalence Studies

Using a postal questionnaire, Holmstrom et al. (1992) conducted a study of 1,733 construction workers in order to detect severe LBP, Manual Materials Handling (MMH) and psychosocial factors. Their results showed that the prevalence rate for LBP in 1 year was 54 % and for severe LBP it was 7 %. Relation to heavy MMH differed according to age in such a way that it could be interpreted as a healthy-worker effect. A dose-response relationship was found between severe LBP and crouching or kneeling positions. Principal psychosocial factors associated with an increase in LBP were the stress index, psychosomatic and psychic indices.

Jelcid et al. (1993) studied 101 health professionals and found that the incidence rate of work-related risk factors had a magnitude of 200 % in comparison with non-occupational risk factors. Physical and sports activities were identified as aggravating factors in patients with a history of LBP and healthy subjects. 26 % had a temporary disability.

Burdorf et al. (1993) carried out a study to identify LBP history, sedentary work, postural load on the back and whole-body vibration in crane operators (n = 94), straddle-carrier drivers (n = 95) and a reference group of office workers (n = 86). A standardized interview and observation of non-neutral postures of the trunk during normal work were utilized for data collection. They reported the following findings: the 12-months prevalence of LBP among crane operators was 50 %; among straddle-carrier drivers, 44 %; and among office workers, 34 %. After adjustment (age and confounders), the odds ratio (OR) among crane operators was 3.29 (95 % Confidence Interval (CI), 1.52–7.12), and among straddle-carrier drivers 2.51 (95 % CI, 1.17–5.38). In neither occupation was daily exposure to whole-body vibration considered an important risk factor for LBP in this study. The observations showed that non-neutral postures of the trunk were frequently adopted among all workers.

Croft and Rigby (1994) reported results from a large-scale study of 9,003 adults aged 18 years and up whose data came from the Health and Lifestyle Survey. The main variables studied were LBP, socioeconomic status, cigarette smoking, weight and height, anxiety and depression. Their results showed that women with households in the lowest income category were more likely to report back pain than those in the highest income group (OR = 1.6; 95 % CI 1.2–2.1). Women with no formal educational qualification were more likely to report back pain than women who had a degree (OR = 1.5; 95 % CI 1.0–2.1).

Masset and Malchaire (1994) researched the history of LBP, personal characteristics and past and present working conditions using a verification list of 618 blue-collar workers in two metalworking industries. They found that lumbar symptoms occurred in 66 % at some time during a subject's life, 53 % within the past year and 25 % 7 days prior to the interview. The multivariate logistic regression showed an association with heavy shoulder effort (OR = 1.62) and driving a vehicle (OR = 1.15). Physical workload, trunk postures and movements, or exposure to whole-body vibration were not associated with increase in prevalence of LBP.

Zheng et al. (1994) studied the prevalence of LBP in 10,087 textile, machine-shop, transportation, apparel and aquaculture workers, finding 13.66 % prevalence in men and 8.75 % in women. Risk factors found via a relative risk analysis were: lifting, bending back posture, rotation back posture and prolonged sitting plus vibration.

2,405 nurses were studied by Smedley et al. (1995): their objective was to determine history of LBP, occupational activities and non-occupational risk factors for back symptoms. Results showed that women and men had similar rates: the lifetime prevalence of back pain was 60 %, and 10 % had been absent from work because of back pain for a cumulative period exceeding 4 weeks. In women back pain during the prior 12 months was weakly associated with height, and was significantly more common in those who reported frequent non-musculoskeletal symptoms such as headache and low mood. After adjustment for height and non-musculoskeletal symptoms, significant associations were found with frequency of manually handling patients around/on the bed, manually transferring patients between bed and chair, and manually lifting patients from the floor. No clear increase in risk was found related to transfer of patients with canvas and poles, manually lifting patients in and out of the bath, or lifting patients with mechanical aids.

Prevalence of musculoskeletal complaints of the back, arms or neck, and legs, and MMH among nurses (n = 846) were studied by Engels et al. (1996). Most of the respondents were women (89.8 %), whose mean age was 29.0 years. Sixty-three percent were state enrolled nurses, 18 % were state enrolled nursing students, and 19 % had other functions such as kitchen assistant. Sixteen percent of the subjects had managerial tasks. The mean duration of employment in their profession was 9.2 years. They worked 32.7 h a week.

Back problems, back-related disability, status, occupation (if applicable) and residence location were the principal variables studied by Hurwitz and Morgenstern (1997) in 84,572 subjects based on data from the 1989 National Health Interview Survey. They found that the prevalence of a disabling back condition, versus no back condition, was relatively higher among subjects aged 25–64 years, male, non-high-school graduates, unemployed, living in the West, with disabling non-back morbidities, and with body mass index and weight above the 50th percentile. The occupations that more frequently reported disabling back conditions were technical, sales, clerical, private household, service, precision production and repair or transportation, compared to workers in professional occupations. The characteristics of adults associated with chronic back disability were: more than 34 years old, weight above the 50th percentile, and history of back trauma.

Nahit et al. (2001) described a study of 1,081 subjects from 12 occupational groups: its principal objective was to research the influence of mechanical factors on regional musculoskeletal pain. Their results showed 24 % reporting LBP, 20 % shoulder pain, 9 % wrist or forearm pain, and 21 % knee pain. Some specific manual handling activities were found to be associated with low back, shoulder, and knee pain. Carrying weights of more than 50 lbs on one shoulder was the factor most strongly associated with LBP (OR = 2.4; 95 % CI 1.5–3.8) and knee pain (OR = 3.5; 95 % CI 2.2–5.5). On the other hand, forearm pain was most strongly associated with repetitive wrist movements (OR = 1.8; 95 % CI 1.04–3.1). Few postures were associated with regional pain, but examples included bending forward in an uncomfortable position for at least 15 min, associated with shoulder pain (OR = 1.6; 95 % CI 1.2–2.2) and kneeling for at least 15 min, associated with knee pain (OR = 1.8; 95 % CI 1.2–2.6).

Elders and Burdorf (2001) identified risk factors for LBP by using a questionnaire of physical load via frequent observations in 229 construction (scaffolding) workers and 59 supervisors. A strong interrelation was found between self-reported determinants and physical load, but an inverse tendency for both age and total job experience could indicate the presence of healthy-worker effect. Relation was observed to be weak between psychosocial variables and physical load. Multivariate analysis showed a significant relation between MMH, perceived health and job demands and LBP within the past 12 months. Chronic back pain was significantly related to high perceived effort and the general state of health. Severe LBP correlated significantly with risky back postures, high recovery need and high job demands. The results of this study suggest construction workers to be a high-risk group for developing persistent forms of LBP.

A British population was studied by Webb et al. (2003) to identify prevalence and predictors for neck and LBP. Using the modified Health Assessment Questionnaire and a questionnaire on multiple joint symptoms, plus a questionnaire regarding the predominant pain site, they found that the 1-month-period prevalence of all reported spinal pain was 29 %. Most people with back (75 %) or neck (89 %) pain also reported pain at other sites. The significant predictor of LBP with disability were: age, high body mass index, living in an area of high material deprivation, and South Asian ethnicity.

Alexopoulos et al. (2003) reported on a study with nursing personnel (n = 351) to identify risk factors for several musculoskeletal disorders such as neck, shoulder and LBP. Using a self-reported questionnaire, they found physical load to be associated with occurrence of back pain (OR = 1.85), neck pain (OR = 1.88), and shoulder pain (OR = 1.87). A trend with physical load was shown with the number of musculoskeletal complaints: for two complaints (OR = 2.47) and for three (OR = 4.13).

Risk factors for LBP in machinery manufacturing were studied by Xiao et al. (2004) using interviews, postural analysis and the revised National Institute for Occupational Safety & Health Lifting equation. The subjects studied were 69 workers involved in MMH (Job A) and 51 machinery workers less involved with MMH tasks (Job B). Their results showed that for LBP with at least one episode lasting for 24 h or more in the past 12 months, prevalence rates were 63.8 and 37.3 % for Jobs A and B, respectively. Prevalence rates of LBP every day for a week or more attributed to lifting were 26.09 % and 5.88 % for Jobs A and B, respectively. Multiple regression analysis suggested that lifting repetitiveness and work age contributed to the occurrence of LBP. Object weight as well as activity repetitiveness had significant adverse effects on LBP.

Chen et al. (2005) reported on the results of analyzing occupational factors associated with LBP based on data from the Taxi Drivers' Health Study (n = 1,242) in Taiwan. The principal risk variables were: job dissatisfaction, demographic features (age, gender), marital status, socioeconomic positions, lifestyle factors, driving time profiles, and average frequency of physical activities (lifting tasks and bending/twisting activities) while driving at work or during leisure time.

Fifty one percent of urban taxi drivers reported LBP over the previous 12 months, significantly higher (P, 0.001) than other professional drivers (33 %). After adjusting for confounding factors (demographic characteristics, lifestyle factors, anthropometric measures and socioeconomic positions) they found that 4 h/day driving time (OR = 1.78; 95 % CI 1.02–3.10), frequent bending/twisting activities while driving (OR = 1.86; 95 % CI 1.15–2.99), self-perceived job stress (OR = 1.75; 95 % CI 1.20–2.55), job dissatisfaction (OR = 1.44; 95 % CI 1.05–1.98) were the major occupational factors significantly associated with higher LBP in taxi drivers.

Eight hundred and forty four Japanese nurses were the sample for a study by Smith et al. (2006). They identified musculoskeletal disorders (MSDs); demographic items such as tobacco smoking, alcohol, etc.; along with workplace factors such as work hours, shift work, physical tasks, posture and psychosocial factors. Data collection was done by a questionnaire based on Standardised Nordic, and a workplace questionnaire. The 12-month period-prevalence of MSDs at any body site was 85.5 %. The first place for MSDs was shoulder pain (71.9 %), followed by LBP (71.3 %), neck pain (54.7 %), with the last being upper back (33.9 %). After adjusting, alcohol consumption (OR = 1.87; 95 % CI 1.17–2.96), tobacco smoking (OR = 2.45; 95 % CI 1.43–4.35), and having children (OR = 2.53; 95 % CI 1.32–4.91) were significant risk factors for MSDs.

Mitchell et al. (2008) conducted a prevalence study for LBP with 897 undergraduate nursing students (years 1, 2 and 3) and 111 graduate nurses recruited by personal invitation during lectures. Using a modified version of the Nordic Low Back Questionnaire, they found mean age was consistent across all groups (26.7 years) and had no significant effect on lifetime lLBP prevalence (p = 0.30). The prevalence of LBP was very high for lifetime (79 %). For 12 months it was 71 % and for 7 days it was 31 %. LBP prevalence rates were consistent across all 3 year-groups of undergraduate nursing students, but were significantly higher after 12 months of full-time employment. Nursing students and graduate nurses believed the causes of their LBP were bending, or lifting despite recent efforts to reduce manual (lifting) demands on nurses in the workplace.

LBP and related factors among 1,436 Iranian office workers was the study undertaken by Rezaee et al. (2010). Risk variables were occupational factors such as prolonged sitting, waist rotating over hip and forward bending. Non-occupational factors: age, sex, years of employment, awareness of back care, participation in instructional courses, and existence of regular exercise program. For data collection they used a direct interview, a body discomfort assessment tool that consisted of a 10-cm color Visual Analogue Scale (VAS) and a questionnaire. 79.8 % of respondents were male: results showed that over 60 % had at least one episode of LBP during their working lives. Lifetime prevalence of LBP was 92.1 % and for the prior 12 months it was 37.3 %. Age up to 40 years, high weight, sitting work style (more than 4 h), computer use (more than 5 h a day) and past history of LBP, all had positive association with LBP.

Alperovitch-Najenson et al. (2010) presented results of studying 384 male full-time urban bus drivers. Information was collected by using the Standardized Nordic Questionnaire and a questionnaire seeking data on regular physical activity, work-related ergonomic and psychosocial stressing factors. Forty-five percent of subjects had experienced LBP in the previous 12 months. Ergonomic factors associated with LBP were an uncomfortable seat (OR = 2.6; 95 % CI 1.4–5.0) and an uncomfortable back support (OR = 2.5; 95 % CI 1.4–4.5). Participation in regular physical activities was higher in the group with LBP (48.5 %) than in the group without LBP (67.3 %; $p < 0.01$). Several psychosocial stressing factors showed significant association with LBP: "limited rest period during a working day" (OR = 1.6; 95 % CI 1.0–2.6), "traffic congestion on the bus route" (OR = 1.8; 95 % CI 1.2–2.7), "lack of accessibility to the bus stop for the descending and ascending of passengers" (OR = 1.5; 95 % CI 1.0–1.5), and "passengers' hostility" (OR = 1.8; 95 % CI 1.1–2.9).

14.3.2 Case-Control Studies

Mundt et al. (1993a) composed their study of 287 people diagnosed with disc hernia, matched for sex, age, medical attention source and geographic area; a control group had no disc hernia or other back conditions. Their objective was to find non-occupational lifting risk factors for LBP. As in the majority of the studies reviewed, they used a questionnaire to collect the data. Their results showed significant association for lifting more than 25 times per day (RR = 3.95). Frequent lifting with arms extended (RR = 1.87) and twisting while lifting (RR = 1.90).

In the same year, Mundt et al. (1993b) found sports and weight-lifting to be possible risk factors for herniated lumbar and cervical discs. Two hundred and eighty-seven patients with lumbar disc hernia and 63 with cervical hernia, were matched by sex, type of hospital and age (by decade); with a control that had no disc hernia nor other neck or back conditions. The majorities of sports were not associated with risk of hernia and could be protective. Relative risk (RR) was generally less than 1.0. They found a weak association between bowling and lumbar or cervical hernia. Use of weight-lifting equipment was not associated with herniated lumbar or cervical disc, but a possible association was indicated between using free weights and risk of cervical herniation (RR = 1.87; 95 % CI 0.74–4.74).

Zwerling et al. (1993) made a study of 154 postal workers with injuries and 942 control subjects. They sought data on job type, gender, age, pre-existing disabilities, and Quetelet Index via a questionnaire. Multivariate logistic regression showed that a previous history of disability (OR = 2.9; 95 % CI 1.88–4.48) and heavy lifting work (OR = 1.91; 95 % CI 1.32–2.76) are associated with occupational low back injuries.

A case (174 soldiers diagnosed with lumbosacral strain and unable to continue working) control (173 controls with no disability) study was done by Feuerstein et al. (1999). Their objective was to find possible predictors of occupational low back disability and their implications for secondary prevention. The most important predictors for disability from LBP were: age (OR = 1.13); infrequent aerobic exercise (OR = 2.2), high job stress (OR = 2.71) and low social support (OR = 5.07).

Mortimer et al. (2001) studied the relation between LBP and sports activities, smoking and weight. They found neither low intensity training for many hours per week ($\geq$5 h), high intensity training for few hours (1–2 h), nor moderate training for many hours ($\geq$5 h) per week were LBP risk factors among men. For women, however, few hours of high intensity training increased relative risk of LBP (RR = 1.6; CI 1.1–2.4). A risk for LBP was observed in men (but not women) with high body weight (RR = 2.2; CI 1.2–3.9). In this study, smoking contributed no risk for LBP.

Prado et al. (2005) carried out a case- control study in a Mexican population. The cases were 77 workers with lumbar spondyloarthrosis matched with 154 "healthy" workers (controls). They collected data about MMH tasks: load weight, postures, task duration and frequency; Quetelet index, smoking, pregnancies, sport activities, sedentary and standing posture, and non-occupational MMH tasks. Data collection was by interview. Lifting tasks, combined with driving tasks, were associated with lumbar spondyloarthrosis (OR = 7.3; 95 % CI 1.7–31.4). Daily lifting frequency as it interacted with work as a driver resulted in a greater risk (OR = 10.4; 95 % CI 2.0–52.5). Load weight, daily task-hours and cumulative time showed a dose-response relationship.

14.3.3 Longitudinal or Cohort Studies

Manninen et al. (1995) conducted a longitudinal study of 537 farmers with 12 years of follow-up. During this time, they collected data on several possible risk factors for pain in different sites, including: smoking, mental stress, height and body mass index. Results showed that full-time farmers had a significant prevalence, high in comparison with those who were part-time or retired. Logistic regression model analysis showed a high risk for current smokers (OR = 9.6; 95 % CI 2.7–65.2) and for ex-smokers (OR = 13.1; 95 % CI 1.7–53.0), as compared with non-smokers. Mental stress, body height and body mass index were not predictors of LBP.

Josephson et al. (1997) carried out a study of 565, 553, 562, and 419 subjects who answered a questionnaire at the first, second, third, and fourth surveys, respectively with 3 years of follow-up. Neck, shoulder, and back musculoskeletal symptoms and job strain data were collected by a 10-point (0-9) scale with verbal end points of "no symptoms" and "very intense symptoms". A Swedish version of Karasek was also applied. Risk for musculoskeletal symptoms was weak (RR = 1.1–1.5) when comparing the group manifesting job strain and the group without job strain in the four measurements. For the combination of job strain and perceived high physical exertion the estimated RR was between 1.5 and 2.1.

From a cohort of 285 construction workers in the Hamburg construction worker study Latza et al. (2000) analyzed LBP by job history, work organization, demographic information, education, psychosocial factors, lifestyle factors, health status and work tasks during 3 years of follow-up. They used a questionnaire and a standardized orthopedic examination for collecting data. Their results showed prevalence of LBP of 80.7 %, determined through a year of self-reporting. Results indicated that differences in brick features (size and type of stone) and temporal aspects of construction work (average hours per shift) could predict future prevalence of LBP.

The relationship Return-to-Work (RTW) after compensated LBP with organizational and psychosocial working conditions, were studied by Krause et al. (2001) in 433 LBP workers with 1–4 years follow-up interview. They found associations for LBP with high physical and psychological job demands and low supervisory support during all disability phases. High job control, especially control over work and rest periods, is associated with over 30 % higher RTW rates, but only during the subacute/chronic disability phase starting 30 days after injury. Job satisfaction and co-worker support are unrelated to time to RTW.

Miranda et al. (2002) made a study of 2077 "healthy" workers (controls) and 327 workers with severe LBP, with 1 year of follow-up. By applying a questionnaire they measured LBP, individual characteristics such as age, smoking and mental stress; occupational loading and participation in different sports. Greater age, mental stress, long-term smoking and trunk rotation on the job were risk factors for LBP. Physical workload factors seem to present more risk in the incidence of LBP, while psychosocial factors were more related to persistence or chronicity of LBP.

Bergström et al. (2007) studied 2,187 employees with data from Work and Health in the Processing and Engineering Industries. Information was collected by self-reported questionnaires, the General Nordic Questionnaire and Questionnaire for Psychological and Social Factors at Work, during 3 years of follow-up. At 18 months, 151 participants reported at least one episode of sick-listing due to neck or back pain during the previous year. Risk factors assessed were blue-collar work, back pain one or several times during the previous year, 1–99 days of cumulative sickness absence during the previous year (all causes except neck or back pain), uncertainty of one's own working ability in 2 years' time and the experience of few positive challenges at work. At the end of the study, 127 participants reported at least one episode of neck pain or LBP during the year prior to follow-up. The principal risk factors for these pains were blue-collar work, several earlier episodes of neck pain, no everyday physical activities during leisure time (cleaning, gardening and so on), and lower physical functioning.

And lastly, Plouvier et al. (2008) researched how LBP related to awkward postures, driving and MMH tasks in 2,218 men 383 women; with 5 years of follow-up using self-administered questionnaires. Significant associations could be observed between LBP and durations of driving (OR = 1.24) and bending/twisting for men (OR = 1.37); LBP for more than 30 days and exposure to bending/twisting for men (OR = 2.20) and women (2.00); driving for women (OR = 3.15); LBP radiating to the leg and duration of driving for men (OR = 1.43) and bending/twisting for women (OR = 1.95), and LBP radiating below the knee and duration of exposure to pulling/pushing/carrying for men (OR = 1.88).

As may be observed, the majority of findings related to LBP risk factors come from cross-sectional studies (19), many of them conducted with a specific population (postal workers, health service workers, construction workers, etc.) or in a specific work setting, for example, in a hospital. Case-control studies, on the other hand, are less often seen (6), with only a slightly greater number of cohort studies (7). Generally, these studies entail more cost, time and effort, which is probably a reason why they are less often conducted, compared with cross-sectional or prevalence studies. Still, they have greater advantages because they permit more dependable causal inferences. It could be observed that cross-sectional as well as longitudinal studies carried out analyses using databases from prior large-scale studies (Croft and Rigby 1994; Hurwitz and Morgenstern 1997; Krause et al. 2001; Bergström et al. 2007).

Thus, it is important to note that studies concerning LBP have basically included the widest category of LBP, which only indicated pain in that anatomical region, and described different prevalences, such as lifetime prevalence, point prevalence or 1-year prevalence.

Referring to exposure measurements, most of them were done through interviews or questionnaires, some even via mail or e-mail, and only some studies done within a particular industry added observation and work tasks records.

Individual risks to which the studies refer most often are: poor physical condition, poor general health, smoking, pregnancy, age—elderly, higher risk (Feuerstein et al. 1999; Miranda et al. 2002; Rezaee et al. 2010); gender—with women appearing to present higher prevalence (Zheng et al. 1994); weight—higher body mass index, higher risk (Mortimer et al. 2001; Rezaee et al. 2010).

Psychological factors most frequently found were: stress, low mood and psychosocial load. The relationship among them is unclear, but they appear to delay recovery and lead to chronic LBP (Holmstrom et al. 1992; Smedley et al. 1995; Feuerstein et al. 1999; Elders and Burdorf 2001).

Socioeconomic factors were not much studied, but a greater proportion of LBP could be observed in people with lower salaries and lower education levels (Croft and Rigby 1994; Prado et al. 2005).

The most recurrent occupational factors found were physical load, mainly MMH tasks and awkward working postures; often observed in such occupations as warehouse, blue collar or construction work, or nursing (Zheng et al. 1994; Smedley et al. 1995; Nahit et al. 2001; Plouvier et al. 2008; Mitchell et al. 2008).

Also, prolonged static postures plus vibration i.e. driving a motor vehicle for a prolonged period is associated with a high incidence of LBP (Zheng et al. 1994; Elders and Burdorf 2001; Plouvier et al. 2008; Alperovitch-Najenson et al. 2010).

In general sports activities appear to be protective, though this depends on the physical load involved. For example, the practice of bowling may be risky for its bent-back posture and the weight of the ball (Mundt et al. 1993b; Mortimer et al. 2001; Alperovitch-Najenson et al. 2010).

14.4 Conceptual Model of LBP

The graphic rendering (Fig. 14.1) attempts to explain that the most relevant occupational factors, as well as workers' individual characteristics, socioeconomic and psychological factors, can increase the probability of LBP fatigue. However, if recovery does not occur, other short-range changes will be produced at an anatomic-physiological level in the spine, which will manifest as inflammation and restriction of spinal movement, and if this continues, added to accumulation over time, LBP will become chronic with anatomic-physiological changes growing more serious, and medium- to long-range degenerative processes manifesting in the spine.

There follow explications of three concepts relevant to the literature review, and essential for understanding the model derived from it: MSDs, MMH and LBP.

14.4.1 Concept of Musculoskeletal Disorders

From an ergonomic point of view the principal occupational causes of MSDs are highly repetitive activities, often undertaken from inadequate postures, with movement of the involved corporal segments, and pressure from work equipment upon the body. Putz-Anderson (1998) also underscores the importance of non-existent or insufficient rest/recuperation.

LBP constitutes one of the most important MSDs. The initial supposition is that everyone does things that are potentially damaging to the back, but if these actions take place repetitively, there arises a cumulative process of damage arises over weeks, months or years. This situation causes the damage range to exceed the recuperation range, producing degenerative damage to the lumbosacral spine, which manifests in one context or another, although the context may not be directly provoked by the damage, but by prior antecedents (Pheasant 1991).

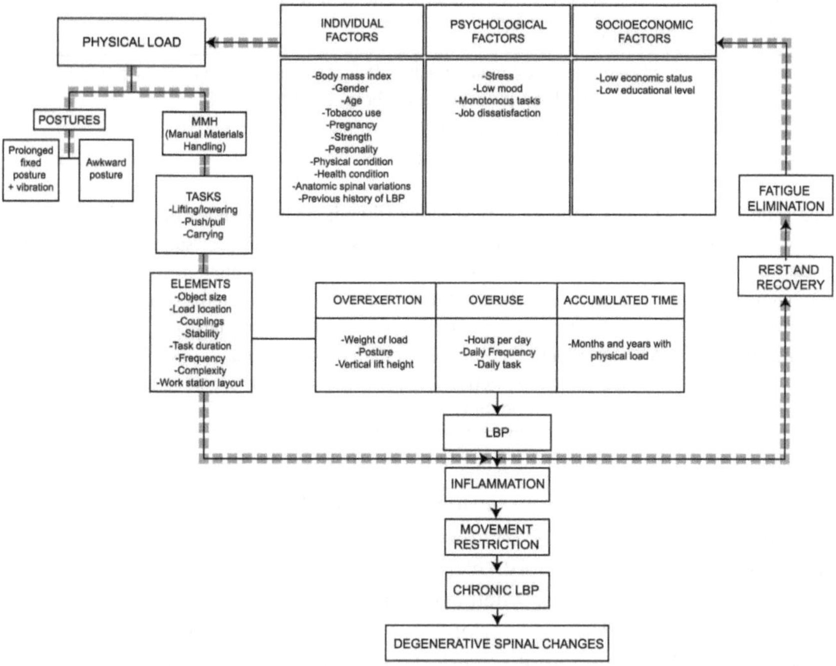

Fig. 14.1 Risk factors for low back pain

14.4.2 Concept of Manual Materials Handling

Is an occupational activity undertaken very often, and in various work settings. It is characterized by the presence of one or more of the following activities that involve excess effort or overexertion, and movements in extreme postures (Kumar and Mital 1992; Pheasant 1991): lifting, lowering, pushing, pulling and carrying animals, people, heavy objects, equipment or tools, without the assistance of mechanical devices.

From a systemic point of view, MMH includes 4 components (Ayoub 1992).

- Worker
- Task
- Tools and equipment
- Setting.

Table 14.1 presents the elements of the task and its indicators 1 (Ayoub and Mital 1989).

MMH y LBP Each MMH task requires a dynamic and static muscular effort. Damage may occur when the effort produced by these tasks exceed the viscoelastic capacity of the ligaments, tendons, bones and discs (Pheasant 1991). This excess

Table 14.1 MMH: component task, its elements and indicators

Elements	Indicators
Load dimensions	Measurement of mass, push/pull force requirements, mass moment of inertia, load measurement: height, width, depth
Load distribution	Localized measure of load unit (one- or two-handed)
Coupling	Measure of hand-object interface, or method of gripping or fastening the load
Load stability	Localization and consistency (liquids or voluminous materials)
Layout of work station	Measurement of the work area's spatial properties, such as distance of movement, obstacles and nature of the destination
Frequency/duration/step	Measurement of time dimensions for the task: frequency, duration and dynamics of the activity (short-term or long-term)
Complexity	Measurement of combined demands such as manipulation, movement requirements, activity objective, precision tolerance and number of kinetic components

may arise from overuse, overexertion or temporal accumulation. By overuse is understood repetitive activity (frequency) of some MMH tasks within the work day, though this doesn't imply overload when applied to characteristics of an isolated task. In other words, when a worker carries out the task of lifting 20 kg, the weight does not qualify as overload, but doing this task with a frequency of one lift per minute throughout an 8-h work day does count as overload. Overload thus encompasses the remaining MMH elements: load weight, lifting posture, work station layout, complexity, load distribution and stability, etc., and finally temporal accumulation, which refers to the total period during which the person has been performing MMH activities.

14.4.3 Concept of Low Back Pain

LBP refers to those pathological conditions that constitute MSDs presenting pain in the lower part of the back, and significantly related torelating significantly to tasks performed in the workplace (Pheasant 1991). The pain often radiates toward the thighs or buttocks, restricting mobility in the back; possibly causing muscular spasms due to incorrect functional use of the lumbosacral spine (Cailliet 1990; La Dou 1993; Crenshaw and Campbell 1988).

14.5 Conclusion

Based on data about the high incidence of diseases related to LBP in the world population, it can be deduced that this represents one of the most widespread work-related health problems to generate both temporary and permanent incapacities, and disability, in such a way that the socioeconomic cost it represents is relevant for our society.

Although technological advances have affected all areas of production, even product handling and transport via the use of automatic or semi-automatic mechanisms as full or partial substitutes for manual handling, yet in many countries such as Mexico, MMH practices, the transfer and loading of countless products, remain in the hands of stevedores, adding to the population at risk of suffering LBP. The concept of LBP which is considered in this work is analyzed from an ergonomic perspective, considering it as MSDs, of multifactorial nature, with some relevant risk factors due to work exposure.

The contribution of the ergonomic conception thus supposes that LBP is, in large measure, of a cumulative origin and that manifestation of a serious LBP episode in the workplace, or any other place, does not therefore mean that a work incident has occurred: it rather means continued and prolonged exposure to inadequate work conditions from an ergonomic perspective.

Identifying risk factors that may be modified, such as the nature of the task, its duration and repetitiveness, could lead to establishing preventive measures that reduce the incidence of spondyloarthrosis; which, as one of the high-incidence diseases involving LBP, has great individual, economic, and social repercussions. In this sense it may be noted that epidemiologic studies have supported a conclusion that back injuries may be prevented or reduced by 33 %, if the work station is redesigned. (Snook 1978; quoted by Kumar and Mital 1992).

References

Alexopoulos, E. C., Burdor, F. A., & Kalokerinou, A. (2003). Risk factors for musculoskeletal disorders among nursing personnel in Greek hospitals. *International Archives of Occupational and Environmental Health, 76*, 289–294. doi:10.1007/s00420-003-0442-9

Alperovitch-Najenson, D., Santo, Y., Masharawi, Y., Katz-Leurer, M., Ushvaev, D., & Kalichman, L. (2010). Low back pain among professional bus drivers: ergonomic and occupational-psychosocial risk factors. *The Israel Medical Association Journal, 12*(1), 26–31.

Ayoub, M., & Mital, A. (1989). *Manual materials handling*. Newyork: Taylor & Francis.

Ayoub, M. (1992). Problems and solutions in manual materials handling: The state of the art. *Ergonomics, 35*(7/8), 713–728.

Bergström, G., Bodin, L., Bertilsson, H., & Irene, B. (2007). Risk factors for new episodes of sick leave due to neck or back pain in a working population. A prospective study with and a three-year follow-up. *Occupational and Environmental Medicine, 64*(4), 279–287.

Bombardier, C., Kerr, M., & Shannon, H. (1994). Guide to interpreting epidemiologic studies on the etiology of back pain. *Spine, 19*(185), 20475–20562.

Burdof, A., Naaktgeboren, B., & De Groot, H. (1993). Occupational risk factor for low back pain among sedentary workers. *Journal of Occupational Medicine, 35*(12), 1213–1220.

Cailliet, R. (1990). *Síndromes dolorosos*. Ciudad de México: El Manual Moderno.

Chen, J. C., Chang, W. R., & Chang, W. (2005). Occupational factors associated with low back pain in urban taxi drivers. *Ocupational Medicine, 55*, 535–540.

Crenshaw, A., & Campbell, A. (1988). *Cirugía Ortopédica* (7a. edición). Buenos Aires, Argentina: Médica Panamericana.

Croft, P., & Rigby, A. (1994). Socioeconomic influences on back problems in the community in Britain. *Journal of Epidemiology and Community Health, 48*, 166–170.

Dionne, C. E., Dunn, K. M., & Croft, P. R. (2006). Does back pain prevalence really decrease with increasing age? A systematic review. *Age and Ageing, 35*(3), 229–234.

Elders, L., & Burdorf, A. (2001). Interrelations of risk factors and low back pain in scaffolders. *Occupational and Environmental Medicine, 58*(9), 597–603.

Engels, J. A., van der Gulden, J., & Senden, T. F. (1996). Work related risk factors for musculoskeletal complaints in the nursing profession: Results of a questionnaire survey. *Occupational and Environmental Medicine, 53*, 636–641.

Feuerstein, M., Berkowitz, S. M., & Huang, G. D. (1999). Predictors of occupational low back disability: Implications for secondary prevention. *Occupational and Environmental Medicine, 41*(12), 1024–1031.

Holmstrom, E., Lindewll, J., & Moritz, U. (1992). Low back and neck/shoulder pain in construction workers: Occupational workload and psychosocial risk factors. *Spine, 17*(6), 663–671.

Hoy, D., Brooks, P., & Blythc, F. (2010). The epidemiology of low back pain. *Best Practice & Research Clinical Rheumatology, 24*(6), 769–781.

Hurwitz, E., & Morgenstem, H. (1997). Correlates of back problems and back-related disability in the United States. *Journal of Clinical Epidemiology, 50*(6), 669–681.

Instituto Mexicano del Seguro Social (Mexican Social Security Institute). (2010). *Memoria estadística de salud en el trabajo* (Statistical memory of work health). Retrieved from http://www.imss.gob.mx/estadisticas/financieras/pages/memoriaestadistica.aspx

Jelcid, A., Culjak, M., & Horvacic, B. (1993). Low back pain in health personnel. *Reumatizam, 40*(2), 13–20.

Josephson, M., Lagerstrom, M., & Hagberg, M. (1997). Musculoskeletal symptoms and job strain among nursing personnel: A study over a three year period. *Occupational and Environmental Medicine, 54*, 681–685.

Krause, N., Dasinger, L., & Deegan, L. (2001). Psychosocial job factors and return-to-work after compensated low back injury: A disability phase-specific analysis. *American Journal of Industrial Medicine, 40*, 374–392.

Kumar, S., & Mital, A. (1992). Margin of safety for the human back: a probable consensus based on published studies. *Ergonomics, 35*(7/8), 769–781.

La Dou, J. (1993). *Medicina Laboral* (1ª Edición). Ciudad de México: El Manual Moderno.

Latza, U., Karmaus, W., & Stammer, T. (2000). Cohort study of occupational risk factors of low back pain in construction workers. *Occupational and Environmental Medicine, 57*(1), 28–34.

Manninen, P., Riihimak, H., & Heliovaara, M. (1995). Incidence and risk factors of low—back pain in middle-aged farmers. *Ocupational Medicine, 45*(3), 141–146.

Masset, D., & Malchaire, J. (1994). Epidemiologic aspects and work-related factors in the steel industry. *Spine, 9*(2), 143–146.

Miranda, H., Viikari-Juntura, E., Martikainen, & R. (2002). Individual factors, occupational loading, and physical exercise as predictors of sciatic pain. *Spine (Phila Pa 1976), 27*(10), 1102–1109.

Mitchell, T., O'Sullivan, P., & Burnett, A. (2008). Low back pain characteristics from undergraduate student to working nurse in Australia: A cross-sectional survey. *International Journal of Nursing Studies, 45*, 1636–1644.

Mortimer, M., Wiktorin, C., & Pernol, G. (2001). Weight and smoking in relation to low-back pain: A population-based case-referent study. *Scandinavian Journal of Medicine and Science in Sports, 11*(3), 178–184.

Mundt, D., Kelsey, J., & Golden, A. (1993a). An epydemiologic study of non-occupational lifting as a risk factor for herniated lumbar invertebral disc. *Spine, 18*(5), 595–602.

Mundt, D., Kelsey, J., & Golden, A. (1993b). An epidemiologic study of sports and weight lifting as possible risk factors for herniated lumbar a cervical discs. *American Journal of Sports Medicine, 21*(6), 854–860.

Nahit, E., Macfarlane, G., Pritchard, C., et al. (2001). Short term influence of mechanical factors on regional musculoskeletal pain: A study of new workers from 12 occupational groups. *Occupational and Environmental Medicine, 58*(6), 374–381.

Occupational Safety & Health Administration. (2004). Retrieved June 20, 2013 from http://www.osha.gov/publications/osha3173.pdf

Pheasant, S. (1991). *Ergonomics, work and health*. Hong Kong: McMillan Press, Scientific & Medical.

Plouvier, S., Renahy, E., & Chastang, J. (2008). Biomechanical strains and low back disorders: Quantifying the effects of the number of years of exposure on various types of pain. *Occupational and Environmental Medicine, 65*, 268–274.

Putz-Anderson, V. (1998). Cumulative trauma disorders: A manual for musculoskeletal diseases of the upper limbs. Bristol, PA: Taylor & Francis.

Prado, L., Celis, A., & Avila, R. (2005). Occupational lifting tasks as a risk factor in low back pain: A case-control study in a Mexican population. *Work, 25*, 107–114.

Rapoport, J., Jacobs, P., & Bell, N. R. (2004). Refining the measurement of the economic burden of chronic diseases in Canada. *Chronic Diseases in Canada, 25*(1), 13–21.

Rezaee, M., Ghasemi, M., & Jonaidi-Jafari, N. (2010). Low back pain and related factors among Iranian office workers. *International journal of occupational hygiene, 3*, 23–28.

Rothman, K., & Greenland, S. (1998). *Modern epidemiology* (2nd ed.). Philadelphia, USA: Lippincott-Raven Publishers.

Smedley, J., Egger, P., & Cooper, C. (1995). Manual handling activities and risk of low back pain in nurses. *Occupational and Environmental Medicine, 52*, 160–163.

Smith, D., Mihashi, M., & Adachi, Y. (2006). A detailed analysis of musculoskeletal disorder risk factors among Japanese nurses. *Journal of Safety Research, 37*, 195–200.

Thelin, A., Holmberg, S., & Thelin, N. (2008). Functioning in neck and low back pain from a 12-year perspective: A prospective population-based study. *Journal of Rehabilitation Medicine, 40*(7), 555–561.

Webb, R., Brammah, T., & Lunt, M. (2003). *Prevalence and predictors of intense, chronic, and disabling neck and back pain in the UK general population*. Baltimore: Lippincott Williams & Wilkins, Inc.

Xiao, G., Dempsey, P., & Lei, L. (2004). Study on musculoskeletal disorders in a machinery manufacturing plant. *Journal of Occupational Environmental Medicine, 46*, 341–346.

Zheng, Y., Hu, Y., & Shou, B. (1994). An epidemiologic study of workers with low back pain. *Chinese Journal of surgery, 32*(1), 43–58.

Zwerlin, C., Ryan, J., & Schootman, M. (1993). A case-control study of risk factors for industrial low back injury. *Spine, 18*(9), 1242–1247.

Chapter 15
Lean-Six Sigma Framework for Ergonomic Compatibility Evaluation of Advanced Manufacturing Technology

Aide Maldonado-Macías, Jorge Luis García-Alcaraz, Rosa María Reyes-Martínez and Juan Luis Hernández-Arellano

Abstract In this chapter an Ergonomic Compatibility Evaluation Model for the selection of Advanced Manufacturing Technology (AMT) into a Lean Six Sigma Manufacturing Framework is proposed. The approach was applied on the selection of CNC plastic molding machines for an AMT training center. The purpose of this model is to constitute a decision aid for decision makers to perform evaluations of AMT in a more complete manner, while considering human factors and ergonomic aspects. These aspects are found to be scarce in actual AMT evaluation and selection models. Model was created using a multi-attribute axiomatic design perspective supporting the selection of AMT. In this chapter a theoretical Lean Six Sigma Framework was proposed for its application. A numerical example is presented by the application of the model with the participation of three experts. A procedure including phases with several steps was deployed using the Define, Measure, Analyze, Improve and Control (DMAIC) stages for AMT evaluation. The alternative which best meet established Design Ranges in terms of Ergonomic Compatibility was selected among three alternatives, according to the Ergonomic Incompatibility Content in a fuzzy environment.

A. Maldonado-Macías (✉) · J. L. García-Alcaraz
Department of Industrial and Manufacturing Engineering, Autonomous University of Ciudad Juarez, Av. Del Charo 450 Norte. Col. Partido Romero, Ciudad Juarez, Chihuahua, Mexico
e-mail: amaldona@uacj.mx

J. L. García-Alcaraz
e-mail: jorge.garcia@uacj.mx

R. M. Reyes-Martínez
Undergraduate Studies Division, Technological Institute of Juarez City, Ave. Tecnológico No. 1340, C.P 32500 Ciudad Juárez, Chihuahua, Mexico
e-mail: rosyreyes2001@yahoo.com.mx

J. L. Hernández-Arellano
Department of Industrial Design, Autonomous University of Ciudad Juarez, Av. Del Charo 450 Norte. Col. Partido Romero, Ciudad Juarez, Chihuahua, Mexico
e-mail: luis.hernandez@uacj.mx

J. L. García-Alcaraz et al. (eds.), *Lean Manufacturing in the Developing World*,
DOI: 10.1007/978-3-319-04951-9_15,

Keywords Fuzzy multiple attribute decision making (FMADM) · Advanced manufacturing technology (AMT) · Lean six-sigma framework · Ergonomic compatibility (EC) · Human factors and ergonomics (HFE) · Axiomatic design (AD)

15.1 Introduction

Advanced Manufacturing Technology (AMT) has brought major changes in manufacturing systems in the world's industrial scenery. It is considered one of the main elements towards efficiency and competitiveness of enterprises. It generally includes Computer Numerical Controlled (CNC) equipment, Computer Aided Design/Computer Aided Manufacturing (CAD/CAM), Flexible Manufacturing Systems (FMS), robotics, rapid prototyping, environmentally sustainable technologies, etc. (Venkata Rao 2007; García et al. 2009). Nowadays, aiming to optimize manufacturing systems, AMT selection plays an important role in decision making, this implies a large amount of information and uncertainty. Decision Makers (DMs) continuously face the problem of evaluation and selection of equipment among a wide variety of alternatives involving multiple attributes and conflicting criterion. AMT has been broadly used in modern industries around the world and there are evaluation tools and models available to support equipment selection processes, even though the publications on this subject are limited (Kulak et al. 2005; García et al. 2008). Among these authors is accepted that decision making and management of AMT constitutes a complex problem that involves multiple aspects, which are sometimes difficult to consider in their totality. In this way, evaluation models regarding planning and selection of AMT equipment are found to be scarce of the adequate and desirable Human Factors and Ergonomics (HFE) aspects (attributes) and their importance is under estimated among DMs and Decision Making Models (DMM) as well. In this chapter a Lean Six Sigma theoretical framework was applied to develop a fuzzy multi attribute axiomatic design procedure following the DMAIC stages.

A numerical example is presented using this approach; in this case DMs started a project where the main goal was selecting Plastic Molding Machines among three alternatives for a university's manufacturing laboratory. It was important for them to consider ergonomic and safety aspects additionally of economic and technological ones. The main purpose of this chapter is to present the theoretical framework of Lean Six Sigma that was used to assess the selection of the equipment taking into account ergonomic attributes of equipment. An Ergonomic Compatibility Survey (ECS) was applied in order to measure the Ergonomic Compatibility (EC) of these artifacts. The EC construct was validated to be used and a fuzzy axiomatic design approach was used to obtain Ergonomic Incompatibility Content (EIC); the alternative which has the minimum EIC was selected as the best for this purpose.

15.2 Lean-Six Sigma Manufacturing Framework

Lean manufacturing focuses on the systematic elimination of wastes from organization operations through a set of synergistic work practices to produce products and services at the rate of demand (Womack et al. 1990; Fullerton et al. 2003; Simpson and Power 2005; Shah and Ward 2007). Lean manufacturing represents a multifaceted concept that may be grouped together as distinct bundles of organizational practices (McLachlin 1997; MacDuffie 1995). A list of bundles of lean practices includes JIT, total quality management (TQM), total preventative maintenance, and human resource management, pull, flow, low setup, controlled processes, productive maintenance and involved employees (McKone et al. 1999; Swink et al. 2005; Linderman et al. 2006; Shah and Ward 2007). Decision Making about Advanced Manufacturing Technology (AMT) is a very common organizational practice that offers opportunities for management to adopt a Lean Manufacturing position.

The main two practices of Lean Manufacturing are quality systems and human resource management. The first practice, quality systems, includes quality improvement strategies, such as TQM and Six Sigma. According to Carayon and Smith (2000) and Drury (1997), there exists an interaction of common benefits between ergonomics and TQM. These interactions are: the use of ergonomics to improve the performance of quality control activities; applications of TQM to safety aspects of ergonomics; relations between TQM and macro-ergonomics or socio-technical systems; open systems strategic problems; systems approaches to organization design and leadership; measurement-based operations; appropriate use of technology; individuals, teams and the change process.

Six sigma, is a statistical problem-solving methodology, quality improvement strategy and a management philosophy which is flexible in terms of being able to apply it to any functional business area, including manufacturing. It contains five distinct problem-solving stages known as the DMAIC approach (Shaffie and Shahbazi 2012):

- *Define* the problem statement, the goal, and the financial benefits.
- *Measure* the current performance of the process and collect the required data.
- *Analyze* the root cause of the problem.
- *Improve* the process to eliminate errors and instability.
- *Control* the performance of the process, ensuring that the improvements are sustained.

15.3 Fuzzy Axiomatic Design Methodology

Fuzzy and Crisp Axiomatic Design approaches for AMT evaluation and selection are found in literature in the state of the art of the evaluation and selection models. A literature review of these applications can be found in Maldonado-Macías et al. (2008) evaluation and selection processes using the Information Axiom seem to offer several advantages for authors, emphasizing its capability to evaluate designs following the designers', judges' or experts' appraisal necessities stated as Functional Requirements (FR) with their correspondent Design Ranges (DR). The alternative which best meet such requirements is selected as the best for the particular purposes. Nevertheless, the Ergonomic Compatibility Evaluation proposal using a Fuzzy Multi-Attribute Axiomatic Design approach for AMT is considered innovative. EC is a construct used in this work evoking the concepts of human-system and human-artifact compatibility introduced by Karwowski (1997, 2000, 2001, and 2005) which offer comprehensive treatment of compatibility in human factors discipline. It intends to measure in a subjective way the probability of a design to satisfy ergonomic requirements using the EIC in a fuzzy environment. For this, the Theory of Axiomatic Design extended by Helander (1995), Helander and Lin (2002) and adopted by Karwowski (2001, 2005) was also evolved. For this approach a Hierarchical Fuzzy Axiomatic Design Survey for Compatibility Evaluation of AMT was designed based on a pragmatic perspective and a large amount of literature that was reviewed and presented in Maldonado et al. (2009), Maldonado et al. (2013). This evaluation approach can be adapted to be immersed into a Lean-Six Sigma framework. The methodology is explained in the following sections.

15.4 Methods

15.4.1 Lean Six Sigma Framework for Model Application

The Lean Six-Sigma framework for the Ergonomic Compatibility Evaluation Model implicates all stages of DMAIC methodology. At the *Define* stage Ergonomic Compatibility attributes are defined and Ergonomic Functional Requirements (EFR) are expressed as desirable ergonomic attributes of the equipment, the main goal is define an ergonomic oriented project for selection or evaluation of AMT.

At the *Measure* stage the Ergonomic Compatibility Survey is applied and the alternative's ratings of each attribute are evaluated by each expert. At the *Analyze* stage the importance of each attribute is determined using the experts' opinions. Appropriate aggregation procedures to determine the weight of importance for each attribute and the analysis and determination of Design Ranges and System Design Ranges (SDR) are conducted; graphically these ranges can be represented to determine a Common Area (CR). For the *Improve* stage the Equations for the Weighted Ergonomic Incompatibility Content (WEIC) and Membership Functions are needed to determine the highest ergonomic incompatibility content among all attributes, this can lead us to improve decision making processes about AMT and compare alternatives. At the *Control* stage the procedure focuses on ensuring long term sustainability of improvement once the decision about AMT was made.

15.4.2 Stage of Define

This section describes the Fuzzy Axiomatic Design Procedure under Lean-Six Sigma Framework for this stage. In this stage several steps are proposed. The main goal at this stage is define a project where AMT must be subject to evaluation for selection or improvement taking into consideration ergonomic attributes. Also, the Ergonomic Compatibility Attributes involved in the evaluation are defined.

15.4.2.1 Fuzzy Axiomatic Design Procedure Under Lean-Six Sigma Framework for the Stage of Define

In this stage the first phase of the procedure is developed under a Lean-Six Sigma Framework

Phase 1:

Step 1: Determine the alternatives to consider in the evaluation, define the project. Where $A_i = (1, 2,\ldots,N)$ number of alternatives

Step 2: Determine the attributes to evaluate, establishing the EFR's. Where $B_j = (1, 2,\ldots,M)$ number of attributes

Step 3: Constitute the group of experts. Where $k = (1, 2,\ldots,k)$ number of experts

Step 4: Choose appropriate linguistic variables for the importance weights of the attributes for each alternative and the linguistic ratings according Table 15.1.

15.4.2.2 Definition of Ergonomic Compatibility Attributes

Ergonomic Compatibility Attributes are divided into main attributes and sub attributes. This attributes were defined by an extensive literature review shown in Maldonado et al. (2009) and Maldonado et al. (2013). Table 15.2 shows these attributes and their correspondent description. Main attributes are divided into five main attributes (ECMA): compatibility with human skills and training (A11), physical work space compatibility (A12), usability (A13), equipment emissions requirements (A14) and organizational requirements (A15). The main attribute A11 includes two sub-attributes: human skills compatibility (A111) and training compatibility (A121). The main attribute A12 includes five sub-attributes: access to machine and clearances (A121), horizontal and vertical reach zones (A122), adjustability of design (A123), postural comfort of design (A124), physical work and endurance of design (A125). The main attribute A13 includes seven sub-attributes: controls' design compatibility (A131), controls' physical distribution (A132), visual work space design (A133), information load (A134), error tolerance (A135), man–machine functional allocation (A136), design for maintainability

Table 15.1 Linguistic scales using fuzzy triangular numbers

Tangible attributes	
Linguistic term	Triangular fuzzy numbers
Very low (VL)	(0, 0, 0.3)
Low (L)	(0, 0.25, 0.5)
Medium (M)	(0.3, 0.5, 0.7)
High (H)	(0.5, 0.75, 1)
Very high (VH)	(0.7, 1, 1)
Intangible attributes	
Linguistic term	Triangular fuzzy numbers
Poor (P)	(0, 0, 0.3)
Regular (R)	(0.2, 0.35, 0.5)
Good (G)	(0.4, 0.55, 0.7)
Very good (VG)	(0.6, 0.75, 0.9)
Excellent (E)	(0.8, 1, 1)

(A137). The main attribute (A14) includes four sub-attributes: temperature (A141), vibration (A142), noise (A143), residual materials (A144). Finally, the main attribute (A15) includes two sub-attributes: compatibility with rate of work (A151) and compatibility with job content (A152).

15.4.3 Stage of Measure

15.4.3.1 Fuzzy Axiomatic Design Procedure under Lean-Six Sigma Framework for the Stage of Measure

Phase 2:

On this phase the ECS proposed by Maldonado et al. (2009) is applied; a group of experts must participate in this phase by responding the survey and measure each attribute and evaluate every AMT alternative.

The Ergonomic Compatibility Survey (ECS) was designed to collect the information of evaluations and also for the determination of the relative importance of the attributes and sub-attributes. The survey includes 95 questions divided in four parts. In the first part, importance is assigned for the attributes and sub-attributes using linguistic scales, in the second part Ergonomic Design Range is determined for each sub-attribute using linguistic scales, in the third part Ergonomic System Range for the alternatives are evaluated using linguistic scales and finally crisp pair wise comparisons using Analytic Hierarchical Process (AHP) proposed by Saaty (1980) were conducted to obtain the importance weights by means of *Expert Choice*^{TM} *software*. The validation of this survey was made applying the Cronbach Alpha Test using crisp values on a Likert Scale. Examples of the Ergonomic Compatibility Survey can be found in Maldonado et al. (2009) and Maldonado et al. (2013) and are recommended for further reading. This stage involves the following steps of the procedure.

Table 15.2 Description of the ergonomic attributes (Maldonado et al. 2013)

Attribute	Description
A11 Human skills and training compatibility: set of attributes that define the compatibility of the equipment with the user's technical skills and training	
A111 Skill level compatibility	Attribute of equipment's design regarding its adaptability to differences on users' technical skills. (Allowing safe and efficient operation for expert and novice users)
A112 Training compatibility	Attribute of equipment's design in terms of the training required (quality and duration) that will be available taking into account users' needs
A12 Physical work space compatibility: set of attributes that define the equipment's compatibility with the physical work space through the allowance of comfortable reaches and postures as well as taking into account the strength and endurance required for its operation promoting safety and effectiveness	
A121 Access to machine and clearances	Attribute of equipment's design concerning the allowance of mobility and secure access to arms, hands, legs, head, trunk, and knees of the operator through its space and clearances
A122 Horizontal and vertical reaches	Attribute of equipment's design concerning the allowance of comfortable, safe and effective human vertical and horizontal reaches (upper and lower extremities)
A123 Adjustability of design	Attribute of equipment's design concerning the allowance of adjustment and/or change on its physical structure (size, position, etc.) or on its components that would be satisfactory to operator
A124 Postural comfort of design	Attribute of equipment's design regarding the allowance of neutral and diverse body postures for a safe and effective operation
A125 Physical work and endurance of design	Attribute of the equipment's design concerning the level of physical work and endurance that will required of operator during interaction
A13 Usability: set of attributes that promote easiness of use of equipment's design	
A131 Controls design compatibility	Attribute of the equipment's design regarding the type and design of controls and sensors (as buttons, knobs, levers, switches, sensors stoppage of movement, etc.) providing an effective and safe operation

(continued)

Table 15.2 (continued)

Attribute	Description
A132 Control's physical distribution	Attribute of the equipment's design regarding the physical distribution (location) of the controls (buttons, knobs, levers, switches, etc.) providing a safe and effective manipulation
A133 Visual work space design	Attribute of the equipment's design concerning the size and location of screens and displays of information: size and type of characters used, colors, contrast, resolution and brightness facilitating human visual tasks during human–machine interaction
A134 Information load	Attribute of the equipment' design which allows and facilitates a safe and effective operation through a satisfactory human understanding, learning and processing of the information (visual, auditory, sensory) during human–machine interaction
A135 Error Tolerance of design	Attribute of the equipment's design which allows and facilitates to the operator the management and prevention of errors, through simple and clear messages and dialogues on the human–machine interface
A136 Man–Machine functional allocation of design	Attribute of equipment's design concerning that difficult tasks for operator such as quick response, short term storing information, high accuracy and repeatability, among others are allocated in the equipment design preferable to the machine
A137 Design for maintainability	Attribute of equipment's design considering whether a simple, rapid, effective and safe maintenance tasks will be allowed, during repairing, installation and dismantling, transportation, loading, cleaning, assembling and disassembling among other maintenance activities
A14 Equipment emissions requirements: set of attributes related to temperature, vibration, noise and residual materials generated by the equipment and may adversely affect operator and/or environment	
A141 Temperature	Attribute of equipment's design which assesses the temperature emitted by the equipment and its components and that may adversely affects operator and/or environment
A142 Vibration	Attribute of equipment's design which assesses the vibration emitted by the equipment and that may adversely affects operator and/or environment

(continued)

Table 15.2 (continued)

Attribute	Description
A143 Noise	Attribute of equipment's design which assesses the noise emitted by the equipment and its components and that may adversely affects human operators and/or environment
A144 Residual materials	Attribute of equipment's design which assesses the amount and kind of residual materials generated by the equipment and its components and that may adversely affect the operator and or environment
A15 Equipment design organizational requirements: set of attributes that define the equipment's compatibility with the pace and speed of work as well as with the total content of the work according human limitations and capabilities	
A151 Rate of work machine compatibility	Attribute of the equipment's design considering that it avoids or prevents inappropriate pace and speed of work for operator
A152 Total work content machine compatibility	Attribute of the equipment's design considering that it support the prevention of excessive force application, long term awkward postures, repetitive tasks and high-risk task (i.e. manual handling of loads) on complementary tasks of the machine operation (total content of work)

Step 1: Assess the importance of each attribute using linguistic terms

Step 2: The experts evaluate subjectively each attribute of each alternative establishing the Design Range (DR) and System Range (SR) for each attribute. Where m_{ij} = (for i = 1, 2,…,N, j = 1, 2,…,M, k = 1, 2,…,k) are ratings the alternatives and attributes.

15.4.4 Stage of Analyze

15.4.4.1 Fuzzy Axiomatic Design Procedure Under Lean-Six Sigma Framework for the Stage of Analyze

At this stage three phases of the procedure are required: Phase 3, 4 and 5. On Phase 3, opinions from experts must be converted to numerical values and aggregated to conform decision matrices. On Phase 4 Membership Functions for every Ergonomic Functional Requirement for AMT are defined using triangular fuzzy numbers. On Phase 5 the Ergonomic Incompatibility Content (EIC) is determined by establishing Ergonomic System Range, Ergonomic Design Range and Common Area for every EFR. Theses phases are explained in the following sections.

Phase 3:

Step 1: Convert the linguistic terms of importance and the ratings assigned to each attribute to their numeric value

Step 2: Aggregate experts' opinions about the importance to obtain the weight of each attribute from pair wise comparisons of Analytical Hierarchical Process (AHP) methodology created by Saaty (using the geometric means expressed by Eq. (15.1)

Step 3: Determine DR for each attribute from experts' opinions and from Corollary number 6 of Axiomatic Design Theory

Step 4: Aggregate experts' opinions on the assigned rating of each attribute and sub attribute

$$\mathrm{m_{ij}} = \frac{1}{k}\left[m_{ij1} \oplus m_{ij2} \oplus \ldots\ldots \oplus m_{ijk}\right] \tag{15.1}$$

Step 5: Construct decision matrices to the assigned importance, DR, and SR for each attribute and each alternative.

Phase 4: Definition of the Membership Functions (MF) or $\mu(x)$ for every Ergonomic Functional Requirement. Figures 15.1 and 15.2 show the MF used for the proposed method in this Chapter. Membership functions were obtained by Eq. 15.2, where X_i, α, and θ are shown in Fig. 15.3.

$$\mu(\mathrm{x}) = \frac{\mathrm{X_i} - \alpha}{\theta - \alpha},\ \text{for benefit attributes}\quad \mu(\mathrm{x}) = \frac{\alpha - \mathrm{X_{i,}}}{\theta - \alpha}\ \text{for cost attributes} \tag{15.2}$$

Phase 5: Assess the EIC of each attribute for each alternative using the Information Axiom with weight. Figure 15.3 shows the SR, DR, and the Common Area. Eqs. 15.3–15.6 are used for calculate EIC.

$$\boldsymbol{EIC} = \mathbf{log_2} \frac{\textbf{Area of Ergonomic System Design (Triangular Fuzzy Number)}}{\textbf{Common Area}} \tag{15.3}$$

$$\mathit{Weighted\ EIC} = \sum \mathrm{w_i EIC_i}\ \text{Weighted Ergonomic Incompatibility Content} \tag{15.4}$$

$$\mathit{Common\,Area}\,(\mathit{CA}) = [\mu(\mathrm{x})^*(\mathrm{c}\ - \alpha)]^{*}1/2 \tag{15.5}$$

$$\text{System Area(SA)} = (\mathrm{c} - \mathrm{a})^{*}1/2 \tag{15.6}$$

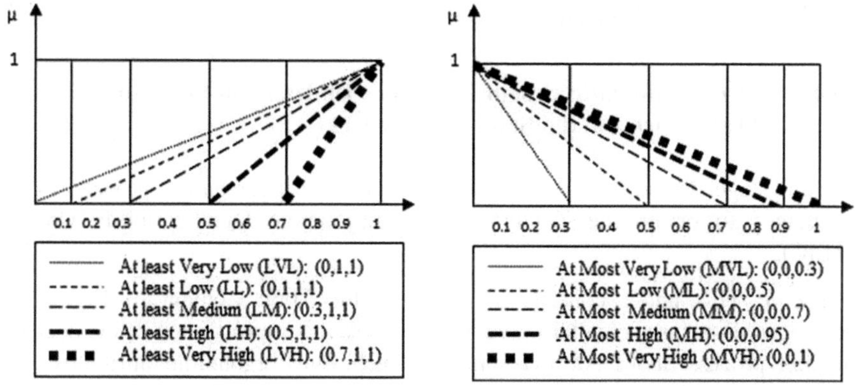

Fig. 15.1 Membership functions for intangible attributes

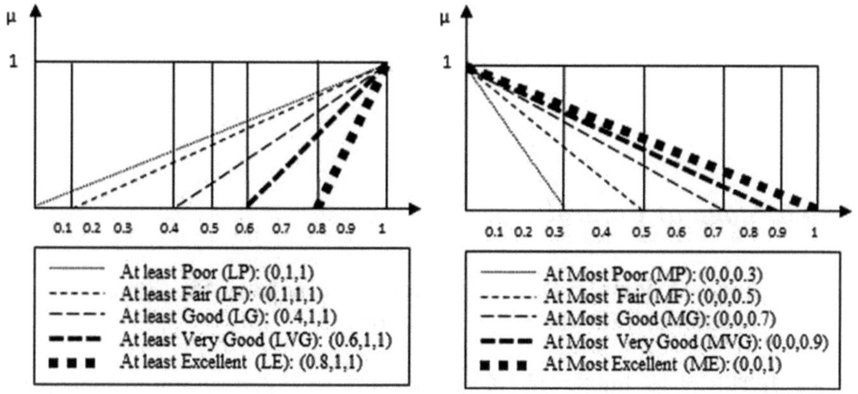

Fig. 15.2 Membership functions for tangible attributes

Fig. 15.3 Ergonomic system range, ergonomic design range and common area

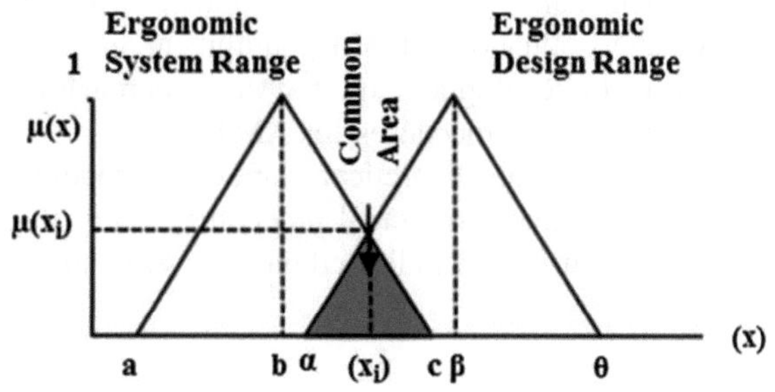

15.4.5 Stage of Improve

15.4.5.1 Fuzzy Axiomatic Design Procedure Under Lean-Six Sigma Framework for the Stage of Improves

On this stage, Phase 6 of the procedure implies the aggregation procedures are conducted to obtain the Total Weighted Ergonomic Incompatibility Content (TEIC).

Phase 6: This phase consists in obtaining the Total Content of Ergonomic Incompatibility for each alternative. In this phase the improvement of the decision making process about AMT is pursued integrating ergonomic attributes. The alternative which has the minimum Ergonomic Incompatibility Content (EIC) is chosen as the best choice. Decision makers in this phase are provided of the EIC of each alternative, also EIC for every attribute and subattribute are provided to lead to more complete decision about AMT.

$$TWEIC = \sum_{i=1}^{w} w_i EIC_i \tag{15.7}$$

15.4.6 Stage of Control

15.4.6.1 Fuzzy Axiomatic Design Procedure under Lean-Six Sigma Framework for the Stage of Control

Phase 7: This phase implies the development of an expert system to support this methodology application and look for long term sustainable improvement once decision of AMT has been made. Expert knowledge can be used every time a project for ergonomic compatibility evaluation is needed. Safety and Ergonomics department can develop in collaboration with Ergonomics and AMT experts' systematic evaluations of AMT equipment and AMT workstations; this idea have guided us to develop an expert system approach to enable future evaluations and decision making procedures for AMT.

To illustrate this theoretical framework proposal, Fig. 15.4 shows the Lean-Six Sigma deployment for actual procedure. Every stage of DMAIC methodology is developed, phases are indicated into this structure. The steps are included for every stage.

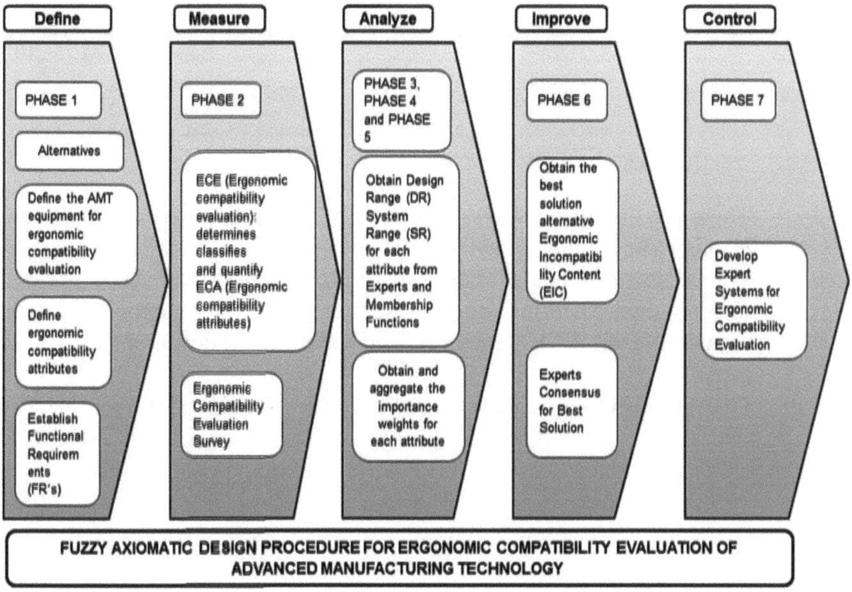

Fig. 15.4 Lean-six sigma framework for ergonomic compatibility evaluation of AMT

15.5 Results

15.5.1 Numerical Example

To illustrate the proposed methodology under this Lean-Six Sigma framework a numerical example for three alternatives of plastic molding machines has been presented. The procedure is resumed as following:

15.5.2 Define

Phase 1:

Step 1: Five attributes and twenty sub-attributes (explanation of attributes was described above). Corollary 6 of Axiomatic Design was used to establish the EFR's, which were: EFR_{A111}: At least good, EFR_{A112}: At least good, EFR_{A121}: At least excellent, EFR_{A122}: At least regular, EFR_{A123}: At least good, EFR_{A124}: At least regular, EFR_{A125}: Low, EFR_{A131}: At least good, EFR_{A132}: At least good, EFR_{A133}: At least good, EFR_{A134}: At least good, EFR_{A135}: At least good, EFR_{A136}: At least very good, EFR_{A137}: At least very good, EFR_{A141}: Low, EFR_{A142}: Low, EFR_{A143}: Low, EFR_{A144}: Low,FR_{A151}: At least good, and EFR_{A152}: At least very good

Table 15.3 Plastic molding machine alternatives

Alternative	Plastic molding machine
X	Battenfeld TM 75/210
Y	Vandorn 75
Z	New bury 75

Step 2: There are three alternatives of Plastic Molding Machines for evaluation in this case of study (Table 15.3)

Step 3: Three experts evaluated the alternatives; all experts had vast experience in the manufacturing and academic fields. Ergonomic Attributes included in the ECS were explained individually during a face to face interview

Step 4: Five linguistic terms were chosen according Tables 15.1 and 15.2.

15.5.3 Measure

Phase 2:

Step 1: The importance of each attribute was obtained via pairwise comparisons of AHP methodology

Step 2: Experts subjective evaluations were made using ECS. Further reading of Maldonado et al. (2009) is recommended.

15.5.4 Analyze

Phase 3:

Step 1: Convert the linguistic terms of the ratings assigned to each attribute (Tables 15.4, 15.5 and 15.6) to their numeric value (Tables 15.7, 15.8 and 15.9). Table 15.4 shows the evaluation of each attribute for the Alternative X in linguistic terms

Step 2: AHP was used to obtain the importance of each attribute. The results are shown in Table 15.10

Step 3: Determine DR for each attribute from experts' opinions and Corollary number 6 of Axiomatic Design Theory

Step 4: Add the experts' opinions on the assigned rating of each attribute to each alternative obtaining the SR. For example, for only one attribute A125 with Alternative X it is calculated as following: (Complete ratings of alternatives are shown in Tables 15.7, 15.8 and 15.9)

Table 15.4 Evaluation of alternative X for each attribute in linguistic terms by experts

Attributes	A111	A112	A121	A122	A123	A124	A125	A131	A132	A133
E1-X	G	VG	VG	VG	G	VG	H	VG	VG	VG
E2-X	VG	VG	VG	G	VG	VG	L	E	VG	VG
E3-X	VG	G	VG	VG	VG	VG	L	VG	VG	VG
Attributes	A134	A135	A136	A137	A141	A142	A143	A144	A151	A152
E1-X	E	VG	VG	G	M	L	L	L	G	G
E2-X	VG	VG	VG	VG	M	L	L	H	G	VG
E3-X	VG	VG	VG	G	VL	VL	VL	L	VG	VG

Table 15.5 Evaluation of alternative Y for each attribute in linguistic terms by experts

Attributes	A111	A112	A121	A122	A123	A124	A125	A131	A132	A133
E1-Y	P	VG	VG	G	R	VG	H	R	R	VG
E2-Y	G	G	G	R	P	G	L	R	R	R
E3-Y	R	R	G	G	G	G	L	R	G	R
Attributes	A134	A135	A136	A137	A141	A142	A143	A144	A151	A152
E1-Y	VG	R	G	R	M	L	M	L	G	R
E2-Y	R	R	R	R	M	H	VH	H	G	G
E3-Y	G	R	R	R	L	VL	VL	L	VG	G

Table 15.6 Evaluation of alternative Z for each attribute in linguistic terms by experts

Attributes	A111	A112	A121	A122	A123	A124	A125	A131	A132	A133
E1-Z	VG	VG	VG	G	G	VG	M	G	G	R
E2-Z	G	G	G	R	R	G	L	R	R	R
E3-Z	R	R	G	R	R	R	M	R	R	R
Attributes	A134	A135	A136	A137	A141	A142	A143	A144	A151	A152
E1-Z	G	P	R	R	H	L	M	L	G	G
E2-Z	R	G	G	R	M	H	VH	H	G	G
E3-Z	R	R	R	G	L	VL	L	L	VG	G

$$X_{A125\,(11)} = 1/3\,(0.5 + 0.0 + 0.0) = 0.167$$

$$X_{A125\,(12)} = 1/3\,(0.75 + 0.25 + 0.25) = 0.417$$

$$X_{A125\,(13)} = 1/3\,(1.0 + 0.5 + 0.5) = 0.667$$

So,

$$X_{A125} = [0.167,\ 0.417,\ 0.667]$$

Step 5: Construct decision matrices to the assigned importance, recommended rating (DR), and assigned ratings of each attribute to each alternative (SR). The results are shown in Table 15.11.

Table 15.7 Evaluation of alternative X for each attribute by experts (SR for each attribute)

Attributes	A111	A112	A121	A122	A123
E1-X	(0.4, 0.55, 0.7)	(0.6, 0.75, 0.9)	(0.6, 0.75, 0.9)	(0.6, 0.75, 0.9)	(0.4, 0.55, 0.7)
E2-X	(0.6, 0.75, 0.9)	(0.6, 0.75, 0.9)	(0.6, 0.75, 0.9)	(0.4, 0.55, 0.7)	(0.6, 0.75, 0.9)
E3-X	(0.6, 0.75, 0.9)	(0.4, 0.55, 0.7)	(0.6, 0.75, 0.9)	(0.6, 0.75, 0.9)	(0.6, 0.75, 0.9)
Attributes	A124	A125	A131	A132	A133
E1-X	(0.6, 0.75, 0.9)	(0.5, 0.75, 1)	(0.6, 0.75, 0.9)	(0.6, 0.75, 0.9)	(0.6, 0.75, 0.9)
E2-X	(0.6, 0.75, 0.9)	(0, 0.25, 0.5)	(0.8, 1.00, 1.0)	(0.6, 0.75, 0.9)	(0.6, 0.75, 0.9)
E3-X	(0.6, 0.75, 0.9)	(0, 0.25, 0.5)	(0.6, 0.75, 0.9)	(0.6, 0.75, 0.9)	(0.6, 0.75, 0.9)
Attributes	A134	A135	A136	A137	A141
E1-X	(0.8, 1.00, 1.0)	(0.6, 0.75, 0.9)	(0.6, 0.75, 0.9)	(0.4, 0.55, 0.7)	(0.3, 0.5, 0.7)
E2-X	(0.6, 0.75, 0.9)	(0.6, 0.75, 0.9)	(0.6, 0.75, 0.9)	(0.6, 0.75, 0.9)	(0.3, 0.5, 0.7)
E3-X	(0.6, 0.75, 0.9)	(0.6, 0.75, 0.9)	(0.6, 0.75, 0.9)	(0.4, 0.55, 0.7)	(0.0, 0.0, 0.3)
Attributes	A142	A143	A144	A151	A152
E1-X	(0.0, 0.25, 0.5)	(0.0, 0.25, 0.5)	(0.0, 0.25, 0.5)	(0.4, 0.55, 0.7)	(0.4, 0.55, 0.7)
E2-X	(0.0, 0.25, 0.5)	(0.0, 0.25, 0.5)	(0.5, 0.75, 1.0)	(0.4, 0.55, 0.7)	(0.6, 0.75, 0.9)
E3-X	(0.0, 0.00, 0.3)	(0.0, 0.00, 0.3)	(0.0, 0.25, 0.5)	(0.6, 0.75, 0.9)	(0.6, 0.75, 0.9)

Table 15.8 Evaluation of alternative Y for each attribute in fuzzy numbers by experts (SR for each attribute)

Attributes	A111	A112	A121	A122	A123
E1-Y	(0.0, 0.00, 0.3)	(0.6, 0.75, 0.9)	(0.6, 0.75, 0.9)	(0.4, 0.55, 0.7)	(0.2, 0.35, 0.5)
E2-Y	(0.4, 0.55, 0.7)	(0.4, 0.55, 0.7)	(0.4, 0.55, 0.7)	(0.2, 0.35, 0.5)	(0.0, 0.0, 0.3)
E3-Y	(0.2, 0.35, 0.5)	(0.2, 0.35, 0.5)	(0.4, 0.55, 0.7)	(0.4, 0.55, 0.7)	(0.4, 0.55, 0.7)
Attributes	A124	A125	A131	A132	A133
E1-Y	(0.6, 0.75, 0.9)	(0.5, 0.75, 1.0)	(0.2, 0.35, 0.5)	(0.2, 0.35, 0.5)	(0.6, 0.75, 0.9)
E2-Y	(0.4, 0.55, 0.7)	(0.0, 0.25, 0.5)	(0.2, 0.35, 0.5)	(0.2, 0.35, 0.5)	(0.2, 0.35, 0.5)
E3-Y	(0.4, 0.55, 0.7)	(0.0, 0.25, 0.5)	(0.2, 0.35, 0.5)	(0.4, 0.55, 0.7)	(0.2, 0.35, 0.5)
Attributes	A134	A135	A136	A137	A141
E1-Y	(0.6, 0.75, 0.9)	(0.2, 0.35, 0.5)	(0.4, 0.55, 0.7)	(0.2, 0.35, 0.5)	(0.3, 0.50, 0.7)
E2-Y	(0.2, 0.35, 0.5)	(0.2, 0.35, 0.5)	(0.2, 0.35, 0.5)	(0.2, 0.35, 0.5)	(0.3, 0.50, 0.7)
E3-Y	(0.4, 0.55, 0.7)	(0.2, 0.35, 0.5)	(0.2, 0.35, 0.5)	(0.2, 0.35, 0.5)	(0.0, 0.25, 0.5)
Attributes	A142	A143	A144	A151	A152
E1-Y	(0.0, 0.25, 0.5)	(0.3, 0.5, 0.7)	(0.0, 0.25, 0.5)	(0.4, 0.55, 0.7)	(0.2, 0.35, 0.5)
E2-Y	(0.5, 0.75, 1.0)	(0.7, 1.0, 1.0)	(0.5, 0.75, 1.0)	(0.4, 0.55, 0.7)	(0.4, 0.55, 0.7)
E3-Y	(0.0, 0.00, 0.3)	(0.0, 0.0, 0.3)	(0.0, 0.25, 0.5)	(0.6, 0.75, 0.9)	(0.4, 0.55, 0.7)

Phase 4: Definition of the Membership Functions for DR and SR; for attribute A125 in Alternative X is shown as an example as following. Complete Common Area calculations are shown in Tables 15.12, 15.13 and 15.14.

Table 15.9 Evaluation of alternative Z for each attribute in triangular fuzzy numbers by experts (SR for each attribute)

Attributes	A111	A112	A121	A122	A123
E1-Z	(0.6, 0.75, 0.9)	(0.6, 0.75, 0.9)	(0.6, 0.75, 0.9)	(0.4, 0.55, 0.7)	(0.4, 0.55, 0.7)
E2-Z	(0.4, 0.55, 0.7)	(0.4, 0.55, 0.7)	(0.4, 0.55, 0.7)	(0.2, 0.35, 0.5)	(0.2, 0.35, 0.5)
E3-Z	(0.2, 0.35, 0.5)	(0.2, 0.35, 0.5)	(0.4, 0.55, 0.7)	(0.2, 0.35, 0.5)	(0.2, 0.35, 0.5)
Attributes	A124	A125	A131	A132	A133
E1-Z	(0.6, 0.75, 0.9)	(0.3, 0.50, 0.7)	(0.4, 0.55, 0.7)	(0.4, 0.55, 0.7)	(0.2, 0.35, 0.5)
E2-Z	(0.4, 0.55, 0.7)	(0.0, 0.25, 0.5)	(0.2, 0.35, 0.5)	(0.2, 0.35, 0.5)	(0.2, 0.35, 0.5)
E3-Z	(0.2, 0.35, 0.5)	(0.3, 0.50, 0.7)	(0.2, 0.35, 0.5)	(0.2, 0.35, 0.5)	(0.2, 0.35, 0.5)
Attributes	A134	A135	A136	A137	A141
E1-Z	(0.4, 0.55, 0.7)	(0.0, 0.0, 0.3)	(0.2, 0.35, 0.5)	(0.2, 0.35, 0.5)	(0.5, 0.75, 1.0)
E2-Z	(0.2, 0.35, 0.5)	(0.4, 0.55, 0.7)	(0.4, 0.55, 0.7)	(0.2, 0.35, 0.5)	(0.3, 0.5, 0.7)
E3-Z	(0.2, 0.35, 0.5)	(0.2, 0.35, 0.5)	(0.2, 0.35, 0.5)	(0.4, 0.55, 0.7)	(0.0, 0.25, 0.5)
Attributes	A142	A143	A144	A151	A152
E1-Z	(0.0, 0.25, 0.5)	(0.3, 0.5, 0.7)	(0.0, 0.25, 0.5)	(0.4, 0.55, 0.7)	(0.4, 0.55, 0.7)
E2-Z	(0.5, 0.75, 1.0)	(0.7, 1.0, 1.0)	(0.5, 0.75, 1.0)	(0.4, 0.55, 0.7)	(0.4, 0.55, 0.7)
E3-Z	(0.0, 0.00, 0.3)	(0.0, 0.25, 0.5)	(0.0, 0.25, 0.5)	(0.6, 0.75, 0.9)	(0.4, 0.55, 0.7)

$$\mu(x) = \frac{0.570 - 0.300}{1 - 0.300} = 0.385$$

Phase 5: Assess the EIC of each attribute for each alternative using the Information Axiom with weight. In order to obtain the EIC a sample calculation for only one attribute (A125 in Alternative X) is given as an example:

$$Common\,Area\,(CA) = [0.385^{*}\,(0.667 - 0.300)]\,{}^{*}\,1/2 = 0.070$$
$$System\,Area\,(SA) = (0.667 - 0.167)\,{}^{*}\,1/2 = 0.25 \text{ and}$$
$$EIC_{A125} = \log_2(0.25/0.070) = 1.820$$

This is a preliminary result, since importance weight at each level has not been applied.

Table 15.12 shows the EIC of each attribute for Alternative X (Battenfeld TM 75/210).

Table 15.13 show the EIC of each attribute for Alternative Y (van Dorn 75).

15.5.5 Improve

Phase 6: This phase consists of obtaining the Total Ergonomic Incompatibility Content (TEIC) for each alternative, using Eq. 15.3 at each level of the hierarchy. The alternative which has the minimum EIC is chosen as the best choice.

Table 15.10 Importance of the attributes by experts obtained by AHP

	A11	A12	A13	A14	A15	A111	A112	A121	A122	A123	A124	A125	A131
w AHP	0.355	0.175	0.236	0.113	0.122	0.712	0.288	0.175	0.266	0.269	0.167	0.123	0.128

	A132	A133	A134	A135	A136	A137	A141	A142	A143	A144	A151	A152
w AHP	0.132	0.128	0.203	0.120	0.156	0.135	0.320	0.094	0.286	0.300	0.590	0.410

Table 15.11 Fuzzy decision matrix for the assigned rating to each alternative by experts (SR)

Alternative	A111	A112	A121	A122	A123
X	(0.53, 0.68, 0.83)	(0.53, 0.68, 0.83)	(0.60, 0.75, 0.90)	(0.53, 0.68, 0.83)	(0.53, 0.68, 0.83)
Y	(0.20, 0.30, 0.50)	(0.40, 0.55, 0.70)	(0.47, 0.62, 0.77)	(0.33, 0.48, 0.63)	(0.20, 0.30, 0.50)
Z	(0.40, 0.55, 0.70)	(0.40, 0.55, 0.70)	(0.47, 0.62, 0.77)	(0.27, 0.42, 0.57)	(0.27, 0.42, 0.57)
W AHP	0.094	0.286	0.3	0.59	0.41
Alternative	A124	A125	A131	A132	A133
X	(0.60, 0.75, 0.90)	(0.17, 0.42, 0.67)	(0.67, 0.83, 0.93)	(0.60, 0.75, 0.90)	(0.60, 0.75, 0.90)
Y	(0.47, 0.62, 0.77)	(0.17, 0.42, 0.67)	(0.20, 0.35, 0.50)	(0.27, 0.42, 0.57)	(0.33, 0.48, 0.63)
Z	(0.40, 0.55, 0.70)	(0.20, 0.42, 0.63)	(0.27, 0.42, 0.57)	(0.27, 0.42, 0.57)	(0.20, 0.35, 0.50)
w AHP	0.167	0.123	0.128	0.132	0.128
Alternative	A134	A135	A136	A137	A141
X	(0.67, 0.83, 0.93)	(0.60, 0.75, 0.90)	(0.60, 0.75, 0.90)	(0.47, 0.62, 0.77)	(0.20, 0.33, 0.57)
Y	(0.40, 0.55, 0.70)	(0.20, 0.35, 0.50)	(0.27, 0.42, 0.57)	(0.20, 0.35, 0.50)	(0.20, 0.42, 0.63)
Z	(0.27, 0.42, 0.57)	(0.20, 0.3, 0.50)	(0.27, 0.42, 0.57)	(0.27, 0.42, 0.57)	(0.27, 0.5, 0.73)
w AHP	0.203	0.12	0.156	0.135	0.32
Alternative	A142	A143	A144	A151	A152
X	(0.00, 0.17, 0.43)	(0.00, 0.17, 0.43)	(0.17, 0.42, 0.67)	(0.47, 0.62, 0.77)	(0.53, 0.68, 0.83)
Y	(0.17, 0.33, 0.6)	(0.33, 0.50, 0.67)	(0.17, 0.42, 0.67)	(0.47, 0.62, 0.77)	(0.33, 0.48, 0.63)
Z	(0.17, 0.33, 0.6)	(0.33, 0.58, 0.73)	(0.17, 0.42, 0.67)	(0.47, 0.62, 0.77)	(0.40, 0.55, 0.70)
W AHP	0.094	0.286	0.3	0.59	0.41

Table 15.12 EIC for plastic molding machines, alternative X

Alternative X				
Attribute	Design range		System range	C.A.
	Design range	TFN (α, β, θ)	TFN (a, b, c)	
A111	At least good	(0.400, 1.000, 1.000)	(0.533, 0.683, 0.833)	0.105
A112	At least good	(0.400, 1.000, 1.000)	(0.533, 0.683, 0.833)	0.105
A121	At least good	(0.400, 1.000, 1.000)	(0.600, 0.750, 0.900)	0.122
A122	At least good	(0.400, 1.000, 1.000)	(0.533, 0.683, 0.833)	0.105
A123	At least good	(0.400, 1.000, 1.000)	(0.533, 0.683, 0.833)	0.105
A124	At least good	(0.400, 1.000, 1.000)	(0.600, 0.750, 0.900)	0.122
A125	At least medium	(0.300, 1.000, 1.000)	(0.167, 0.417, 0.667)	0.07
A131	At least good	(0.400, 1.000, 1.000)	(0.667, 0.833, 0.933)	0.121
A132	At least good	(0.400, 1.000, 1.000)	(0.600, 0.750, 0.900)	0.122
A133	At least good	(0.400, 1.000, 1.000)	(0.600, 0.750, 0.900)	0.122
A134	At least good	(0.400, 1.000, 1.000)	(0.667, 0.833, 0.933)	0.121
A135	At least good	(0.400, 1.000, 1.000)	(0.600, 0.750, 0.900)	0.122
A136	At least good	(0.400, 1.000, 1.000)	(0.600, 0.750, 0.900)	0.122
A137	At least good	(0.400, 1.000, 1.000)	(0.467, 0.617, 0.767)	0.085
A141	At most medium	(0.000, 0.000, 0.700)	(0.200, 0.333, 0.567)	0.13
A142	At most medium	(0.000, 0.000, 0.700)	(0.000, 0.167, 0.433)	0.201
A143	At most low	(0.000, 0.000, 0.500)	(0.000, 0.167, 0.433)	0.178
A144	At most medium	(0.000, 0.000, 0.700)	(0.167, 0.417, 0.667)	0.148
A151	At least good	(0.400, 1.000, 1.000)	(0.467, 0.617, 0.767)	0.085
A152	At least fair	(0.200, 1.000, 1.000)	(0.533, 0.683, 0.833)	0.126

(continued)

Table 15.12 (continued)

Alternative X (continued)								
Attribute	System area	Log SA	Log CA	Incompatibility content				
				I	w (2nd. level)	(I)(w) (2nd. level)	w (1st. level)	(I)(w) (1st. level)
A111	0.15	−2.73	−3.24	0.509	0.712	0.362	0.355	0.128
A112	0.15	−2.73	−3.24	0.509	0.288	0.146	0.355	0.052
A121	0.15	−2.73	−3.03	0.296	0.175	0.051	0.175	0.009
A122	0.15	−2.73	−3.24	0.509	0.266	0.135	0.175	0.023
A123	0.15	−2.73	−3.24	0.509	0.269	0.136	0.175	0.024
A124	0.15	−2.73	−3.03	0.296	0.167	0.049	0.175	0.008
A125	0.25	−2.00	−3.82	1.820	0.123	0.223	0.175	0.039
A131	0.13	−2.90	−3.04	0.142	0.128	0.018	0.236	0.004
A132	0.15	−2.73	−3.03	0.296	0.132	0.039	0.236	0.009
A133	0.15	−2.73	−3.03	0.296	0.128	0.037	0.236	0.008
A134	0.13	−2.90	−3.04	0.142	0.203	0.028	0.236	0.006
A135	0.15	−2.73	−3.03	0.296	0.120	0.035	0.236	0.008
A136	0.15	−2.73	−3.03	0.296	0.156	0.046	0.236	0.010
A137	0.15	−2.73	−3.56	0.823	0.135	0.111	0.236	0.026
A141	0.18	−2.44	−2.94	0.498	0.320	0.159	0.113	0.018
A142	0.21	−2.20	−2.31	0.112	0.094	0.010	0.113	0.001
A143	0.21	−2.20	−2.49	0.285	0.286	0.081	0.113	0.009
A144	0.25	−2.00	−2.75	0.753	0.300	0.226	0.113	0.025
A151	0.15	−2.73	−3.56	0.823	0.590	0.485	0.122	0.059
A152	0.15	−2.73	−2.99	0.256	0.410	0.105	0.122	0.012
						Ergonomic incompatibility content X =		0.486

Table 15.13 EIC for plastic molding machines, alternative Y

Alternative Y				
Attribute	Design range		System range	C.A.
	Design range	TFN (α, β, θ)	TFN (a, b, c)	
A111	At least good	(0.400, 1.000, 1.000)	(0.200, 0.300, 0.500)	0.006
A112	At least good	(0.400, 1.000, 1.000)	(0.400, 0.550, 0.700)	0.060
A121	At least good	(0.400, 1.000, 1.000)	(0.467, 0.617, 0.767)	0.085
A122	At least good	(0.400, 1.000, 1.000)	(0.333, 0.483, 0.633)	0.036
A123	At least good	(0.400, 1.000, 1.000)	(0.200, 0.300, 0.500)	0.006
A124	At least good	(0.400, 1.000, 1.000)	(0.467, 0.617, 0.767)	0.085
A125	At least medium	(0.300, 1.000, 1.000)	(0.167, 0.417, 0.667)	0.071
A131	At least good	(0.400, 1.000, 1.000)	(0.200, 0.350, 0.500)	0.007
A132	At least good	(0.400, 1.000, 1.000)	(0.267, 0.417, 0.567)	0.019
A133	At least good	(0.400, 1.000, 1.000)	(0.333, 0.483, 0.633)	0.036
A134	At least good	(0.400, 1.000, 1.000)	(0.400, 0.550, 0.700)	0.060
A135	At least good	(0.400, 1.000, 1.000)	(0.200, 0.350, 0.500)	0.007
A136	At least good	(0.400, 1.000, 1.000)	(0.267, 0.417, 0.567)	0.019
A137	At least good	(0.400, 1.000, 1.000)	(0.200, 0.350, 0.500)	0.007
A141	At least medium	(0.000, 0.000, 0.700)	(0.200, 0.417, 0.633)	0.132
A142	At least medium	(0.000, 0.000, 0.700)	(0.167, 0.333, 0.600)	0.148
A143	At least low	(0.000, 0.000, 0.500)	(0.333, 0.500, 0.667)	0.021
A144	At least medium	(0.000, 0.000, 0.700)	(0.167, 0.417, 0.667)	0.148
A151	At least good	(0.400, 1.000, 1.000)	(0.467, 0.617, 0.767)	0.085
A152	At least regular	(0.200, 1.000, 1.000)	(0.333, 0.483, 0.633)	0.085

(continued)

Table 15.13 (continued)

Alternative Y (continued)								
Attribute	System area	Log SA	Log CA	Incompatibility content				
				I	w (2nd. level)	(I)(w) (2nd. level)	w (1st. level)	(I)(w) (1st. level)
A111	0.15	−2.73	−7.32	4.585	0.712	3.264	0.355	1.158
A112	0.15	−2.73	−4.05	1.322	0.288	0.380	0.355	0.135
A121	0.15	−2.73	−3.56	0.823	0.175	0.144	0.175	0.025
A122	0.15	−2.73	−4.78	2.049	0.266	0.545	0.1750	0.095
A123	0.15	−2.73	−7.32	4.585	0.269	1.233	0.175	0.215
A124	0.15	−2.73	−3.56	0.823	0.167	0.137	0.175	0.024
A125	0.25	−2.00	−3.82	1.820	0.123	0.223	0.175	0.039
A131	0.15	−2.73	−7.22	4.492	0.128	0.575	0.236	0.135
A132	0.15	−2.73	−5.75	3.015	0.132	0.398	0.236	0.093
A133	0.15	−2.73	−4.78	2.049	0.128	0.262	0.236	0.061
A134	0.15	−2.73	−4.05	1.322	0.203	0.268	0.236	0.063
A135	0.15	−2.73	−7.22	4.492	0.120	0.539	0.236	0.127
A136	0.15	−2.73	−5.75	3.015	0.156	0.470	0.236	0.111
A137	0.15	−2.73	−7.22	4.492	0.135	0.606	0.236	0.143
A141	0.21	−2.20	−2.92	0.718	0.320	0.229	0.113	0.026
A142	0.21	−2.20	−2.75	0.547	0.094	0.051	0.113	0.005
A143	0.16	−2.58	−5.58	2.998	0.286	0.857	0.113	0.096
A144	0.25	−2.00	−2.75	0.753	0.300	0.226	0.113	0.025
A151	0.15	−2.73	−3.56	0.823	0.590	0.485	0.122	0.059
A152	0.15	−2.73	−3.55	0.818	0.410	0.335	0.122	0.040
						Ergonomic incompatibility content Y =		2.684

Table 15.14 EIC of each attribute for alternative Z

Alternative Z				
Attribute	Design range		System range	C.A.
	Design range	TFN (α, β, θ)	TFN (a, b, c)	
A111	At least good	(0.400, 1.000, 1.000)	(0.400, 0.550, 0.700)	0.060
A112	At least good	(0.400, 1.000, 1.000)	(0.400, 0.550, 0.700)	0.060
A121	At least good	(0.400, 1.000, 1.000)	(0.467, 0.617, 0.767)	0.085
A122	At least good	(0.400, 1.000, 1.000)	(0.267, 0.417, 0.567)	0.019
A123	At least good	(0.400, 1.000, 1.000)	(0.267, 0.417, 0.567)	0.019
A124	At least good	(0.400, 1.000, 1.000)	(0.400, 0.550, 0.700)	0.060
A125	At least medium	(0.300, 1.000, 1.000)	(0.200, 0.417, 0.633)	0.061
A131	At least good	(0.400, 1.000, 1.000)	(0.267, 0.417, 0.567)	0.019
A132	At least good	(0.400, 1.000, 1.000)	(0.267, 0.417, 0.567)	0.019
A133	At least good	(0.400, 1.000, 1.000)	(0.200, 0.350, 0.500)	0.007
A134	At least good	(0.400, 1.000, 1.000)	(0.267, 0.417, 0.567)	0.019
A135	At least good	(0.400, 1.000, 1.000)	(0.200, 0.300, 0.500)	0.006
A136	At least good	(0.400, 1.000, 1.000)	(0.267, 0.417, 0.567)	0.019
A137	At least good	(0.400, 1.000, 1.000)	(0.267, 0.417, 0.567)	0.019
A141	At most medium	(0.000, 0.000, 0.700)	(0.267, 0.500, 0.733)	0.148
A142	At most medium	(0.000, 0.000, 0.700)	(0.167, 0.333, 0.600)	0.148
A143	At most low	(0.000, 0.000, 0.500)	(0.333, 0.583, 0.733)	0.019
A144	At most medium	(0.000, 0.000, 0.700)	(0.167, 0.417, 0.667)	0.148
A151	Al least good	(0.400, 1.00, 1.00)	(0.460, 0.610, 0.76)	0.085
A152	At least fair	(0.200, 1.00, 1.00)	(0.400, 0.55, 0.70)	0.101

(continued)

Table 15.14 (continued)

Alternative Z (Continued)								
Attribute	System area	Log SA	Log CA	Incompatibility content				
				I	w (2nd. level)	(I)(w) (2nd. level)	w (1st. level)	(I)(w) (1st. level)
A111	0.15	−2.73	−4.05	1.322	0.712	0.941	0.355	0.334
A112	0.15	−2.73	−4.05	1.322	0.288	0.380	0.355	0.135
A121	0.15	−2.73	−3.56	0.823	0.175	0.144	0.175	0.025
A122	0.15	−2.73	−5.75	3.015	0.266	0.802	0.175	0.140
A123	0.15	−2.73	−5.75	3.015	0.269	0.811	0.175	0.141
A124	0.15	−2.73	−4.05	1.322	0.167	0.220	0.175	0.038
A125	0.21	−2.20	−4.04	1.838	0.123	0.226	0.175	0.039
A131	0.15	−2.73	−5.75	3.015	0.128	0.385	0.236	0.091
A132	0.15	−2.73	−5.75	3.015	0.132	0.398	0.236	0.093
A133	0.15	−2.73	−7.22	4.492	0.128	0.575	0.236	0.135
A134	0.15	−2.73	−5.75	3.015	0.203	0.612	0.236	0.144
A135	0.15	−2.73	−7.32	4.585	0.120	0.550	0.236	0.129
A136	0.15	−2.73	−5.75	3.015	0.156	0.470	0.236	0.111
A137	0.15	−2.73	−5.75	3.015	0.135	0.407	0.236	0.096
A141	0.23	−2.10	−2.75	0.654	0.320	0.209	0.113	0.023
A142	0.21	−2.20	−2.75	0.547	0.094	0.051	0.113	0.005
A143	0.20	−2.32	−5.75	3.431	0.286	0.981	0.113	0.110
A144	0.25	−2.00	−2.75	0.753	0.300	0.226	0.113	0.025
A151	0.15	−2.73	−3.56	0.823	0.590	0.485	0.122	0.059
A152	0.15	−2.73	−3.31	0.573	0.410	0.235	0.122	0.028
						Ergonomic incompatibility content Z =		1.911

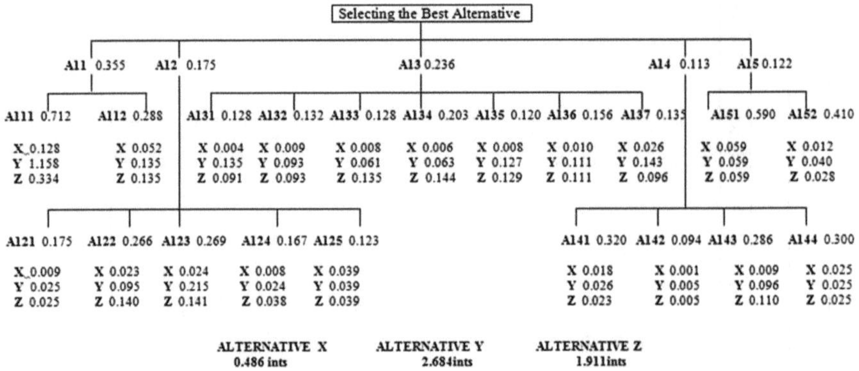

Fig. 15.5 Ergonomic incompatibility content diagram

Figure 15.5 shows the weights for each alternative and the Total Ergonomic Incompatibility Content.

*Note: Some attributes are shown below second level attributes due to the limited space of this chapter.

The alternative X is chosen as the best alternative for our goal. The university manufacturing laboratory personnel agreed in selecting it as the one which has more probability of accomplish the EFR's.

15.5.6 Control

Phase 7: Safety and Ergonomics personnel collaborated with Ergonomics and AMT experts' to develop systematic evaluations of AMT equipment and AMT workstations and documented the project. This idea helped propose an expert system approach to enable future evaluations and decision making procedures for AMT assuring that the measurement system may encourage the attention on ergonomic problems arisen by AMT interaction and may help formulate and implement ergonomic interventions.

15.6 Conclusions

As conclusions, this Lean-Six Sigma Theoretical Framework for ergonomic evaluation of AMT offers the structure where ergonomic compatibility evaluation can be implemented. Six Sigma is about improving the quality and accuracy of processes by reducing variation, while Lean focuses on achieving quicker response times by eliminating waste. These two methodologies assist finding the root cause of diverse business challenges. One of these challenges is based in the lack of effective methods for evaluate ergonomics and safety attributes for AMT selection. The main goal in this case is conducting an ergonomic oriented project for selection or evaluation of AMT and a Lean-Six Sigma (LSS) approach may help encouraging more complete decision making processes considering safety and ergonomics aspects. In this way, the DMs could conveniently regard an ergonomic perspective in their final decision having a robust reference framework. Some conclusions can be made about the effectiveness of a multi-attribute approach, given that the structured hierarchy for the multiple ergonomic requirements for AMT selection helps understanding this complex problem. Also, the proposed fuzzy axiomatic design methodology for ergonomic evaluation of AMT becomes feasible under the LSS framework and promotes the importance of ergonomic and safety aspects for reducing waste and ergonomic risk the manufacturing industry.

Acknowledgments We like to thanks the contribution of the CNC experts of Centro de Entrenamiento en Alta Tecnología (CENALTEC), we appreciate your help and enthusiasm to develop this project.

References

Carayon, P., & Smith, M. J. (2000). Work organization and ergonomics. *Applied Ergonomics, 31*, 649–662.

Drury, C. G. (1997). Ergonomics and the quality movement. *Ergonomics, 40*(3), 249–264.

Fullerton, R. R., McWatters, C. S., & Fawson, C. (2003). An examination of the relation- ships between JIT and financial performance. *Journal of Operations Management, 21*(4), 383–404.

García, J. L., Noriega, S. A. & Ventura, R. A. (2008). Multicriteria methodology for advanced manufacturing technology (AMT) evaluation. *International Journal of Industrial Engineering*, Special Issue, 499-509.

García, J. L., Noriega, S. A., Martínez, E. A. (2009). A multicriteria approach for the location of product warehouse. *International Journal of Industrial Engineering*, Special Issue, 409-417.

Helander, M. G. (1995). Conceptualizing the use of axiomatic design procedures in ergonomics. In *Proceedings of the IEA World Conference, Associação Brasileira de Ergonomia*, Rio de Janeiro, Brazil (pp. 38–41).

Helander, M. G., & Lin, L. (2002). Axiomatic design in ergonomics and extension of information axiom. *Journal of Engineering Design, 13*(4), 321–339.

Karwowski, W. (1997). Ancient wisdom and future technology: The old tradition and the new science of human factors/ergonomics. In *Proceedings of the Human Factors and Ergonomics Society 4th Annual Meeting, Albuquerque, NM, USA* (pp 875–877). Santa Monica, CA, USA: Human Factors and Ergonomics Society.

Karwowski, W. (2000). Simvatology: The science of an artifact-human compatibility. *Theoretical Issues in Ergonomics Science, 1*(1), 76–91.

Karwowski, W. (2001). *International Encyclopedia of Ergonomics and Human Factors*. London: Taylor & Francis.

Karwowski, W. (2005). Ergonomics and human factors: The paradigms for science, engineering, design, technology, and management of human-compatible systems. *Ergonomics, 48*(5), 436–463.

Kulak, O., Bulent, D. M., & Kahraman, C. (2005). Fuzzy multi-attribute equipment selection based on information axiom. *Journal of Materials Processing Technology, 169*, 337–345.

Linderman, K., Schroeder, R. G., & Choo, A. (2006). Six sigma: The role of goals in improvement teams. *Journal of Operations Management, 24*(6), 779–790.

MacDuffie, J. P. (1995). Human resource bundles and manufacturing performance: Organizational logic and flexible production systems in the world auto industry. *Industrial and Labor Relations Review, 48*(2), 197–221.

McKone, K. E., Schroeder, R. G., & Cua, K. O. (1999). Total productive maintenance: A contextual view. *Journal of Operations Management, 17*(2), 123–144.

McLachlin, R. (1997). Management initiatives and just-in-time manufacturing. *Journal of Operations Management, 15*(4), 271–292.

Maldonado-Macías, A., De la Riva J., Noriega S., Díaz J.J. (2008). Aplicaciones del Axioma de Información en Procesos de Evaluación y Selección de Instalaciones y Equipamiento. In *Proceedings of the 1st. International Congress of Undergraduate Studies and Research*, Ciudad Juarez Technology Institute, (pp. 380–388).

Maldonado, A., Sánchez J., Noriega S., Díaz J.J., García J.L., Vidal L. (2009). A hierarchical fuzzy axiomatic design survey for ergonomic compatibility evaluation of advanced manufacturing technology—AMT. In *Proceedings of the XXIst. Annual International Conference of Occupational Safety and Ergonomics, International Society for Occupational Ergonomics and Safety*, (pp 18–23).

Maldonado, A., García, J., Alvarado, A., & Balderrama, C. (2013). A hierarchical fuzzy axiomatic design methodology for ergonomic compatibility evaluation of advanced manufacturing technology. *International Journal of Advanced Manufacturing Technology, 66*(1–4), 171–186.

Saaty, T.L. (1980). *The analytic hierachy process*. New York: Mc Graw Hill Inc.

Shaffie,S., Shahbazi, S. (2012). *Lean six sigma. The McGraw-Hill 36-hour course*. New York: The McGraw-Hill Companies.

Shah, R., & Ward, P. T. (2007). Defining and developing measures of lean production. *Journal of Operations Management, 25*(4), 785–805.

Simpson, D. F., & Power, D. J. (2005). Use the supply relationship to develop lean and green suppliers. *Supply Chain Management: An International Journal, 10*(1), 60–68.

Swink, M., Narasimhan, R., & Kim, S. W. (2005). Manufacturing practices and strategy integration: effects on cost efficiency, flexibility, and market-based performance. *Decision Sciences, 36*(3), 427–457.

Venkata, R. R. (2007). *Decision making in the manufacturing environment. Using graph theory and fuzzy multiple attribute decision making methods*. London: Springer.

Womack, J., Jones, D., & Roos, D. (1990). *The machine that changed the world*. New York: Macmillan.

Chapter 16
Expert System Development Using Fuzzy If–Then Rules for Ergonomic Compatibility of AMT for Lean Environments

Arturo Realyvásquez-Vargas, Aidé Aracely Maldonado-Macías, Jorge Luis García-Alcaraz and Alejandro Alvarado-Iniesta

Abstract Lean Manufacturing is a revolution that is impacting the manufacturing process—it isn't just about using tools, or changing a few steps in our manufacturing processes—it's about the complete change of businesses—how the supply chain operates, how the directors direct, how the managers manage, how employees—people—go about their daily work. Advanced Manufacturing Technology (AMT) is a field that offers opportunities to the enterprises to adopt a Lean Manufacturing posture. This chapter provides a description of the development of an expert system for the ergonomic compatibility evaluation of AMT for Lean Environments. This system allows the evaluation of tangible and intangible ergonomic compatibility attributes of a variety of alternatives of AMT based on a multiattribute fuzzy axiomatic design approach rules, so the evaluation is by mean of linguistic terms. As result, system offers the ergonomic compatibility content of all the alternatives and the expert system direct them to the alternative that best satisfies ergonomic attributes. The results of the expert system were validated by the ergonomic compatibility evaluation of AMT carried out by experts.

Keywords Ergonomic compatibility · AMT · Fuzzy logic · Axiomatic design

16.1 Introduction

Lean manufacturing was developed by Toyota motor company to address their specific needs in a restricted market in times of economic trouble (Manotas and Rivera 2007). The Toyota Production System was based around the desire to

A. Realyvásquez-Vargas (✉) · A. A. Maldonado-Macías · J. L. García-Alcaraz · A. Alvarado-Iniesta
Department of Industrial and Manufacturing Engineering, Autonomous University of Ciudad Juarez, Avenue Del Charro 450 Norte, Col. Partido Romero, Ciudad Juárez, CHIH, Mexico
e-mail: arturo.realyvasquez@uacj.mx

J. L. García-Alcaraz et al. (eds.), *Lean Manufacturing in the Developing World*,
DOI: 10.1007/978-3-319-04951-9_16,

produce in a continuous flow which did not rely on long production runs to be efficient; it was based around the recognition that only a small fraction of the total time and effort to process a product added value to the end user (Melton 2005). Lean manufacturing is a multi-dimensional approach that encompasses a wide variety of management practices, including just-in-time, quality systems, work teams, cellular manufacturing, supplier management, total preventative maintenance, human resource management, etc., in an integrated system (MacDuffie 1995; McLachlin 1997; Shah and Ward 2003).

The focus of lean manufacturing is the systematic elimination of wastes from organization's operations through a set of synergistic work practices to produce products and services at the rate of demand (Fullerton et al. 2003; Shah and Ward 2007; Simpson and Power 2005; Womack et al. 1990; Yang et al. 2011). This ensures that every activity and process step adds *value* to the end product or service (Shaffie and Shahbazi 2012).

Two important practices of lean manufacturing are quality systems and human resource management. The first practice, quality systems, includes quality improvement strategies, such as total quality management (TQM) and six sigma. According to Carayon and Smith (2000) and Drury (1997), there exists an interaction of mutual benefits between ergonomics and TQM. These interactions are:

- the use of ergonomics to improve the performance of quality control inspectors;
- applications of TQM to safety aspects of ergonomics;
- linkages between TQM and macro-ergonomics or socio-technical systems;
- open systems strategic issues;
- systems approaches to organization design and leadership;
- measurement-based operations;
- appropriate use of technology;
- Individuals, teams and the change process.

Six sigma as a quality improvement strategy is a statistical problem-solving methodology and a management philosophy which is flexible in terms of being able to apply it to any functional business area, including manufacturing. It contains five distinct problem-solving phases known as the DMAIC approach (Shaffie and Shahbazi 2012):

- *Define* the problem statement, the goal, and the financial benefits.
- *Measure* the current performance of the process and collect the required data.
- *Analyze* the root cause of the problem.
- *Improve* the process to eliminate errors and instability.
- *Control* the performance of the process, ensuring that the improvements are sustained.

Human resource management is defined as a strategic and coherent approach to the management of an organization's most valued asset- the people working there who individually and collectively contribute to the achievement of its objectives (Armstrong 2006).

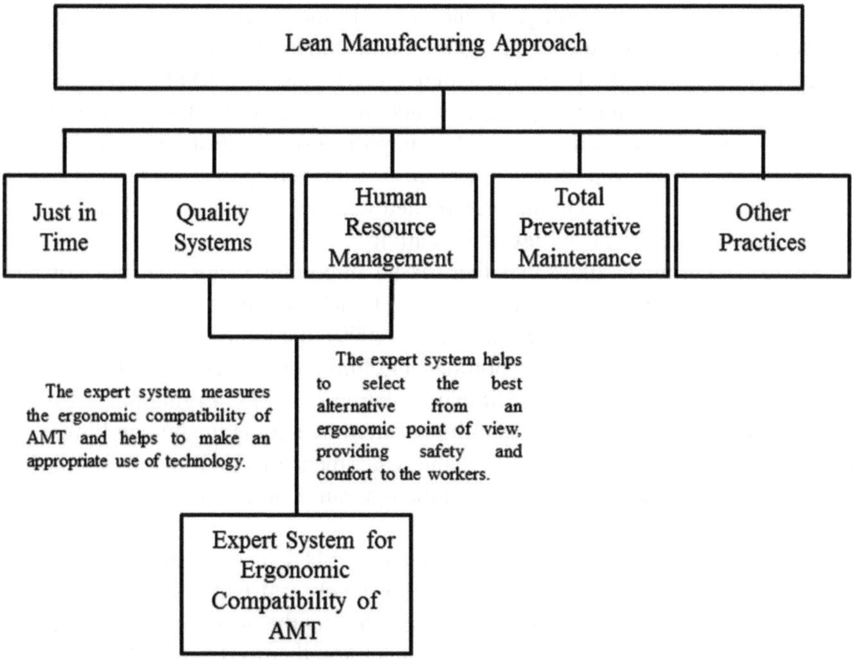

Fig. 16.1 Space of the expert system for ergonomic compatibility of AMT

This project describes the development of an expert system to propose the measurement of the ergonomic compatibility of Advanced Manufacturing Technology (AMT) in order to select the best alternative of AMT from an ergonomic point of view. Thus, it can be said that the expert system falls within of the practices of quality system by helping to *measure* ergonomic compatibility for an appropriate use of technology, and in the human resource management practices by providing a safe and comfortable environment to the workers and also improving their performance. All of these effects are in the lean manufacturing approach. This is graphically shown on Fig. 16.1.

16.2 Literature Review

AMT is a topic that since the end of the 20th century has demanded an efficient methodology for its evaluation and selection. Several models have been proposed along the time to achieve this goal. However, some shortcomings have been found either on the application or on the attributes that they take into account.

AMT has become into a tool for the manufacturing companies that provides a greater competitiveness level in a global market (Chuu 2009a; Karsak and Tolga 2001). Historically, AMT assessment has been based on attributes such as cost, quality, productivity, flexibility, inventory level, floor space requirements and life of the equipment (Beskese et al. 2004; Chuu 2009b; Karsak and Tolga 2001). Corbett (1988), and Mital and Pennathur (2004) state that AMT development is based on a technocentric approach, which means that its design is based on technical aspects, excluding those like simplification, control and rate of work. These authors point out that the technical approach generates high levels of attention demand, stress and diseases related with stress. Technical approach also causes loss of worker capacities and creativity.

These backgrounds show that there is an important factor that has not been considered on the assessment and selection of AMT: the human factor. According to Karwowski et al. (2003) and Maldonado (2009), the human factor considers the human being with regard the capabilities and limitations integrating them into the design, evaluation, selection and implementation of AMT. Ergonomic attributes, which help to improve human capabilities and to overcome the human limitations, have not been taken into account on the assessment of AMT due to the lack of timely and appropriate information and the ignorance of their benefits (Maldonado 2009).

Failure to consider ergonomic attributes on AMT causes greater investment in training time, more errors, lower production levels and bad quality by the companies. It also generates injuries and accidents to the workers, which leads to economical disasters for the companies that face them (Bandrés 2001; Bridger 1995; Maldonado 2009; Mital and Pennathur 2004). This problem reflects the need of carry out an investigation that promotes the usage of ergonomic attributes on the assessment and selection of AMT. This project integrates the knowledge of 8 experts in AMT assessment (Maldonado 2009) to achieve the main objective: develop an expert system based on fuzzy rules for the ergonomic compatibility of AMT.

16.2.1 Models of Assessment and Selection of AMT

Justifying investment on AMT by mean of economic models such as discounted cash flow, net present value and return of investment is not the most correct (Chuu 2009a, b; Karsak and Tolga 2001; Kulak and Kahraman 2005; Mohanty and Deshmukh 1998; Ordoobadi and Mulvaney 2001). Mohanty and Deshmukh (1998) proposed a model based on learning organization theory to the evaluation of AMT. This model applies the methodology of Nominal Group Technique, which allows taking into account intangible attributes. The main deficiency of this model is its high demand of time and effort to the decision makers.

O'Kane et al. (2000) developed a methodology based on simulation to evaluate changes and their effects in a manufacturing automatized system. This model measures factors such as productivity, inventory in process, delivery times and utilization of equipment. On other hand, Talluri and Yoon (2000) applied the method of one-ratio data envelopment analysis to analyze attributes such as cost, repeatability, speed and load capacity. Kengpol and O'Brien (2001) developed a decision making tool that proposes an integration of three models: cost/benefit analysis, decision making effectiveness, and a common criteria model to select AMT. This model uses technical, economic and manufacturing attributes. The main problem with these models is that they are based on exact measures and evaluations (Chuu 2009b), which do not reflect la qualitative and subjective nature of many attributes (Abdel-Kader and Dugdale 2001).

In order to evaluate tangible and intangible attributes simultaneously incorporating vague and incomplete information, evaluation models of AMT have developed methodologies based on fuzzy logic (Chuu 2009b). The fuzzy method developed by Karsak and Tolga (2001) for the assessment of AMT applies a fuzzy analysis of discounted cash flow and linguistic evaluations of attributes such as flexibility and quality. Ordoobadi and Mulvaney (2001) developed a process known as system-wide benefits value analysis combined with a fuzzy expert system (FES) to evaluate and select AMT. This method requires the decision maker performs several settings, and also it become cyclic and not able to find if the investment is justifiable in a first iteration. Abdel-Kader and Dugdale (2001) developed a fuzzy model to assess AMT by mean of factors such as flexibility, customer requirements, delivery times, quality, saves in net cost, initial investment and others financial factors. According to Chuu (2009b), these methods do not consider a group decision making perspective. This author presented a model of evaluation and selection of AMT which applies a fuzzy information fusion methodology to measure intangible attributes by mean of linguistic evaluations and a group decision approach.

The models exposed in this section show at least one of the following shortcomings: (1) they require exact measures to assess intangible attributes, and (2) they require expert knowledge to perform the steps. In addition, none of them takes into account ergonomic attributes.

16.2.2 Ergonomic Attributes for the Selection of AMT

Human beings have limitations on the interaction with AMT (Maldonado 2009; Mital and Pennathur 2004). These limitations must be taken into account when designing, evaluating and selecting AMT, since failure to do so lead to significant human and equipment downtime, human error, injuries and accidents that affect the production time and human wellbeing (Maldonado 2009). A modern manufacturing approach centered on human factor may be more effective based on the real productivity gains, economy, technical feasibility and the capacity and reliability of equipment (Mital and Pennathur 2004).

Based on this idea, Maldonado (2009) classified 20 ergonomic sub-attributes into five main groups corresponding to the attributes. This classification is called as Ergonomic Compatibility Evaluation Model (ECEM) and it is showed on Fig. 16.2. This model uses a hierarchical structure where at first level is the goal to

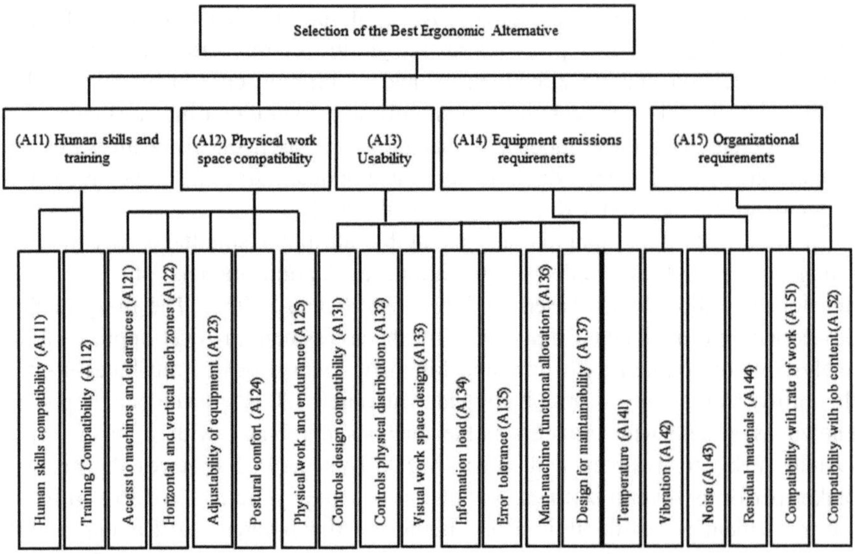

Fig. 16.2 Ergonomic compatibility evaluation model. Maldonado (2009)

achieve, on the second and third level are the attributes and sub-attributes, respectively, and on the fourth and last level are the alternatives of AMT with which a selection will be made. This author uses a key that identifies attributes and sub-attributes. For the attributes the key is made by a letter and number "A1", followed by the number of the group (1, 2, 3, 4 or 5); and for the sub-attribute the key is made by the key of the corresponding attribute followed by the number of the sub-attribute. The ECEM, with the groups of ergonomic attributes, was used to developed the FES to evaluate ergonomic compatibility of AMT, and all the attributes were considered as beneficial and intangibles, with exception of the sub-attributes with the keys A125, A141, A142, A143, A144 (Maldonado et al. 2011).

16.2.3 Fuzzy Logic

Fuzzy logic aims to model inherent impreciseness present in our natural language; it captures, through the process of inference, the uncertainty, ambiguity, and complexity of the human cognitive processes (He et al. 1998). Fuzzy logic is employed to represent and manipulate inferences though the use of fuzzy if–then rules, which are based on linguistic variables (Prasad et al. 2003). Despite traditional computer programs, fuzzy logic helps computers to think instead of computing a series of operations. It is designed to make fuzzy choices when solving problems. In the process of mapping input spaces into proper outputs spaces, fuzzy logic has shown numerous advantages in contrast with other technologies.

These advantages include being faster, cheaper, and more attractive to implement due to its simplicity. In general, fuzzy logic attempts to emulate and approximate human reasoning capabilities. Some reasons for using fuzzy logic are when a system is unknown, when parameters are dynamic, when there are several interactions, conditions, and constraints within the process and the environment, which are not easily or feasible to modeling mathematically (He et al. 1998).

16.2.3.1 Fuzzy Inference

The process of fuzzy inference or fuzzy reasoning consists in obtaining an appropriate output space from a fuzzy rule. Fuzzy rules are commonly expressed in a form such as "IF x is A_i THEN y is B_i", where A_i and B_i are fuzzy sets representing fuzzy concepts (linguistic variables). A number of techniques have been proposed for fuzzy reasoning such as Mamdani model. Mamdani combines fuzzy rules as,

$$\mathrm{R}(x,y) = \bigvee_{i=1}^{n} A_i(x) \wedge B_i(y) \tag{16.1}$$

and for an extended set of rules,

$$\mathrm{R_i} : \text{If } A_{i1} \text{ and } A_{i2} \text{ and...and } A_{ik} \text{ then } B_i, i = 1,..,n \tag{16.2}$$

$$\begin{aligned} \mathrm{R}(x,y) &= \mathrm{R}(x_1, x_2, .., x_k, y) \\ &= \bigvee_{i=1}^{n}((A_{i1}(x_1)) \wedge (A_{i2}(x_2)) \wedge \ldots \wedge (A_{ik}(x_k)) \wedge (B_i(y))) \end{aligned} \tag{16.3}$$

Fuzzy inference process aims to combine fuzzy rules to produce a fuzzy set (output space). Generally, in most of the applications a crisp number is desirable; this involves a process named "defuzzification". Defuzzification is the process to obtain a crisp value from a fuzzy set. As well as fuzzy reasoning, there are different methods for carrying out the defuzzification process (composite moments and composite maximum). Center of area (COA) method (composite moments) is one of the most employed techniques; thus, a crisp number is obtained using COA by

$$y_0 = \frac{\sum_{j=1}^{n} \mathrm{y_j} \mathrm{R}(x, y_j)}{\sum_{j=1}^{n} \mathrm{R}(x, y_j)} \tag{16.4}$$

16.3 Methodology

As it was mentioned above, Maldonado (2009) developed the ECEM (see Fig. 16.2) to assess AMT from an ergonomic approach. This model was adopted as basis to develop the FES. This author delivered the weights for each attribute and sub-attribute by applying the Analytic Hierarchy Process (AHP) and using the linguistic terms *Poor (P), Regular (R), Good (G), Very Goo (VG),* and *Excellent*

(E) for intangible attributes and sub-attributes; *Very Low (VL), Low (L), Medium (M), High (H)* and *Very High (VH)* for tangible attributes and sub-attributes. Figure 16.3 shows the weights delivered by Maldonado (2009).

For the development of the FES, software like Matlab 2010, Minitab 16, Word and Excel were used.

16.3.1 Methods

The method to develop the FES was divided into two stages. The first stage belongs to the fuzzy rules (IF–THEN rules), and the second stage refers to the development of the FES with Matlab and its validation.

16.3.1.1 Stage 1: Fuzzy Rules Obtainment

Different fuzzy sets were adopted to deliver the fuzzy rules (Celik et al. 2007). For the tangible attributes five fuzzy sets (Very Low, Low, Medium, High, and Very High) were distributed on a scale with range 0–1. For the intangible attributes another five fuzzy sets (Poor, Regular, Good, Very Good, and Excellent) were also distributed on a scale with range 0–1. Figures 16.4, 16.5 show the fuzzy sets for tangible and intangible attributes respectively.

The scale for the Ergonomic Incompatibility Content (EIC) was developed based on the fact that membership functions can be assigned to linguistic terms by mean of the intuition delivered of the experts' judgment. This scale uses the same linguistic terms than those used for tangible attributes, and it comprises the range 0–4 because the CIE delivered by Maldonado (2009) is nearly to this scale. Figure 16.6 shows the EIC scale.

The method of Azadeh et al. (2008) was applied in order to decrease the number of fuzzy rules derived by all possible combinations of linguistic terms for attributes and sub-attributes. This method organizes the attributes in a hierarchical way classifying into a same group those attributes with common specifications. The attributes were regrouped according to specifications pointed out by Corlett and Clark (1995). For example, sub-attributes A121, A122 and A123 were classified into the group Equipment Spatial Design (A12123) which is in an intermediate level between attributes and sub-attributes. Figure 16.7 shows the final regrouping.

Once fuzzy sets and linguistic terms were defined, fuzzy rules were derived by following the next steps:

Step 1: Defuzzification. At this step, a precise value was associated with each fuzzy set by mean of the centroid method applying the next equation

$$c = \frac{\sum_0^n f(x)_i x_i}{\sum_0^n f(x)_i} \tag{16.5}$$

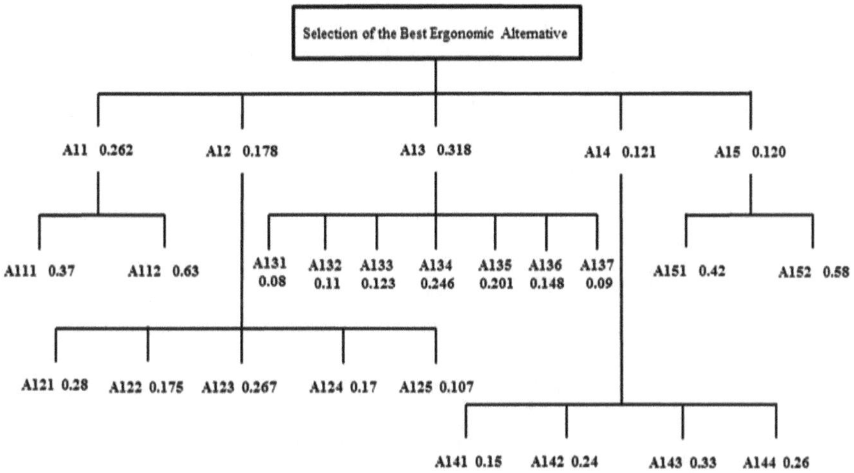

Fig. 16.3 Weights of the attributes and sub-attributes. Maldonado (2009)

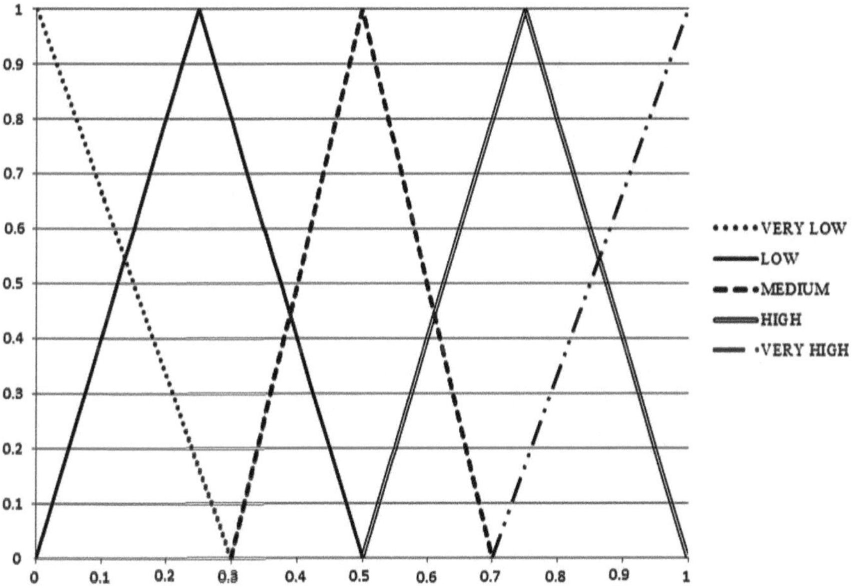

Fig. 16.4 Fuzzy sets for tangible attributes. Celik et al. (2007)

Step 2: Applying the Human Incompatibility Axiom. This axiom states that a design with less human incompatibility content has a greater success probability; it mean, the alternative with lower EIC is the best ergonomic alternative. This axiom was applied by mean of the next equation:

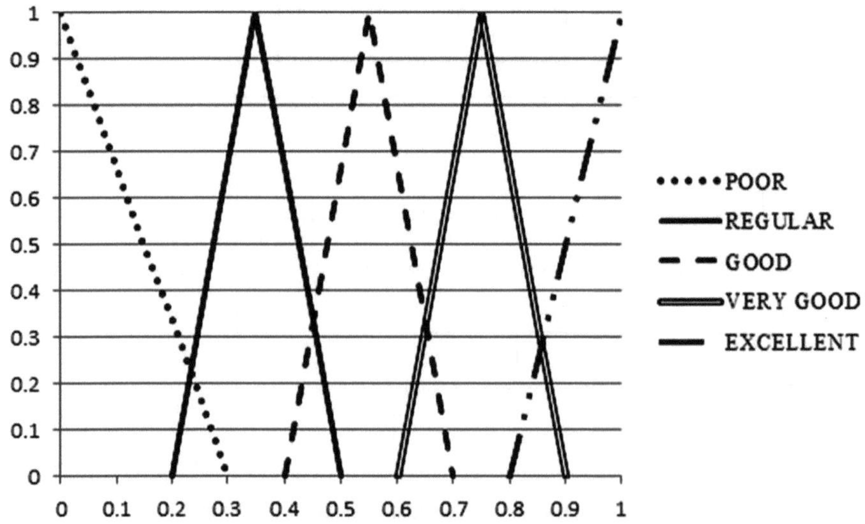

Fig. 16.5 Fuzzy sets for intangible attributes. Celik et al. (2007)

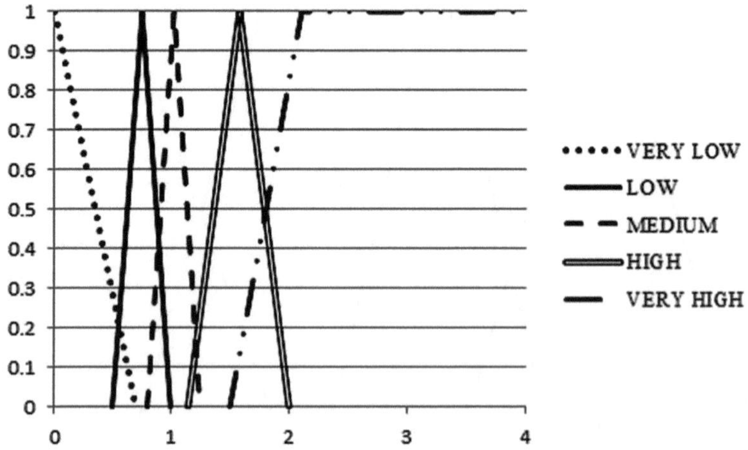

Fig. 16.6 Fuzzy sets for the EIC scale

$$EIC_i = \log_2(1/c_i) \tag{16.6}$$

where EIC_i is the ergonomic incompatibility content for the attribute i on a defined alternative, and c_i is the centroid value—compatibility content—for the linguistic term given to the attribute i on the defined alternative. This step applies only for the sub-attributes, because the qualifications for the attributes are already in terms of incompatibility.

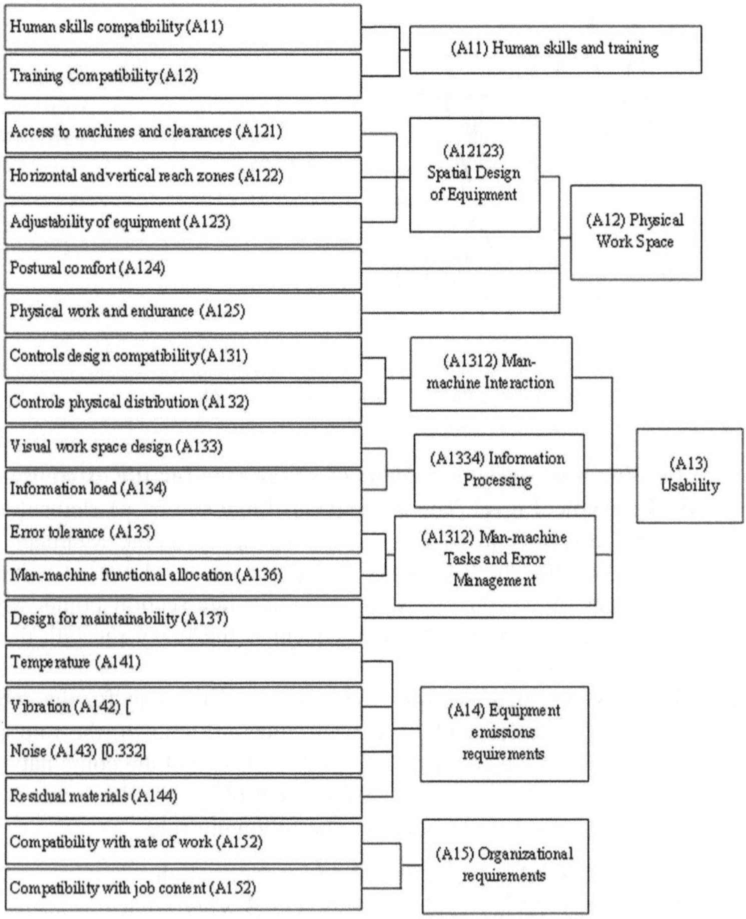

Fig. 16.7 Regrouping of sub-attributes

Step 3: Applying the importance weights of attributes. The weights of the attributes on the Fig. 16.3 are the original weights, however, with the regrouping many weights changed. To keep a weighing, the new weights were normalized. The new weights were derived by dividing the original weight of each attribute by the sum of the weights on a new group. The sum of new weights are equal to one, and the weight for the new group is the sum of the original weights of the attributes regrouped into it. Once the new weights were defined for each group, the EIC of each attribute was multiplying by its corresponding weight.

Step 4: Adding and obtaining the weighted EIC. The weighted EIC were added in order to get a total EIC for each group. This step applies to all the groups taking into account all the possible combinations of qualifications for the sub-attributes.

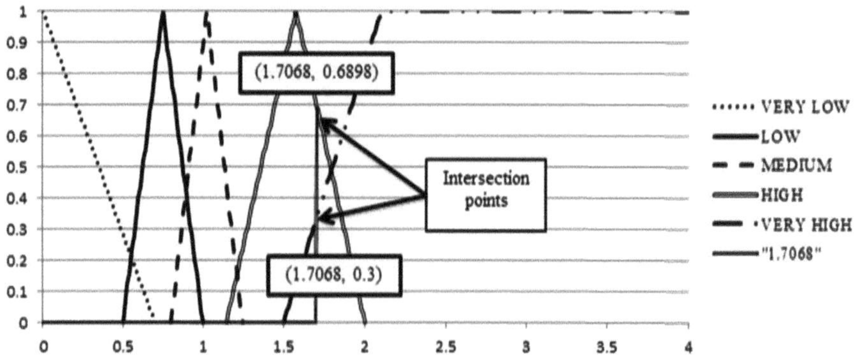

Fig. 16.8 Determining the consequent for an EIC equal to 1.7068

Step 5: Finding the consequent element of the rule for the attribute of the subsequent hierarchical level. The consequent (linguistic term) element for the attribute of the subsequent level was derived by applying the Mamdani fuzzy inference system. For this, the EIC valued, computed on step 4, was introduced on the fuzzy sets scale shown on Fig. 16.8. After the intersection points with the different fuzzy sets are found and the consequent is the fuzzy set with the point with the higher membership.

Step 6: Fuzzy rules formulation. Now that linguistic term for the sub-attributes and the consequent element for the attributes are known, the rule can be made on this manner: *IF x is A and y is B, THEN z is* C. All the combinations are taken into account. In order to decrease the number of fuzzy rules, some of these are summarized on one rule. For example, the following rules

IF *A111* is Poor and *A112* is Poor, THEN *A11* is Very High
IF *A111* is Poor and *A112* is Regular, THEN *A11* is Very High
IF *A111* is Poor and *A112* is Good, THEN *A11* is Very High
IF *A111* is Poor and *A112* is Very Good, THEN *A11* is Very High
IF *A111* is Poor and *A112* is Excellent, THEN *A11* is Very High

Can be stated in only one rule: IF *A111* is Poor, THEN *A11* is Very High. The adverbs "at least" and "at most", were used to summarize other rules when the range of qualifications for an attribute may use the extreme value of some linguistic term or vice versa respectively, on a continuous way and the consequent is the same in all the rules. For example, the following rules for the attribute *A11*:

IF *A111* is Poor and *A112* is Poor, THEN *A11* is Very High
IF *A111* is Poor and *A112* is Regular, THEN *A11* is Very High
IF *A111* is Poor and *A112* is Good, THEN A*11* is Very High
IF *A111* is Poor and *A112* is Very Good, THEN *A11* is High
IF *A111* is Poor and *A112* is Excellent, THEN *A11* is High

Table 16.1 Example of EIC qualification for alternatives of AMT

Alternative	Qualification
X	Low
Y	High
Z	Very high

Can be summarized in only two rules:

IF *A111* is Poor and *A112* is at most Good, THEN *A11* is Very High
IF *A111* is Poor and A*112* is at least Very Good, THEN *A11* is High

16.3.1.2 Selection of the Best Ergonomic Alternative

Once the evaluators assign qualifications to each of the sub-attributes for a specific number of alternatives, the expert system must provide a final EIC qualification for each one of the alternatives, which give place to several cases. For example, suppose the qualifications shown on Table 16.1 were obtained for a set of three alternatives X, Y and Z. At this case it is easy to make any decision, because of the three alternatives have different linguistic qualifications, and being X the best alternative since it has a minor ***EIC***.

There are cases on which the decision making is not easy. For example, when there is equality on the final linguistic qualifications for all the alternatives. To solve this problem the final *EIC* obtained for each alternative are compared. So, alternatives can have different values of *EIC*, but all of them belong to a same category. At this case, the best alternative is that with the mininum *EIC*. Figure 16.9 shows graphically this case. Note that all the alternatives have a different EIC value, but all of them belong to the category of *High.* At this figure, the alternative Z has a minor *EIC*, so it is the best alternative; alternative X has a greater *EIC,* then it is the worst alternative to choose.

16.3.1.3 Stage 2: Development and Validation of the Expert System

At this stage, the expert system development process with Matlab and the functions that it contains are described. The methods applied to the validation of the system are also described.

Development of the Expert System with Matlab

For the application of the expert system on the evaluation and selection of AMT from an ergonomic approach, a program was developed in Matlab 2010^{TM}. This program contains the methodology exposed above. It was necessary to create

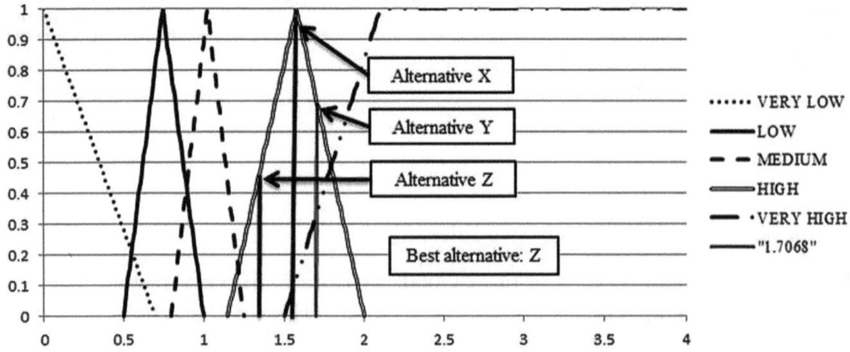

Fig. 16.9 Tiebreaker for alternatives with the same linguistic value

several *for* loops. Some of these loops allow the evaluator to assess all the alternatives and then continue with the following evaluator. Other loops allow to the same evaluator to continue with the assessment to the following alternative. Logic operations were also developed by using the operator *if*. These logic operations help to assign a precise value to each fuzzy qualification for each alternative regarding each one of the ergonomic sub-attributes and assigning a linguistic qualification (Very Low, Low, Medium, High, Very High) to the *EIC* depending on its correspondent value range. Other mathematical operations were indicated on the steps 1 to 4, and the geometric mean value was used in the case when the alternatives are assessed by a group of evaluators.

Validation of the Expert System by mean of Sensibility Analysis

An expert system can be validated against historical results despite the number of cases against which the system is validated (O'Keefe et al. 1986). So, the expert system was validated against the results obtained by Maldonado (2009) on three study cases. O'Keefe et al. (1986) also point out that it is not correct to validate the system on critical application, such as manufacturing applications, since this can generate economic loss. They state there are two ways of performance validation for an expert system, according to the time when the validation is carried out: formal validation and informal validation. To validate the expert system developed at this work, the informal validation was used, since it is carried out at the end of the expert system development. They also point out that there exist qualitative and quantitative validations, according to the validation techniques applied. Within the qualitative validation there are methods such as sensibility analysis. This method complemented with the study cases, was applied to validate the expert system.

Sensibility analysis is carried out by changing the values of the system input variables on an interest range observing the effect on the system performance (O'Keefe et al. 1986). For example, suppose the following fuzzy rules were obtained:

IF *A111* is *Poor* and *A112* is *Poor*, THEN *A11* is *Very High*
IF *A111* is *Poor* and *A112* is *Regular*, THEN *A11* is *Very High*
IF *A111* is *Poor* and *A112* is *Good*, THEN *A11* is *Very High*
IF *A111* is *Poor* and *A112* is *Very Good*, THEN *A11* is *Very High*
IF *A111* is *Poor* and *A112* is *Excellent*, THEN *A11* is *Very High*

As it is known, the qualification for *A112* does not have any effect on the qualification for *A11*, always that *A111* is *Poor*. Then, giving any value to *A112* must not affect intermediate and final results. To perform the sensibility analysis to the expert system some summarized rules were randomly selected. On these rules, the values of the sub-attributes, that according to the rule do not have any effect within a range of qualification, were changed. For example, suppose that the following rule was selected:

IF *A111* is *Poor* and *A112* is at most *Good*, THEN *A11* is *Very High*

When the sensibility analysis is performed *A111* was constant with a qualification equal to *Poor*, while *A112* changed its qualifications to *Poor*, *Regular* and *Good*. With these changes the intermediate and final results must keep constant.

16.4 Results

This section provides the results derived by applying the methodology exposed above. Among the results presented are the linguistic terms defuzification, the EIC denoted by each linguistic value, the new weights for every attribute on the regrouping, and the fuzzy rules. Other important results are the expert system validation and its development in Matlab.

16.4.1 Defuzzification of the Linguistic Terms

The following precise values (centroids), for each of the linguistic terms, resulted by applying the step 1 described on the methodology. Tables 16.2, 16.3 show the linguistic terms centroids for intangible and tangible attributes.

Table 16.2 Linguistic terms for intangible attributes and their centroids

Linguistic term	Poor	Regular	Good	Very good	Excellent
Centroid	0.0833	0.35	0.55	0.75	0.95

Table 16.3 Linguistic terms for tangible attribute and their centroids

Linguistic term	Very low	Low	Medium	High	Very high
Centroid	0.9167	0.75	0.5	0.25	0.0833

Table 16.4 *EIC* for each linguistic term

Linguistic term	Very low	Low	Medium	High	Very high	Poor	Regular	Good	Very good	Excellent
EIC	0.125	0.415	1	2	3.585	3.585	1.514	0.862	0.415	0.074

16.4.2 Ergonomic Incompatibility Content of the Fuzzy Sets

At step 2, Eq. 16.6 was applied for each one of the centroids shown on Tables 16.2, 16.3. With this equation the *EICs* for the linguistic terms were derived. Table 16.4 shows the *EIC* for each linguistic term.

16.4.3 Weighting of the Sub-Attributes on the Regrouping

The subattributes new weights are shown in brackets on Fig. 16.10. Note that on each group the sum of the weights is 1.

16.4.4 Fuzzy Rules for the Assessment of AMT

Most of the fuzzy rules are presented in a summarized way. Because of the large number of fuzzy rules, only few of them are presented in this chapter. Fuzzy rules are presented in tables. Table 16.5 shows a sample of fuzzy rules for each one of the subattributes and a rule for the final *EIC*. Fuzzy rules 1–5 correspond to the attributes *A11, A12, A13, A14*, and *A15*, respectively. Fuzzy rule 6 shows the final *EIC* based on the *EICs* of the attributes. Summarized rules were used to validate the system by mean of sensibility analysis.

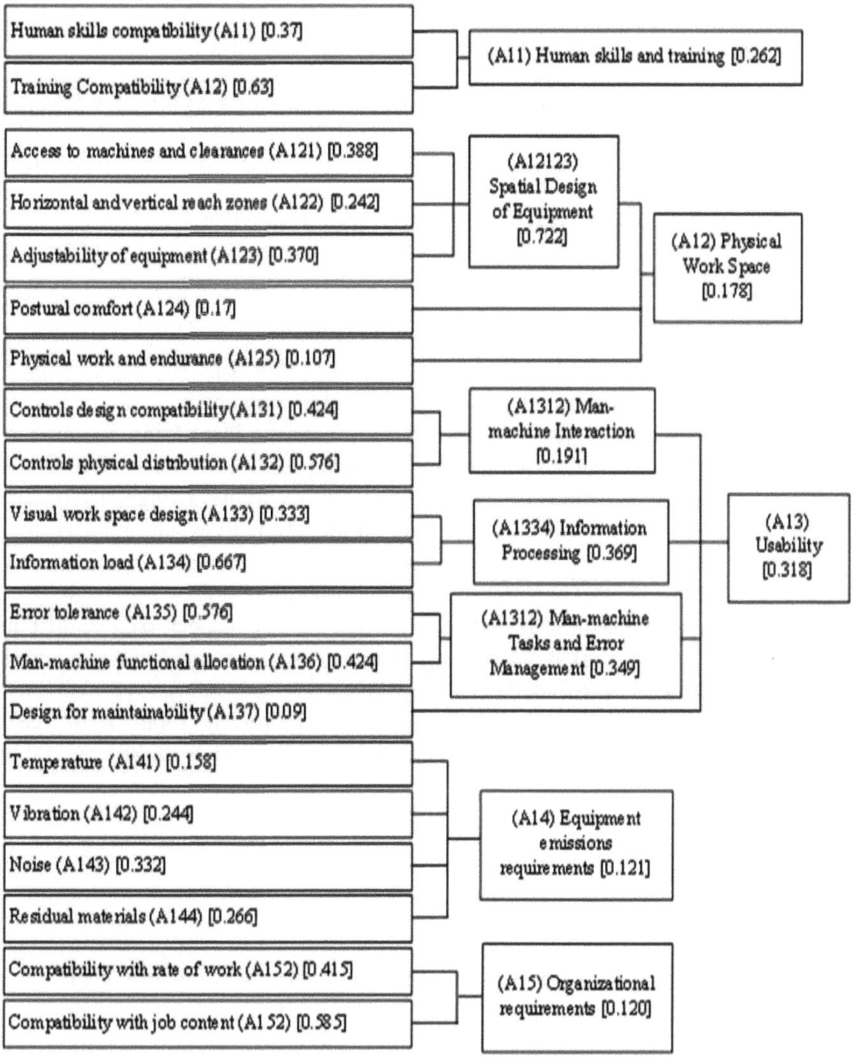

Fig. 16.10 Subattributes weights on regrouping

16.4.5 Development of the Expert System with Matlab

In order to apply the expert system for the evaluation and selection of AMT from an ergonomic approach, a program in Spanish was developed by applying the methodology exposed on this work. The software Matlab® 2010 was used to develop the program. The program allows assessing several alternatives either by one or more evaluators. Figure 16.11 shows program's starting screen. On this

Table 16.5 Examples of fuzzy rules for ergonomic compatibility of AMT

1	IF A111 is Poor and A112 is at most Good, Then A11 is Very High
2	IF A121 is Poor and A122 is Poor, Then A12 is Very High
3	IF A131 is Poor, and A132 is at most Good, and A134 is Poor, and A135 is Poor, Then A13 is Very High
4	IF A141 is High, and A142 is Very Low, and A143 is Very Low, and A144 is at most Low, Then A14 is Very Low
5	IF A151 is Very Good, and A152 is at least Very Good, Then A15 is Very Low
6	IF A11 is Very Low, and A12 is Very High, and A13 is Very High, and A14 is at least Low, and A15 is High, Then EIC is Very High

starting screen the title of the program is shown, and it asks for the number of alternatives to be evaluated and for the number of evaluators.

During the evaluation process, the program allows identifying the current evaluator and provides and explanation about what each attribute is. The evaluators assess each subattribute for each alternative by giving a linguistic qualification. When an evaluator finishes assessing all the alternatives with respect to all subattributes, the program gives the place to the next evaluator automatically. Figure 16.12 shows these characteristics of the program.

The program provides the EIC for each alternative with respect to each attribute (*A11, A12, A13, A14,* and *A15*) both in terms of number and linguistic terms. Figure 16.13 show the *EIC* of the five attributes for each alternative evaluated. Finally, as it is shown at Fig. 16.14, the program provides the total *EIC* for each alternative and also it indicates what alternative is the best from an ergonomic approach.

16.4.6 Expert System Validation

This section shows the results of the system validation. A validation with sensibility analysis is presented.

16.4.6.1 Validation of the Expert System with Sensibility Analysis

Validation with sensibility analysis was performed by randomly selecting five fuzzy rules. On these rules, the values of the linguistic terms were changed on the range indicated by the rules. With these changes the results must be the same. So, validation with sensibility analysis consists of comparing the results obtained with different linguistic term values. A rule and its changes are presented below as example.

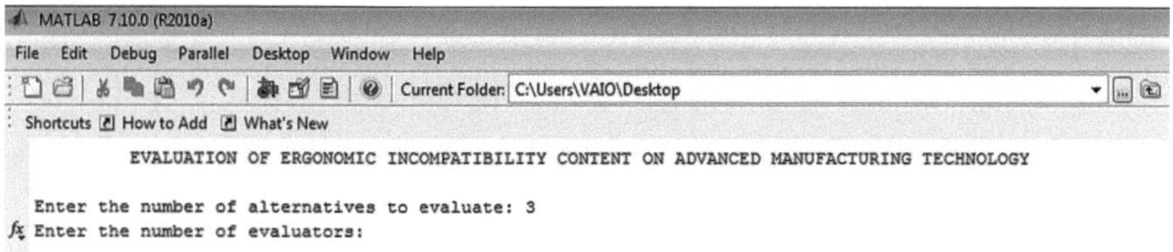

Fig. 16.11 Starting screen of the expert system for assessment and selection of AMT on Matlab

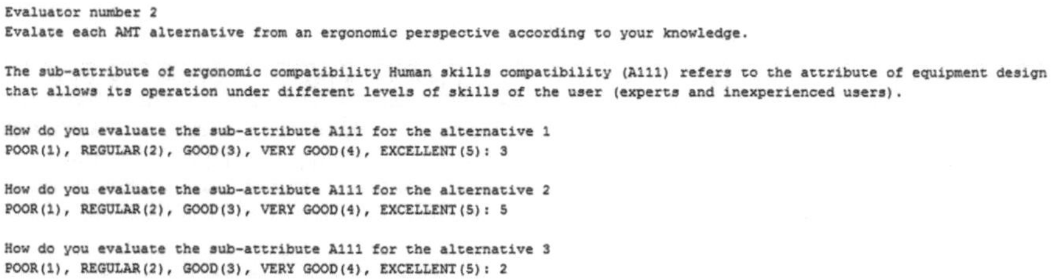

Fig. 16.12 Assessment of the ergonomic subattributes for different alternatives by different evaluators

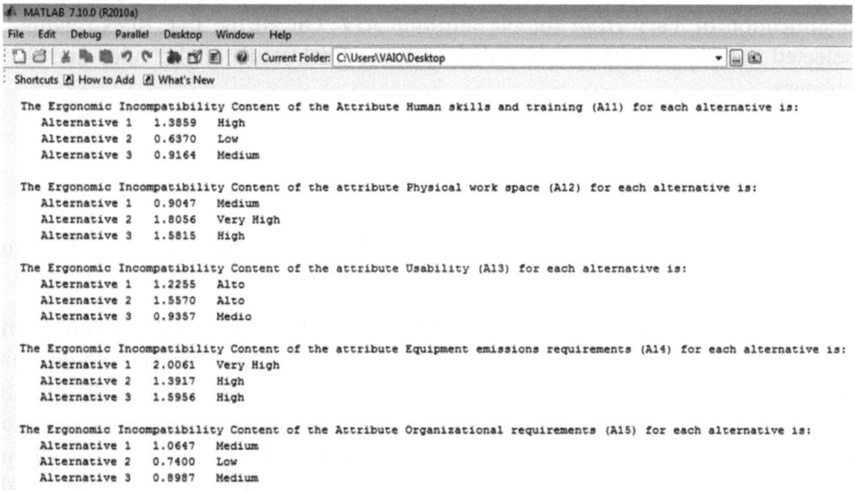

Fig. 16.13 EIC of the ergonomic attributes for each alternative

Rule for the attribute A13:

IF *A131* is Good and *A132* is Regular, and *A133* is Poor, and *A134* is at most Regular, and *A135* is Good, and *A136* is Good, and *A137* is Poor, Then *A13* is Very High.

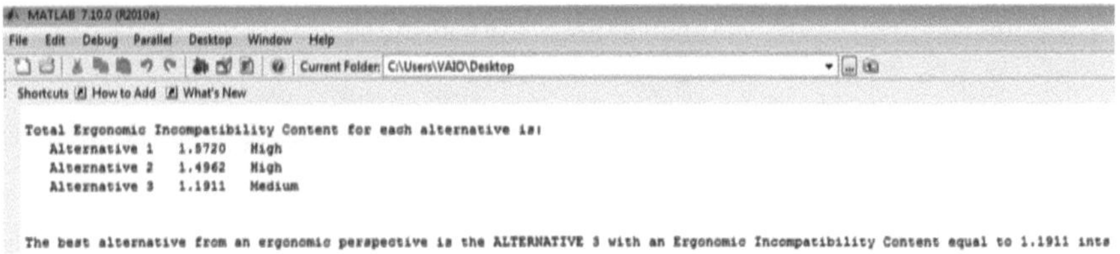

Fig. 16.14 Total EIC for ach alternative and selection of the best alternative

Table 16.6 Sensibility analysis for fuzzy rules

A131	A132	A133	A134	A135	A136	A137	A13
Good	Regular	Poor	Poor	Good	Good	Good	Very high
Good	Regular	Poor	Regular	Good	Good	Good	Very high

At this fuzzy rule *A134* can take the values of *Poor* and *Regular*, the other subattributes must keep constant and the final result for *A13* must be always the same within these values. Table 16.6 shows the changes to the subatribute A134 and the result for *A13* given by the expert system developed with Matlab.

As can be seen, for all the changes on this rule the expert system gave a qualification of *Very High* for the attribute *A11.* On all the five rules randomly selected the expert system coincided with the final result of the rules and their changes.

16.4.6.2 Example

An example corresponding to selection of AMT in Computer Integrated Manufacturing (CIM) is presented in this section.

Example 4.1 Ergonomic assessment of CIM. Table 16.7 shows the qualifications given by the experts to the CIM alternatives. At this table the alternatives are indicated by the letters X and Y, the experts by E1, E2 and E3, and the qualifications are abbreviated on the following way: *P* = *Poor, R* = *Regular, G* = *Good, VG* = *Very Good,* and *E* = *Excellent, VL* = *Very Low, L* = *Low, M* = *Medium, H* = *High, VH* = *Very High.* At this example, three experts evaluate three CIM alternatives. Qualifications given by the experts are introduced in the Matlab program. When there are several evaluators, as in these examples, the expert system applies a geometric mean to the centroid of the qualifications of a specific subattribute for each alternative. At the end, the expert system provides the final EIC for each alternative. Table 16.8 shows the results of the expert system in the case study of CIM.

At this case the best alternative is Y according to the expert system, since Y has a minor *EIC.*

Table 16.7 Qualifications given by the experts to the CIM alternatives

Subattributes		A111	A112	A121	A122	A123	A124	A125	A131	A132	A133
E1	X	G	R	R	G	G	R	M	G	G	G
	Y	G	G	G	G	G	R	M	G	G	R
E2	X	VG	G	G	G	G	G	L	G	VG	VG
	Y	VG	VG	E	VG	VG	VG	L	VG	VG	VG
E3	X	G	R	VG	VG	VG	VG	L	G	VG	VG
	Y	G	R	R	G	G	G	L	G	G	G
Subattributes		A134	A135	A136	A137	A141	A142	A143	A144	A151	A152
E1	X	G	G	R	R	L	L	M	M	G	G
	Y	G	R	G	R	L	L	M	L	G	G
E2	X	VG	G	VG	G	L	M	M	M	VG	VG
	Y	VG	G	VG	VG	VL	L	M	M	VG	VG
E3	X	G	G	G	P	VL	L	M	L	G	VG
	Y	G	G	G	P	VL	L	M	L	G	G

Table 16.8 *EIC* for the CIM alternatives

Alternative	EIC with the expert system
X	0.9117
Y	0.8383

This example was taken from Maldonado (2009) and it was also used to perform another validation of the expert system. Maldonado (2009) and the expert system coincided in most of the ranking of the alternatives. On this validation it was concluded that the expert system has an acceptable performance.

16.5 Conclusions and Suggestions

Based on the validation and on the examples, it is concluded the expert system is an efficient and effective tool that helps evaluates AMT from an ergonomic approach since the users can save time and effort when they compute the outcomes. This expert system may promote the development of additional expert system approaches that will contribute to evaluate AMT on a more general way (i.e. including several approaches: ergonomics, flexibility, productivity, etc.) and on a specific way (i.e. on a specific approach).

The following suggestions will help to improve the expert system and will make it applicable to more areas:

- It is suggested that the expert system is available online in order to it has an easier access and expanded application.

- Another suggestion is include other approaches (quality, productivity, flexibility, etc.) into the assessment of the AMT in order to have a general evaluation and facilitate decision-making.
- A third suggestion is that the expert system may be available in several languages.

Currently, this expert system can contribute to develop a lean manufacturing perspective in a specific field, but, if suggestions are carried out, it can contribute in a more general way to have a lean manufacturing culture in modern companies.

References

Abdel-Kader, M. G., & Dugdale, D. (2001). Evaluating investments in advanced manufacturing technology: A fuzzy set theory approach. *British Accounting Review., 33*, 455–489.

Armstrong, M. (2006). *A handbook of human resource management Practice* (10th ed., p. 3). London: Kogan Page Limited.

Azadeh, A., Fam, I. M., Khoshnoud, M., & Nikafrouz, M. (2008). Design and implementation of a fuzzy expert system for performance assessment of an integrated health, safety, environment (HSE) and ergonomics system: The case of a gas refinery. *Information Sciences, 178*, 4280–4300.

Bandrés, M. E. (2001). *Ergonomía. 20 preguntas para aplicar la Ergonomía en la empresa* (Segunda edición, p. 25). Madrid: MAPFRE.

Beskese, A., Kahraman, C., & Irani, Z. (2004). Quantification of flexibility in advanced manufacturing systems using fuzzy concept. *International Journal of Production Economics, 89*, 45–56.

Bridger, R. S. (1995). *Introduction to Ergonomics* (p. 350). New York: McGrawHill.

Carayon, P., & Smith, M. J. (2000). Work organization and ergonomics. *Applied Ergonomics, 31*, 649–662.

Celik, M., Kahraman, C., Cebi, S., & Er, I. D. (2007). Fuzzy axiomatic design-based performance evaluation model for docking facilities in shipbuilding industry: The case of Turkish shipyards. *Expert Systems with Applications, 36*(1), 599–615.

Chuu, S. J. (2009a). Group decision-making model using fuzzy multiple attributes analysis for the evaluation of advanced manufacturing technology. *Fuzzy Sets and Systems, 160*, 586–602.

Chuu, S. J. (2009b). Selecting the advanced manufacturing technology using fuzzy multiple attributes group decision making with multiple fuzzy information. *Computers and Industrial Engineering, 57*, 1033–1042.

Corbett, J. M. (1988). Ergonomics in the development of human-centred AMT. *Applied Ergonomics, 19*, 35–39.

Corlett, E. N., Clark, T. S. (1995). *The Ergonomics of Workspaces and Machines. A Design Manual* (2nd ed., pp. 1–122). Library Network.

Drury, C. G. (1997). Ergonomics and the quality movement. *Ergonomics, 40*(3), 249–264.

Fullerton, R. R., McWatters, C. S., & Fawson, C. (2003). An examination of the relation- ships between JIT and financial performance. *Journal of Operations Management, 21*(4), 383–404.

He, W., Zhang, Y. F., Lee, K. S., Fuh, J. Y. H., & Nee, A. Y. C. (1998). Automated process parameter resetting for injection molding: a fuzzy-neuro approach. *Journal of Intelligent Manufacturing, 9*, 17–27.

Karsak, E. E., & Tolga, E. (2001). Fuzzy multi-criteria decision-making procedure for evaluating advanced manufacturing systems investments. *International Journal of Production Economics, 69*, 49–64.

Karwowski, W., Rizzo, F., Rodrick, D. (2003). Ergonomics in Information Systems. In H. Bidgoli (Ed.), *Encyclopedia of information Systems* (pp. 185–201). San Diego: Academic.

Kengpol, A., & O'Brien, C. (2001). The development of a decision support tool for the selection of advanced technology to achieve rapid product development. *International Journal of Production Economics, 69*, 177–191.

Kulak, O., & Kahraman, C. (2005). Multi-attribute comparison of advanced manufacturing systems using fuzzy vs. crisp axiomatic design approach. *International Journal of Production Economics, 95*, 415–424.

MacDuffie, J. P. (1995). Human resource bundles and manufacturing performance: organizational logic and flexible production systems in the world auto industry. *Industrial and Labor Relations Review, 48*(2), 197–221.

Maldonado, A. A. (2009). *Modelo de Evaluación Ergonómica para la Planeación y Selección de Tecnología de Manufactura Avanzada.* Disertación Doctoral, Instituto Tecnológico de Ciudad Juárez.

Maldonado, A. A., Guillén, L., Barrón, L., García, J. L. (2011). Evaluación Ergonómica para la Selección de Tecnología de Manufactura Avanzada: una Propuesta de Software. *Revista de la Ingeniería industrial, 5*, 154–164.

Manotas, D. F., & Rivera, L. (2007). Lean manufacturing measurement: the relationship between lean activities and lean metrics. *Estudios gerenciales, 23*(105), 69–83.

McLachlin, R. (1997). Management initiatives and just-in-time manufacturing. *Journal of Operations Management, 15*(4), 271–292.

Melton, T. (2005). The benefits of lean manufacturing. What lean thinking has to offer the process industries. *Chemical Engineering Research and Design, 84*(6), 662–673.

Mital, A., & Pennathur, A. (2004). Advanced technologies and humans in manufacturing workplaces: an interdependent relationship. *International Journal of Industrial Ergonomics, 33*, 295–313.

Mohanty, R. P., & Deshmukh, S. G. (1998). Advanced manufacturing technology selection: A strategic model for learning and evaluation. *International Journal of Production Economics, 55*, 295–307.

O'Kane, J. F., Spenceley, J. R, Taylor, R. (2000). Simulation as an essential tool for advanced manufacturing technology problems. *Journal of Materials Processing Technology, 107*, 412–424.

O'Keefe, R. M., Balci, O., Smith, E. P. (1986). Validation of expert system performance. *IEEE Expert*, *2*(4), 81–90.

Ordoobadi, S. M., & Mulvaney, N. J. (2001). Development of a Justification Tool for advanced manufacturing technologies: system-wide benefits value analysis. *Journal of Engineering and Technology Management, 18*, 157–184.

Prasad, N. R., Nguyen, H. T., Walker, E. A., & Walker, C. L. (2003). *A first course in fuzzy and neural control.* Boca Raton, Florida: Chapman & Hall.

Shaffie, S., Shahbazi, S. (2012*). Lean Six Sigma. The McGraw-Hill 36-hour course*. United States: The McGraw-Hill Companies.

Shah, R., & Ward, P. T. (2003). Lean manufacturing: context, practices bundles, and performance. *Journal of Operations Management, 21*(2), 129–149.

Shah, R., & Ward, P. T. (2007). Defining and developing measures of lean production. *Journal of Operations Management, 25*(4), 785–805.

Simpson, D. F., & Power, D. J. (2005). Use the supply relationship to develop lean and green suppliers. *Supply Chain Management: An International Journal, 10*(1), 60–68.

Talluri, S., & Yoon, K. P. (2000). A cone-ratio DEA approach for AMT justification. *International Journal of Production Economics, 6*, 119–129.

Womack, J., Jones, D., & Roos, D. (1990). *The machine that changed the world.* New York, NY: Macmillan.

Yang, M. G., Hong, P., & Modi, S. B. (2011). Impact of lean manufacturing and environmental management on business performance: An empirical study of manufacturing firms. *International Journal of Production Economics, 129*(2), 251–261.

Chapter 17
Assessment of Human Fatigue: A Comparison Between Machining and Assembly Tasks

Juan-Luis Hernández-Arellano, Jorge Luis García-Alcaraz, Gabriel Ibarra-Mejia and Aidé Aracely Maldonado-Macías

Abstract Research on human fatigue in industrial environments is limited, especially in developing countries. Currently, with the inclusion of new and more complex technologies; levels of mental effort performed by workers have increased while levels of physical exertion have decreased. This research presents the comparison of fatigue scores of two processes including Advanced Manufacturing Technologies (AMT). The first one involves machining of steel pieces using Computerized Numerical Control (CNC) lathes and the second one involves Constant Velocity (CV) Joints assembly using hydraulic presses. For the assessment of fatigue, the Swedish Occupational Fatigue Inventory Spanish version (SOFI-S) was applied. The study was conducted in a manufacturing company of CV Joints located in Central Mexico. A sample of 75 operators of CNC lathes and 66 operators of hydraulic presses were surveyed. As results, Lack of energy was the fatigue dimension that obtained the highest score of (SOFI-S) in both processes analyzed. Nonparametric comparisons show no difference in scores of fatigue dimensions between both processes analyzed.

Keywords Fatigue in work · Work load · SOFI survey

J.-L. Hernández-Arellano (✉)
Department of Design, Autonomous University of Ciudad Juarez, Av. Del Charo 450 Norte. Col. Partido Romero, Ciudad Juarez, Chihuahua, Mexico
e-mail: luis.hernandez@uacj.mx

J. L. García-Alcaraz · G. Ibarra-Mejia · A. A. Maldonado-Macías
Department of Industrial and Manufacturing Engineering, Autonomous University of Ciudad Juarez, Av. Del Charo 450 Norte. Col. Partido Romero, Ciudad Juarez, Chihuahua, Mexico

J. L. García-Alcaraz et al. (eds.), *Lean Manufacturing in the Developing World*, DOI: 10.1007/978-3-319-04951-9_17, © Springer International Publishing Switzerland 2014

17.1 Introduction

In manufacturing processes, there are tasks where the worker performs only one kind of effort (physical or mental). For example, office work, control room work, and some supervisory tasks where the tasks performed by workers are limited to operate a computer and/or observe activities with relatively low physical activity level. In these examples, is demonstrated that the level of mental effort is significantly greater than physical effort (Bridger, 2003). Inversely, activities performed by construction workers, operators performing the loading–unloading of parts and/or materials; physical exertion is significantly higher than mental effort. The above activities can be considered extreme because they only demand one kind of effort of worker performance. However, there are plenty of tasks where the worker develops both kind of efforts, physical and mental simultaneously.

With the advancement of technology, machines and tools included in industrial processes are more complex and demands of worker greater amount of mental effort, while the physical effort has decreased, creating a stability between the physical and mental efforts. This is the case of the operation of machinery of Computer Numerical Control (CNC) which is classified as Advanced Manufacturing Technology (AMT) where operators perform three main tasks: (1) loading–unloading parts into the machines, (2) control panel operation of machines, and (3) inspection of pieces after processing. According to the classification proposed by Small and Chen (1997), CNC machines are classified as AMT in manufacturing, machining, and assembly group.

In the context of the AMT, research about fatigue among workers who operate it in industrial environments is scarce, especially in developing countries (Gonzalez and Gutierrez 2006). In the case of Mexico, few studies have been identified where working conditions and their impact on fatigue of workers in industrial environments are analyzed. Some of the most relevant are: the cutoff validation of the Subjective Symptoms of Fatigue Test (SSFT) in a sample of Mexican workers (Barrientos-Gutierrez et al. 2004), the influence of mental workload as a factor of stress risk in workers in the electronics industry (Gonzalez and Gutierrez 2006), validation of the Scale of Points Estimates Fatigue-Energy (SPEFE; Juarez-Garcia 2007), the determination of fatigue curves in women packing tomatos in Jalisco, Mexico (Hernández-Arellano et al. 2010) and the construction of a survey for assessing workload and fatigue in AMT operators in Mexico (Hernandez-Arellano et al. 2012).

17.2 Assessment of Fatigue

The assessment of fatigue has been developed mainly using statistically validated questionnaires and surveys. One of the first questionnaires developed is the Fatigue Related Symptoms Questionnaire (F-RSQ; Yoshitake 1978), which allows assessing fatigue in three dimensions: Sleepiness and apathy, inability to concentrate and physical discomfort. Each dimension includes 10 items that are assessing using a dichotomous scale. Another method was developed in France by the Laboratory of Economics and Sociology of Work (LEST; Guelaud et al. 1977) which includes five dimensions: physical load, mental load, physical environment, psychosocial environment, and work time. Each dimension are assessed qualitatively and quantitatively to obtain a 0–10 scale rating. The method Cooper Harper (Wiewille and Casali 1983) is a scale of 10 divisions with a linguistic expression where the worker mentioned their fatigue perception.

The Swedish Occupational Fatigue Inventory (SOFI; Åshberg et al. 1997) is an instrument for the assessment of fatigue that includes five dimensions: lack of energy, physical exertion, physical discomfort, lack of motivation, and Sleepiness, each dimension includes 5 items that are assessed with a visual analog scale. The Fatigue Assessment Scale (FAS) was developed in order to obtain the total fatigue assessment, includes the dimensions mental fatigue and physical fatigue that are assessed using a likert scale (Michielsen et al. 2003). The Spanish version of the questionnaire SOFI (SOFI-Spanish; Gonzalez et al. 2005) includes the same dimensions of Ahsberg (2000), but only includes 3 items for the assessment of the five dimensions of fatigue. The Occupational Fatigue Exhaustion Recovery Scale (OFER; Winwood and Casalli 2005) includes three fatigue dimensions: acute fatigue, chronic fatigue, and inter-shifts recovery, which are assessed using a likert scale.

Finally, 3 instruments for assessment of fatigue developed in Mexico are described. The first is the construct evaluation, the internal reliability and the establish the best cuttoff point of the Subjective Symptoms of Fatigue Test (SSFT) in a sample of Mexican workers (Barrientos-Gutierrez et al. 2004). The second is the Fatigue Energy Point Estimate Scale (FEPES; Juarez-Garcia 2007) that is a method based on perceived exertion scale of Borg 1–10 (Borg 1985) and relates the perception of fatigue-energy available to perform a task. The third is the Survey for assessment of workload and fatigue among AMT operators in Mexico (Hernandez et al. 2012). Due to limited research about fatigue, these three methods are the only identified for fatigue assessment of Mexican workers.

17.3 SOFI Questionnaire

This instrument is one of the tools that has proven to be more efficient in assessing fatigue, since it has obtained high levels of reliability and internal consistency and has had applications in diverse environments; however, it has been used mostly in healthcare environments. The development of this instrument was mainly aimed for assessing fatigue in more than two dimensions (physical and/or mental). The research that developed SOFI included a variety of tasks and activities as those conducted by teachers, bus drivers, firefighters, working outdoors, nurses, nuclear plant operators, and so on (Åhsberg et al. 1997). All activities included were significantly high in (physical or mental) workload levels.

The first version of the questionnaire includes five dimensions of fatigue, lack of energy, physical exertion, physical discomfort, lack of motivation and Sleepiness. Each dimension is assessed by 5 items. The revision of this instrument by Åhsberg (2000), the activities included teachers, firefighters, supermarket cashiers, bus drivers and production engineers. In this new version, the items for assessing fatigue dimensions were reduced from 5 to 4. Two translations from English have been developed. The first one was into Chinese language where the work done by computer users was investigated (Leung et al. 2004), and the second was into Spanish language, where the work of special care nurses was investigated (Gonzalez et al. 2005). The latest version is identified by Sebastian et al. (2008), which added the irritability dimension to the questionnaire. Table 17.1 shows the dimensions and items that have been obtained for this survey.

17.4 Case Study

The case study presented herein reports the application of SOFI-S among workers of machining and assembly process in a manufacturer company of Constant Velocity (CV) Joints located in Central Mexico. Three production areas are found in the company: forging, machining, and assembly. Different AMT machines are used for each production area in the company. However, the machining and assembly areas have the greatest number of semi-automated machinery, therefore, only these two processes were analyzed in this case. In machining process are CNC lathes while in the assembly process are CNC hydraulic presses. Figure 17.1 shows the two types of machines mentioned above.

There are three main differences between machining and assembly processes. The first is the type pieces, in machining process three pieces are manufacturing: "bell", "semi-axis" and "tulip". In assembly process two pieces are manufacturing, CV Joint-short and CV Joint-long. The second difference is the process cycle time: in machining process is 40 s (±5) while in assembly process is 30 s (±5). The third difference is the weight of the pieces, in machining process the pieces have an average weight of 4.5 kg., while in assembly process have an average weight of 7 kg. Movements and tasks performed by operators are similar in both processes. First task is the loading–unloading of pieces in the machines, second task is control panel operation, and inspection is the final task.

Due to increased demand of CV Joints, the company implemented extended 12-h shifts throughout 4 days with 2 days of rest. After that, workers have expressed perceiving situations of "excessive tiredness or fatigue" at the end of their working day, and feeling an increase in musculoskeletal discomfort generated by their work. Due to these circumstances, the study of fatigue is developed with the following objectives considering both processes mentioned:

- Apply the SOFI-S among AMT workers to determine the fatigue scores
- Compare fatigue scores intra and between processes.

Table 17.1 Dimensions and items of SOFI

Dimensions	Åhsberg (1997); Leung et al. (2004)	Åhsberg (2000)	Gonzalez (2005)	Sebastian et al. (2008)	Hernandez et al. (2012)
Lack of energy	Over worked	Over worked,	Worn out	Worn out	Worn out
	Spent	Spent	Exhausted	Exhausted	Exhausted
	Worn out	Worn out	Drained	Drained	Drained
	Exhausted	Drained			
	Drained				
Physical exertion	Breathing heavily	Breathing heavily	Breathing heavily	Breathing heavily	Breathing heavily
	Out of breath	Out of breath	Palpitations	Palpitations	Palpitations
	Taste of bloody,	Palpitations	Warm	Warm	Warm
	Palpitations	Sweaty			
	Sweaty				
Physical discomfort	Aching	Aching	Stiff joints	Stiff joints	Stiff joints
	Hurting	Numbness	Numbness	Numbness	Numbness
	Numbness	Stiff joints	Aching	Aching	Aching
	Stiff joints	Tense muscles.			
	Tense muscles				
Lack of motivation	Uninterested	Uninterested	Passive	Passive	Passive
	Passive	Passive	Listless	Listless	Listless
	Listless	Indifferent	Indifferent	Indifferent	Indifferent
	Indifferent,	Lack of initiative			
	Lack of initiative.				
Sleepiness	Sleepy	Sleepy	Sleepy	Sleepy	Sleepy
	Yawns	Yawning	Falling asleep	Falling asleep	Falling asleep
	Drowsy	Drowsy	Yawning	Yawning	Yawning
	Lazy	Falling asleep			
	Falling asleep				
Irritability				Irritable	
				Angry	
				Furious	

Fig. 17.1 Machining and assembly task. **a** Worker operating a CNC lathe. **b** Worker operating a hydraulic press

17.5 Methodology

17.5.1 Study Design

The assessment of fatigue in semi-automated machinery operators was performed once. The descriptive results and the comparison intra and between processes are reported. Therefore, it is a cross-sectional, descriptive and comparative study.

17.5.2 Sample

Operators of machining and assembly processes were surveyed. For both processes, workers were responsible of loading and unloading the machines, operating the machine control panel and final inspection of pieces after processing time. Inclusion criteria to include participants in the research were the following:

- Have worked at least 6 months in the company as operator machining and/or assembly
- Have received training to operate CNC lathes and hydraulic presses
- At least 18 years of age
- Have no physical injuries in the past 6 months

The sample of workers was chosen according to the convenience of the researcher for the inclusion criteria fulfillment mentioned before.

Table 17.2 Dimensions and items analyzed in the study

Dimensions	Items
Lack of energy	Worn out, exhausted and drained
Physical exertion	Breathing heavily, palpitations, warm
Physical discomfort	Stiff joints, numbness, aching
Lack of motivation	Passive, listless, indifferent
Sleepiness	Sleepy, falling asleep, yawning

17.5.3 Materials and Methods

For the assessment of fatigue, the SOFI-Spanish version developed by Gonzalez (2005) and validated for Mexican workers by Hernandez et al. (2012) was applied. This version considers five dimensions of fatigue: lack of energy, physical exertion, physical discomfort, lack of motivation and Sleepiness. Items included in each of the factors are shown in Table 17.2.

For the valuation of the items, a visual analog scale of 5-point response including the terms "low" on the left and "high" on the right was used. The workers were asked to reply to the survey according to their perception of fatigue at the end of a 12 h typical workday.

The survey was delivered to the group of supervisors of the machining and assembly areas. They were responsible for distributing the survey to participant workers. Prior to the distribution, workers and supervisors were properly trained to minimize the doubts about the survey. They were taught about the kind of questions included and how each section must be answered. Participants were informed about the objectives of the research, emphasizing the confidentiality of the information obtained from them. The project was supported and approved by management of the company.

17.5.4 Data Analysis

The internal consistency of the data was analyzed using Cronbach's alpha index (Nunally 1995 ; Levy et al. 2003). The adequacy of the sample was analyzed with the Keiser Meyer Olkin (KMO) index (Rodriguez-Salazar et al. 2001) and (Levy et al. 2003). Because ordinal and categorical nature of data, two nonparametric tests for comparisons of data were applied: for comparison intra groups, Wilcoxon Rank test for related samples was applied, and for comparison between groups the Mann–Whitney U test for independent samples was applied. Analysis of Variance (ANOVA) and Tukey Post Hoc test were applied to determine differences between the analyzed variables and dimensions. For all statistical analyzes, a significance level of $\alpha = 0.05$ was used. All data analyzes were performed using SPSS ® statistical software version 17.

Table 17.3 Demographic data

Gender	Mach	Assem	Sleeping time	Mach	Assem
Men	73	48	8–9	8	6
Women	2	18	7–8	20	17
			6–7	25	21
			5–6	15	16
			<5 h	7	5
Education			*Marital status*		
Middle school	17	37	Single	28	19
Preparatory	30	18	Married	47	46
Technical	28	11	Common law mariage	0	1

Mach machining, Assem assembly

17.6 Results

A total of 75 workers of machining area and 66 of assembly area were surveyed. Table 17.3 shows the demographic information of the participants.

17.6.1 Internal Consistency

The results of reliability and internal consistency analysis are presented in Table 17.4. The Cronbach alpha values are greater than 0.7, with the exception of the dimensions physical exertion and physical discomfort that obtained values between 0.500 and 0.609. In case the dimension physical exertion, one of the most important dimensions for the generation of fatigue, historically has shown Cronbach's alpha values lower than 0.55 (Åhsberg et al. 1997, 2000; Leung et al. 2004; Barker and Nusbaum 2011; Hernandez et al. 2012). Despite this, the Cronbach's alpha and KMO indices for all items analyzed were higher than 0.764. According to Nunally (1995) and Levy and Varela (2003), data show reliability and internal consistency.

17.6.2 Assessment of Fatigue

The descriptive results of the items analyzed in the SOFI-S (mean, median, and standard deviation) in the machining and assembly areas are shown in Table 17.5. In both processes, the item with the highest score of fatigue was "warm" with 3.76 and 3.55 points on average for the machining and assembly processes, respectively. This is because the processes include heat-treating furnaces increasing ambient temperature and workers perceive more heat. In contrast, the lowest level of fatigue was obtained by the variable "Breathing heavily" with 1.41 and 1.58

Table 17.4 Internal consistency analysis

Dimensions	Cronbach alpha	
	Machining	Assembly
Lack of energy	0.827	0.839
Physical exertion	0.558	0.500
Physical discomfort	0.742	0.609
Lack of motivation	0.754	0.858
Sleepiness	0.793	0.798
	C. A.: 0.881	C. A.: 0.923
	KMO: 0.813	KMO: 0.764

KMO Keiser Meyer Olkin, *C. A.* Cronbach alpha

Table 17.5 Descriptive data of variables of fatigue

Item	Machining			Assembly		
	Mean	Median	SD	Mean	Median	SD
Worn out	2.92	3	0.93	3.06	3	1.05
Exhausted	2.70	3	0.80	2.83	3	1.06
Drained	2.61	3	0.94	2.50	3	1.06
Breathing heavily	1.41[b]	1	0.68	1.58[b]	1	0.88
Palpitations	1.73	1	0.92	1.79	2	0.98
Warm	3.76[a]	4	1.23	3.55[a]	4	1.22
Stiff joints	1.86	2	0.99	2.09	2	1.19
Numbness	1.80	2	0.90	2.00	2	1.08
Aching	2.73	3	1.06	3.02	3	1.10
Passive	2.29	2	0.89	2.27	2	1.12
Listless	2.02	2	0.90	1.86	2	1.02
Indifferent	2.00	2	0.93	1.70	1	0.99
Sleepy	2.06	2	0.84	2.09	2	1.09
Falling asleep	1.73	2	0.90	2.02	2	0.98
Yawning	2.14	2	1.04	2.36	2	1.03

SD standard deviation
[a] Fatigue highest score. [b] Fatigue lowest score

points on average for the machining and assembly processes, respectively. This is because in both processes work energy demand is low.

Descriptive results of fatigue dimensions (mean, median, and standard deviation) are shown in Table 17.6. In both processes, lack of energy dimension obtained the highest score of fatigue with 8.22 and 8.39 average-points for machining and assembly process, respectively. This result contrasts with the only one reference identified where the application of the SOFI report the operation of machinery (Åhsberg et al. 1997), where the dimension "drowsiness" obtained the highest fatigue score followed by "lack of energy". In this case, sleepiness obtained the lowest fatigue score in the machining process, while in assembly process lack of motivation obtained the lowest score.

Table 17.6 Descriptive data of fatigue dimensions

Dimensions	Machining			Assembly		
	Mean	Median	SD	Mean	Median	SD
Lack of energy	8.22	8	2.27	8.39	8	2.75
Physical exertion	6.89	7	2.12	6.90	7	2.25
Physical discomfort	6.39	6	2.50	7.10	6	2.80
Lack of motivation	6.30	6	2.24	5.83	5	2.60
Sleepiness	5.92	5.5	2.45	6.46	6	2.75

SD standard deviation

Table 17.7 Fatigue dimensions post hoc analysis

Fatigue dimensions	Group 1	Group 2
Lack of energy	8.227	
Physical exertion		6.893
Physical discomfort		6.393
Lack of motivation		6.303
Sleepiness		5.924

Machining process
Means for groups in homogeneous subsets are displayed. $\alpha = 0.05$

Table 17.8 Fatigue dimensions post hoc analysis

Fatigue dimensions	1	2	3
Lack of energy	8.393		
Physical discomfort		7.106	
Physical exertion		6.909	6.909
Sleepiness		6.469	6.469
Lack of motivation			5.833

Assembly process
Means for groups in homogeneous subsets are displayed
$\alpha = 0.05$

17.6.3 Intragroup Comparison

The ANOVA conducted to the machining process data shows that at least one dimension of fatigue is different to the others; therefore, there are significant differences between the fatigue dimensions compared ($p < 0.05$). The Tukey Post Hoc test (see Table 17.7) shows the two groups generated after this analysis. Lack of energy dimension is located in-group 1; the four remaining dimensions are located in- group 2.

The ANOVA conducted to the assembly process data shows that at least one dimension of fatigue is different to the others; therefore, there are significant differences between the fatigue dimensions compared ($p < 0.05$). The Tukey Post Hoc test (see Table 17.8) shows the three groups generated after this analysis.

Table 17.9 Nonparametric comparison between fatigue dimensions

Fatigue dimensions	Comparison		¿Difference?
	U-value	P-value	
Lack of energy	2154	0.914	No
Physical exertion	2145	0.881	No
Physical discomfort	1875	0.165	No
Lack of motivation	1843	0.124	No
Sleepiness	1909	0.216	No

Lack of energy dimension is located in-group 1, discomfort physical, physical exertion, and sleepiness are located in-group 2, physical exertion, sleepiness, and lack of motivation are located in-group 3.

17.6.4 Comparison Between Groups

Non-parametric comparison of the fatigue dimensions analyzed between two processes (machining and assembly) is shown in Table 17.9. As noted, there were no significant differences in any of the five dimensions compared.

17.7 Conclusion

The SOFI-S, on their second test of the adapted version for fatigue assessment among Mexican workers obtained values greater than 0.7 (considered good) on Cronbach's alpha test and KMO indexes, with the exception of the physical exertion and physical discomfort dimensions that historically have obtained internal consistency scores considered as medium.

The fatigue score obtained for each dimension was similar in both processes. The highest fatigue score was obtained by lack of energy dimension (8.22 and 8.39, for machining and assembly process, respectively) while sleepiness dimension obtained the lowest score (5.92) in machining process, and lack of motivation dimension obtained the lowest score (5.83) in the assembly process. This order was confirmed by Tukey Post Hoc analysis where two groups of items were obtained for the machining process and three groups of items for the assembly process. For both processes, lack of energy dimension was placed in a group as the sole dimension.

The nonparametric comparison of the five fatigue dimensions between the two processes analyzed showed no significant differences, therefore, after this case study, it can be concluded that the use of CNC lathes and CNC hydraulic presses produce the same fatigue to workers.

References

Åhsberg, E. (2000). Dimensions of fatigue in different working populations. *Scandinavian Journal of Psychology, 41*(3), 231–241.

Åhsberg, E., & Fürst, C. J. (2001). Dimension of fatigue during Radiotherapy-an application of the Sewdich Occupational Fatigue Inventory (SOFI) on cancer patients. *Acta Oncologica, 40*(1), 37–43.

Åhsberg, E., & Gamberale, F. (1998). Perceived fatigue during physical work: An experimental evaluation of a fatigue inventory. *International Journal of Industrial Ergonomics, 21*, 117–131.

Åhsberg, E., Gamberale, F., & Kjellberg, A. (1997). Perceived quality of fatigue during different occupational task. Development of a questionnaire. *International Journal of Industrial Ergonomics, 20*, 121–135.

Barker M., L., & Nussbaum, M. (2011). Fatigue, performance and the work environment: A survey of registered nurses. *Journal of advanced nursing, 67*(6): 1370–1382.

Barrientos-Gutierrez, T., Mártinez-Alcántara, S., & Méndez-Ramírez, I. (2004). Validez de constructo, confiabilidad y punto de corte de la Prueba de Síntomas de Subjetivos de Fatiga en Trabajadores Mexicanos. *Salud Publica de México, 46*(6), 516–523.

Borg, G. (1985). *An introduction to Borg's RPE-scale.* (Ithaka, Ed.) New York: Movement publications.

Bridger, R. (2003). *Introduction to ergonomics* (2nd ed.). New York: Taylor & Francis.

Fürst, C., & Ahsberg, E. (2001). Dimensions of fatigue during Radiotherapy. An application of the multidimensional fatigue inventory. *Supportive Care in Cancer, 9*(5), 355–360.

Gonzalez, E. L., & Gutierrez, R. E. (2006). La carga de trabajo mental como factor de riesgo de estrés en trabajadores de la industrial electrónica. *Revista Latinoamericana de Psicología, 38*(2), 259–270.

Gonzalez, J. L., Moreno, J. B., Garrosa, H. A., & López, L. A. (2005). Spanish version of the Swedish Occupational Fatigue Inventory (SOFI): factorial replication and validity. *Introducción Journal of Industrial Ergonomics, 35*, 737–746.

Guelaud, F., Beauchesne, M., Gautrat, J., & Roustang, G. (1977). *Pour une analyse desd contions du travail ouvrier dans l'entrerprice.* Paris: A Collin.

Hernandez-Arellano, J. L., Ibarra-Mejía, G., Serratos-Pérez, J. N., García-Alcaraz, J. L., & Brunette, M. J. (2012). Construction of a survey to assess workload and fatigue among AMT operators in Mexico. *Work, 41*, 1790–1796.

Hernandez-Arellano, J., Hernandez, V., & Serratos-Perez, J. (2010). Fatigue curves in tomato packers in Jalisco state Mexico. *Proceedings of the 2010 Industrial Engineering Research Conference.* Cancun: IIE.

Juarez-Garcia, A. (2007). La dimensión de fatiga-energía como indicador de presentismo: Validez de una escala en trabajadroes mexicanos. *Ciencia y trabajo, 9*(24), 55–60.

Leung, A., Chan, C., & He, J. (2004). Structural stability and reliability of the swedish occupational fatigue inventory among chinese VDT workers. *Applied Ergonomics, 35*, 233–241.

Levy, J., Varela, J., & Gonzalez, J. (2003). *Análisis multivariable para las ciencias sociales.* Prentice Hall-Pearson: Madrid.

Michielsen, H., De Vries, J., & Van H., G. (2003). Psychometrics qualities of a brief self-rared fatigue measure. The fatigue assessment scale. *Journal of Psychosomatic Research, 54*: 345–352.

Nunally, J. (1995). *Teoría psicométrica.* México D. F.: Mc. Graw Hill.

Rodriguez-Salazar, M. E., Álvarez-Hernández, S., & Bravo-Núñez, E. (2001). *Coeficientes de Asociación.* México: Plaza y Valdez.

Sebastian, C., Idoate G., V., Llano, L., & Almanzor, E. F. (2008). SOFI-SM: cuestionario para el análisis de la fatiga laboral, física y psíquica. Revista digital de seguridad y salud en el trabajo, (2): 1–22.

Small, M., & Chen, I. (1997). Economica and strategic justification of AMT inferences from industrial practices. *International Journal of Production Economics, 49*, 65–75.
Wiewille, W. W., & Casali, J. G. (1983). A Validated Rate Scale for Global Mental Workload Measurement in a Test and Evaluation Environment. In H. F. Society (Ed.), *Proceedings of the Human Factors Society 27th Annual Meeting*, (pp. 129–133).
Winwood, P., & Casalli, J. (2005). Development and validation of a scale to measure work-related fatigue and recuvery: The occupational fatigue exhaustion recovery scale (OFER). *Journal of Occupational and Environmental Medicine, 47*, 594–606.
Yoshitake, H. (1978). Three characteristics patterns of subjective fatigue symptoms. *Ergonomics, 21*, 231–233.

Chapter 18
Theoretical Approach for Human Factors Identification and Classification System in Accidents Causality in Manufacturing Environment

Rosa María Reyes-Martínez, Aidé Aracely Maldonado-Macías, Rodolfo de la O and Jorge de la Riva-Rodríguez

Abstract Assessment of workplace risk is generally accepted as the basis of any effort to manage it. Researchers have emphasized on the need to incorporate human factors in risk assessment, due to the increasing relevant role of humans in the modern work environment. The aim of this chapter is to offer a theorical approach for human factors identification and classification system in accidents' causality in manufacturing environment. This chapter emphasizes in the study of human error as a relevant aspect in the human factors approach. It presents an introduction to human error taxonomies. Also, a new taxonomy is proposed and developed in a case study. The Swiss Cheese defense model developed by Reason is treated in detail because it has built the theoretical fundaments for the proposed taxonomy applied in the manufacturing industry for automotive harnesses. Finally, an application of the proposed taxonomy is shown, using accident report found in the database of an international automotive wire harness manufacturing corporation.

Keywords Accident causality · Human errors · Risk management

18.1 Introduction

The Health and Safety Executive (HSE) describes human factors as the perceptual, mental and physical capabilities of people and interactions of individuals with their job and work environment, and the influence of equipment design and system

R. M. Reyes-Martínez (✉) · R. de la O · J. de la Riva-Rodríguez
Undergraduate Studies Division, Technological Institute of Ciudad Juarez, Ciudad Juarez, Chihuahua, Mexico
e-mail: rosyreyes2000@uacj.mx

A. A. Maldonado-Macías
Department of Industrial and Manufacturing Engineering, Autonomous University of Ciudad Juarez, Ciudad Juarez, Chihuahua, Mexico

J. L. García-Alcaraz et al. (eds.), *Lean Manufacturing in the Developing World*,
DOI: 10.1007/978-3-319-04951-9_18,

design on human performance. It also notes characteristics of the organization which influence safety-related behavior at work (HSE 1989). The identification and study of human factors is of particular interest for the field of safety as these factors are involved in human errors. Their field is vast as it "studies the intersection between people, technology and work, with the major aim to find areas where design and working conditions produce human error" (Woods and Cook 2006).

Analysis of human error and their role in accidents is an important part of developing systematic methods for reliability in the industry and risk prediction. To obtain data for predictive analysis is necessary to analyze accidents and incidents to identify its causes in terms of component failures and human errors (Leplat and Rasmussen 1984). Therefore, a proper understanding of human factors in the workplace is an important aspect in the prevention of accidents (Harris 2004), and human factors should be considered in any program to prevent those that are caused by human error.

Human error in accidents has been studied in two different environments. The first one is in manufacturing and the second in high-risk complex systems. The complexity of research in relation to human error is different according to each approach. In manufacturing is oriented to individual accidents (occupational) while high-risk systems approach is focused on organizational accidents. Research in high-risk systems tends to be more complex as organizational accidents, due to it has multiple causes without clear causal links (Saurin et al. 2008). In the case of occupational accidents, they are associated with unsafe acts that have a big impact, and generally a person or group of persons is the agent and the victim. In addition, the person or group at risk may be known (Reason 1990). One drawback of the current studies about identification and classification of human error and its contributing factors, is that occupational accidents in the manufacturing industry have been neglected to give attention to high-risk complex systems, such as medicine, aviation, military, rail industry and shipping. This implies that there is very little empirical data on human error in the field of manufacturing and it limits the empirical validation of their concepts and classification, in this domain of knowledge.

To prevent and/or reduce occurrence of accidents and incidents, the organizations must work towards reducing human error or making the work system to be more error tolerant. The process of management of human error accidents involves prevention, recovery from errors and containment of the consequences resulting from its occurrence (Cacciabue 2000). The first step in this process is error identification, which may allow appropriate prevention and mitigation strategies to be developed for this purpose (Baysari et al. 2008).

The causes of occupational accidents from the human factors perspective have received little attention in the field of scientific research. Available studies related to the topic have been developed mainly for aerospace applications and they have been found insufficient to explain accidents causalities in the manufacturing industry.

18.2 Human Error

The human reliability is defined as: the body of knowledge related to the prediction, analysis and reduction of human error, focusing on the role of the person in operations design, maintenance, use and management of a socio-technical system. (Arquer and Nogareda 1988). The human reliability study aims to human error. Human error is a complex construct that has received constant attention among researchers of human factors, has been consistently identified as a contributing factor in a high proportion of incidents in complex and dynamic systems. The dominant definition of human error is raised by Reason who defines it as "a generic term that encompasses all those occasions in which a sequence of physical or mental activities, fails to achieve its desired result and when these failures cannot be attributed to the intervention of some chance" (Reason 1990).

According to Cañas and Waerns (2001), the human error has been studied from three different approaches. The first, taken from the field of engineering has developed a number of techniques known by the generic name of techniques of Human Reliability Analysis (HRA), its basic assumption is that the actions of a person in the workplace can be considered from the same point of view that the operation of a machine. HRA goal is to predict the human error probability and evaluate how the whole system work degrades as a result of mistake alone or in connection with the operation of the machines, the characteristics of the task or the person and the job design system. The engineering methods have been applied in human error analysis in healthcare systems and occupational accidents among which are the method of Failure Mode and Effects Analysis (FMEA), Root Cause Analysis (RCA), Fault Tree Analysis (FTA), Cause Effect Diagram, Study Hazard Operability (HAZOP), Probability Tree Method, Analysis of Human–Machine System (MMSA) and Markov Analysis Dhillon (2003). Although these techniques have made considerable progress in efforts to predict the occurrence of human error, they have been criticized as inadequate. In this regard Reason (1990), points out that the main difficulty of their application is the estimation of the probability of error. In addition, experts have difficulty making accurate estimates of past or future events.

A second approach is adopted from cognitive psychology; in this case the interest is focused on knowing the mental processes responsible for the occurrence of errors (Norman 1988; Reason 1990). The authors state that error are not irresponsible behavior nor occur for poor mental functioning, rather, they may be the result of having ignored how a person perceives, attends, remember and make decisions during the design of the work system. In this perspective, the investigation of causes of human errors is made by analyzing the characteristics of human information processing.

In this approach, the first step to explain the causes of human error, has been classified according to the mental processes involved in those behaviors that led to the error. According to Rasmussen (1987) who have studied cognitive factors, it is possible to distinguish three types of errors depending on the sequence in the processing of information during the execution of the task of solving problems or decision making. Reason (1990) extends the Rasmussen classification by the model known as the Generic Modeling System Error (GEMS). However, although researches made by Reason and Rasmussen have succeeded in finding psychological explanations based on their proposed cognitive models, they have been considered insufficient to facilitate the prediction and prevention of error.

The third approach is recently adopted by the cognitive ergonomics and is a combination of HRA, developed by engineers and cognitive models. In the field of cognitive ergonomics, it is recognized that the cause of human errors often be sought in the design of the machine, the computer signals or forms of communication between people. For this reason, explain, predict and prevent such errors, is a central topic of cognitive ergonomics.

The cognitive ergonomists have departed from the basic idea that a good design is one in which we take into account that the person who will operate the machine, has a number of cognitive features that enforce limitations on their ability to process information and make decisions. In the field of cognitive ergonomics, is recognized that the cause of human errors often be sought in the design of the machine, the computer signals or forms of communication between people; where a good design is developed taken into account a group of cognitive characteristics that impose restrictions on information processing capacity and make decisions. The cognitive ergonomics approach is based on two assumptions:

- Human error cannot be explained only from the cognitive models. The person and the system where it works should be viewed as a joint cognitive system where their interaction occurs.
- The conduct of a person is determined by the context in which it occurs and since working system is dynamic, it is necessary to consider the context of the task. It is not enough to estimate errors from the perspective of human information processing.

18.2.1 Human Error in Manufacturing Environments

For Liu et al. (2009) a manufacturing process can be conceptualized, as a sequence of physical actions and information processing activities that transforms raw materials into customers' specified products. The ultimate goal of this process is to create profits and sustain industry growth. This goal cannot be achieved without a stable manufacturing environment. Through carefully designed safety mechanisms, it is expected that this process is robust to preserve the firmly coupled activities remain connected under the interferences of occupational stresses, such as human errors. Even though modern technology is incorporated during the process design and installation stages, the inconsistent performance of human behaviors and unstable quality of hardware often create problems in the manufacturing chain. Unexpected human errors and hardware failures disturb the system activities and initiate fluctuation. Once the magnitude of the fluctuation exceeds the limits of system safety capacity, the manufacturing process is interrupted and industrial accidents occur. Investigators have shown that the human errors have been generally recognized as the major cause of industrial accidents (Hale and Glendon 1987; Runciman and Sellen 1990; Reason 1990; Hollnagel 1993; Maurino et al. 1995; Salminen and Tallberg 1996; Feyer and Williamson 1996; Feyer et al. 1997; Reyes et al. 2012). This problem is aggravated by increasing mental workload of human in modern manufacturing environment and insufficient company resources spending due to the lack of understanding the losses of human errors (Strobhar 1995; Spintzer 1996; Cacciabue 2000).

When human errors occur, actions should be taken to control resulting impacts. It is practical to provide factory managers with a tool to assess the potential human error threats from daily operations (Strater and Bubb 1999). A taxonomy developed to identify accidents' causes can be useful when is combined with diagnostic methods to prevent such accidents. This chapter presents a literature review of several human error taxonomies and a case study where a new taxonomy for accidents affecting hands in a manufacturing process of automotive harnesses was proposed.

18.2.2 Human Error Taxonomies

The concept of "taxonomy" was originated in biological classification. In this field of knowledge, it is constructed based on the degree of similarity and difference between qualities, which are attributes or characteristics that distinguish one individual and provide evidence for taxonomic classification. The qualities are chosen by their creators for use in discrimination or group. Taxonomy is defined as a system that divides a body of knowledge in taxonomic units and defines the relationships between them. Its aim is to classify the body and understanding of knowledge (ANSI/IEEE STD 1002 1987). Taxonomic units should be hierarchically nested at different levels and cover the entire body of knowledge (Blackwelder 1967; Gordon 1999; Li et al. 2006). The classification is made by arranging the objects into categories that share a specific set of properties and can be divided into a number of subcategories, with each subset of the original category. This process can be repeated and subcategories can be divided into lower levels. Similarly categories can be grouped into higher-level categories, whose structure is known as a hierarchy. A classification system is commonly represented by a hierarchical tree, allowing arbitrary levels of sophistication (Li et al. 2006).

Taxonomies are used for identification and classification of human error and are widely used in science. Taxonomy is a prerequisite for the creation of empirical science, to a deep understanding of the nature, origins and causes of human error. Therefore, it is necessary to have a clear classification system to describe the phenomenon under study (Senders and Moray 1991). Taxonomies provide a theory that stimulates the classification system to enable analysts and researchers categorize operator errors and the factors contributing to the accident/incident. The classification of these errors and contributing factors based on their theoretical nature provides a great benefit in allowing that the accident paths are identified through the forms of error (Reinach and Viale 2006). Also taxonomies allow to build an overview over a large number of accidents and incidents, allowing the identification of the dominant failure factors and recurrent (Baysari et al. 2008), and causal factors and concurrent over time.

Some models and human error frameworks have been developed from the late seventies, to help understand how humans make mistakes and how accidents and incidents occur in the context of systems. These include, the Information Processing Model of Four Stages (Wickens and Flach 1988), the Skill-Rule-Knowledge (Rasmussen 1982), Wheel of Misfortune taxonomy (O'Hare 2000) the socio-technical error system (Moray 2000), the Generic Modeling System Error (Reason 1990). The Swiss cheese model (Reason 1990) and the Human Factors Classification System (Shapell 1997) which is considered by experts as the system with the best taxonomy. These models and their structures include taxonomies.

18.2.2.1 The Skill-Rule-Knowledge Model

In order to propose a theoretical reference regarding the human reliability, the cognitive model proposed by Rasmussen (1987) is presented. The model identifies cognitive functioning levels involved in the tasks and the different phases in diagnostic activity and problem solution (detection, identification, evaluation, planning and execution). Rasmussen's model distinguishes eight stages in information processing and decision making: activation, observation, identification, interpretation, evaluation, target selection, choice of procedure and enforcement. Also, the three levels of functioning of the person are differentiated depending on their activity in automation, rules or procedures and knowledge.

These levels correspond to decreasing degrees of familiarity with the environment and the task, when a person behaves based on automation, will stage "activation and observation" to "execution". When this is based on the rules and procedures, the leap to the "execution" will be from the stage of "identification" and finally, when the behavior is based on knowledge, the person goes through the eight stages mentioned. (Arquer and Nogareda 1988).

The advantage of Rasmussen's cognitive model is that it combines the identification of the functional origin of the error (processing step), with the implied performance level (based on knowledge, rules or automatic) and allows separate treatment stages errors; additionally, the specificity of some errors, allows to propose preventive measures adapted to the level of performance person. The model proposes three levels of cognitive activity:

- The skill-based level, refers to the knowledge called automatic actions, it is hard to raise awareness and has little control by operators. It is extremely economical from the point of view of mental workload.

- The level of knowledge based in rules refers to conscious order, whose characteristic is the form of rules applied to new situations, with a level of control action greater than skill level.
- The level of knowledge also refers to knowledge of consent order, but unlike the other two levels, they build new rules allow novel situations solution, developed diagnostic activities and problem solving. Declarative knowledge possesses a high level of action control and demand many cognitive resources.

According to Rasmussen (1983), the *skill based* behavior is represented by sensory-motor performance during activities, which following a declaration of an intention occur without conscious control as smooth, automated and highly integrated patterns of behavior. Experimental tracking task are common examples about the skill based level and is used rarely and only for slow, very accurate movements such as assembly task or drawing. In the case of *rule based* behavior, a sequence of subroutines in a familiar work situation is controlled by a store ruled or procedure which may have been derived empirically during previous occasions, communicated from other person' know-how as instruction, or it may be prepared on occasion by conscious problem solving and planning. The *knowledge based* behavior occurs during unknown situations, faced with an environment for which no know-how or rules for control are available from previous encounters. Thus the goal is explicitly formulated, based on an analysis of the environment and the aims of the person.

One aspect of the categorization of human performance founded in behavior is the role of the information observed from the environment, which is based in three levels: skill/rule/knowledge. At the *skill-based* level the perceptual motor system acts as a multivariable continuous control system synchronizing the physical activity such as navigating the body through the environment and manipulating external objects in the time–space domain. The performance at the skill based level may be released or guided by value features involved with previous experience of certain patterns of information not taking part at the time–space control, but acting as cues or signs activating the organism. At the *rule-based* level, the information is typically perceived as signs. The information perceived is defined as a sign when it serves to activate or modified predetermined actions or manipulations. A sign refers to situations or proper behavior by convention or prior experience; it does not refer to concept or represent a functional property of the environment.

18.2.2.2 The Swiss Cheese Model

Reason (2000) mentions that human error problem is viewed in two different perspectives: the person approach and the system approach. Each has its own model of error causation arising in quite different philosophies of error management and control. Understanding the differences between these approaches, have important implications for dealing with risk, which is always present. The person approach focuses on unsafe acts (errors) and procedural violations in direct contact with the production system. It considers these unsafe acts primarily occur due to abnormal mental processes such as memory failure, inattention, poor motivation, carelessness, negligence and recklessness. Methods used in this approach include poster campaigns that appeal to people′s sense of fear, writing another procedure, disciplinary measures, litigation threat, retraining, blaming and shaming. Followers of this approach tend to threat errors as moral issues, assuming that bad things happen to bad people.

The basic premise in the systemic approach is that humans are fallible and errors are to be expected even in the best organizations. Errors are seen as consequences rather than causes, having their origins not so much in the perversity of human nature, but in the early stages of systemic factors. These include recurrent error traps in the workplace and organizational processes that give rise to them. Countermeasures are based on assumptions that cannot change the human condition, but it is possible to change the conditions of human performance. A central idea is that the system's defenses, all technologies have dangerous obstacles and safeguards. When an adverse event occurs, the important thing is not who did it, but how and why the defenses failed.

James Reason presents the cognitive model of great relevance to the psychology of Work Safety, Defence Model Swiss Cheese (Reason 1990; Romera 2007). Also known as the Model of Unsafe Acts, which has become one of the most used in the study of human error in cognitive ergonomics. This model was initially developed as the system's Generic Error Classification comprises the Skill-Rule-Knowledge model of Rasmussen and the dichotomy of slip/mistake of Norman (1988) and includes rules violations as a distinct form of unsafe acts (Hobbs and Williamson 2003; Reason 1990).The theoretical framework of Reason provides the following: the taxonomy of human error, the simultaneous failure model "Swiss Cheese" failures distinction depending on the immediacy of their consequences (active failures and latent conditions), the metaphor of "pathogens", accident investigation and human behavioral factors and their levels.

In the Swiss Cheese model, Reason states that the condition for an accident is that the holes produced by active failures (errors and violations), are aligned with the holes of the latent failures and then there is a window of opportunity for the occurrence of an accident. This model has led to the explanation of an accident as the "overlap or coincidence of failures at different levels of the organization at the same time." In this model, the term "defense" is used as different means applied to guarantee people safety and company goods. Automated systems, physical barriers, personal protective equipment are hard applications while the soft are legislation, regulations, rules and procedures. The holes of the defences arise for two reasons.

Active failures are unsafe acts committed by people who are in direct contact with the system and have a direct and short-lived impact on the integrity of the defences. They take a variety forms such as slips, lapses, mistakes and procedural violations. Latent condition is the resident pathogen within the system. They arises from decisions made by designers, builders, procedure writers, and top level management. Such decisions may be mistaken, but they need not be. All such strategic decisions have the potential for introducing pathogens into the system. Latent conditions have two kinds of adverse effect: they can translate into error provoking conditions within the local workplace (for example, time pressure, understaffing, inadequate equipment, fatigue, and inexperience) and they can create longlasting holes or weaknesses in the defences (untrustworthy alarms and indicators, unworkable procedures, design and construction deficiencies). Latent conditions, as the term suggests may lie dormant within the system for many years before they combine with active failures and local triggers to create an accident opportunity. Unlike active failures, whose specific forms are often hard to foresee, latent conditions can be identified and remedied before an adverse event occurs. Understanding this leads to proactive rather than reactive risk management (Reason 2000).

According to Reason's taxonomy human errors can be caused by intentional and unintentional actions. Errors by intentional actions can be of two types, mistakes and violations. Lapses and slips are unintentional actions. Lapses are linked to observable facts and are commonly associated with attention deficit such as omission and inversion steps, grown given orders and out of time actions. Predisposing conditions may be psychological or circumstantial. The psychological type refer to capture attention due to distraction or concern for things unrelated to immediate task and therefore poor capacity attention to control the process of current actions. Circumstantial conditions often occur when there is a change in the nature of the task and/or the environment in which it performs.

Lapses are internal phenomena related to memory failures, such as omission, repetition of items planned, location losses, forgetting of intent. In this case occurs a deviation for not respecting the sequence of a plan. The error occurs at a higher level in the mental process involved in determining the available information, the planning and formulation of intent. The mistakes are conceptions errors, are based on knowledge and rules established, its detection is difficult because it can remain dormant over time and can be on rules and knowledge. The case of mistakes based on rules comprises applying good and bad rules. While the case of mistakes based on knowledge occurs when you run out of ready-made solutions and new thinking is required. Violations are deviations from operating procedures, standards and existing rules regarding safety, are presented in three categories: routine, exceptional and sabotage. The routine presented in the level of skill-based behavior and takes the least effort to accomplish the task. The exceptional arise in working conditions and are interpreted as necessary for the fulfillment of the task. Sabotage is intentions to harm people or equipment.

Organizational issues have always been the most neglected aspect of the analysis and accident data collection. Reason mentions "pathogens metaphor", indicating that factors related to the organization and management, contribute to latent conditions of work systems, similar to the "resident pathogens" in biological systems. As pathogens, latent conditions such as design defect, failure in supervision, undetected manufacturing defects, maintenance failures, wrong procedures of tools and equipment unsuitable may be present for many years before combined with local factors and active failures to penetrate different layers of defense system. Latent conditions are present in sick systems are an inevitable part of organizational life (Harris 2004; Romera 2007). The organizational failures are linked to the organizational structure so that any change of it, affects the processes of selection, training, career planning, compensation systems, leadership style, work environment, communication, quality of equipment, planning, maintenance, operational and commercial pressures, supervision and management.

According to the Swiss Cheese Model, accidents are the result of the simultaneous occurrence of windows of opportunity at the system level, allowing security threats remain undetected and uncorrected. Since the error is simply a consequence of failures, is not sufficient to identify it. Preferably are proposed effective safety programs, after priority events and characteristics situational allowed humans to cause the error have been identified. Besides the information about these systems-level causes, must be communicated to the safety experts who can identify ways to reduce or eliminate error, if it is possible (Reason 2000).

In relation to the accident investigation, the model presents a sequential path in the development of the accident causal history. Reason presents the thesis that human error is inevitable and says: "It is extremely crucial that staff and managers in particular, to become aware of the human potential for error and the activities of the workplace and the factors organizational will shape their possibilities and their consequences" (Romera 2007 p.19). Even though the Swiss cheese model provides a unifying framework for studying root causes, researchers and practitioners have questioned the effectiveness of applying this model in the study of accidents. In this regard Weigmann and Shapell (2003) noted that Reason's model fails to identify the exact nature of the holes in the cheese, the researchers argued that it tends to be theoretical and less analytical, which makes difficult its application. Another limitation is that it not contains explicitly human fallibility and ignored the physical, psychological and social aspects of the person. In order to facilitate the implementation of the Swiss cheese model, Weigmann and Shapell during the period 1997–1999, defined cheese holes through the development Human Factors Analysis and Classification System (HFACS).

18.2.2.3 The Human Factors Analysis and Classification System

The HFACS was originally developed as an evaluation framework to analyze and classify human error in naval accidents and mishaps. However, the advanced version of HFACS based upon Reason'model of latent and active failures has provided an applicable taxonomy to investigate human error in accidents, and it is considered as one of the most powerful tool for reconstructing human contributions to various types of accidents.

The HFACS taxonomy was developed by reviewing hundreds of accident reports in military environment and has been studied extensively; studies have confirmed reliability, comprehensiveness and utility for investigating military and general aviation accidents. Also small studies have demonstrated the efficacy of

HFACS for commercial aviation accidents, but raters observed that factors included are appropriate for military aviation and it is difficult for raters to use if they lack military experience. Although the generic framework of HFACS was designed for use in military and aviation environment, it had being applied in the identification and classification of human error associated with accidents in other areas, such as the railroad industry (Baysari et al. 2008; Reinanch and Viale 2006), to reduce occupational accidents in Turkish shipyards (Celik and Cebi 2009).

HFACS described four levels of failure, according to the structure of Reason (1990). These levels were: unsafe acts, unsafe supervision, preconditions for unsafe acts and organizational influences (Baker and Krokos 2007; Shappell et al. 2007). The unsafe acts category can be classified in: errors and violations. In general errors represent the mental or physical activities of individuals fail to achieve their intended outcome. Violations refer to the willful disregard for the rules and regulations; occur much less frequently since they often involve fatalities. Preconditions for unsafe acts category are important because accidents investigators must dig deeper into why the unsafe acts took place, the category consists of two mayor subdivisions: substandard conditions of operators and the substandard practices they commit. Unsafe supervisor category identified four subdivisions: inadequate supervision planned inappropriate operations, failure to correct a known problem and supervisory violations. Organizational influences often are ignored by safety professionals, due in large part to the lack of a clear framework from which to investigate them, the most elusive of latent failures revolve around issues related to resource management, organizational climate and operational processes. A brief description of each casual category is provided in Table 18.1.

18.2.2.4 Human Factors Analysis and Classification System for Harnesses Manufacturing Process

HFACS was developed to identify, analyze and classify human error in naval accidents and mishaps. The taxonomy has been applied successfully to accidents in high risk systems. However apply this system to occupational accidents in the manufacturing industry has been difficult; for this reason the Human Factors Analysis and Classification System for Harnesses Industry (HFACSH) was developed. In this sense, O'Connor (2007) showed that researchers must be careful when adjusting taxonomies that were developed for specific industries, although they show similarities in the general level of the categories, it is likely that this does not happen at the level subcategories, as they depend on the characteristics of the industry and the types of accidents that occur. The theoretical framework of the proposed taxonomy is consistent with the "Swiss Cheese Model" of Reason (1990), in relation to: the category of unsafe acts corresponds to the active failures, the unsafe conditions, personal factors, supervision and organizational factors are latent failures. In relation to the taxonomy categories there is a complete agreement with those developed for the HFACS, specifically, the category of unsafe

Table 18.1 Brief description of human factors analysis and classification system (reproduced from Shappell et al. 2007)

Organizational influences

Organizational climate: Prevailing atmosphere/vision within the organization, including such things as policies, command structure, and culture

Operational process: Formal process by which the vision of an organization is carried out including operations, procedures, and oversight, among others

Resource management: How human, monetary, and equipment resources necessary to carry out the vision are managed

Unsafe supervision

Inadequate supervision: Oversight and management of personnel and resources, including training, professional guidance, and operational leadership, among other aspects

Planned inappropriate operations: Management and assignment of work, including aspects of risk management, crew pairing, operational tempo, etc.

Failed to correct known problems: Those instances in which deficiencies among individuals, equipment, training, or other related safety areas are "known" to the supervisor yet are allowed to continue uncorrected

Supervisory violations: The willful disregard for existing rules, regulations, instructions, or standard operating procedures by managers during the course of their duties

Preconditions for unsafe acts

Environmental factors

Technological environment: This category encompasses a variety of issues, including the design of equipment and controls, display/interface characteristics, checklist layouts, task factors, and automation

Physical environment: Included are both the operational setting (e.g., weather, altitude, terrain) and the ambient environment (e.g., as heat, vibration, lighting, toxins)

Condition of the operator

Adverse mental states: Acute psychological and/or mental conditions that negatively affect performance, such as mental fatigue, pernicious attitudes, and misplaced motivation

Adverse physiological states: Acute medical and/or physiological conditions that preclude safe operations, such as illness, intoxication, and the myriad pharmacological and medical abnormalities known to affect performance

Physical/mental limitations: Permanent physical/mental disabilities that may adversely impact performance, such as poor vision, lack of physical strength, mental aptitude, general knowledge, and a variety of other chronic mental illnesses

Personnel factors

Crew resource management: Includes a variety of communication, coordination, and teamwork issues that impact performance

Personal readiness: Off-duty activities required to perform optimally on the job, such as adhering to crew rest requirements, alcohol restrictions, and other off-duty mandates

Unsafe acts

Errors

Decision errors: These "thinking" errors represent conscious, goal-intended behavior that proceeds as designed, yet the plan proves inadequate or inappropriate for the situation. These errors typically manifest as poorly executed procedures, improper choices, or simply the misinterpretation and/or misuse of relevant information

(continued)

Table 18.1 (continued)

Skill-based errors: Highly practiced behavior that occurs with little or no conscious thought. These "doing" errors frequently appear as breakdown in visual scan patterns, inadvertent activation/deactivation of switches, forgotten intentions, and omitted items in checklists. Even the manner or technique with which one performs a task is included
Perceptual errors: These errors arise when sensory input is degraded, as is often the case when flying at night, in poor weather, or in otherwise visually impoverished environments. Faced with acting on imperfect or incomplete information, aircrew run the risk of misjudging distances, altitude, and descent rates, as well as of responding incorrectly to a variety of visual/vestibular illusions
Violations
Routine violations: Often referred to as "bending the rules," this type of violation tends to be habitual by nature and is often enabled by a system of supervision and management that tolerates such departures from the rules
Exceptional violations: Isolated departures from authority, neither typical of the individual nor condoned by management

acts, which describe human error from Reason's taxonomy, based on the model Skill-Rule-Knowledge of Rasmussen (1982). The category of unsafe conditions is equivalent to the preconditions for unsafe acts, while organizational factors category corresponds to HFACS organizational influences. The scientific methodology used in the developed of HFACSH corresponds to a combined Cognitive Anthropology approach, with the application of an informal version of the Cultural Consensus Theory. The data collection techniques such as free listing and pile sort (card sorting) were used with cultural domain methods analysis described by Weller and Romney (1988) and Weller (2007). These techniques were applied sequentially, that is, each subsequent data collection step was based on the findings from the preceding step, the categories of unsafe acts, unsafe conditions, personal factors and organizational factors were generated from the knowledge of the members of the safety group, composed mostly by supervisors. The supervision category was developed from the knowledge of multifunctional operators, and theoretical foundations of HFACS insecure supervision. Consequently, categories and human factors that make up the taxonomy reflect the knowledge that participants have regarding human error and safety system failures. HFACSH framework was integrated by five categories: human error, unsafe supervision, unsafe conditions, personal factors and organizational factors. A brief description of each casual category is provided in Table 18.2. Human error types described in HFACSH, slips, lapses and mistakes match of Reason's taxonomy predominant types of errors are deviations from operating procedures, standards and existing rules regarding safety and correspond to routine and exceptional violations. The first presented in the skill-based level and occur when the person makes the slightest effort to accomplish the task, the latter are necessary for the fulfillment of the task.

The unsafe conditions category presents a solid theoretical foundation in the conceptual structure of human error, developed by Sharit and Gables (2006). This conceptual framework provides an integrative approach from the perspectives of

Table 18.2 Brief description of human factors analysis and classification system harnesses industry

Organizational factors
Organizational climate. Safety policies, procedures, practices, and the overall importance and the true priority of safety at work
Operational process. The formal process by which the vision of an organization is carried out, consisting of operations, procedures and working methods
Resource management. Human and economic resources, equipment and facilities
Unsafe supervision
Planned inappropriate operations. Management and assignment, including aspects of risk management and the pace of operations
Failed to correct known problems. Those cases where deficiencies among individuals, equipment, training and other security related areas are "known" to the supervisor but are allowed to continue uncorrected
Supervisory violations. The willful failure by the administrators during the course of their duties, in relation to the rules, regulations, instructions or standard operating procedures related to safety
Inadequate supervision. Oversight of management of personnel and resources, including professional guidance, training, supervision resources, motivation, leadership regarding operational safety
Unsafe conditions
Machinery and equipment. Conditions in the machinery and equipment that do not allow the execution of the task safely
Safeguards. The different ways in which human error can be contained
Tools. Instruments that help workers to perform their task
Environmental conditions. Features of the physical environment that may influence occupational accidents such as lighting, noise, vibration, temperature
Personal factors
Physical factors. Physical characteristics of the operators may influence occupational accidents
Social factors. Behaviors resulting from workers' social life that influence occupational accidents
Psychological factors. Behavior characteristics workers that influence in occupational accident
Human error
Mistakes. Errors of conception are based on knowledge and rules established, its detection is difficult, as they can remain dormant over time, can be about rules and knowledge
Slips. Errors that are related to observable facts and are commonly associated with attention deficits, such as the introduction, omission, investment, given orders wrong and untimely actions
Violations. This kind of human error are deviations are operating procedures, standards and existing rules regarding safety, may be: routine, exceptional and sabotage
The routine violations are presented in the level of skill-based behavior and take the least effort to accomplish the task
Outstanding violations are generated in the working conditions and are interpreted as necessary for the fulfillment of the task
Sabotage is intentions to harm people or equipment

ergonomics and human factors, cognitive engineering and socio-technical approach. Most factors describing the class, and part of the context corresponding to the technology system, which is part of a socio-technical approach regarding environmental conditions and equipment design workspace. According to Reason are known unsafe conditions and underlying causes include poor design of equipment, tools and maintenance failures. The category also presents factors related to barriers. According to Sharit and Gables (2006), barriers function is to prevent the propagation of errors that result in an accident. Category factors are located with physical barriers, specifically safeguards machinery. In the model of Swiss cheese defense, this term refers to the various means used to ensure the safety of people and facilities. Accident structure proposed by Raouf (1998) presents similarities with that of the category and the author mentions ineffective safety devices as well as its absence, which despite being necessary correspond to physical barriers such as safeguards.

Personal Factors category is integrated by physical, psychological and social issues. In contrast to HFACS, the taxonomy does not show matches within the category preconditions for unsafe acts of HFACS, although it presents a subcategory called personal factors; it is specific for the aviation industry. A new classification within this taxonomy are social factors because such factors are not included in HFACS The category of unsafe supervision was adapted from HFACS taxonomy, the names of the categories are preserved, but their definitions correspond to harnesses industry needs. Supervision is part of the administrative task and could be studied as a subcategory within organizational factors; however, given its importance in preventing risks, it was decided to give the importance of general category. Finally, in the case of the category called organizational factors described failures resulting from administrative tasks of the organization and are theoretically grounded in the model of Swiss cheese defense. These factors are considered by Reason (1990) as pathogens in reference to the metaphor with biological systems, because they contribute to latent conditions in work systems. The category was structured with three subcategories: safety climate, operational process and resource management. The conceptual structure of human error and Sharit and Gables (2006) provides theoretical support to these subcategories.

18.2.3 Taxonomy Application in Harnesses Manufacturing Process

The taxonomy was applied in a corporation that manufactures automotive harnesses for the purpose of identifying and classifying the types of errors and their contribution factors whose outcomes are occupational accidents with hand injuries. The harness is an exportation product, assembled in Mexico and Ciudad Juárez is a city where the mayor companies in this sector (United Technology Automotive, Yazaki, Delphi, Alcoa Fijikura, Packard Electric System, Electric Wire Products, and Sumitomo) are located. In 1997, the United States imported 5.9 billion and Mexico supplied 63.4 % of the total U.S. market harnesses. The assembly is done in 58 plants in some states of Mexico: Baja California, Chihuahua, Coahuila,

Nuevo Leon, Tamaulipas, Sinaloa, Guanajuato, Zacatecas, San Luis Potosi, Queretaro City, Mexico State, Tlaxcala, and Colima (Carrillo and Hinojosa 2001; Reyes et al. 2011). Harnesses are multiple assemblies with insulated electrical conductors, which are mounted in terminals, connectors, sockets and other wiring products used to connect to a power source (batteries and motors) various electrical components such as lights, instruments and motors and/or care for selected high voltage ignition parts (starters, generators, distributors and spark plugs) in vehicles such as cars, aircraft and ships (USITC 1997).

The data collection was obtained by reports of occupational accidents with lost of work days that occurred during the period from 2001 to 2009 in a Mexican Harnesses Corporation. The amount of occupational accident reports recorded in the database was 368, of which only 225 reports were useful. In order to facilitate data collection form was designed to summarize information about causal factors reported by safety personnel during accident analysis. The category, subcategory, contributing factors and frequency were identified, classified and analyzed using the interpretive method of content analysis. Only four of the five categories that integrates Taxonomy were used, due reports did not present data related to the supervision category. During the analysis of the reports identified the types of injuries caused by accidents and their relationship to human error types (Table 18.3). The three types of injuries most frequently were: cuts, contusions and fractures with 109, 43 and 23 respectively. Among the types of errors, violations were the most frequent with 141 cases.

Table 18.4 shows several types of human error, their contributing factors, and their respective frequency. Regarding the three types of violations, the most relevant factors found were: "*do not respect rules and safety procedures*" with a frequency of 63, "*failure to follow work instructions (method)*" with 55 and finally "*do not use the right tool*" with frequency of 24. Slips showed as the most important factor "*distraction or carelessness of the operator to perform its task*" with 44. It is interesting to mention that the total contribution factors do not match the total accident reports because; during the analysis of the accident reports identified more than a factor of contribution by accident.

The causal contribution factors for unsafe conditions, personal factors and organizational factors categories are shown by Table 18.5. In the category of unsafe conditions data shows that the most frequent important factors were: "*poor condition equipment for lack of maintenance*" with a frequency of 47 and "*poor ergonomic design in work station*" 30, both belonging to subcategory of machinery and equipment, while in the subcategory of barriers "*Lack of safety*".

According to the results obtained by applying the taxonomy, it can be concluded that accident analysis in the harnesses corporation shows a noticeable tendency to point the worker as the main cause of the accident. In this sense, results coincide with Reason (2000) who establishes the human error approach centered on the person. This approach focuses on unsafe acts (errors) and violations of established procedures, and performed by workers who work in direct contact with the production system.

Table 18.3 Relationships among injuries types and human error

Human error					
Injuries types	Slips	Violations	No error	Mistakes	Total
Cuts	12	75	20	2	109
Bruises	10	16	17	0	43
Fractures	2	23	4	0	29
Finger amputations	5	5	1	1	12
Injuries and sprains	7	15	2	0	24
Burns	0	7	1	0	8
Total	36	141	45	3	225

Table 18.4 Frequency of human error and contribution factors in industry automotive harnesses

Category	Subcategory	Contribution causal factors	Frequency
Unsafe acts	Violations	Perform tasks without personal protective equipment (gloves)	11
		Failure to follow work instructions (method)	55
		The operator does not respect rules and procedures safety	63
		Trying to save time in developing their operation	17
		Do not respect rules and safety procedures	11
		Two workers operating equipment	8
		Manage the sharp terminal cable without gloves	1
		Work without safety guards in machines	2
		Do not use the right tool	24
		Remove with fingers stuck terminals	5
		Unauthorized use of knives	1
		Workers play in workspaces	2
	Mistakes	Operating equipment without knowledges	1
		Overconfidence	2
	Lapses	Distraction or carelessness of the operator to perform its task	44
		Negligence	20

"*Guard in equipment*" and "*Improper design of safety guards in press*" with frequencies of 15 and 18 respectively. In the category of personal factors showed very low frequencies, while organizational factors for the higher frequencies were presented at the factors: "*The Procedures and methods do not consider safety recommendations*" and "*Lack of training in Jobs with risk of accidents*", corresponding to the subcategory of operational and organizational climate process whose frequencies were respectively 15 and 13.

Table 18.5 Frequency of causal contribution factors in Industry Automotive Harnesses

Category	Subcategory	Contribution causal factors	Frequency
Unsafe conditions	Machinery and equipment	Long, hard and misplaced bolts	2
		Poor condition equipment for lack of maintenance	47
		Poor ergonomic design of the workstation	30
		Sharp edges in Workstation	4
		Workspaces reduced	4
	Barriers	Lack of safety guard in equipment	15
		Improper design of safety guards in press	18
		Lack of visual and auditive displays	5
	Tools	Damaged tools	7
Personal factors	Physical factors	Drowsiness	2
	Social factors	Personal problems	1
	Psychological factors	Workers inappropriate behaviour	3
		Recklessness	4
Organizational factors	Organizational climate	Lack of training in Jobs with risk of accidents	13
		Lack of communication in aspects of accidents prevention	3
		Lack of focus in ergonomics	2
		Lack motivation to comply with the Procedures and rules to prevent accidents	6
		The needs of production cause the release of unsafe equipment	8
	Operational process	The working methods are confusing and without safety instructions	1
		The Procedures and methods do not consider safety recommendations	15
	Resource management	Poor attention to correct Unsafe conditions	1
		Lack of personal protective equipment (gloves and thimbles)	7

References

ANSI/IEEE STD 1002 (1987). *IEEE standard taxonomy for software engineering standards.* New York: Institute of Electrical and Electronics Engineers.

Arquer MI y Nogareda C (1988). NTP: *Fiabilidad humana: Conceptos básicos.* Instituto Nacional de Seguridad e Higiene en el Trabajo. Recuperado el 12 de febrero de 2008 del sitio Web del Ministerio de Trabajo y Asuntos Sociales de España: http://www.insht.es/InshtWeb/Contenidos/Documentacion/FichasTecnicas/NTP/Ficheros/301a400/ntp_360.pdf.

Baker, P. D., & Krokos, J. K. (2007). Development and validation of aviation causal contributors for error reporting system (accers). *Human Factors, 49*(2), 185–199.

Baysari, T. M., McIntosh, S. A., & Wilson, J. R. (2008). Understanding the human factors contribution to railway accidents in Australia. *Accidents Analysis and Prevention, 40*, 1750–1757.

Blackwelder, R. E. (1967). *Taxonomy: A text and reference book*. New York: Wiley.

Cacciabue, P. C. (2000). Human factors impact on risk analysis of complex systems. *Journal of Hazardous Materials, 71*, 101–116.

Cañas, J., & Waerns, Y. (2001). *Ergonomía Cognitiva Aspectos psicológicos de la interacción de las personas con la tecnología de la información*. España: Editorial médica panamericana.

Carrillo, J., & Hinojoza, R. (2001). Cableando el norte de México: la evolución de la industria maquiladora de arneses. *Región y sociedad.XII, 21*, 79–114.

Celik, M., & Cebi, S. (2009). Analytical HFACS for investigating human errors in Shipping accidents. *Accident Analysis and Prevention, 41*, 66–75.

Dhillon, B. (2003). Methods for performing human reliability analysis in health care. *International Journal of Health, 16*, 306–317.

Feyer, A. M., & Williamson, A. M. (1996). *Accident models: Human factors in accidents* (4th ed.). Geneva: ILO Encyclopedia of Occupational Health and Safety International Labour Office.

Feyer, A. M., Williamson, A. M., Cairns, & David R. (1997). The involvement of human behavior in occupational accidents: Errors in context. *Safety Science, 25*, 55–65.

Gordon, A. D. (1999). *Classification* (2nd ed.). Boca Raton: Chapman & Hall/CRC.

Hale, A. R., & Glendon, A. I. (1987). *Individual behavior in the control of danger*. Amsterdam: Industrial Safety Series.

Harris, A. (2004). Erring of the side of danger. *Occupational Health, 56*, 24–27.

Health and Safety Executive. (1989). *Human factors in industrial safety*. London: HMSO.

Hobbs, A., & Williamson, A. (2003). Associations between errors and contributing factors in aircraft maintenance. *Human Factors, 45*(2), 186–201.

Hollnagel, E. (1993). *Human reliability analysis: Context and control*. London: Academic Press.

Leplat, J., & Rasmussen, J. (1984). Analysis of human errors in industrial incidents and accidents for the improvement of work safety. *Accident Analysis and Prevention, 16*, 77–88.

Li, B., Li, M., Chen, K., & Smidts, C. (2006). Integrating software into PRA: A software-related failure mode taxonomy. *Risk Analysis, 4*, 997–1012.

Liu, H., Huang, S. L & Liu, T. H. (2009). Economic assessment of human errors in manufacturing environment. *Safety Science, 47*, 170–182.

Moray, N. (2000). Culture, politics and ergonomics. *Ergonomics, 43*, 858–868.

Maurino, D. E., Reason, J., Johnston, N., & Lee, R. B. (1995). *Beyond aviation human factors*. Brookfield: Ashgate Publishing Company.

Norman, D. A. (1988). *The psychology of everyday things*. New York: Basic Books.

O'Connor, P., O'Dea, A., & Melton, J. (2007). A methodology for identifying human error in U.S. Navy diving accidents. *Human Factors, 49*(2), 214–226.

O'Hare, D. (2000). The wheel of misfortune: A taxonomic approach to human factors in accident investigation and analysis in aviation and other complex systems. *Ergonomics, 43*(12), 2001–2019.

Raouf, A. (1998). Prevención de accidentes: teoría de las causas de los accidentes. En Enciclopedia de Salud y Seguridad en el Trabajo (Vol. 2, pp. 56.6–56.8).Organización Internacional del Trabajo. Disponible en la página Web, del INSHT España, http://www.insht.es/portal/site/Insht/menuitem.1f1a3bc79ab34c578c2e8884060961ca/?vgnextoid=5f5b4cf5a69a5110VgnVCM100000dc0ca8c0RCRD&vgnextchannel=9f164a7f8a651110VgnVCM100000dc0ca8c0RCRD.

Rasmussen, J. (1982). Human errors: A taxonomy for describing human malfunction in industrial installations. *Journal of Occupational Accidents, 4*, 311–333.

Rasmussen, J. (1983). Skill, rules and knowledge: Signals, signs and symbols and other distinctions in human performance models. *IEEE Transactions on Systems, Man and Cybernetics, 13*, 257–266.

Rasmussen, J. (1987). Risk and information processing. In W. T. Singleton & J. Hovden (Eds.), *Risk and decisions*. New York: Wiley.
Reason, J. T. (1990). *Human error*. Cambridge: Cambridge University Press.
Reason, J. T. (2000). Education and debate: Human error: Models and management. *British Medical Journal, 320*, 768–770.
Reinanch, S., & Viale, A. (2006). Application of the human error framework to conduct train accident/incident investigations. *Accident Analysis and Prevention, 38*, 396–406.
Reyes, M., Prado, L., Aguilera, V., & Soltero, A. (2011). Descripción de los Conocimientos sobre Factores Humanos que Causan Accidentes en una Industria Arnesera Mexicana. *E-Gnosis, 9*, 1–17.
Reyes, R. M., Maldonado, A. A., & Prado, L. R. (2012). Human factors identification and classification related to accidents 'causality on hand injuries in the manufacturing industry. *WORK A Journal of Prevention, Assessment and Rehabilitation, 41*(1), 3155–3163.
Romera, J. (2007).Causalidad del error humano en los accidentes laborales (Modelo psicológico de Queso Suizo). *Seguridad y Salud en el Trabajo, Revista del INSHT, prevención trabajo y salud, 48*, 10–18.
Runciman, W. B., & Sellen, A. (1990). Errors, incidents and accidents in anesthetic practice. *Anesthetic and Intensive Care, 21*, 506–519.
Salminen, S., & Tallberg, T. (1996). Human errors in fatal and serious occupational accidents in Finland. *Ergonomics, 39*, 980–988.
Saurin, T. A., Buarque, M. L., Fabiano, C. M., & Ballardin, L. (2008). An algorithm for classifying error types of front-line workers based on the SRK framework. *International Journal of Industrial Ergonomics, 38*, 1067–1077.
Senders, J. W., Moray, N. P. (1991). *Human error: Cause, prediction and reduction*. Lawrence Erlbaum: Hillsdale.
Shappell, S., & Wiegmann, D. (1997). A human error approach to accident investigation: the taxonomy of unsafe operations. *International Journal of Aviation Psychology, 7*, 269–291.
Wiegmann, D. A., & Shappell, S. A. (2003). Assessing the reliability of the human factors analysis and classification system (HFACS). *Aviation, Space and Environmental Medicine, 72*, 1006–1016.
Shappell, S., Detwiler, C., Holcomb, K., Hackworth, C., Boquet, A., & Wiegmann, D. A. (2007). Human error and commercial aviation accidents: An analysis using the human factors analysis and classification system. *Human Factors, 43*(2), 227–242.
Sharit, J., & Gables, C. (2006). Human error. In G. Salvendy (Ed.), *Handbook of human factors and ergonomics* (pp. 708–760). U.S.A: Wiley.
Spintzer, C. (1996). Review of probabilistic safety assessment:insigth and recomendations regarding further developments, Reliability Engineering. *System Safety, 52*, 153–163.
Strobhar, David A. (1995). Evaluation of operator decision-making. *ISA Transactions, 34*, 405–409.
Strater, O., & Bubb, H. (1999). Assessment of human reliability based on evaluation of plant experience: Requirements and implementation. *Reliability Engineering and System Safety, 63*(2), 199–219.
USITC. (1997). *Production sharing: Use of U.S components and materials in foreign assembly operations*. Washington: United States International Trade Commission. (Publication 3032).
Weller, S. C., & Romney, A. K. (1988). *Systematic data collection*. Newbury Park: Sage.
Weller, S. R. (2007). Cultural consensus theory: Applications and frequently asked questions. *Field Methods, 19*(4), 339–368.
Wickens, C., & Flach, J. (1988). Information processing. In E. L. Wiener & D. C. Nagel (Eds.), *Human factors in aviation* (pp. 111–155). San Diego, CA: Academic.
Woods, D. D., & Cook, R. I. (2006). Incidents—markers of resilience or brittleness? In E Hollnagel D. D. Woods & N. Levenson (Eds.), *Resilience engineering: Concepts and precepts* (pp. 69–76). Aldershot: Ashgate.

Part IV
Alternative Methodologies for Lean Manufacturing

Chapter 19
Alternatives Methodologies for Lean Manufacturing: Genetic Algorithm

Alejandro Alvarado-Iniesta, Jorge Luis García-Alcaraz and Luis Pérez-Domínguez

Abstract Global competition has forced industries to be in a constant search of technology; however, any technology that involves an advantage is temporary. Nowadays, a computer is seen in every enterprise mostly with the objective of being an advantage and as a tool to facilitate tasks (besides entertaining). Thus, a question arises, is it the most modern and expensive computer a significant advantage by default? I guess not today. The development of computers has arguably been the most radical achievement in the history of science and technology. The first computer scientists were motivated in large part by visions of creating computers programs with intelligence, mainly focusing to modeling the brain, imitating human behavior, and simulating biological evolution. This matter had its takeoff in the early 1980s, mostly by the advances in computational power; with evolutionary computation as one of the first ones, where "genetic algorithm" is the most prominent example (Mitchell 1998). Therefore, the "best" technologies are not limited to the acquisition of modern equipment or expensive machinery, but also include the progress and implementation of intelligent computational approaches such as the Genetic Algorithm. Therefore, the aim of this chapter is to show some applications of Genetic algorithm in the manufacturing sector in order to obtain a lean environment. This chapter begins with a brief definition of what a Genetic Algorithm is, its basic elements, and some numerical examples in order to facilitate its application. Next, two cases taken from the manufacturing industry of Juarez, Mexico are illustrated.

Keywords Genetic algorithms · Production optimization · Computational optimization

A. Alvarado-Iniesta (✉) · J. L. García-Alcaraz · L. Pérez-Domínguez
Department of Industrial and Manufacturing Engineering, Autonomous University of Ciudad Juarez, Av. Del Charo 450 Norte. Col. Partido Romero, Ciudad Juárez, Chihuahua, Mexico
e-mail: alejandro.alvarado@uacj.mx

J. L. García-Alcaraz et al. (eds.), *Lean Manufacturing in the Developing World*,
DOI: 10.1007/978-3-319-04951-9_19,

19.1 Genetic Algorithm

Fortunately, there is an extensive variety of literature available about Genetic Algorithm (GA). Nonetheless, there is limited information that explains a numerical example easily and in detail. The intention of this section is precisely cover that issue, and not explain all the concepts related to GA into depth. A Genetic Algorithm (GA) is an optimization and search technique based on the principles of genetics and natural selection (Haupt and Haupt, 2004). Figure 19.1 shows a diagram of the simple GA.

Before starting with GA, let's define what is optimization? It can be defined as finding the best possible alternative (from a set of alternatives) given some circumstances, as time or resources (Bertsimas and Tsitsiklis 1997). Another definition of optimization is that it can be seen as a search task, where the objective is to find the best choice among a group of options, or candidate solutions, within a feasible region. In general, an optimization problem consists of three basic parts: design variables, cost function, and a set of constraints (which define a feasible region). Hence, mathematically, an optimization problem can be formulated as:

Find $X = [x_1, x_2, \ldots, x_n]$ (where "n" represents the numbers of design variables)
That minimizes $f(X)$ (cost function)
Subject to a set of constraints that defines a feasible region such as

$$x_i \geq 0$$

The most important aspect in any optimization problem is to establish an exact cost function, and it is considered the most difficult part in an optimization problem (Erwin Diewert 2008). A cost function is a performance index of the problem and represents the relationship among the design variables.

In general, any optimization problem consists of at least a constraint. Whichever method is employed, conventional or genetic algorithm, it must deal with constraints. The constraints define the feasible region as a system of equalities or inequalities.

19.1.1 Variable Encoding

In general, there are three types of encoding (Haupt and Haupt 2004),

Binary encoding
Real-Valued encoding
Permutation encoding

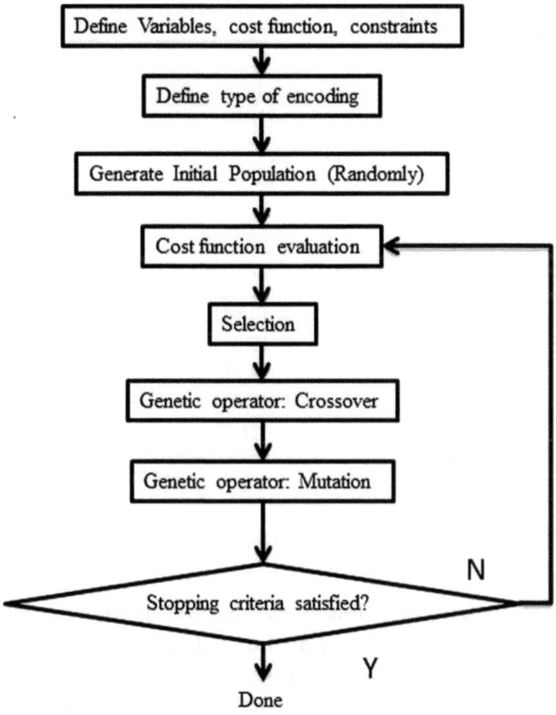

Fig. 19.1 Simple genetic algorithm

19.1.1.1 Binary Encoding

Assume the following optimization problem (Goldberg 1989).

Find x

Maximizes $f(x) = x^2$

s.t.

$$0 \leq x \leq 31$$

A set of feasible solutions (candidate) are presented in Table 19.1 with its respective binary encoding.

19.1.2 Real Valued Encoding

Assuming the same optimization problem, a set of candidate solutions is presented in Table 19.2 with real-valued encoding.

Table 19.1 Binary representation

x	Binary encoding
0	[00000]
8	[01000]
10	[01010]
15	[01111]
30	[11110]

Table 19.2 Real-valued representation

X
[2.64]
[0.5]
[3.1]
[4.78]
[26.5]

19.1.3 Permutation Encoding

This type of encoding is mostly used in combinatorial optimization problems. For instance, assume the traveling salesman problem. Figure 19.2 shows a graphical representation of the problem.

A set of four candidate solutions is presented in Table 19.3.

19.2 Initial Population (Randomly Generated)

The GA starts with a group of solutions known as the population. Table 19.1–19.3 presents different types of encoding but also illustrates a population of solutions to the given problems. This size of the population is fixed in advance at the start of the algorithm, and must be maintained in all the algorithm run. With a small population the chances of finding the global optimum are less; on the other hand, a large population may result in a significant computational time. It is recommended that the initial population is generated randomly; otherwise, there is a big chance of occurring premature convergence to a local optimum.

19.3 Cost Function Evaluation

The Genetic algorithm works in two types of spaces alternatively: encoding space and solution space (cost function evaluation). Genetic operators work on the encoding space, and evaluation and selection work on solution space (Fig. 19.2)

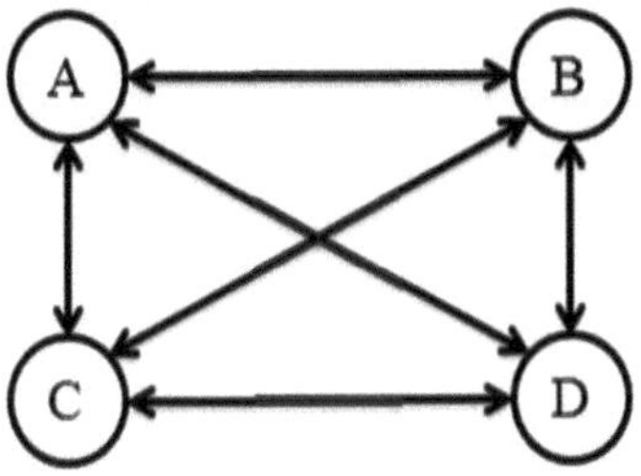

City	A	B	C	D
A	-	20	15	35
B		-	40	20
C			-	25
D				-

Fig. 19.2 Symmetric travelling salesman problem

Table 19.3 Permutation encoding representation

X
[A-B-C-D-A]
[B-A-C-D-B]
[D-B-C-A-D]
[C-A-B-D-C]

(Gen and Cheng 2000). Thus, the mapping from encoding space to solution space impacts considerably the performance of a GA (Fig. 19.3).

From the three types of encoding given above, the cost function evaluation results are given in Table 19.4.

19.4 Selection

There are several selection operators proposed so far for the GA. To mention some, roulette-wheel selection, tournament selection, steady-state reproduction, ranking and scaling, sharing, etc. (Gen and Cheng 2000). Roulette wheel selection, proposed by Holland, is the best known selection type (Holland 1992). The main idea is to determine a selection probability for each solution proportional to its fitness value, such the sum of all percentages is equal to one. The selection probability can be computed with the formula 19.1.

$$p_i = \frac{\mathrm{CostFcn}_i}{\sum_{n=1}^{\mathrm{PopSize}} \mathrm{CostFcn}_n} \tag{19.1}$$

To illustrate, consider the case of permutation encoding. An example is shown in Table 19.5.

The principle behind selection operators is to imitate natural selection. Thus, once best solutions are selected, there are other genetic operators that may be applied. For instance, elitism, in which a certain number of the best solutions are

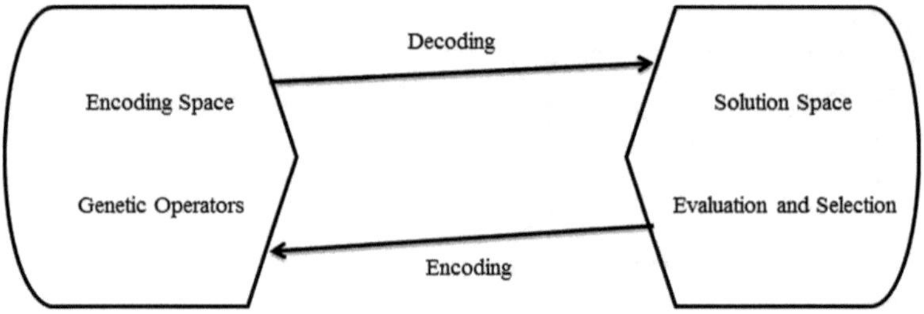

Fig. 19.3 Encoding space and solution space

Table 19.4 Cost function evaluation

Binary encoding	Real-valued encoding	Permutation encoding
f(x)	f(x)	f(x)
0	6.97	120
64	0.25	80
100	9.61	110
225	22.85	80
900	702.25	–

Table 19.5 Roulette-wheel selection

x	f(x)	Probability
[A-B–C-D-A]	120	0.31
[B-A-C-D-B]	80	0.21
[D-B–C-A-D]	110	0.28
[C-A-B-D-C]	80	0.21
Sum	390	1.00

passed directly to the next generation. Similarly, in order to apply the following operator, crossover, one of the solutions can be chosen from one of the selection operators and the second randomly.

19.5 Genetic Operator: Crossover

Crossover operator allows to create one or more offspring from the solutions selected in the selection process. Commonly, two solutions are selected (two parents) to create two offspring (two new solutions). In some cases, the offspring replace directly the parents, or four solutions are evaluated (cost function evaluation) and the two strongest are maintained for the next generation. The manner in which crossover occurs depends directly in the encoding type. There are several crossover operators depending on the nature of the encoding.

19.5.1 Binary Encoding

Typically, there are some ways of doing crossover with binary encoding (Goldberg 1989).

Single point crossover: A single crossover point is selected from the solution.

$$Parent\ 1\text{: } [11100|00111]$$
$$Parent\ 2\text{: } [10101|11001]$$

$$Offspring\ 1\text{: } [11100|11001]$$
$$Offspring\ 2\text{: } [10101|00111]$$

Two point crossover: Two crossover points are selected.

$$Parent\ 1\text{: } [111|0000|111]$$
$$Parent\ 2\text{: } [101|0111|001]$$

$$Offspring\ 1\text{: } [111|0111|111]$$
$$Offspring\ 2\text{: } [101|0000|001]$$

19.5.2 Real-Valued Encoding

There exist different crossovers operator real-valued encoding as well. To mention some: arithmetic, heuristic, simple, two point, uniform, among others (Gwiazda 2006).

Arithmetic crossover: The offspring is generated by the following formulas,

$$Offspring_i = \alpha x_i + (1 - \alpha)x_j \tag{19.2}$$

$$Offspring_j = \alpha x_j + (1 - \alpha)x_i \tag{19.3}$$

where α is a random number between 0 and 1, x_i is the solution i, and x_j is solution j.

Heuristic crossover: The offspring is generated by the following formula,

$$Offspring_i = \alpha(x_{best} - x_i) + x_{best} \tag{19.4}$$

$$Offspring_j = \alpha x_{best} + (1 - \alpha)x_i \tag{19.5}$$

where x_{best} is the best solution of the current generation and x_i is the solution i.

19.5.3 Permutation Encoding

A crossover operator is applied in order to obtain a modification in the current solution.

Cycle operator: Consider parents (solutions) x_i and x_j. (Davis 1991).

$$x_i = [1, 2, 3, 4, 5]$$

$$x_j = [3, 5, 4, 1, 2]$$

To create the first offspring, the first element from the first solution (parent) is taken,

$$Offspring_i = [1, ., ., ., .]$$

To continue

$$Offspring_i = [1, ., ., 3, .]$$

$$Offspring_i = [1, 2, ., 3, .]$$

$$Offspring_i = [1, 2, 4, 3, .]$$

Finally

$$Offspring_i = [1, 2, 4, 3, 5]$$

The same principle is applied to obtain offspring 2.

$$Offspring_j = [3, 5, 1, 4, 2]$$

Order operator: It creates offspring by transferring a randomly chosen subsequence of random length and position from one solution, and filling the remaining positions according to the order form from the other solution (Oliver et al. 1987). From both parents, a cut point is defined,

$$x_i = [1, 2, |3, 4, 5, 6|, 7, 8]$$

$$x_j = [3, 7, |4, 6, 2, 8|, 5, 1]$$

$$Offspring_i = [., ., 3, 4, 5, 6, ., .]$$

$$Offspring_j = [., ., 4, 6, 2, 8, ., .]$$

Thus, starting in the second cut point and getting back to the beginning, the order of x_j results as,

$$[5, 1, 3, 7, 4, 6, 2, 8]$$

Removing, 3, 4, 5, 6 (since are already placed in offspring 1), the resulting subsequence is,

$$[1, 7, 2, 8]$$

Hence,

$$Offspring_i = [2, 8, 3, 4, 5, 6, 1, 7]$$

Similarly,

$$Offspring_j = [3, 5, 4, 6, 2, 8, 7, 1]$$

19.6 Genetic Operator: Mutation

This operator refers to a simple change in the structure of the solution, once again depending on the type of encoding. Generally, the most common type is a random replacement.

19.6.1 Binary

For binary encoding, mutation results quite simple. In this case a 0 is mutated into a 1, and vice versa.

$$x_i = [0|0|110] -\!\!\rightarrow x_i = [0|1|110]$$

19.6.2 Real-Valued

Uniform mutation: This operator is defined with upper and lower bonds. In order to apply this operator, a position j is randomly chosen within the solution, and that element is modified into a random value between the upper and lower bounds.

$$x_i = [34.2, |65|, 2, 45, 101.1]$$

$$x_i = [34.2, |79.4|, 2, 45, 101.1]$$

19.6.3 Permutation

Random swaps: This operator randomly chooses two positions from a possible solution, $i \neq j$, and swaps the positions i and j (Alvarado et al. 2013). Figure 19.4 shows an example, $i = 2, j = 10$.

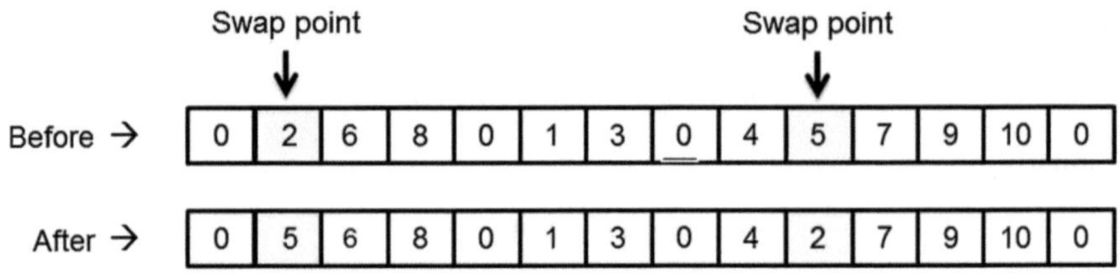

Fig. 19.4 Random swaps

Random swaps of subsequences: This operator is an extension of random swaps. Two subsequences of points and depot are selected and swapped. The length of the subsequences may be different. Figure 19.5 shows an illustration.

Reversing a subsequence: This operator selects a subsequence of consecutive customers and depot of random length, and then the order is reversed. Figure 19.6 illustrates an example.

Random swaps of reversed subsequences: This operator is a combination of two operators: random swaps and reversing a subsequence. Two subsequences of points and depot are chosen and swapped in reversing order. Figure 19.7 shows an instance.

19.7 Stopping Criteria

Stopping criteria decides whether the GA continues searching or stop the search. Each time the criterion is checked to determine if it is time to stop. Next, there are mentioned some types of termination (Haupt and Haupt 2004).

Generations: Maximum number of generations.

Time Limit: maximum time in seconds.

Cost function Limit: If best cost function value is less than or equal to specified fitness limit (minimize objective).

Cost function convergence: If cumulative change in the cost function value is less than specified fitness tolerance.

19.8 Examples Step by Step by Type of Encoding

This section pretends to show to the reader the simple genetic algorithm (as shown in Fig. 19.1) with different type of encoding and by use of some genetic operators presented in previous section.

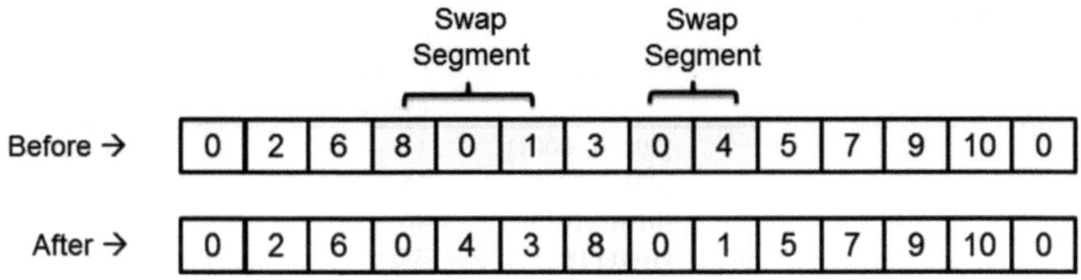

Fig. 19.5 Random swaps of subsequences

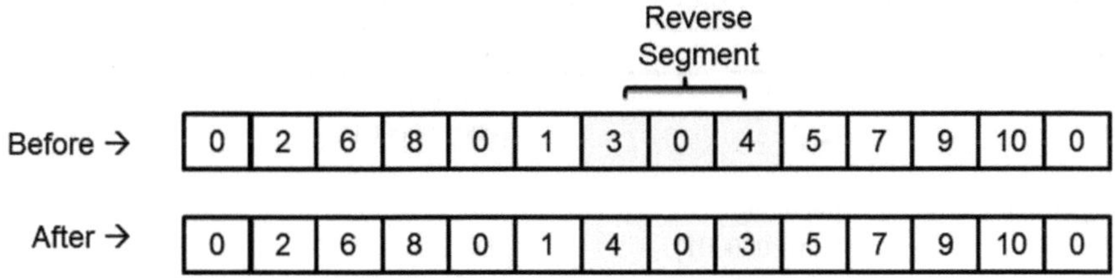

Fig. 19.6 Reversing a subsequence

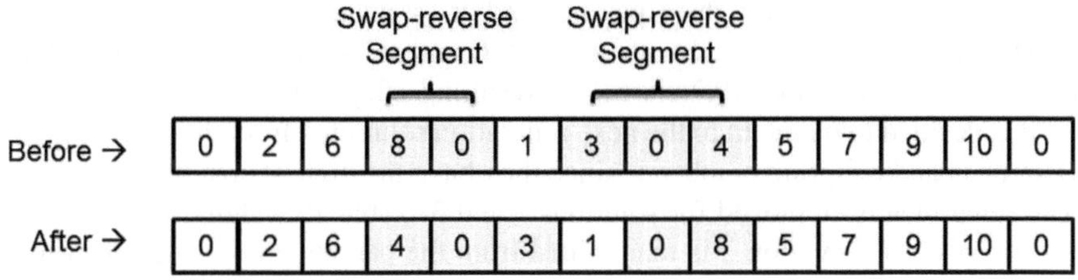

Fig. 19.7 Random swaps of reversed subsequences

19.8.1 Binary Encoding Example

Assume the following optimization problem,

Find x

Maximizes $f(x) = x^2 - 3y$

s.t.

$$0 \leq x, y \leq 31$$

The population size is fixed to be 4. Hence, initial population is generated, and cost function is evaluated. Results are shown in Table 19.6.

From Table 19.6, it can be seen the best solution is the number 4 so far. If elitism operator were applied, this solution would pass directly to the next generation, or maybe the best two. For our case, the two best solutions are used to

Table 19.6 Initial population and cost evaluation (Binary encoding case)

Solution	x	y	Encoding	Cost fitness evaluation	Probability
1	2	1	[00010,00001]	1	0.002
2	5	0	[00101,00000]	25	0.060
3	10	20	[01010,10100]	40	0.096
4	19	3	[10011,00011]	352	0.842
			Sum	418	1

Table 19.7 First single point crossover operator (Binary encoding case)

Solution	x	Encoding	Cost evaluation
4	[19, 3]	[1001|100011]	352
1	[2, 1]	[0001|000001]	1
Offspring	*Encoding*	*x*	
1	[1001|000001]	[18, 1]	321
2	[0001|100011]	[3, 3]	0

apply crossover operator with the other two solutions. Thus, each offspring generated is compared with their parents and it is maintained the best pair. To illustrate, solution 4 is mated with solution 1, and solution 3 with 2. Single point crossover is used. Table 19.7 shows the resulting offspring.

It is decided to maintain to the next generation solution 4 from the parents group (Parent 1) and offspring number 1, since they have the highest cost function value. The same idea is employed for solutions 2 and 3. Table 19.8 shows the results.

For this case, solution 3 is maintained from the parents group, and offspring 2 after applying crossover operator. Hence, the new population is obtained and it is shown in Table 19.9.

In order to illustrate the mutation operator, a mutation probability is fixed in advance. That is, each solution has a small probability to be mutated, typically 1 % (Goldberg 1989). If a solution is selected to be mutated, a bit randomly elected is mutated according with the type of encoding used. In our example, solution 2 is chosen to be mutated, and position 6 (from left to right) is selected as the bit to be mutated.

	Encoding	x	Cost evaluation
Before	[01010,**1**0000]	[10, 16]	52
After	[01010,**0**0000]	[10, 0]	100

Therefore, the final population (after genetic operators) in generation one is shown in Table 19.10.

Table 19.8 Second single point crossover operator (Binary encoding case)

Solution	x	Encoding	Cost evaluation
2	[5, 0]	[00101001000]	25
3	[10, 20]	[01010101100]	40
Offspring	*Encoding*	*x*	
1	[00101001100]	[5, 4]	13
2	[01010101000]	[10, 16]	52

Table 19.9 Current population after crossover (Binary encoding case)

Solution	x	y	Encoding	Cost function evaluation
1	18	1	[10010,00001]	321
2	10	16	[01010,10000]	52
3	10	20	[01010,10100]	40
4	19	3	[10011,00011]	352
			Sum	765

Table 19.10 Current population after mutation (Binary encoding case)

Solution	x	y	Encoding	Cost function evaluation
1	18	1	[10010,00001]	321
2	10	0	[01010,00000]	100
3	10	20	[01010,10100]	40
4	19	3	[10011,00011]	352
			Sum	813

After genetic operators are employed, the stopping criteria is checked to see if it is satisfied; otherwise, the current population after mutation is selected as the initial population and the algorithm begins the same way. This is done until the stopping criterion is satisfied. For our example, solution 4 is the best solution, with a cost function value equals to 352.

If we make a change in the optimization problem, and instead of maximizing the function, it is desired to minimize it. We can convert a minimization problem into a maximization problem, and vice versa. A way to do it is replacing the cost function by a fitness function. That is,

Assume $f(x)$ is our function to be minimized. If we want to treat the problem the same way the previous example, we can use the following conversion.

$$g(x) = \begin{cases} \frac{1}{1+f(x)} & f(x) \geq 0 \\ 1 + abs(f(x)) & f(x) < 0 \end{cases} \tag{19.6}$$

Thus, while $g(x)$ is the fitness function to be maximized, $f(x)$ is the cost function for being minimized, and the algorithm can perform the same manner.

19.8.2 Real-Valued

Consider the same optimization problem.

Find x

Maximizes $f(x) = x^2 - 3y$

s.t.

$$0 \leq x, y \leq 31$$

The population size is fixed to be 4. Initial population is generated and evaluated. Table 19.11 shows both.

Arithmetic crossover operator is applied to the resulting pairs. In this case, solution 2 is mated with solution 3, and solution 1 with solution 4. $\alpha = 0.3$

$$Offspring_1 = 0.3 * [30\ 5] + (1 - 0.3) * [15\ 25]$$

$$Offspring_1 = [19.5\ 19]$$

$$cost\,function\,value = 323.25$$

$$Offspring_2 = 0.3 * [15\ 25] + (1 - 0.3) * [30\ 5]$$

$$Offspring_2 = [25.511]$$

$$cost\,function\,value = 617.25$$

As a result, parent 1, solution 3, and offspring 2 are retained in the population. For the other pair, the procedure is exactly the same. Resulting population after crossover operator is shown in Table 19.12.

Mutation operator is applied to the solution with the lowest cost function value. In this case, solution 1 is mutated, where the value of a variable (randomly chosen) is changed uniformly.

	x	Cost evaluation
Before	[17, **30**]	199
After	[17, **20**]	229

Current population after mutation operator is shown in Table 19.13.

This is done until stopping criteria is satisfied.

Table 19.11 Initial population and cost evaluation (Real-valued encoding case)

Solution	x	y	Cost function evaluation	Probability
1	17	30	199	0.143
2	30	5	885	0.636
3	15	25	150	0.108
4	13	4	157	0.113
		Sum	1391	1

Table 19.12 Current population after crossover (Real-valued encoding case)

Solution	x	y	Cost function evaluation
1	17	30	199
2	30	5	885
3	25.5	11	617.25
4	15.8	22.2	183.04
		Sum	1884.29

Table 19.13 Current population after mutation (Real valued encoding case)

Solution	x	y	Cost function evaluation
1	17	20	229
2	30	5	885
3	25.5	11	617.25
4	15.8	22.2	183.04
		Sum	1914.29

19.8.3 Permutation

For the permutation example, let's consider the traveling salesman problem (TSP). Table 19.14 shows the symmetric matrix of distances considering 4 cities.

The objective is to find the tour with the minimum distance. Table 19.15 shows the initial population and the cost function evaluation. Formula 6 is applied to continue the same manner we did in previous exercises.

In this case, the population is fixed to 3. The best solution found is the number 3, and it is maintained to the next generation. Thus, solution 1 and 2 are used for applying order crossover operator.

$$x_1 = [1 - 2 - 3 - 4 - 1]$$

$$x_2 = [2 - 4 - 1 - 3 - 2]$$

$$Offspring_1 = [. - . - |3 - 4| - .]$$

$$Offspring_2 = [. - . - |1 - 3| - .]$$

Table 19.14 Symmetric matrix TSP (Permutation encoding case)

City	1	2	3	4
1	–	8	7	9
2	8	–	10	11
3	7	10	–	7
4	9	11	7	–

Table 19.15 Initial population and cost evaluation (Permutation case encoding case)

Solution	x	Cost function evaluation	Fitness function evaluation	Probability
1	[1-2-3-4-1]	34	0.0286	0.339
2	[2-4-1-3-2]	37	0.0263	0.312
3	[3-1-2-4-3]	33	0.0294	0.349
	sum	104	0.0843	1

$$Offspring_1 = [2 - 1 - |3 - 4| - 2]$$

$$Offspring_2 = [2 - 4 - |1 - 3| - 2]$$

$$CostFcnOffS_1 = 33$$

$$CostFcnOffS_2 = 37$$

Therefore, solution 1 is retained, and solution 2 is replaced by offspring 1. Table 19.16 shows current population

For mutation operator, solution 2 is chosen and random swaps operator is selected.

	x	Cost function evaluation
Before	[2-**4**-**1**-3-2]	37
After	[2-1-4-3-2]	34

Current population after mutation operator is shown in Table 19.17.

Once again, this is done until stopping criteria is satisfied.

Table 19.16 Current population after crossover (Permutation encoding case)

Solution	x	Cost function evaluation	Fitness function evaluation	Probability
1	[2-1-3-4-2]	33	0.0294	0.345
2	[2-4-1-3-2]	37	0.0263	0.309
3	[3-1-2-4-3]	33	0.0294	0.345
	Sum	103	0.0851	1

Table 19.17 Current population after mutation (Permutation encoding case)

Solution	x	Cost function evaluation	Fitness function evaluation
1	[2-1-3-4-2]	33	0.0294
2	[2-1-4-3-2]	34	0.0286
3	[3-1-2-4-3]	33	0.0294
	Sum	100	0.0874

19.9 Manufacturing Cases

19.9.1 Optimization of Material Flow in a Manufacturing Facility by Genetic Algorithm

Material flow planning organizes and optimizes production systems in order to increase productivity and reduce waste. This section presents a case in a manufacturing plant located in Ciudad Juarez, Chihuahua, Mexico, where the main objective is to optimize the material flow from the warehouse to the different production lines. This manufacturing company is focused in telecommunication products. The firm is expanding and works closely with other plants in USA and China. Due to insufficient space for an upcoming new product line, the facility had to change locations. The new facilities' layout was different from the previous one in that all product numbers were not considered correctly in the new arrangement. Therefore, some production lines were not considered in the planning of the material flow through the whole plant. Thus, the need arose to plan the routes to deliver the raw material to the different production lines for each operator in charge of this task, with the aim of minimizing time and human effort. A common problem this company faced was the time these operators used to take to come back to the warehouse to perform the next duty cycle. As a result, a standard time was established for each operator delivering material to a specific number of production lines.

This section presents a single case from the plan implemented in this company. A unique operator is considered, $M = 1$, and 10 production lines, $N = 10$. The demands are given in the Table 19.18. In this case, material is homogeneous for each one of the lines, and operator capacity $q = 15$. The distances between warehouse and lines are known in advance. Figure 19.8 shows a graphical representation of the warehouse and production lines, considering 0 as warehouse.

Table 19.18 Demand of each production line

Line	1	2	3	4	5	6	7	8	9	10
Demand	7	5	4	6	7	6	5	4	3	3

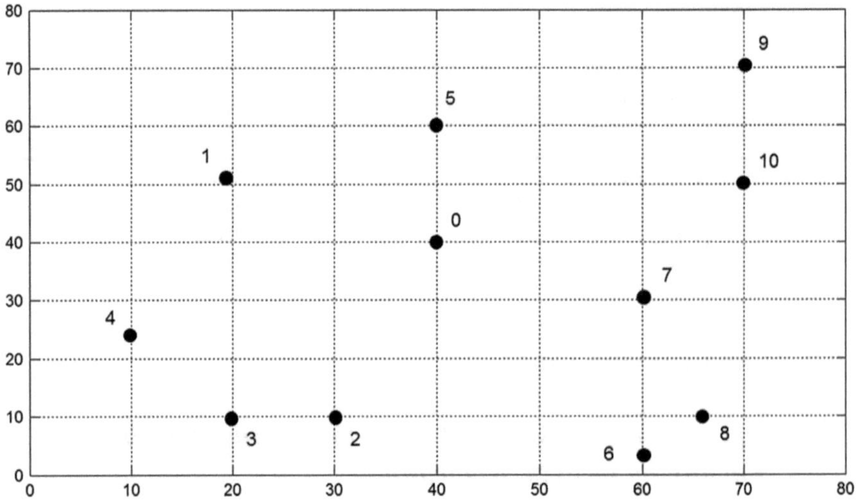

Fig. 19.8 Warehouse and production lines

Observing Table 19.18 and Fig. 19.8, the question arises: What should be the optimal route to deliver the raw material to each one of the production lines, satisfying the demand to each one, without violating the capacity constraint of the operator?

In order to solve this, a Genetic Algorithm with permutation encoding is programmed and applied to obtain the best optimal route. All experiments were performed on a modern quad core CPU, and the Genetic Algorithm was coded in Java. Figure 19.9 illustrates the best route found by the GA. Thus, the optimal route is: 0-5-0-6-2-3-0-8-10-9-7-0-4-1-0 with a cost function value of 18820 inches.

According with results, the total traveled distance should be 18820 inches. Therefore, it is consider the standard time in (Meyers and Stewart 2001), which establishes that takes 1 min to walk 3168 inches (3 miles per hour), the standard time to walk 18820 inches should be approximately 6 min. It is added a tolerance of 10 %; thus, it is established a standard time of 7 min approximately for walking. Of course, this time must be added to other standard times such as the ones for picking and delivering material, which may be easily calculated by doing a time study or by using a predefined time standard system.

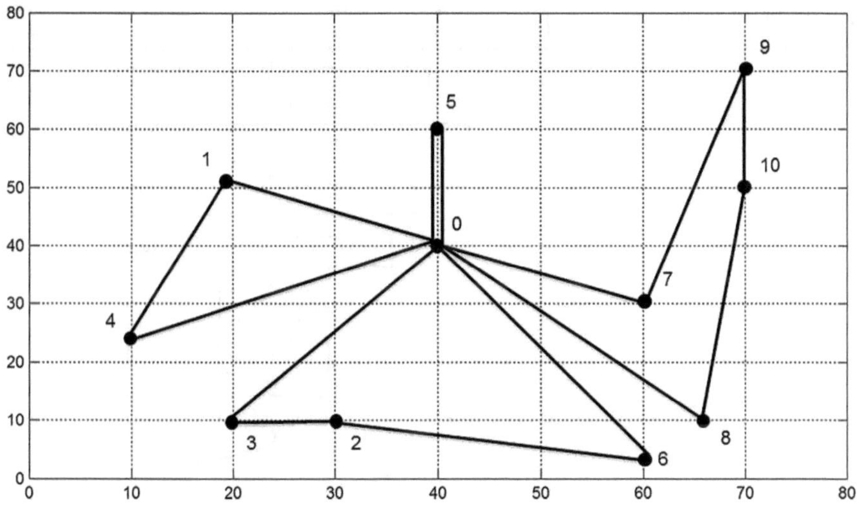

Fig. 19.9 Best route found by ABC

19.9.2 Workstation Optimization by Genetic Algorithm

It is said that one of the key factors in manufacturing facilities that are in constant improvement is the optimization of the process flow. It is considered that the flow is the most important parameter of a production system (Meyers and Stewart 2001); almost all improvements are reflected directly into the flow through a plant which directly impacts the costs of the organization. The flow may refer to information, raw material, or work in progress. It is for this reason that a workstation plays a major role within a manufacturing company, and directly impacts the flow of the final product. The present case aims to develop a tool for optimizing a workstation by the use of a Genetic algorithm. This tool aims to optimizing a workstation setup, whose cost function is given by a time study in relation to the basic motions of reach, grasp, move and release, which represent almost 50 % of all work in a workstation (Meyers and Stewart 2001). The purpose of the algorithm is to find the best setup of equipment and material at the station; therefore, obtaining a better flow in the station which directly impacts the flow of the production line. As a result, it is obtained a decrement in terms of time and unnecessary motions which leads to increase productivity in the process.

Figure 19.10 shows a diagram that illustrates the methodology used in this work in order to facilitate the understanding thereof.

Step 1: To start with the optimization of a workstation, the first step is to detect it in order to improve it and analyze the current state, i.e. the configuration of raw material, tools, etc.

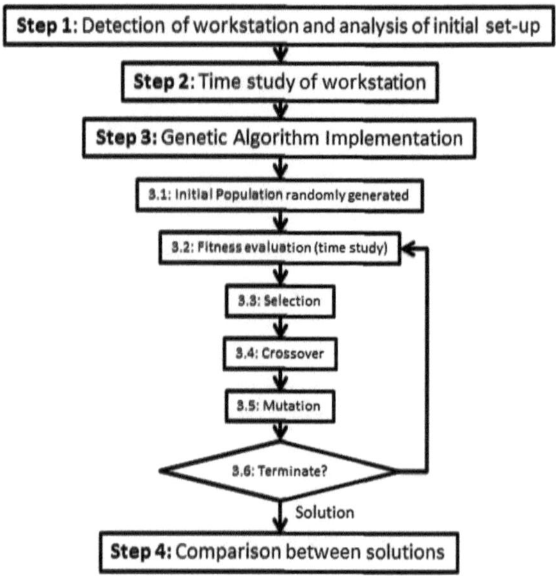

Fig. 19.10 Flow diagram of methodology

Step 2: A time study is carried out based on the basic motions of reach, grasp, move and release, by using a predetermined time standard system (PTSS) (Meyers and Stewart 2001).

Reach: Basic motion used to move the hand to a location or destination. The formula for reaching is 2 inches per 0.001 plus 0.003 min, up to 48 inches. For instance,

Reach 1″ = 1/2 + 0.003 = 0.004 min

Reach 15″ = 15/2 + 0.003 = 0.011 min

Move: Basic motion used to move an object to a location or destination. There are three causes for move time to change: distance measured in inches, beginning or ending in motion, and weight or force required. The first two caused for move time changes are exactly the same as those for reaches. For items that weigh over 5 pounds, it is added 25 % more time for every 10 pounds over 5 pounds. If both hands are used, weight is divided by 2. For instance,

Move 18″ = 18/2 + 0.003 = 0.012 min

Move 18″ weight 15 pounds = (18/2 + 0.003)*125 % = 0.015 min

Move 18″ weight 20 pounds = (18/2 + 0.003)*150 % = 0.018 min

Grasp: Basic motion used to secure sufficient control of an object to perform the next motion. There are five types of grasps:

Contact grasp: A contact grasp is the fastest motion, 0.001 min. When something must be moved without picking it up, a contact grasp is used.

Large parts grasp (G1): Grasp used when picking up something that measures at least 1 inch at the point of grasp, 0.003 min.

Medium parts grasp (G2): Grasp used when picking up parts between ¼ inch and 1 inch at the point of grasp, 0.006 min.

1	5	2	4	8	9	3	10	6	11	7	14	13	12

Fig. 19.11 Permutation encoding

Small parts grasp (G3): Grasp used when picking up parts under ¼ inch at the point of grasp, 0.009 min.

Regrasp (G4): Used in many different situations and is called the contingency element, 0.004 min.

Release: Used when control is relinquished, 0.001 min.

Step 3: Genetic algorithm implementation.

3.1 Initial population: The type of encoding used is the permutation encoding. Therefore, a solution may be represented as shown in Fig. 19.11.

where the first box represents the position one in the workstation layout. Similarly, the number one in the first box represents that item 1 is located in that position. Alike, item 5 is located in position two of the layout, and so on. Items may be raw material, tools, or any item in the workstation.

3.2 Fitness evaluation: This step refers to the sum of all motions described in step 2 according with the reaches, movements, grasps and releases of each item according with the corresponding position in the layout of the workstation.

3.3 Selection: Roulette selection technique is used to rank the best solutions.

3.4 Crossover: Order operator is used for crossover.

3.5 Mutation: Random swap operator is used for mutation.

Step 4: The initial solution is compared with the current solution given by the genetic algorithm in terms of standard time.

19.9.3 Illustrative Case

Step 1: Consider the following setup for a workstation.

where $j = 1, 2, 3, \ldots.., 9$ indicates that any item might be located at any of the positions of the workstation (Fig. 19.12).

Table 19.19 shows the items and their characteristics as well as the current position in the workstation.

This indicates that item *a* is located in position *1*, item *b* in position 2, and so on.

Step 2: For this setup, the standard time taken for the basic motions previously mentioned is equal to 30.60 s. To explain,

Consider the case of item *a* in position *1*. The distance, from the center of the work area, to the position 1 is 20 inches diagonally. Thus,

Reach: 20″ = 20/2 + 0.003 = 0.013 min
Grasp: 0.003 min

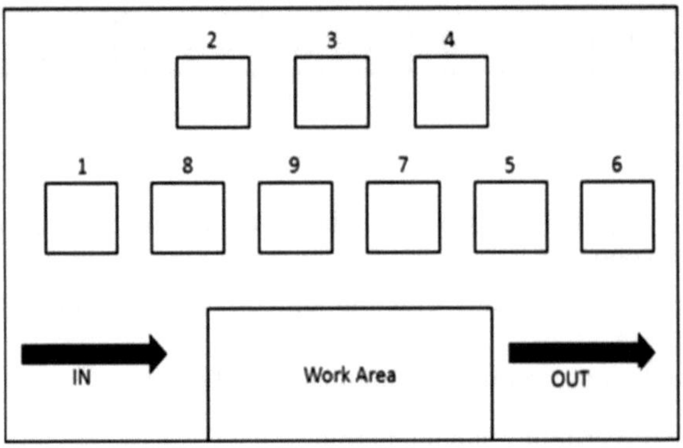

Fig. 19.12 Initial setup

Table 19.19 Items and their characteristics

Position	Item	Point of grasp (inches)	Weight (lbs.)	Both hands? Y	N
1	a	1.96	4.377	*	
2	b	2.36	0.747		*
3	c	1.77	0.170		*
4	d	2.75	0.625		*
5	e	3.14	3.160		*
6	f	3.14	2.350		*
7	g	1.37	0.110		*
8	h	1.39	0.065		*
9	i	1.37	0.110		*

Move: 20″ weight 4.38/2 pounds = (20/2 + 0.003)*125 % = 0.013 min
Release: 0.001 min

Total + 1.8 s

The sum of all reaches, grasps, moves, and releases for all items, gives the current solution that corresponds to the actual standard time (30.60 s)for that setup of the workstation.

Step 3: Genetic algorithm parameters:

Population: 20
Iterations: 2500
Selection: Roulette
Crossover: Order operator
Mutation: Random swap operator
Best solution after 10 runs:
c-b-h-i-d-a-g-f-e
with a standard time: 30.1 s.

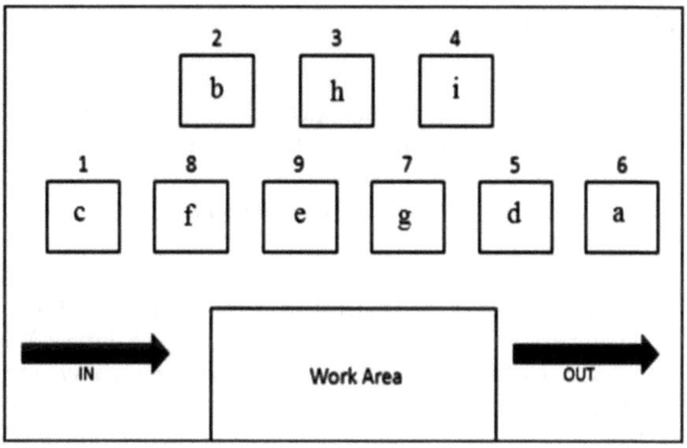

Fig. 19.13 Final setup solution

Step 4: Comparison between solutions
Initial solution: a-b-c-d-e–f-g-h-i
Standard time = 30.6 s
New solution: c-b-h-i-d-a-g-f-e
Standard time = 30.1 s
New solution is better by 0.5 s.

Figure 19.13 shows the final setup.

Results show a small difference between the initial solution and the final solution. No matter how significant the savings are or not, while they are done continuously. The case presented in this study is only a representative example with the aim of illustrating how it may be implemented at any workstation with different number of items. Clearly, it is more beneficial if the number is considerable large. Similarly, it is a first attempt for optimizing stations, the goal is to design completely which may be the best design starting from the location of the positions.

19.10 Conclusions

This chapter intended to show the applicability of Genetic Algorithms in continuous improvement programs for the industrial sector of Juarez, Mexico. Similarly, it is intended to give an overview of its components with the aim the reader might be able to employ it in a project. Today, most of the companies worldwide seek to implement methodologies in order to establish a lean environment in their systems. The Genetic Algorithm has gained attention in several fields of applicability, as the industrial field. The idea is that these types of techniques could be used more often.

A general recommendation is that this methodology, among other common ones, might be taken as an alternative for continuous improvement programs. In addition, it is encouraged getting deeper in this kind of techniques in order to extend the knowledge which may offer an advantage in the business world we live.

References

Alvarado-Iniesta, A., García-Alcaraz, J. L., Rodríguez, M., & Maldonado, A. (2013). Optimization of the material flow in a manufacturing plant by use of artificial bee colony algorithm. *Expert Systems with Applications, 40*, 4785–4790.

Bertsimas, D., & Tsitsiklis J. N. (1997). *Introduction to linear optimization.* Belmont: Athena Scientific.

Davis, L. (1991). *Handbook of genetic algorithms*. New York: Van Nostrand Reinhold.

Erwin, D (2008). In S. N. Durlauf & L. E. Blume (Eds.) *Cost functions, the new palgrave dictionary of economics.* (2nd ed.). London: Palgrave Macmillan.

Gen, M., & Cheng, R. (2000). *Genetic Algorithms and engineering optimization*. New York, NY: Wiley.

Goldberg, D. (1989). *Genetic algorithms in search, optimization, and machine learning*. Boston, MA: Addison-Wesley.

Gwiazda, T. D. (2006). *Genetic Algorithms Reference: Crossover for single-objective numerical optimization problems.* Polongne: Tomas Gwiazda.

Haupt, R.L. & Haupt, S.E. (2004). *Practical genetic algorithms.* (2nd ed.). New York: Wiley.

Holland, J. (1992). *Adaptation in natural and artificial systems*. Cambridge, MA: MIT Press.

Meyers, F. E., & Stewart, J. R. (2001). *Motion and Time study for lean manufacturing* (3rd ed.). New Jersey: Prentice Hall.

Mitchell, M. (1998). *An introduction to genetic algorthms*. Cambridge, MA: MIT Press.

Oliver, I. M., Smith, D. J., Holland J.H. (1987). A study of permutation crossover operators on the travelling salesman problem, In *Proceedings of the Third International Conference on Genetic Algorithms*, London, Lawrence Eribaum Associates, 1987.

Chapter 20
System for Improving Productivity Using the Techniques of 5'S

Wilson Adarme-Jaimes, Camilo Alvarez-Payon, Martin Darío Arango-Serna and Julian Andrés Zapata-Cortes

Abstract In a lean manufacturing environment there are lot techniques, methodologies and philosophies been applied. One of the most important is the 5'S, that is focused to improve the work site order for mechanical tools, keep a clean work site, standardized work among others, that are important for a good maintenance program. This chapter is focused to describe a 5'S program, the implementation process, and illustrates an example that was applied to an industry in Colombia. The chapter finalize discussing results and proposing some task for an effective implementation.

Keywords 5'S programs · Work standardization · Maintenance preface

20.1 5'S Program

The program is structured in the order listed in Fig. 20.1

W. Adarme-Jaimes (✉)
Universidad Nacional de Colombia Sede Bogotá, Bogotá, Colombia
e-mail: wadarmej@unal.edu.co

C. Alvarez-Payon
Universidad Nacional de Colombia Sede Palmira, Palmira, Colombia

M. D. Arango-Serna · J. A. Zapata-Cortes
Universidad Nacional de Colombia Sede Medellín, Medellín, Colombia

J. L. García-Alcaraz et al. (eds.), *Lean Manufacturing in the Developing World*,
DOI: 10.1007/978-3-319-04951-9_20,

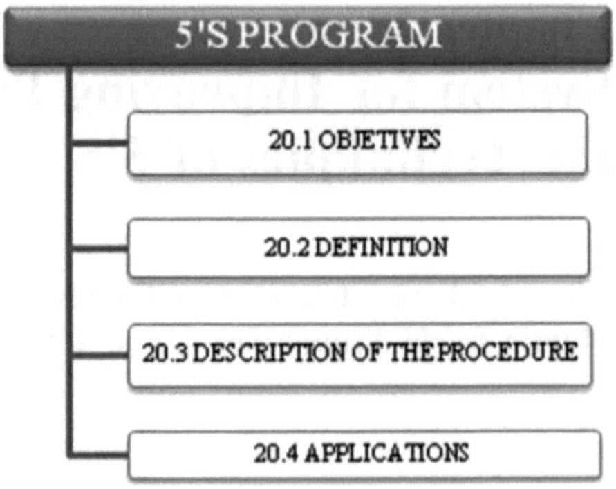

Fig. 20.1 Thematic structure—5'S Program. *Source* Authors

20.2 Objectives

The following are some goals that serve to indicate the importance of the *5'S* program in an organization.

- Create safe and healthy work environments.
- Generate a culture of commitment and continuous improvement based on respect and order.
- Facilitate the simplification and standardization of processes.
- Seek improvement in productivity, through the elimination of waste.
- Offering goods and services, consistent with the customer needs.
- Support the occupational health programs.

20.3 Definition

5'S: This is a simple program for management applicable to any organization that wants to initiate, maintain and improve the way they compete and survive in this changing world. Its application extends to home viewed this as the first link in the entire macro universal system.

The provision of a service that is timely, safe and meets to the needs of customers, along with quality and competitive price, they become key elements to generate sustainable competitive advantages today. But these features and/or attributes that can be viewed as means and not as organizational purposes, found in 5'S level of industrial plants and service organizations a fundamental pillar for its scope.

Customer satisfaction involves work in organizations focusing on human talent, implying a holistic development of people, from their own satisfaction, generating in them, engagement, membership, motivation, self-esteem and as expected as a result the development of their potential for better institutional contribution (Moore 2007).

To start a program of continuous improvement, the ideal way is to improve the working conditions of people in their everyday environment and encourage habits and procedures which enable their overall improvement across the organization. On these premises *5'S*, as 5'S program called Oriental—Japanese, are the condition and basis for initiating the implementation of quality assurance programs and improving productivity in any company. Its reach is important and well regarded primarily demand drastic attitude changes with the passage of time are expected become usual, take the unconscious of the individual autonomously make your application.

The *5'S* are therefore a response to the need to improve the work environment, which prepares the way for implementation of any management tool. It speaks of the workstation because it a very important part of development in life, in many cases the time at work is greater than the time given household or social occupations and responsibilities assumed by each person (Lareau 2003).

A pleasant working environment ensures commitment, personal development and quality care for users. 5'S terms is coming from Japanese words and beginning with S:

- SEIRI
- SEITON
- SEISO
- SEIKETSU
- SHITSUKE

The first four terms are prefixed SEI, this signifies the feeling of, honesty, respect and fairness. Rescue the fact of self-esteem as the axis on which turns the proposal.

The first three relate directly to the Workspace, the office, the workshop, the clinic and ancillary services. The last two relate directly to the person.

Each Japanese word described previously associated with terms in English that follow a logical procedure for application, such as improvement program through the order.

1. ***SEIRI:*** Inventor, Sort, Group, Choose, Register.
2. ***SEITON:*** Organize, Identify, Store, Archive.
3. ***SEISO:*** Cleanliness, Tidiness, Hygiene, Maintenance (Sacristán 2005).

As was pointed out, the first three will refer to the work environment, the last two points to the person, the individual, it is he who is ultimately responsible for the changes and where you should start.

4. ***SEIKETSU:*** Personal Health (physical and mental), Personal Wellness
5. ***SHITSUKE:*** Self Discipline, Lifestyle, habit, custom.

The following is a detailed explanation of each one of the program components.

20.3.1 SEIRI: Inventory, Sort, Group, Choose, Register

Make an inventory, sort, group, select and register properly leads to simplification and standardization so that is available on each workstation is absolutely necessary.

For example located in the office or workplace and/or located in the home and with reference to a particular area (work: office, desk, filing cabinet; house: living room, bedroom, closet) (Alukal and Manos 2006).

Once in the workplace, look around: Maybe mention something there are sheets corresponding to memorandums, internal correspondence, formats that are no longer in use (all this yellow so old), there are newspapers, magazines, six ashtrays, brochures, advertising, lottery seventies, a cabinet with five previous years papers, a library with books and objects that nobody uses four pots (no garden) that occupies half of the space, a collection of dolls (thirty four), three typewriters damaged (they are in a corner on the floor), a damaged chair with which stumble at least three people daily.

So it is worth asking.
Do these documents and/or items are essential? (e.g. damaged machines).
If used, how often is it done?,
Is the current location the best?
Is it justified to keep still?
If you definitely need to get rid of them?

20.3.2 SEITON: Organize, Identify, Store—Archive

After the classification, through the second S, which implies the organization, identification and storage, consisting of available information, work items, materials, machines and others means of production where we find easily, so that everyone knows how they have been saved and where to look. This coupled with an efficient plant layout ensures a special atmosphere for work (Asef-Vaziri and Laporte 2005).

They have classified the elements, and other information, but you have to locate the best possible site. For example, all the information about complaints the previous semester will save which is possibly on the floor. It should be kept in a place where it will be cataloged so that if someone is not dependency or someone new arrives, you can see where and how the respective claims are saved.

Today the archive information, store supplies, spare parts and other foremost demand proper identification, therefore it is important to apply appropriate techniques which highlights barcodes and microchips among others.

20.3.3 SEISO: Cleanliness, Tidiness, Hygiene Maintenance

This third S is permanently maintain hygienic and pleasant environment to labor, tending towards the elimination of the root *causes* that cause dirt and grime (Talty 1998; Terpstra 1998).

Analyze the following typical example: When entering a customer, is a co-worker, a relative to visit us, a user; you think your visitor will feel quiet and comfortable to see: ashtrays full of cigarette butts, chewing gum at different sites including adhesives on the walls, dusty and stained glass also used for fly cobwebs and say something irrelevant information system, remembering that the guest could not sit for the state of their chairs.

What is your reaction? Tranquility and comfort? Or rather that the situation would be different.

In industrial plants is considered relevant to handle maintenance programs at TPM (Total Productive Maintenance and cleaning) which in its purposes is to ensure the adequate development of labor conditions (Kumiega and Van Vliet 2008).

These are all components of the third S, the feeling of having the information or work items which are easily found, so that everyone knows how they have been saved and where to look, is perhaps one of the most important feelings to work at home.

20.3.4 SEIKETSU Personal Health (Physical and Mental), Personal Welfare

No one shall be able to satisfy their family and social goals, as like performing on the job, but is clear that everything must start from yourself, your health both bodily and emotional level constitutes the initial step to achieve comprehensive development (Barrios Casas and Paravic Klijn 2006).

Within the occupational health programs, specifically those related to preventive medicine, should include campaigns that seek to apply good work habits, as sports and cultural practice, not smoking, and regular practice of medical and psychological nature. Programs Safety and Industrial Hygiene should ensure the use of personal protective equipment to develop the safest possible work, as well as waste treatment and/or management of pollutants (Shi and Shiichiro 2012).

A free work environment, organized, neat and clean, along with the implementation of an appropriate program of Occupational Health, in some organizations governed by the rules of industrial hygiene and safety and foremost a personal disposition that encourages attitudes and behaviors aimed to personal health ensure the fourth S (Cortes 2007).

20.3.5 SHITSUKE: Self Discipline, Way of Life, Habit, Tradition, Status Quo. The Feeling of Self-Esteem and Self-Discipline

It is the commitment to compliance, ethical standards, moral and technical organization and the pursuit of continuous improvement to personal and organizational level.

Play a key role in the process of implementing techniques such as 5'S, engineering methods, where achieving relate the different resources and human potential through simplified methods, standardized, understandable, permanently improved, cause sustained increases in levels of competitiveness, based on a premise that "there will always be a better way of doing things".

The sum of the 5'S, comes when turning the habit, into custom, a way of life, this applies as simple as leaving the keys in their respective site each time you use examples, leaving the marker, eraser, pencil in the right place once they are used, the board let it not as is but maybe better (clean), leave the phone in the normally position always more comfortable to use, leaving ordered workstation when you leave it and generally make of the 5'S as a habit, which unconsciously things are where they were found or is no longer where they should be. This allows each person improves and of course as part of an organization and a community (Tozawa and Bodek 2001).

The program must be designed for each organization because there is no magic inflexible model that can be used interchangeably, each organization also receives a special name, memorable, to invite the action, as in a business is called 5'S, in others prefer to call 5'S program, in other is called Health, Life, Success, Future, Improve, hygiene, energy, among others.

20.4 Description of the Procedure

For example, the procedure to develop the art of 5'S is presented in Table 20.1.

20.5 Applications

Applications are now at a general level in any independent organization of the form of association, the size, micro, small, medium or large business, regardless of their activity belongs to the primary, secondary or tertiary sector undoubtedly has applicability and is primarily a means to improve productivity and competitiveness companies (Radharamanan et al. 1996). Namely:

Table 20.1 Procedure 5'S

Manufactures WMA SAS	
PROCEDURE: 5'S	CODE:
MAIN CHARGE: Ruby Zapata	
CONTINUOUS IMPROVEMENT COORDINATOR: Ing Gabriel Torres	CODE:
AREA: Manufactures WMA SAS	DATE: July 2013
DESCRIPTION	
A series of activities is presented below in sequence, with corresponding actions, accountable and resources required to develop a systematic and organized way the program 5'S in an organization	
STEP	ACTIVITY—ACTIONS
1	ACTIVITY
	The sensitization process starts with trainings for leaders and sub-leaders of each area
	ACTIONS
	As part of the commitment of the Manufactures WMA managers and program planning is proposed to conduct training workshops for leaders and subleaders the following days:
	Monday, July 25
	Thursday, July 28. Trainers: 7:00 to 9:00 a.m.
	Monday, August 1
	Subleaders Thursday, August 4 2:00 to 4:00
	Leaders and subleaders: 4:00 to 5:00
	The training will last for 4 h and will be held in the auditorium of the company. Within the training will make use of available resources in the auditorium plus to billboards, flyers, among others
	On Monday 8 August, a meeting with leaders and their staff to define group name, meeting dates, contract commitment to the implementation of the program will be held
	Guests to the workshop are 15 leaders and 18 leaders of each sub-area, which will be responsible for the implementation of the 5'S in their respective job and sensitization to other members of the company

(continued)

Table 20.1 (continued)

Manufactures WMA SAS		
2	ACTIVITY Creating committees to implement the program ACTIONS Working groups or committees are created by the Commons, that is, each point workspace has a leader and one, maximum two sub-leaders who will support the implementation process and contribute to awareness and set an example to the rest staff Every leader and sub-leader is responsible for the implementation of the 5'S in them workspace under the direction of the heads of each S throughout the plant	
3	ACTIVITY Implementation of the tool—overall assessment ACTIONS 5'S generation projects by areas: working groups or committees, who based on prioritization, trace achievable goals and a plan of implementation, evaluation, maintenance and continuous improvement Committees carries out the characterization of the area, they present the respective diagnosis, where are documented which are particular problems in each area Then the program of 5'S is implemented. In each case following a rigorous sequence including inventory, identification and proper storage, a maintenance program suitable housekeeping. Also the development of occupational health programs to ensure individual health worker Overall, the plan must involve everyone in the institution from the messenger, operator, assistant, to the manager and should include the what, who, how, where, why and when each of the S project	
4	ACTIVITY Monitoring and evaluation ACTIONS As a method of feedback booklets to be sent to the family of each of the members of the company, in order that one or two members of the same notify compliance with the 5'S in each house will be created. A mailbox unique suggestions for the program is created and photographs of the company showing the before and after the implementation process of the 5'S will be used The monitoring process is vital in Manufactures WMA SAS because it is the best way to ensure the success of the program. It is important to make it a lifestyle, where each of the members of the company is related to the same	
Elaborated by: Mr. Wilton Cardenas	Observed by: Ing Fredy Morante	Authorized by: Date.____________

Source Authors

Table 20.2 Worksheet—5'S program

Manufactures WMA SAS	
PROCEDURE: 5'S	CODE:
MAIN CHARGE: Ruby Zapata	CODE:
CONTINUOUS IMPROVEMENT COORDINATOR: Mr. Gabriel Torres	
AREA: Manufactures WMA SAS	DATE:
DESCRIPTION	
A series of activities with their corresponding actions, accountable and resources required to develop systematic and organized program 5'S in a company is presented below in sequence	
PHASE 1. PRELIMINARY	
The committee or group responsible for 5'S program implementation in Manufactures WMA SAS, consists of: Graduate Ruby Zapata Industrial Engineer and Gabriel Torres	Worksheet
Use tool like 5W/2H (what, why, when, where, who, how, how much) (what, why, when, where, who, how, how much)	What Why When Where Who How When
Each of the committee members personally should describe at least two commitments and two benefits they expect from a program "5'S"	
Should be shared commitments and individual benefits among different members of the group and draw conclusions overall	
Additional comments to this phase	
PHASE 2. DEVELOPMENT	
SEIRI: Inventory, Sort, Group, Choose, Register	
Proceed to inventory and the various elements are recorded classes. You could think in four S well	
Class 1. Elements, devices and tools necessary s or very frequent use. (Daily)	

(continued)

Table 20.2 (continued)

Manufactures WMA SAS	
Class 2. Elements arranged positive and tools s needed or frequently used. (Weekly)	
Class 3. Elements, devices, tools and underutilized use of very low frequency. (Monthly)	
Class 4. Elements, devices, and absolutely unnecessary for any use in your section or workspace tools	
Class 1. Items, devices, tools, etc., Absolutely necessary and very frequent use	
Identify and list the means required to develop their work (note that s what means are required and available). Use to the same, worksheets, or computer environment spreadsheets that allow their management by categories or by coding	Use additional sheet
Class 2. Elements, devices, tools needed and often used	
Identify and list the means required to develop their work (note that s what means are required and available). Use to the same, worksheets, or computer environment spreadsheets that allow their management by categories or by coding	Use additional sheet
Class 3. Elements, devices, tools and underutilized use of very low frequency	
Identify and list the means required to develop their work (note that s what means are required and available). Use to the same, worksheets, or computer means, electronic spreadsheets that allow their management by categories or by coding	
Class 4. Elements, devices, and absolutely unnecessary for any use in your section or workspace tools	
Identify and list the resources available in their workplace and are not needed in their dependency. Use to the same, worksheets, or computer environment spreadsheets that allow their management by categories or by coding	Use additional sheet
SEITON: Organize, Identify, Store, Archive	
Classify, organize and identify. The items listed in Class 1, 2 and 3. Elements, devices and tools or very frequent use and often used	Use additional sheet
Set the appropriate place for everything (listed against each item designates the ideal site for storage), given that its use is highly frequent or infrequent. Remember the need to use the methods engineering and techniques of cataloging and classification, then you must group the items by common attributes, such as whether they are office supplies, all belonging to this group apart, if they are tools, whether it is home Menswear for example, within the menswear can be subdivided in lingerie, casual wear, formal wear, etc., lady clothing, children's clothing and so on	Use additional worksheet

(continued)

Table 20.2 (continued)

Manufactures WMA SAS	
Define and identify the items and the proper place for everything, such as libraries, cabinets, cabinets, magnetic media, etc.	Use additional worksheet
All this work is important to develop your work group, as it works to the extent that develops teamwork	
Some identification methods used in classification	Use additional worksheet
Identify the folder or AZ with name and number	
Labeling	
Place ribbons	
Draw silhouettes	
Colors	
Barcode	
Microchips	
Decide to use and the reasons why	
When and how the CLASSIFICATION—SELECTION	
Identify when and how the items will be inventoried, classified and located in the work area.	
Use the questions	Use additional worksheet
What is the sequence of work flow and to optimize the work?—Working method	
What is the most convenient location to minimize the need to bend and reach?	
How best to design the area?—Distribution plan	
Review CLASSIFICATION—SELECTION	Use additional worksheet
Remember that improvements in processes, redesign methods, technological updating, redistribution plan, has implications in the standings, for this reason it requires a continuous evaluation process. Then set the routine evaluation	Use worksheets and/or additional means
Finally, to the elements, materials and other means that do not use (Class 4) identify the location of a storage area far away from the work site and the persons responsible for managing these elements responsible for them (e.g. wineries do, surplus stores, tack stores materials etc.)	
Always include dates to develop each task	
SEISO cleanliness, Tidiness, Hygiene, Maintenance	

(continued)

Table 20.2 (continued)

Manufactures WMA SAS	
Establish maintenance routines and in turn list other cleaning steps that will be needed to keep your work area nice, clean and orderly	Use worksheets and/or additional means
Use suitable planning techniques like 5 W/2H	
List three benefits of cleaning the job site and keep it that way	Use additional sheets work
Design programs for accident prevention, handling unsafe situations to complement the occupational health program	Use additional sheets work
List three ideas to make sure everyone in the work area are accomplishing the cleaning routine and placing the elements used in their respective places	Use worksheets and/or additional means
Write at least two ideas for ways and means of how to offer more training both internally and externally for all	Use worksheets and/or additional means
SEIKETSU Personal Health (physical and mental), Personal Wellness	
Remember that your health is essential, make a list of unsafe conditions and work involving some risks	Use worksheets and/or additional means
Make your own recreation, sports, social and entertainment program	Use worksheets and/or additional means
SHITSUKE: Self Discipline, Lifestyle, habit, custom	
The key is to decide "I COULD DO IT, I'M DOING IT AND I WILL CONTINUE TO DO IT"	Use worksheets and/or additional means
Do this continuously, because the trend is back to zero, i.e., disorder and dirt from your work site and its surroundings	
PHASE 3. Tracing	
It should:	Use worksheets and/or additional means
Take the lead	
Multiply. How?	
Evaluate. How?	

Source Authors

- Absolutely necessary in any organization you want to work depending on the client.
- When seeking to eliminate waste, improve work environments and foremost seek an appropriate organizational climate.

20.5.1 Workshop

Objectives

- Display as a method how to apply the technique.
- Familiarize the reader in the use of the 5'S as a means to increase productivity and establish a healthy and safe environment in the company.

Methodology

- In this format will find sequentially how the proposal must develop.

Means

- One of the main characteristics of 5'S is that in economic terms does not require additional efforts for businesses, it is only the commitment of the members, spend a little time (30 min-week initially) then this should be part daily work.

Results

- Are different outcomes and impacts it has on organizational development, highlighting the followings: improve work attitudes, improved productivity and a healthy and safe environment in which everyone wins in the enterprise.

Below is a worksheet (Table 20.2) is presented as a guide to the implementation of the proposal. In the blanks include the answers, if insufficient space uses continuation sheets.

References

Alukal, G., & Manos, A. (2006). *Lean kaizen: a simplified approach to process improvements.* Milwaukee: ASQ Quality Press.

Asef-Vaziri, A., & Laporte, G. (2005). Loop based facility planning and material handling. *European Journal of Operational Research, 164*(1), 1–11.

Barrios Casas, S., & Paravic Klijn, T. (2006). Health promotion and a healthy workplace. *Revista Latino-Americana de Enfermagem, 14*(1), 136–141.

Cortes, J. M. (2007). *Técnicas de prevención de riesgos laborales: seguridad e higiene del trabajo*. Madrid: Editorial Tebar.

Kumiega, A., & Van Vliet, B. (2008). *30—Kaizen: Continuous improvement.* In A. Kumiega & B. V. Vliet (Eds.), *Quality money management* (pp. 271–277). Burlington: Academic Press. Retrieved from http://www.sciencedirect.com/science/article/pii/B9780123725493000306

Lareau, W. (2003). *Office Kaizen TM: Cómo controlar y reducir los costes de gestión en la empresa* (1a Ed.). Fundación Confemetal (FC) Editorial, Madrid, España.

Moore, R. (2007). 8—Kaizen. In R. Moore (Ed.), *Selecting the Right Manufacturing Improvement Tools* (pp. 159–172). Burlington: Butterworth-Heinemann. Retrieved from http://www.sciencedirect.com/science/article/pii/B9780750679169500096

Radharamanan, R., Godoy, L. P., & Watanabe, K. I. (1996). Quality and productivity improvement in a custom-made furniture industry using Kaizen. *Computers and Industrial Engineering, 31*(1–2), 471–474.

Sacristán, F. R. (2005). *Las 5S: Orden y limpieza en el puesto de trabajo*. Fundación Confemetal (FC) Editorial, Madrid, España.

Shi, G., & Shiichiro, I. (2012). Study on the strategies for developing a safety culture in industrial organizations. *Procedia Engineering, 43*, 535–541.

Talty, J. T. (Ed.). (1998). *8—Principles of Air Cleaning*. In *Industrial Hygiene Engineering (Second Edition)* (pp. 188–197). Park Ridge, NJ: William Andrew Publishing. Retrieved from http://www.sciencedirect.com/science/article/pii/B9780815511755500170

Terpstra, P. M. J. (1998). Domestic and institutional hygiene in relation to sustainability. Historical, social and environmental implications. *International Biodeterioration and Biodegradation, 41*(3–4), 169–175.

Tozawa, B., & Bodek, N. (2001). *The Idea Generator: Quick and Easy Kaizen*. Boston: PCS Inc.: Vancouver

Chapter 21
Performance Measurement in Lean Manufacturing Environments

Leonardo Rivera and Diego F. Manotas

Abstract This chapter presents the main metrics suggested for a lean manufacturing environment, as well as their use and application. The selection of metrics arises from a process that starts with the examination of a lean implementation, followed by the dimensions of improvement that will be examined. Then we suggest a set of metrics for each of the improvement dimensions, and we characterize the a priori impact that tools considered typical of lean manufacturing might have on the metrics. Based on this characterization we perform two types of analyses: A horizontal analysis to uncover the most critical performance indicators and a vertical analysis to present the lean tools that are most influential on the manufacturing system. We then propose different ways to use the indicators, based on their focus (results-oriented or process-oriented) and based on their scope (organizational scope and time scope). Finally, we propose future work to evaluate the impact and cost of implementing and operating a performance measurement system based on the metrics and their application we propose.

Keywords Lean manufacturing · Performance measurement · Measurement cost

21.1 Introduction

The implementation process of lean tools in manufacturing environments has generated significant impacts on the performance in several industries. One of the most important subjects about lean implementation projects is related with the linkage between lean tools and their effects on the overall performance of the production system. The lean concept includes a set of different improvement tools

---

L. Rivera (✉) · D. F. Manotas
School of Industrial Engineering, Universidad del Valle, Ciudad Universitaria Meléndez, Calle 13 No 100-00, Cali, Colombia
e-mail: leonardo.rivera.c@correounivalle.edu.co

J. L. García-Alcaraz et al. (eds.), *Lean Manufacturing in the Developing World*,
DOI: 10.1007/978-3-319-04951-9_21, © Springer International Publishing Switzerland 2014

with specific purposes. These tools are related with different performance indicators such as work-in-process level, availability of equipment and setup time. In this context, it is necessary to establish a framework to measure the different dimensions of performance in lean manufacturing environments.

According to Feld (2001) the main objective of lean manufacturing is to reduce the waste in human effort, inventory, time to market and manufacturing space to become highly responsive to customer demand while producing world-class quality products in the most efficient and economical way. These ambitious goals require the application of performance metrics to measure the implementation level of lean techniques and the impacts derived of this application. The adoption of lean metrics guides the organizations in their transformation process toward lean enterprises.

The problem of performance measurement in lean environments has originated a movement toward the change of finance control systems (Kumar and Meade 2007). Most companies that introduce lean thinking want to have practical methods to control the business, without the hugely wasteful, time-consuming, and misleading costing and measurement systems.

Maskell and Baggaley (2004) propose three important questions to drive the discussion about performance measurement in lean environments. The first question is "What sorts of performance measures can be used in place of the current measures that seem to work against the lean improvements?" The second question is focused around the costing systems: "Are there costing approaches that are lean themselves, that don't require us to track production that now speeds through the plant in a matter of hours or days?" The last question proposed by these authors is related with the financial benefits of lean implementation: "How do we understand the financial benefits of lean efforts?"

In this chapter we propose a framework to measure the performance in lean manufacturing environments. For this, the chapter is divided in four sections. The first section introduces the background about leanness measurement, lean metrics and performance metrics. The second section presents a conceptual model to implement lean manufacturing. To measure the progress of lean implementation projects a set of metrics is presented. Finally, we present some practical guidelines on how to use the indicators to evaluate performance and the progress of lean implementations.

21.2 Background: Lean Metrics

According to Wan and Chen (2008), besides the lean tools, several performance metrics have been developed to evaluate the improvements in lean implementation. These authors confirm the need to evaluate the overall leanness by integrated indicator. The term "leanness" has been interpreted diversely in the literature. Ben Naylor et al. (1999) use the concept "leanness" to describe the process of realizing lean principles simultaneously with the introduction of concept of "leagility".

Several authors cited by Wan and Chen such as Mason-Jones et al. (2000), Comm and Mathaisel (2000) have used the concept "leanness" to describe the relative measure to lean implementation level in a company. The objective is to develop a unique indicator to measure the different implementation levels of lean tools in a company. In the same direction Bayou and De Korvin (2008) developed a leanness measure to compare the performance on lean projects implementation in Ford Motor Company and General Motors.

The relation between performance measures and lean activities has been studied by different authors. Shah and Ward (2003) analyzed the effects of some contextual factors such as plant size, plant age and unionization status on the success of lean implementation process. These authors concluded that lean implementation measures are strongly conditioned by variables such as plant age and plant size. Doolen and Hacker (2005) investigated the implementation level of lean tools in electronics manufacturing companies. They concluded that the companies under studied focused their efforts in a very specific group of lean tools. For this reason the implementation levels of each lean tool were very different. This situation does not permit establish the leanness level of the company. Rivera and Manotas (2007) propose a framework to value qualitatively the effect of lean activities in the performance measures. This work groups the main activities of a lean system in four main categories, according to their focus. These groups are: Industrial engineering activities, physical processes, personnel activities and management support.

Rivera and Chen (2007) developed a methodology to evaluate the impact of lean tools implementation on the cost-time investment of a product using cost-time profiles (CTP). The use of cost-time profiles includes the time dimension in the cost accumulation process. This feature provides a reliable method to estimate the cost of a product based in a value stream mapping analysis. CTP is a very useful tool to evaluate the impacts of lean improvement projects.

Khadem (2008) proposes an indicator to evaluate the efficacy of lean metrics in the production systems. The procedure developed by these authors considers the lean metrics embedded into simulation model. This simulation model was used to forecast the overall performance of a production plant taking into account several improvement opportunities. The use of lean metrics in simulation models provides a very complete framework to analyze the overall performance of a production system.

21.3 Implementation of Lean Manufacturing

Lean Manufacturing is the popularized denomination of the "Toyota Production System" (TPS), developed in the Japanese automotive manufacturer since after World War II (Coriat 2000). It is a system born of the need to manufacture several different models of vehicles in the same factory, without the money or space to acquire vast lots of raw materials. Lean Manufacturing recovers and applies fundamental Industrial Engineering principles and common sense that were obscured by the predominant manufacturing system, Mass Production.

Lean Manufacturing is a continuous improvement philosophy that aims to remove forms of waste hidden in the workplace through the application of well-known technical tools. Some of these tools and techniques have a structure of prerequisites, since they develop capabilities that are required for the next technique (Rivera 2008). In actual implementations we observe that companies tackle several forms of improvement at once. There is not a clear-cut separation between the completion of a technique and the beginning of the next one. These techniques should be used in a continuous improvement environment, therefore their application never really ceases. Figure 21.1 presents a conceptual scheme that includes the prerequisite structure of the techniques and the generic phases to which they belong.

Should the reader be interested in the implementation process and the results achieved by companies in industries, there are ample resources in the literature that cover these topics in detail (Womack and Jones 1996; Womack 2002; Liker and Convis 2011; Nicholas 2010).

21.4 Performance Indicators for Lean Manufacturing

Lean Manufacturing is a production philosophy. Therefore, many of the aspects that are worth measuring are common between Lean and any other production system. At the same time, we must use some indicators that relate to specific aspects of Lean that distinguish it from other systems. We will present a set of indicators that are related to the dimensions of improvement that are expected when implementing Lean Manufacturing.

21.4.1 Performance Indicators for the Dimensions of Improvement

To identify dimensions of improvement, we take the seminal works of Karlsson and Ahlstrom (1996) and Martínez and Pérez (2001). A Lean Manufacturing implementation should generate improvements in the following five dimensions:

1. Elimination of waste
2. Continuous improvement
3. Continuous flow and Pull-driven systems
4. Multifunctional teams
5. Information systems

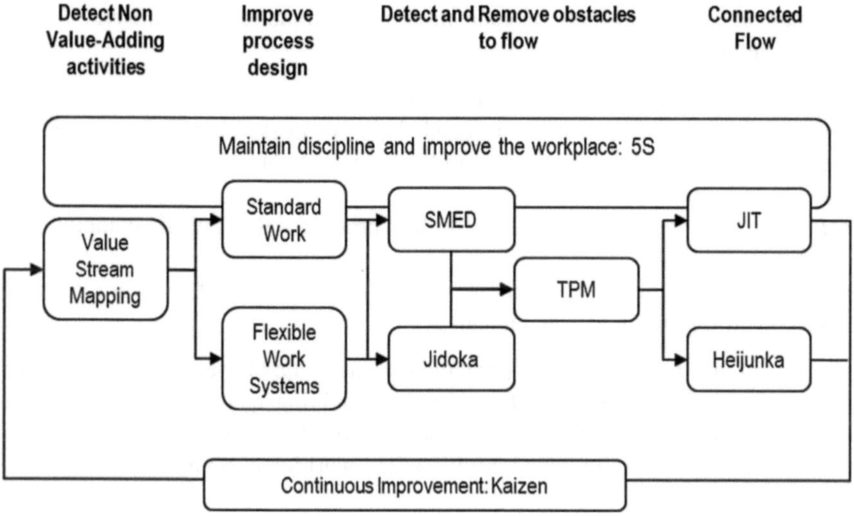

Fig. 21.1 Lean manufacturing implementation model

Each of these five dimensions suggests a set of indicators that could measure their respective improvements. We will enumerate the indicators here (Rivera and Manotas 2007) and then complement them explaining their context and use.

Elimination of waste: Waste is everything that does not add value to the product, such as keeping inventories, devoting time to machine setups, machine downtime, moving parts and generating scrap. The metrics reflect those categories of waste:

- *WIP*: Units of WIP in the line (use the unit load the company employs: Bottles, liters, boxes of product, pallets).
- *Setup time*: Time spent in setups/total scheduled production time (percentage).
- *Machine downtime*: Hours-machine lost due to malfunction/Total machine hours scheduled (percentage).
- *Transportation*: Number of trips moving materials * Distance.
- *Space Utilization*: Area (square footage) required by the line, including WIP and tools.

Continuous improvement: It represents the discipline of considering evolution as the normal state of a system. Some ideas to measure this include:

- *Number of suggestions per employee per year.*
- *Percentage of suggestions that get implemented.*
- *Scrap*: % of the products that need to be scrapped.
- *Rework*: % of the units that need to be sent to rework.

Continuous flow and Pull-driven systems: Lean systems are characterized by smoother flow of products through the line, abandoning the batch mentality and adapting to accept the pull of each process' customers. Some metrics for this dimension are:

- *Lot sizes*: Average lot size for each product.
- *Order flow time*: Time an order spends being processed in the shop floor.
- *Order lead time*: Average time from the placement of an order (by a customer) to its delivery.
- *Pulling processes*: Percentage of the line processes that pull their inputs from their predecessors.
- *Pull value*: % of the total annual value or throughput of the system that is scheduled through pull mechanisms.

Multifunctional teams: In Lean implementations, teams have more responsibility and autonomy, so improvement and problem-solving can happen closer to the source (Niepce and Molleman 1996; Forza 1996). To make flexibility in the line feasible, it is necessary to have a multi-skilled workforce. Some metrics for these aspects:

- *Autonomous control*: % of quality inspection carried out by the team.
- *Workteam task content*: % of the tasks required to make the product performed by the team.
- *Cross training*: Average over team members of Number of skills a team member possesses/Number of skills needed in a team.
- *Number of employees capable of assignment rotation.*

Information systems: The reduction of vertical levels in the structure, and the autonomous operation that teams have to reach, makes necessary that employees have timely access to better information to enable problem-solving and decision making. It does not necessarily mean, but it certainly does not exclude, computerized information systems. Some metrics:

- *Frequency with which information is given to employees.*
- *Percentage of procedures that are documented in the company.*
- *Frequency with which the line or cell progress boards are updated.*

21.4.2 Classification of the Indicators

Not all of the 21 indicators presented in the previous sections are the same. There are two main criteria that we could use to classify them:

- *Result or Process Focus*: A subset of indicators (which are shared by every production system) refer to the tangible results of the operation of the system. They might refer to externally visible results (Order Lead Time) or to internally monitored aspects of the performance of the process (WIP). Other indicators refer to the progress in the implementation of Lean, thus, they are more specific to Lean Manufacturing (Number of employees capable of assignment rotation).

- *Scope*: Indicators have different scopes in terms of who should collect the information, who should report it, how much of a plant must be covered and how often are we interested in the information. Some indicators require collecting information at the workstation level (Setup time), others need averaged information from the whole plant (Order Flow Time). Some indicators need to be monitored daily (Scrap), others can be reviewed monthly (Autonomous control). The frequencies are suggested according to Aragón and Bueno (2010).

Table 21.1 shows the indicators classified according to the categories presented. We will use these classifications in order to propose different treatments for the indicators when configuring a measurement system. For example, indicators such as Cross Training do not need to be collected daily by each operator, whereas Scrap or Rework do.

21.4.3 Impact of Typical Lean Tools on Performance Indicators

Table 21.2 presents the expected impacts of the lean tools presented in Fig. 21.1 (Lean Implementation Model). A number one means that we expect the technique to have a favorable impact on the indicator.

Table 21.2 was filled following the logic presented in the literature (Rivera and Manotas 2007; Hirano 2009), considering the nature of each Lean technique and estimating its impact on the indicators. The indicators are grouped in Process Indicators and Results Indicators. We then proceed to perform horizontal and vertical analysis of the Table.

21.4.3.1 Horizontal Analysis

We added a column to the right of the table, a sum of the ones in each row of the table. In this column we observe that indicators with six or seven techniques affecting them have a black background, indicators with four or five a gray background and all others with a white background. We contend that indicators with the highest number of techniques having an effect on them are critical for the monitoring of the system.

In the case of results indicators, *Order flow time* and *Order lead time* are the ones with more impacts. These indicators are of great interest for decision-makers, since they represent how fast do products move through the system, and how quickly the company is able to fulfill customer's orders.

For process indicators, *Workteam task content* is the one with more impacts. This indicator shows how human-work aspects are crucial to lean implementation, not just the technical tools and work methods (Paez et al. 2004).

Table 21.1 Classification of indicators

	Focus Process	Scope Results	Organizational scope	Frequency
WIP		X	Cell–Plant	Weekly
Setup time		X	Workstation–Cell–Plant	Weekly
Machine downtime		X	Workstation–Cell–Plant	Daily
Transportation		X	Cell–Plant	Monthly
Space utilization		X	Cell	Monthly
Number of suggestions per employee per year	X		Plant–Company	Monthly
Percentage of suggestions that get implemented	X		Plant–Company	Monthly
Scrap		X	Workstation–Cell–Plant	Daily
Rework		X	Workstation–Cell–Plant	Daily
Lot sizes	X		Cell–Plant	Monthly
Order flow time		X	Plant	Weekly
Order lead time		X	Plant	Weekly
Pulling processes	X		Plant	Monthly
Pull value	X		Plant	Monthly
Autonomous control	X		Cell	Monthly
Workteam task content	X		Cell	Monthly
Cross training	X		Cell	Monthly
Number of employees capable of assignment rotation	X		Cell–Plant	Monthly
Frequency with which information is given to employees	X		Plant	Monthly
Percentage of procedures that are documented in the company	X		Plant	Monthly
Frequency with which the line or cell progress boards are updated	X		Cell–Plant	Monthly

Indicators with a gray background (medium number of impacts) are the most commonly discussed as positive expected effects of Lean techniques, such as reduced lot size, increased autonomous control and cross training in the process indicators. Commonly expected results indicators also included reduced WIP, setup times, machine downtimes, scrap and rework. The information presented in the table is consisted with the literature (Womack 2002) related to observed improvements in Lean implementations.

21.4.3.2 Vertical Analysis

When we sum the total of impacts per column, we can make some observations regarding the Lean techniques. In this case, the highest number of impacts resides

Table 21.2 Expected impact of lean tools on the indicators

		Kaizen	5S	VSM	Standard work	FWS (cells)	SMED	JIDOKA	TPM	JIT	HEIJUNKA	Sum
Process indicators	Number of suggestions per employee per year	1						1	1			3
	Percentage of suggestions that get implemented	1	1						1			3
	Lot sizes					1	1	1	1		1	5
	Pulling processes.					1				1	1	3
	Pull value					1				1		2
	Autonomous control	1	1			1	1		1			5
	Workteam task content	1	1		1	1	1		1			6
	Cross training	1			1	1			1			4
	Number of employees capable of assignment rotation	1			1	1			1			4
	Frequency with which information is given to employees	1										1
	Percentage of procedures that are documented in the company	1	1		1	1			1			5
	Frequency with which the line or cell progress boards are updated	1	1		1							3
Results indicators	WIP	1				1	1			1		4
	Setup time		1		1	1	1					4
	Machine downtime		1		1	1			1			4
	Transportation		1			1				1		3
	Space utilization		1			1						2
	Scrap				1	1		1	1			4
	Rework				1	1		1	1			4
	Order flow time		1		1	1	1	1	1	1		7
	Order lead time		1		1	1	1	1	1	1		7
	Sum	10	11	0	11	17	7	6	13	6	2	

(continued)

Table 21.2 (continued)

	Kaizen	5S	VSM	Standard work	FWS (cells)	SMED	JIDOKA	TPM	JIT	HEIJUNKA	Sum
1: Favorable impact. Empty: No impact											
Process	9	5	0	5	8	3	2	8	2	2	
Results	1	6	0	6	9	4	4	5	4	0	

in Flexible Work Systems or Manufacturing Cells. This technique is central to Lean Manufacturing, many of its principles are applied by its implementation. Flexible Work Systems (FWS) connect the technical aspects of Lean work organization and the human components of cross-training, work team contents and others that are crucial in Lean implementations.

The second highest number of impacts corresponds to TPM (Total Productive Maintenance). TPM impacts eight process indicators and five results indicators. TPM is a highly systemized way to ensure machine and system reliability through properly conducted maintenance and operator participation and autonomy.

On the medium-impact tier of techniques (10–12 impacts) we find Kaizen (continuous improvement), 5S (order, cleanliness, discipline and visual controls) and Standard Work. These are primarily techniques devoted to give maintainability, repeatability and consistency to the improvement of the system. They are the core "soft" techniques, which refer more to the way of doing things in the company instead of the specific technical tools that will be implemented.

It was also interesting to observe that Value Stream Mapping (VSM) did not have an impact on any indicator. This is explained by the fact that VSM is a diagnostic tool, a way to take a picture of the current situation and use it for the detection of activities that do not add value and to plan a future desirable state, and then create the plans to reach it (Rother et al. 2009).

21.5 Performance Measurement

21.5.1 Focus of the Indicators

In Table 21.2 we presented the indicators divided into Process and Results categories. This division allows decision makers to use them in different ways.

21.5.1.1 Process Indicators

The reader may observe that these indicators (Number of suggestions per employee per year, Percentage of suggestions that get implemented, Lot sizes, Pulling Processes, Pull Value, Autonomous control, Workteam Task Content, Cross training, Number of employees capable of assignment rotation, Frequency with which information is given to employees, Percentage of procedures that are documented in the company and Frequency with which the line or cell progress boards are updated.) relate to the advance in Lean Manufacturing implementation that the company is achieving. These are indicators that do not have a direct impact on the bottom line of the company by themselves; rather they show the progress in implementing the distinctive features of Lean Manufacturing. However, the discipline and persistence in improving them will take the company further in the Lean road of increasing capabilities and disciplines, and this will finally enhance the ability of the company to obtain the desired results.

Process indicators belong in four of the main dimensions of indicators defined before:

- *Continuous Improvement*: Number of suggestions per employee per year, Percentage of suggestions that get implemented.
- *Pull-driven systems*: Lot sizes, Pulling Processes.
- *Multifunctional Teams*: Autonomous control, Workteam Task Content, Cross training, Number of employees capable of assignment rotation.
- *Information Systems*: Frequency with which information is given to employees, Percentage of procedures that are documented in the company and Frequency with which the line or cell progress boards are updated.

21.5.1.2 Results Indicators

These are more easily identified as indicators that are visible and beneficial for the company. Indicators such as WIP, Setup time, Machine downtime, Transportation, Space Utilization, Scrap, Rework, Order flow time and Order lead time have an impact on the bottom line that is measured more readily. For example, a decrease in WIP frees up space in the company, in the shop floor and in storage rooms. This freed space could lead to savings if we are able to work in a smaller facility and give good use to the space. A decrease in Scrap and Rework will have a direct impact on our bottom line, creating savings in manufacturing costs. Decreasing Setup times makes the system more flexible, enabling smaller lot sizes and smoother production flow. Results indicators come mainly from the dimension of *Elimination of Waste*.

21.5.2 Scope of the Indicators

Indicators also differ in the ways we will monitor them, collect them and use them to plan improvements. There are two main classifications: Organizational Scope and Frequency.

21.5.2.1 Organizational Scope

There are five levels of organizational scope, which differentiate who collects the indicators and who does something with them.

- *Workstation–Cell–Plant*: These indicators will need to be collected by machine operators, then aggregated and reported by Cell Supervisors and Plant Managers.

Table 21.3 Organizational scope of the indicators

Indicator	Organizational scope
Scrap	Workstation–Cell–Plant
Rework	Workstation–Cell–Plant
Setup time	Workstation–Cell–Plant
Machine downtime	Workstation–Cell–Plant
Space utilization	Cell
Autonomous control	Cell
Workteam task content	Cell
Cross training	Cell
WIP	Cell–Plant
Transportation	Cell–Plant
Number of employees capable of assignment rotation	Cell–Plant
Frequency with which the line or cell progress boards are updated	Cell–Plant
Lot sizes	Cell–Plant
Order flow time	Plant
Order lead time	Plant
Pulling processes	Plant
Pull value	Plant
Frequency with which information is given to employees	Plant
Percentage of procedures that are documented in the company	Plant
Number of suggestions per employee per year	Plant–Company
Percentage of suggestions that get implemented	Plant–Company

- *Cell*: These indicators will be collected and reported by Cell Supervisors.
- *Cell–Plant*: These indicators will be collected by Cell Supervisors and reported and discussed by them and also by Plant Managers.
- *Plant*: These indicators will be reported by Plant Managers. At their organizational level, they will probably have a Junior Engineer help with the collection and organization of the indicators.
- *Plant–Company*: These indicators will be reported by Plant Managers and General Managers. At their organizational level, they will probably have a Junior Engineer help with the collection and organization of the indicators.

The discussion of the indicators at their different organizational levels will bring about changes in the implementation procedures, work methods, focus areas and need for kaizen events. Indicators related to workstations and cells will generate rapid and focused action, whereas plant and company indicators help more the discussion of company-wide policies. Table 21.3 presents the indicators sorted by their Organizational Scope.

Table 21.4 Frequency of reporting and discussion of indicators

Indicator	Frequency
Machine downtime	Daily
Scrap	Daily
Rework	Daily
WIP	Weekly
Setup time	Weekly
Order flow time	Weekly
Order lead time	Weekly
Transportation	Monthly
Space utilization	Monthly
Number of suggestions per employee per year	Monthly
Percentage of suggestions that get implemented	Monthly
Lot sizes	Monthly
Pulling processes	Monthly
Pull value	Monthly
Autonomous control	Monthly
Workteam task content	Monthly
Cross training	Monthly
Number of employees capable of assignment rotation	Monthly
Frequency with which information is given to employees	Monthly
Percentage of procedures that are documented in the company	Monthly
Frequency with which the line or cell progress boards are updated	Monthly

21.5.2.2 Frequency

We suggest using three different frequencies for reporting and discussing indicators: Daily, Weekly and Monthly.

- *Daily*: These indicators need to be monitored every day because of their critical nature, because of their closeness to everyday operation and because they can change from one day to the next when we implement changes in the production process.
- *Weekly*: Weekly indicators correspond to operating variables that do not change immediately, that do not respond to minor operating changes. These variables are subjects of more premeditated interventions such as kaizen events.
- *Monthly*: These indicators show cell, plant and company-level changes. Some of them are related to teamwork and information systems, which are wider-reach issues in Lean transformations.

Table 21.4 presents the indicators sorted by the frequency of their discussion and analysis.

21.6 Conclusions and Future Work

We propose a set of 21 indicators that are related to Lean Manufacturing. These indicators vary in their focus (Process or Results), as well as in their scope (both in their spatial and temporal dimensions).

Based on their classifications we also propose different ways to use them. It does not make sense to create indicators that impose unnecessary burdens on the operators for their collection and analysis. At the same time, indicators that measure lean progress and can be collected with less frequency will require that companies devote people and resources to their management.

Lean indicators should be an integral part of a continuous improvement system. In a Deming cycle they should be designed in the Planning stage, deployed in the Do stage, collected in the Check stage and used for the design of improvements in the Act stage.

For future work, studies will be conducted the impact that performance measurement systems may have on performance itself, given that the collection, defense and use of the indicators has the potential to create conflicts between different areas of the company. Dimensions such as individual versus group, results-orientation versus process-orientation, reporting and indicators-based rewards will be examined to identify their impact on daily work and thus present recommendations for their use.

Also, indicators based on work competencies should be developed to monitor and manage the advancement of the workforce towards learning the philosophy and techniques involved in Lean Manufacturing. Since training is a fixture in Lean environments, it is necessary to develop a system of indicators to measure its effectiveness and the critical and creative capabilities that operators develop to continue learning on their own and to improve their working methods and procedures.

References

Aragón, A., & Bueno, M. A. (2010). *Desarrollo e implementación del sistema de Gestión Diaria Operativa (GEDO)*. Master's Thesis, Universidad Icesi, Cali, Colombia.

Bayou, M. E., & de Korvin, A. (2008). Measuring the leanness of manufacturing systems—A case study of Ford motor company and general motors. *Journal of Engineering and Technology Management, 25*(4), 287–304.

Ben Naylor, J., Naim, M. M., & Berry, D. (1999). Leagility: Integrating the lean and agile manufacturing paradigms in the total supply chain. *International Journal of Production Economics, 62*(1–2), 107–118.

Comm, C. L., & Mathaisel, D. F. X. (2000). A paradigm for benchmarking lean initiatives for quality improvement. *Benchmarking: An International Journal*, *7*(2): 118–127.

Coriat, B. (2000). *Pensar al revés: Trabajo y organización en la empresa japonesa*. Siglo XXI de España (Editores), México.

Doolen, T. L., & Hacker, M. E. (2005). A review of lean assessment in organizations: An exploratory study of lean practices by electronics manufacturers. *Journal of Manufacturing Systems, 24*(1), 55–67.
Feld, W. M. (2001). *Lean Manufacturing: Tools, techniques and how to use them.* Florida: St. Lucie Press.
Forza, C. (1996). Work organization in lean production and traditional plants: What are the differences? *International Journal of Operations & Production Management, 16*(2), 42–62.
Hirano, H. (2009). *JIT implementation manual: The complete guide to Just-In-Time manufacturing*. New York: CRC Press.
Karlsson, C., & Ahlstrom, P. (1996). Assessing changes towards lean production. *International Journal of Operations & Production Management, 16*(2), 24–41.
Khadem, M. A., Sk Ahad, A., & Hamid, S. (2008). Efficacy of lean metrics in evaluating the performance of manufacturing systems. *International Journal of Industrial Engineering, 15*(2): 176–184.
Kumar, S., & Meade, D. (2007). *Financial models and tools for managing lean manufacturing.* Florida: Auerbach Publications.
Liker, J., & Convis, G. (2011). *The Toyota way to lean leadership: Achieving and sustaining excellence through leadership development.* New York: McGraw-Hill.
Martinez, A., & Perez, M. (2001). Lean indicators and manufacturing strategies. *International Journal of Operations & Production Management, 21*(11), 1433–1451.
Mason-Jones, R., Naylor, B., & Towill, D. R. (2000). Lean, agile or leagile? Matching your supply chain to the marketplace. *International Journal of Production Research, 38*(17): 4061–4070.
Maskell, B., & Baggaley, B. (2004). *Practical lean manufacturing: A proven system for measuring and managing the lean entreprise*. New York: Productivity Press.
Nicholas, J. (2010). *Production for competitive advantage: A comprehensive guide to lean methodologies and management practices*. New York: Productivity Press.
Niepce, W., & Molleman, E. (1996). A case study: Characteristics of work organization in lean production and socio-technical systems. *International Journal of Operations & Production Management, 16*(2), 77–90.
Paez, O., Dewees, J., Genaidy, A., Tuncel, S., Karwowski, W., & Zurada, J. (2004). The lean manufacturing enterprise: An emerging sociotechnological system integration. *Human Factors and Ergonomics in Manufacturing & Service Industries, 14*, 285–306.
Rivera, L. (2008). Justificación conceptual de un modelo de implementación de lean manufacturing. *Heurística, Revista de Ingeniería Industrial de la Universidad del Valle, 15*, 91–106.
Rivera, L., & Frank Chen, F. (2007). Measuring the impact of lean tools on the cost–time investment of a product using cost–time profiles. *Robotics and Computer-Integrated Manufacturing, 23*(6), 684–689.
Rivera, L., & Manotas, D. F. (2007). Lean manufacturing measurement: The relationship between lean activities and lean metrics. *Estudios Gerenciales, 23*(105), 69–83.
Rother, M., Shook, J., Womack, J., & Jones, D. (2009). *Learning to See: Value Stream Mapping to Add Value and Eliminate MUDA* (Version 1.4). Cambridge, MA: Lean Enterprise Institute.
Shah, R., & Ward, P. T. (2003). Lean manufacturing: context, practice bundles, and performance. *Journal of Operations Management, 21*(2), 129–149.
Wan, H.-D., & Chen, F. F. (2008). A leanness measure of manufacturing systems for quantifying impacts of lean initiatives. *International Journal of Production Research, 46*(23), 6567–6584.
Womack, J. P., & Jones, D. (1996). *Lean thinking: Banish waste and create wealth in your corporation*. New York: Simon & Schuster.
Womack, J. (2002). *Lean Thinking: Where Have We Been and Where Are We Going? Forming & Fabricating*, September 2002, Lean Manufacturing Special Insert, p. L2.

Chapter 22
Layout

J. Blanco-Fernández, E. Martínez-Cámara, E. Jiménez-Macías, A. Cuevas and J. C. Sáenz-Díez

Abstract This chapter analyses the concepts related to plant layout; mainly, the aspects linked to the needs and requirements of space, and emphasizing on the methods of layout generation with its considerations. One specific method has been selected and a methodological proposal is presented for the design of its application for the high quality wine sector, specifically the Qualified Denomination of Origin Rioja. The layout design proposal has been made with the methodology of Systematic Layout Planning (SLP) in order to achieve: (a) a reduction of the handling costs of materials through a reduction of the distances travelled within the production plant; (b) an increase in production through a reduction of the manufacturing time due to the more proper arrangement of the activities within the process; (c) the designation of appropriate and proper spaces for each of the activities of the production process to take place, so that there is no interference between any of the activities.

Keywords Layout · Material flow · SLP · Winery layout

22.1 Introduction

The layout of the facilities is designed with the aim of creating products and services that meet the costumers' needs. This means that they have been created to make efficient products in the required timeframe. In order to achieve that, the physical arrangement of the facilities should be more compact and flexible. In order to save space, some people try to drastically reduce the inventories; they design and put together smaller equipment and the hallways and work places are

J. Blanco-Fernández (✉) · E. Martínez-Cámara · E. Jiménez-Macías · A. Cuevas · J. C. Sáenz-Díez
Department of Mechanical Engineering, University of La Rioja, Logroño, La Rioja, Spain
e-mail: Julio.blanco@unirioja.es

J. L. García-Alcaraz et al. (eds.), *Lean Manufacturing in the Developing World*,
DOI: 10.1007/978-3-319-04951-9_22,

reduced to a minimal space. Furthermore, the workers are trained to be multitask, so that they can use their space in the plant for more than one purpose.

This layout has a strategic effect on the performance of the factories. The materials travel shorter distances, the products go through the plant more rapidly, and clients receive the services in a more efficient way. Likewise, there is a reduction of the costs of space, handling of materials, and keeping of a balanced inventory. This makes the factories and service operations to be more flexible as changes can be made easier and faster.

With this premise, the plant layout is the process of physical arrangement of the industrial elements so that they constitute a productive system capable of fulfilling the goals set, in the most appropriate and efficient way possible.

Plant layout deals with the location of machinery, tools, and support systems in the facilities to achieve an efficient circulation of products. Due to the changes in the demands and clients' needs, the products are no longer unique; therefore, the manufacturing principles require the machinery to be efficient, the tools to be placed in modifiable locations, and the electric connections to be adaptable. To achieve this, the plant must be flexible, in order to adapt to market changes.

With the plant layout a better performance of the facilities is achieved. This can be applied to all cases in which it is necessary to arrange physical means in a determined space, even it has been previously established or not.

Its usefulness extends to both industrial and costumer processes. Plant layout is a foundation of industry which determines efficiency, and sometimes the survival of an enterprise; it also contributes to the minimization of manufacturing costs.

The incidence of production factors is effective when they develop in a proper place (space), controlling and guaranteeing the manufacturing processes. For that reason, it is necessary to adopt a plant layout per work place and operation to allow minimizing the elevated costs of production as well as time losses, in general, an increase in productivity.

22.2 State of the Art

The designing of facilities is one of the key areas of production; it plays an important role in the industrial productivity in terms of cost and consumption in the manufacturing systems. Since 1960, several researchers have developed different models for the issue of designing facilities. The revision of such models is available in several sources. A huge part of the innovative and pioneer work made from 1960 to 1970 has been revised by Francis and Goldstein (1974).

In general, the layout of the facilities is supported in the so-called "life cycle" and has the following phases: design, execution, growth, maturity, and obsolescence (Leung and Suri 1990). In each phase of the life cycle many decisions have to be made: setting, assessment and selection of an efficient design, planning of the production, programming and decision making related to the modifications to be introduced in already existing designs. This last aspect is due especially to the changes in market demands, with a worldwide competence (Bitran and Morabito 1996; Muthaiah and Huang 2006).

On the other hand, few studies have been made about how to measure the efficiency of the layout supported in phases of life cycle (Raoot and Rakshit 1991; Yaman et al. 1993); however, the materials handling cost (MHC) is considered as performance factor. This represents important aspects for efficiency, since sometimes the equipment is moved without an actual handling of materials (MHE), lack of flow of materials (Hatomi 1996), design (Raman et al. 2005), and working area flexibility. Raoot and Rakshit (1991) measure efficiency in terms of a difference obtained between a project and a design according to the desired scores. Tompkins (2003) states that the MHC can be reduced in a 10–30 % with an efficient layout. Benjaafar (2002) says that the layout has a direct impact over the operational performance, measured in the manufacturing time, performance rate, and the work capacity within the process. It is therefore evident that having an efficient layout is a key element to the improvement of the productivity of an enterprise.

22.3 Layout Concepts

22.3.1 Assessment Factors

Nowadays, there is a need of developing a measuring method to determine the level of design efficiency, taking into account several key factors simultaneously.

As a result, Francis et al. (1992) has developed a measuring model that can be adapted to other works by means of the incorporation of three performance factors: facility layout flexibility (FLF), production area utilization (PAU), and proximity (CG).

These factors are developed taking into account the objectives of a layout and include:

22.3.1.1 Facility Layout Flexibility

Flexibility in terms of manufacturing is one of the most important parameters for survival in the current worldwide market. Flexibility can be defined as the manufacturing capability of an enterprise to meet the demands quickly and efficiently with little time, effort, cost, and performance with the changes and uncertainties of the costumers' needs (Koste and Malhotra 1999; Tsourveloudis, and Phillis 1998). This manufacturing flexibility can be classified in different ways: volume, modification, new products, combination of products, expansion, operation, process, route assignment, work, machinery, handling of materials, etc. (Koste and Malhotra 1999; Parker and Wirth 1999).

Yang and Peters (1998) state that layout flexibility can be presented in two ways: (a) design of facilities based on the performance of the variable costs of the production demand, and (b) adaptability of the facilities to new production requirements. However, the objective followed to measure the FLF was once more the minimization of the volume of flow of MHC and the re-layout cost.

22.3.1.2 Productive Area Utilization

The design of the facilities implies the assignment of the required area in the proper location to make the different necessary activities which include production and management activities. Production activities are related to the production in the plant and which is executed and supported by diverse elements such as machinery, MHE, storing of materials and others. Therefore, the area utilized by these elements is defined as productive area.

In general, the level of utilization of the area is measured only in terms of free available area (Hu and Wang 2004). Likewise, Lin and Sharp (1999) use two measures to calculate the utilization area: proportion of free area, and the free distribution area in the workshop or work place. However, these two measures can mislead to a wrong conclusion: the more occupied is an area, the more utilized it is.

Therefore, a true measure of the utilization can only be achieved by an accurate measurement of the area that each element occupies. As a result, the PAU factor (Raman et al. 2005) has been developed and is measured similarly to the concept of lean manufacturing of residue minimization.

In the concept of lean manufacturing, the activities made in an enterprise are grouped, either as an added value or non-added value activity, by means of appropriate methods; the non-added value activities are excluded in order to minimize residues. Likewise, in this approach the productive area utilized for diverse activities or elements is quantified as added or non-added value, in order to minimize utilized area for the n activities or elements of non-added value.

22.3.1.3 Proximity

Based on the characteristics of the production demand, the layout problem of the facilities can be classified into two types:

- Unique layout, when the production demand is nearly constant.
- Flexible/robust layout, when there is a variation in the production demand.

In most available methods, which help analysing the problem of a unique layout period, the effective layout is measured taking into account the possibility of MHC in function of an objective.

In all these statements for the pursuit an efficient layout, the researchers try to locate the departments and facilities that have a huge amount of interaction, as close to each other as possible; a high interaction of departments and facilities can lead to a significant reduction in the total activity in the plant. As a consequence, the reduction of the activities minimizes the corresponding production time and the MHC, which are two of the objectives of an efficient layout (Hu and Wang 2004). However, there are three deficiencies in the methods:

- The MHC does not include the empty movements of MHE, in spite of the fact that empty movements have an important contribution to minimize the MHC, the work in process, and the lead time.
- The total flow between departments, as well as the flow between facilities within a department, have not been considered together.
- For a given layout, only an absolute value of MHC can be obtained, and this absolute value is not relevant in the assessment of the possibilities of new improvements in the current layout.

22.3.2 Layout Generation Methods

Below, a summary of the main layout generation methods in the research sector is presented:

22.3.2.1 ALDEP

It is an automated layout design program; it is more of a construction routine than an improvement routine. This implies that the layouts are determined starting from zero; the program does not require the user to specify an initial layout. This program uses the closeness ratings shown in a relationship diagram.

22.3.2.2 BLOCPLAN

Created by Donaghey and Pire (1990), it is similar to MCRAFT because the departments are arranged in bands. However, there are certain differences between both algorithms. It uses a relation table and a "from-to" table with original data for the "flow". The "cost" of a layout is measured with the objective based on distance or the objective based on adjacency. Besides, the program determines the amount of bands, which is limited to two or three bands.

However, the band width can vary. Likewise, as in BLOCPLAN, every department occupies one band, and all the departments are rectangular. Unlike MCRAFT, BLOCPLAN handles the continuous representation.

22.3.2.3 CORELAP

CORELAP (Computerized Relationship Layout Planning) develops the plant layout by calculating the Total Closeness Rating (TCR) for each department; the TCR is the sum of all the numeric values assigned to the closeness relationships (between a department and the rest). Each department is compared to the rest to obtain a TCR.

The department with the highest TCR is placed in the centre of the layout. If there is a tie on the highest TCR, the following tiebreaking rule is applied: first goes the department with the biggest area, and then the department with the lowest number.

22.3.2.4 CRAFT

CRAFT (Computer Relative Allocation of Facilities Technique) was introduced by Armour and Buffa (1963); it is one of the published layout algorithms. It uses a "from-to" table with original data for the flow.

The cost of the layout is measured by means of the objective function based on distance. The departments are not limited to rectangular shapes and the layout is represented in a discreet way.

22.3.2.5 LOGIC

LOGIC (Layout Optimization with Guillotine Induced Cuts) was developed by Tam (1992). The departments generated by LOGIC are rectangular, as long as the building is rectangular. The disposition is represented in a continuous way.

A construction algorithm can be used, which consists on dividing the building in proportions smaller every time, as successive guillotine cuts are made; that is, straight lines that go from one top of the building to the other. Each cut is either vertical or horizontal. If the cut is vertical, an assignment corresponds to a department, be it on the east or the west side of the cut. Given the width of the building (or a portion of it) and the total area of all the assigned departments, on the east side of a vertical cut, the exact coordinate of the cut (x) is calculated. Likewise, given the longitude of the building (or a portion of it) and the total area of all the assigned departments, on the north side of a horizontal cut, the exact coordinate of the cut (y) is calculated.

22.3.2.6 MCRAFT

The Institute of Industrial Engineers (IEE) distributed a CRAFT establishment for personal computers made by Hosni, Whitehouse and Atkins10 with the name MICRO-CRAFT. This is similar to CRAFT, except in the restrictions, which are lower (for instance, MCRAFT can exchange two departments, whether they are adjacent or not).

22.3.2.7 MULTIPLE

MULTIPLE (Multi-Floor Plant Layout Evaluation) was developed by Meller and Bozer (1996), originally for multi-floor plants. However, it can be used for one-floor plants by just putting the number of floors as one and not including the data requirements related to loads. It is similar to CRAFT; it uses a from-to table with original data for the flow and the function objective is identical to that of CRAFT (that is, an objective based on distances with the distances measured in a rectilinear way between the centroids of the departments). The departments are not limited to rectangular shapes and the layout is represented in a discrete way. It is an improvement layout algorithm that starts with an initial layout specified by the planner.

The layout improvements are achieved by means of two-way exchanges and in every iteration the exchange selected is the one that leads to the higher reduction in the layout cost.

22.3.2.8 PLANET

Unlike CORELAP and ALDEP, PLANET (Plant Layout Analysis and Evaluation Technique) does not use information contained in the relationship diagram to generate a layout. However, unlike CRAFT, PLANET is not an improvement routine. PLANET converts the input data into a diagram of internal flow which provides the cost of sending a flow unit for every two departments. Aside from the entries of the internal flow diagram, the program also requires the user to introduce a priority classification for each department.

22.3.2.9 Systematic Layout Planning

The Systematic Layout Planning (SLP) was first introduced by Muther and Wheeler (1961) as systematic multi-criteria procedure, equally applicable to brand new layouts and layouts of already existing plants. It has been considered as one of the more accepted and commonly used methodologies to solve plant layout problems with qualitative criteria, although it was conceived to design all kinds of plant layout regardless of their nature.

The method has the advantages of the precedent methodological approximations, and incorporates the flow of materials in the layout study, organizing the total planning process in a rational way and establishing a series of phases and techniques that, as Muther and Wheeler (1961) themselves described, allow identifying, assessing, and visualizing all the elements involved in the establishment and the existing relationships between them.

22.4 Case Study

22.4.1 Description of the Problem and Objective

In this section will be carried out the planning process of the facilities of a model winery of the Qualified Denomination of Origin Rioja (DOCa Rioja 2013) for the elaboration of red wine, focusing on the operations and systems of processes the influence its design, without making a thorough analysis of the machinery and technologies used in the wine cellar, nor the enology aspects.

The wine cellar, whose distribution appears in Fig. 22.1, belongs to the DOCa Rioja and is considered to be a big winery in that region. That wine cellar will be used in this chapter as a case study to exemplify the application of the methodology. All the grapes are treated with temperature control in stainless steel tanks, treating approximately 18 million kilograms of grapes per year. It has an average capacity of 14 million bottles between aging, reserve, and great reserve. The 45 % of its production belongs to aging, which represents 6,300,000 bottles of wine.

The objective of this section is to create a layout design proposal through a SLP, applied to the presented winery in order to satisfy the requirements of closeness and space, as well as the handling of materials.

22.4.2 Methodology

The SLP methodology will be used in the present study to find the best layout possible; the hierarchic diagram of the systematic layout planning is shown in Fig. 22.2.

22.4.2.1 Production Flow

The route is the organization of all the operations that allow producing under the best time conditions with a given resources and facilities.

The knowledge of the flow of raw materials, semi-finished products, and finished products is a key factor, mainly in the cases where the maintenance costs are elevated. In each stage of the route the following aspects are examined:

- ELIMINATE: Is the operation necessary or can it be eliminated?
- COMBINE: Can it be combined with another operation?
- CHANGE: Order of operations, place of work, people?
- IMPROVE DETAILS: Can the methods or equipment be enhanced?

In this analysis, the sequence and amount of movement of the products is determined for the different operations during the process. Starting from the

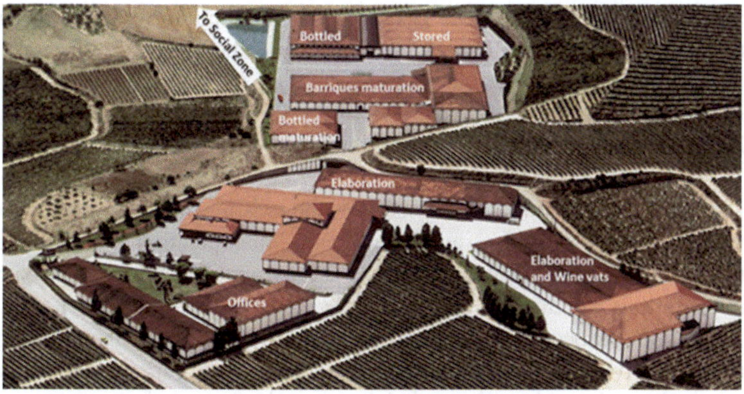

Fig. 22.1 Case study: actual layout

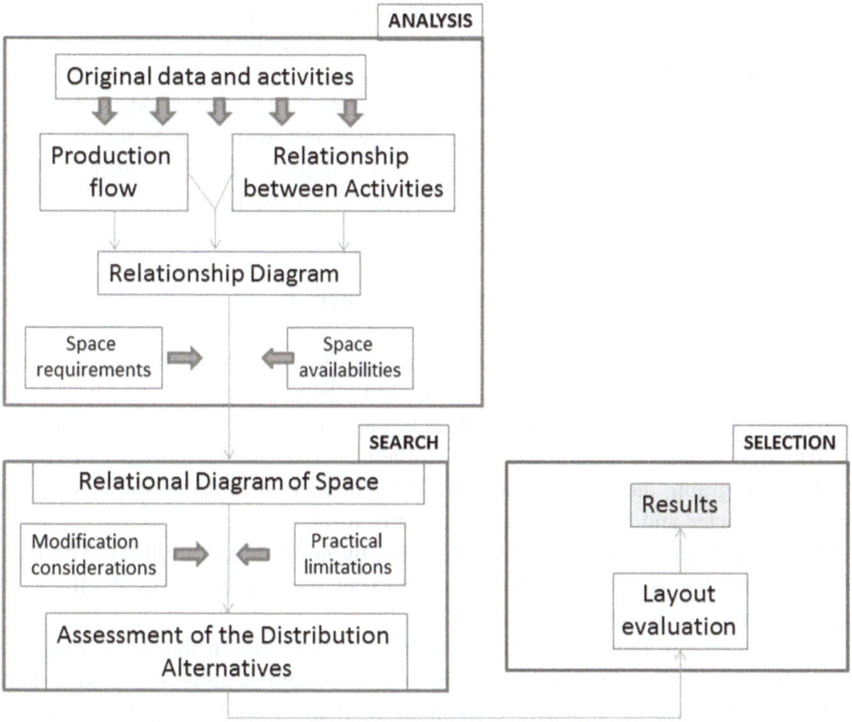

Fig. 22.2 Methodology

information of the production process and the production volumes, descriptive graphs and diagrams of the flow of materials are elaborated.

22.4.2.2 Analysis of the Relationship Between Activities

This section states the kind and intensity of the existing interactions between the different productive activities, auxiliary means, manipulation systems, and different plant services. The flow of materials is only one reason for the proximity of certain operations to others.

Among other aspects, in this stage should be considered the constructive requirements, environmental, safety and hygiene, necessary manipulation systems, energy supply and residue disposal, labour work organization, process control system, information systems, etc.

In order to be able to represent the relationships found in a logical way that allows classifying the intensity of such relationships, the relational activity table is used, consisting of a double entry diagram, in which the proximity needs between each activity and the remaining ones are shown, according the proximity factors defined for such effect.

It is quite common to express these needs by means of a letter code following a scale that decreases following the order of the five vowels: A (absolutely necessary), E (especially important), I (important), O (ordinary importance), and U (unimportant); instability is represented by means of the letter X.

22.4.2.3 Development of the Relationship Diagram

The information gathered so far, regarding both the relationships between activities and the relative importance of the proximity between them, is compiled in the Relational Diagram of Activities. This diagram is intended to gather the topologic distribution based on the information available.

For that graph, the departments that foster the activities do not have a dimensional character and do not possess a definite form.

In the diagram the activities are presented by nodes linked by lines. The latter represent the intensity of the relationship between the activities from the line code (as mentioned with the vowels plus X).

This diagram is adjusted by means of trial and error, which should be made in a way that minimizes the number of crossed lines representing the relationships between activities, or at least between those that represent a higher relational intensity.

22.4.2.4 Analysis of the Needs and Space Availabilities

There is no ideal general procedure to calculate space needs. The most appropriate method should be used, in terms of the level of detail which is being worked on, the amount and accuracy of the information, and previous experience.

The space required for an activity does not depend entirely on factors inherent to itself, it can be conditioned by the characteristics of the global productive process, the management of that process, or of the market.

The area needed for the development of an activity can be affected by the estimated production volume, demand variability, or the kind of management foreseen for warehouses.

22.4.2.5 Development of the Relational Diagram of Space

The Relational Diagram of Space is similar to the aforementioned Relational Diagram of Activities, with the difference that in this occasion the distinctive symbols of each activity are represented in a scale, so that the size occupied by each is proportional to the area needed for the development of the activity.

With the information acquired in this diagram, a group of alternative layouts can be made giving the solution to the problem.

The idea is to transform the ideal diagram into a series of real layouts, considering the conditioning factors and practical limitations that affect the problem.

22.4.2.6 Assessment of the Distribution Alternatives

Once the solutions have been developed, the next step is to select one of them by means of an assessment of proposals. The assessment of the alternative plans will determine which proposal offers the best plant layout.

Some of the methods used to assess the proposals are the following:

- *Comparison of advantages and disadvantages.* It is probably the easiest assessment method. It enumerates the advantages and disadvantages presented by the layout alternatives, that is, a system of pros and cons. However, this kind of method is less accurate; it is thus used in preliminary assessments or in phases where the data is not too specific.
- *Analysis of the weighed factors.* This method consists on the assessment of the layout alternatives with respect to a certain number of factors previously defined and weighed according to the relative importance of each in relation to the other, taking as reference a scale that can range from 1 to 10 or 1 to 100 points. In this way, the highest total score is selected. This increases the objectivity of what could turn into a very subjective decision-making process. Besides, it offers an

excellent way to imply managers in the selection and weighing of factors, and the production and service supervisors in the classification of alternatives of each factor.

22.4.2.7 Comparison of Costs

Comparing costs is the most significant method to assess plant layouts. In most cases, if the cost analysis is not the main basis to make a decision, it is used as support to other assessment methods.

There are two reasons to make a cost analysis:

- *Justifying a particular project.*
- *Comparing the proposed alternatives.*

Preparing a cost analysis implies to consider the total costs involved or just those costs that will be affected by the project.

In spite of the application of novel layout techniques, the final solution usually requires indispensable adjustments based on common sense and experience, according to the specific characteristics of the production process or service that will take place in the projected plant. It is common that in spite of the support of the different software applications currently available, traditional layout techniques are still in use.

22.4.3 Application

In this stage, an analysis of the flow of materials, relationship between activities, and relationship diagram will be made. The space required for each area will be determined, verifying the space available for the new wine cellar.

22.4.3.1 Analysis of the Flow of Materials

The first step of the analysis stage is to define the flow of materials between areas. In this case, the analysis of the flow of materials will be made between the functional areas, where the material that flows between areas is wine. The amount of transported wine will be measured in millions of litres/hours.

Given the fact that in the warehouse the materials of bottling, labelling, corking, and packing are taken prior to the bottling area, those kilograms of materials are represented in equivalent litres, in order to have a single measurement unit pattern.

In this case, to obtain the value of kilograms in equivalent litres, focus is made on the crates moved within the warehouse, which have a storage capacity of 580 bottles. With this data and weight of the bottle, the weight in kilograms of each crate can be obtained, resulting 169.71 kg.

Table 22.1 Flow of materials

From/to	Barriques maturation	Bottled	Bottled maturation	Wine storage	Stored	Stabilization
Elaboration	7			7		7
Barriques maturation		7				7
Bottled			7		7	
Bottled maturation				7		
Wine storage						
Stored		3.2				
Stabilization		7				

To know the amount of crates in the warehouse, the 10 million bottles produced are divided by the bottle capacity of each crate, obtaining in this way the amount of crates of the warehouse, which is of 17,242,00 crates.

Only the areas representative to the process of wine production will be used, and are shown in the from-to diagram of Table 22.1.

22.4.3.2 Relationship of Activities

The activity relationship diagram is made, indicating the existing interaction between the areas of the wine cellar. The relationships between each activity are shown in Fig. 22.3.

22.4.3.3 Relationship Diagram

The Relationship Diagram is the result of the from-to Diagram and the Relational Diagram of Activities. This diagram is presented in Fig. 22.4.

22.4.3.4 Determination of the Space Requirements

To determine the Space Requirements, the considerations followed are similar to those of the current layout, regarding the dimensions of the machinery, space for machine operators, and hallways.

22.4.3.5 Verification of the Available Space

Once determined the total space required for the layout of the winery, it is necessary to verify that the space available is enough to satisfy the needs of the new design.

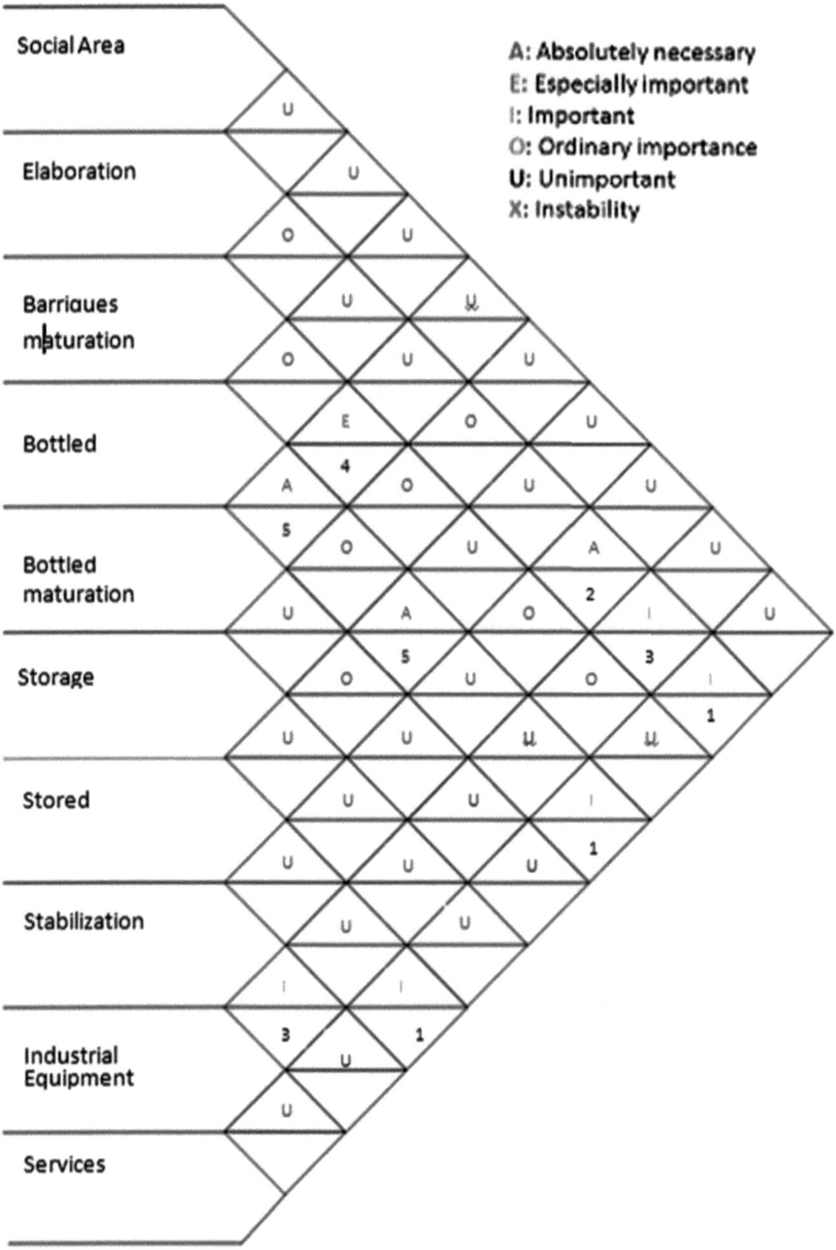

Fig. 22.3 Relationship of Activities

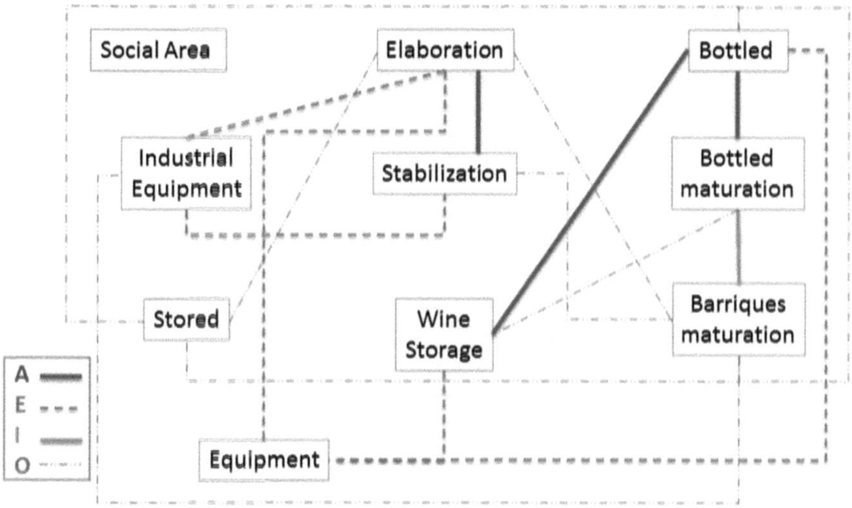

Fig. 22.4 Relationship Diagram

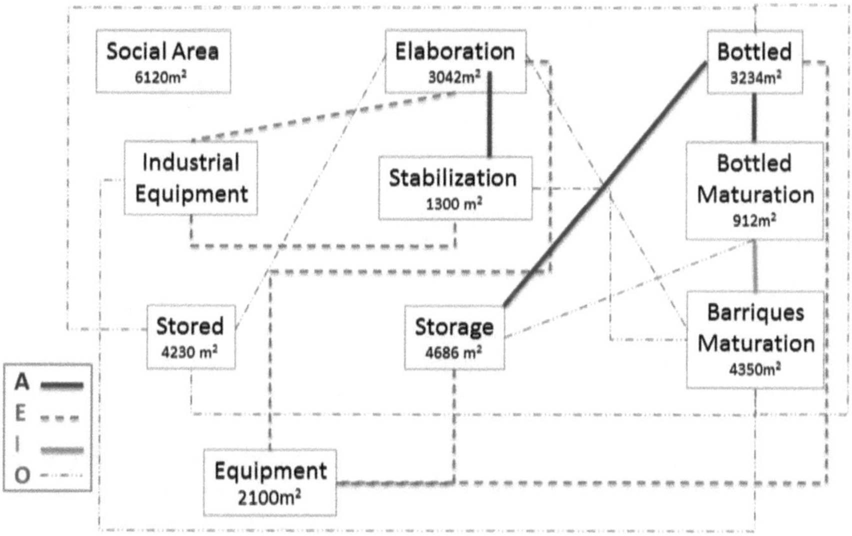

Fig. 22.5 Diagram of Space Relations

22.4.3.6 Diagram of Space Relations

This diagram, just like in the current layout, shows the relationships between the areas that were previously determined, and also their size. The capability of the layout to be fit into the available space is checked, as shown in Fig. 22.5.

22.4.3.7 Development of Alternatives

The layout alternatives are developed once all the requirements of the wine cellar have been determined in the previous sections, and with all the tools created. In this case, three alternatives are assessed, as shown in the next section.

22.4.3.8 Proposal Assessment

The assessment criteria used are the same ones that in the current layout. Three alternatives are assessed in order to select the best layout alternative, according to the results provided by the assessments. Below, the alternatives are assessed by means of these criteria.

Assessment of the adjacency of areas.
This assessment measures to what extent the relations determined in the Relationship Diagram are fulfilled. The kinds of relationships, how many of them are fulfilled, and the total score of the three alternatives are shown in Table 22.2.

Assessment of the Shape of the Areas.
In this case, the assessment is made to know to what extent the shape of each area is proper to the correct performance of its function. The decision criterion is to choose the alternative with the greater number of areas that meet the requirements with the proper shape. For that purpose, the value F is obtained for each area in the three layout alternatives (Table 22.3).

Table 22.4 summarizes the assessment of each area of the three alternatives. The table shows that all the areas of the three alternatives meet the shape requirements for a correct functioning of the wine cellar.

Figure 22.6 shows the possible layouts to implement.

Assessment of the Cost of the Handling of Materials.
The last criterion to assess the alternatives is by means of the cost of handling or movement of materials. The distances between the areas of each alternative will be measured in meters, as shown in Table 22.5.

Once the distances between the areas of each alternative are determined, the cost of the handling of materials is calculated in the following way:

$$C = \sum c_{ij} f_{ij} d_{ij}$$

where:

C = Total cost of the handling of materials

c_{ij} = Cost of the handling of materials between the areas i and j per unit of material and distance.

Table 22.2 Assessment of the adjacency of areas

CASE	Type of relationship	Number of fulfilled relationships	Score per fulfilled relationship	Total
1	A	2	20	40
	E	1	15	15
	I	3	10	30
	0	6	5	30
Total rating	115			
2	A	3	20	60
	E	0	15	0
	I	4	10	40
	0	5	5	25
Total rating	125			
3	A	3	20	60
	E	1	15	15
	I	4	10	40
	0	6	5	30
Total rating	145			

Table 22.3 Assessment of the shape of the areas

AREA	A	P	F	$1 \leq F \leq 1.4$
CASE 1				
Elaboration	3300	280	1.218	Yes
Barriques maturation	1650	170	1.046	Yes
Bottled maturation	1050	130	1.003	Yes
Bottled	1800	180	1.060	Yes
Stored	1500	160	1.032	Yes
Wine storage	2700	240	1.154	Yes
Social area	4200	260	1.003	Yes
CASE 2				
Elaboration	3150	270	1.202	Yes
Barriques maturation	1350	150	1.020	Yes
Bottled maturation	1200	140	1.010	Yes
Bottled	1650	170	1.046	Yes
Stored	1650	170	1.046	Yes
Wine storage	3300	280	1.218	Yes
Social Area	3750	250	1.020	Yes
CASE 3				
Elaboration	3150	270	1.202	Yes
Barriques maturation	1500	160	1.032	Yes
Bottled maturation	1350	150	1.020	Yes
Bottled	1440	156	1.027	Yes
Stored	1650	170	1.046	Yes

Table 22.4 Assessment of the shape of the areas

CASE	Number of acceptable areas	Efficiency (%)
1	7	100
2	7	100
3	7	100

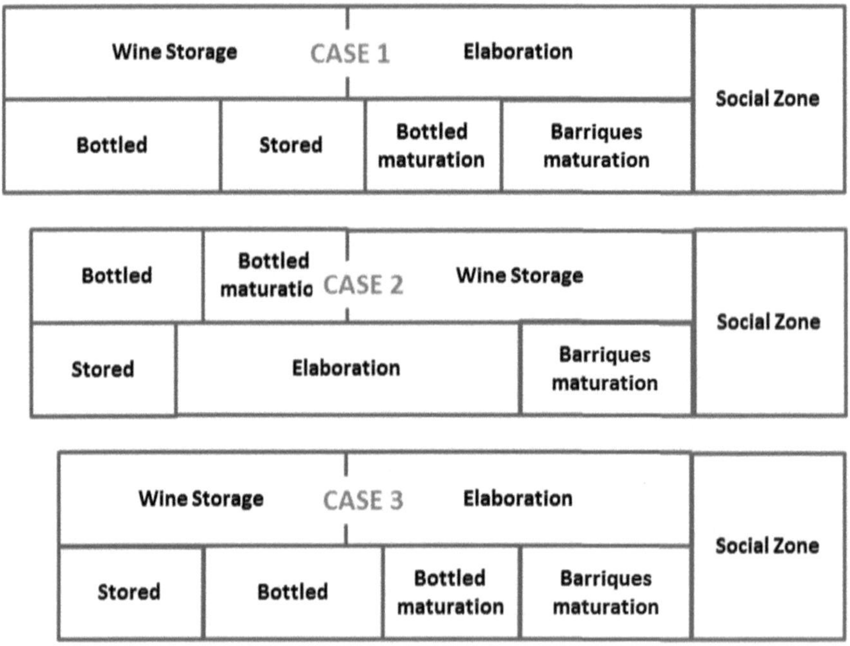

Fig. 22.6 Possible layouts to implement

Results of the Assessments.

Once the alternatives have been assessed under the three criteria, the next step is to determine the most effective alternative according to the assessment criteria used. In Table 22.6, the results of the assessments made are summarized.

22.4.3.9 Analysis of the Results

After the assessment of the alternatives under the three criteria, the next step is to select the most convenient layout for the wine cellar.

According to the results, the current layout is more expensive than the rest of the alternatives and is below the rest in terms of adjacency of the areas. Alternative 3,

Table 22.5 Distances between the areas of each alternative

Fron/to	Barriques maturation	Bottled	Bottled maturation	Wine storage	Stored	Stabilization
CASE 1						
Elaboration	50			45		20
Barriques_maturation		85				35
Bottled			50		40	
Bottled_maturation				60		
Wine Storage						
Stored		40				
Stabilization		65				
CASE 2						
Elaboration	50			45		20
Barriques_maturation		115				30
Bottled			35		40	
Bottled_maturation				35		
Wine storage						
Stored		40				
Stabilization		60				
CASE 3						
Elaboration	50			45		20
Barriques_maturation		45				30
Bottled			35		40	
Bottled_maturation				60		
Wine storage						
Stored		40				
Stabilization		55				

Table 22.6 Results of the assessments made

CASE	Assessment based on the adjacency of the areas	Assessment based on the shape of the areas (%)	Assessment based on the cost of the handling of materials (cost units)
1	115	100	3217.6
2	125	100	3077.6
3	145	100	2727.6

Table 22.7 Comparative of the assessments made

CASE	Assessment based on the adjacency of the areas	Assessment based on the shape of the areas (%)	Assessment based on the cost of the handling of materials (cost units)
1	115	100	3217.6
2	125	100	3077.6
3	145	100	2727.6
Original	110	100	6558

however, has a lower cost than the current layout and the rest of the alternatives, as well as a higher score in terms of adjacency.

The efficiency of the proper shape of the areas had 100 % effectiveness, that is, all the areas have a shape that is proper for a good functioning.

The average in cost of the handling of materials reduced drastically from 6558 to 2788 cost units. This implies that the transport times in the process will also be reduced, thus improving the handling of materials.

Taking that into account, the conclusion is that the best alternative among all the studied ones for the wine cellar layout is alternative 3, as it meets the best value in terms of adjacency of the areas and the lowest in terms of cost of the handling of materials.

All the aspects abovementioned are summarized in Table 22.7.

Once the development of the methodology is finished, it can be said that the objectives stated at the beginning were fulfilled; the main point has been the design of a better plant layout alternative which optimizes the resources of the wine cellar.

22.5 Conclusions

The results show the impact of plant layout design in the wine sector, as well as its importance when designing a winery, but above all, it is an example to show the application of the methodology by means of a case study.

Some relevant aspects of this application are presented below, which can evidently be extrapolated in a qualitative way to a general application:

- The design of the wine cellar layout turns into one of the most important tools when considering an improvement in the production process, as it allows the initial organization of the work places and a more appropriate configuration, in a way that allows the whole wine elaboration process to be more efficient.
- Layout planning plays an important role when designing a wine cellar, as it allows integrating machinery, materials, human resources, industrial equipment and facilities in a big operational unit, which works effectively in tandem.
- The study of layout design of a wine cellar leads to the analysis of the space needs for each work area, total available space, logical relationships within the production process, and costs of the handling of materials generated by the layout. The production process has to be taken into account, as well as the way in which the wine cellar will be made.
- The reduction of the travelled distances helps reducing the wine elaboration time and, therefore, an increase in production takes place.
- When making the design of a wine cellar or a re-layout of it, the main goal should always be an improvement in the process and at the same time an improvement in the working conditions of the employees.

- There is no ideal layout design for a wine cellar, the objective is only to find a better layout that meets the needs of the wine cellar. Every organization tries to fit its layout to the space limitations and the availability of resources. A layout design that is perfect for one wine cellar should not necessarily be effective to another. Multiple variables can change, as well as the conditions and resources.

References

Armour, G. C., & Buffa, E. S. (1963). A heuristic algorithm and simulation approach to relative location of facilities. *Management Science, 9*(2), 294–309.

Donaghety, C. E., & Pire, V. F. (1990). *Solving facility layout problema with BLOCPLAN*. Texas: Industrial Engineering Department, University of Houston.

Benjaafar, S. (2002). Modeling and analysis of congestion in the design of facility layouts. *Management Science, 48*(5), 679–704.

Bitran, G. R., & Morabito, R. (1996). Open queuing networks: Optimization and performance evaluation models for discrete manufacturing systems. *Production Operation Management, 5*(2), 163–193.

DOCa Rioja (2013) Qualified denomination of origin Rioja. Retrieved 2013 from http://es.riojawine.com/en/40-corporation-doca-rioja.html

Francis, R., & Goldstein, J. M. (1974). Location theory: A selective bibliography. *Operations Research, 22*, 400–410.

Francis, R. L., McGinis, L. F., & White, J. A. (1992). *Facility layout and location—an analytical approach*. Englewood Cliffs: Prentice Hall.

Hu, M. H., & Wang, M. J. (2004). Using genetic algorithms on facilities layout problems. *International Journal of Advanced Manufacturing Technology, 23*, 301–310.

Hitomi, K. (1996). *Manufacturing systems engineering*. London: Taylor and Francis.

Koste, L. L., & Malhotra, M. K. (1999). A theoretical framework for analyzing the dimensions of manufacturing flexibility. *Journal of Operation Management, 18*, 75–93.

Leung, Y. T., & Suri, R. (1990). Performance evaluation of discrete manufacturing system. *IEEE Control System Management, 10*(4), 77–86.

Lin, L. C., & Sharp, G. P. (1999). Quantitative and qualitative indices for the plant layout evaluation problem. *European Journal of Operations Research, 116*, 100–117.

Meller, R. D., & Bozer, Y. A. (1996). A new simulated annealing algorithm for the facility layout problem. *International Journal of Production Research, 34*(6), 1675–1692.

Muthaiah, K. M. N., & Huang, S. H. (2006). A review of literature on manufacturing systems productivity measurement and improvement. *International Journal of Industrial and Systems Engineering, 1*(4), 461–484.

Muther, R., & Wheeler, J. D. (1961). *Simplified systematic layout planning*. Boston: Industrial Education Institute.

Parker, R. P., & Wirth, A. (1999). Manufacturing flexibility: Measures and relationships. *European Journal of Operations Research, 118*, 429–449.

Raman, D., Nagalingam, S., Chiu, M. (2005). A fuzzy rule based system to measure facility layout flexibility. *18th International Conference on Production Research*. Fisciano (SA), Italy: University of Salerno.

Raoot, A. D., & Rakshit, A. (1991). A fuzzy approach to facilities lay-out planning. *International Journal of Production Research, 29*(4), 835–857.

Tam, K. Y. (1992). Genetic algorithms, function optimization, and facility layout design. *European Journal of Operational Research, 63*, 322–346.

Tompkins, J. A. (2003). *Facilities planning* (3rd ed.). New York: Wiley.

Tsourveloudis, N. C., & Phillis, Y. A. (1998). Manufacturing flexibility measurement: a fuzzy logic framework. *IEEE Transactions in Robotics and Automation, 14*(4), 513–524.
Yaman, R., Gethin, D. T., & Clarke, M. J. (1993). An effective sorting method for facility layout construction. *International Journal of Production Research, 31*(2), 413–427.
Yang, T., & Peters, B. A. (1998). Flexible machine layout design for dynamic and uncertain production environments. *European Journal of Operational Research, 108*, 49–64.

Chapter 23
Single Minute Exchange of Die and Organizational Innovation in Seven Small and Medium-Sized Firms

António Carrizo-Moreira

Abstract Innovation plays a very important role in businesses competitiveness. As Small and Medium-sized Enterprises (SMEs) stem for more than 95 % of the industrial fabric in the developed world, the improvement of the industrial production or the provision of a service are key to increase their productivity and competitiveness. The Single Minute Exchange of Die (SMED) is a Japanese process-based innovative methodology that involves the separation and conversion of internal setup operations into external ones. The SMED makes it possible for firms to reduce their lead times and to eliminate wastefulness during changeover activities. Although organizational innovation is a very important tool, it plays a silent role in productivity improvement, as it is less tangible than product or process innovation. Moreover, studies about SMED implementation and how teams have managed to achieve their results are still very limited among SMEs. The main objective of this chapter is to provide the results of seven projects involving business-university partnerships addressing this understudied topic: SMED implementation and organizational innovation in SMEs. The main finding of this study is that all firms managed to improve their setup times, although the results vary extensively. From the organizational innovation point of view, only one firm failed to intertwine the initiation and the implementation stages. Although all firms have initiation-implementation routines, there are clear differences among them. As a result, although it is possible to claim that all SMEs analyzed are ambidextrous organizations, their initiation-implementation routines, deserve deeper comprehension.

Keywords SMED methodology · Setup times · Organizational innovation · SMEs

A. Carrizo-Moreira (✉)
DEGEI—University of Aveiro, Campus Universitário de Santiago,
3810-193 Aveiro, Portugal
e-mail: amoreira@ua.pt

J. L. García-Alcaraz et al. (eds.), *Lean Manufacturing in the Developing World*,
DOI: 10.1007/978-3-319-04951-9_23, © Springer International Publishing Switzerland 2014

23.1 Introduction

Single-Minute Exchange of Die (SMED), also known as Quick Changeover of Tools, is a methodology that was developed to reduce the setup time during changeover (Shingo 1985). SMED has been used to respond to fluctuations in demand and leads to lead time reduction, the elimination of wastefulness and the reduction of lot sizes (Shingo 1985; Moreira and Pais 2011).

Organizational innovation has been extensively analyzed to understand how environmental contingencies influence firms. The ambidextrous theory focuses on the adoption of innovation and proposes two stages, the initiation and the implementation, in which firms perform differently.

Based on a set of seven projects from business-university partnerships, the main objectives of this chapter are: firstly, to cover the main results achieved from the implementation of the SMED methodology in the reduction of setup times in firms producing several types of products (dies; polyurethane and polyester foam; sanitary products; corrugated cardboards; rims for bicycles; and bolts and rivets); secondly, to provide examples of the SMED methodology outside the typical exchange of dies; thirdly, to address the importance of organizational innovation, namely the ambidextrous theory, in achieving continuous intangible improvements.

This chapter is divided in seven sections. The first section covers the introduction. The second section addresses the importance of innovation, namely organizational innovation. The third section addresses the literature review regarding the SMED methodology. The fourth section exposes SMED implementation. The research methodology is covered in the fifth section. The sixth section addresses the field work covered. In the seventh section the conclusions are drawn.

23.2 The Importance of Innovation

The increasing competitiveness in global markets unleashed a quest in the industrial business world for the improvement of productivity.

During the last decades the business world has witnessed a great deal of changes in the management and organization of production systems in such a way that industrial firms, whether large or small, need to keep up with their main competitors on a global scale. The technological revolution and the introduction of new organizational techniques played an important role in those changes. The former, led by the introduction of information systems, machinery, telecommunications, pervasive automation and robotics, created unprecedented productivity gains and improved operations planning and control. Most of the firms realized that despite their technological acumen, organizational innovations, based on human capital, sparked their productivity improvement, paving the way for better time to market efforts and improved flexibility (Womack et al. 1990; McIntosh et al. 2001; Holweg 2007).

Several studies have examined organizational innovation in an effort to answer to basic questions such as, "why some organizations are more innovative that others?" or "What organizational structures and management processes facilitate or inhibit innovations?". Most academic research have been on conceiving innovation as an outcome or as process and little attention has been given to organizational innovation.

Although, innovation is an important tool for competitiveness (Dantas and Moreira 2011), two types of innovation processes were distinguished: the firm as provider or user of innovation. As provider of innovation, the firm faces problems and makes decisions that involve the development of new products and processes (Moreira and Pais 2011). The company's success as an innovation provider stems from its exploration capacity and the possibility to profit from such innovation (Gopalakrishnan and Damanpour 1997). With this behavior, the firm follows an outward innovation process. The firm is not a passive entity "importing" concepts, products or processes: it plays the central role in generating change.

As user of innovation, the firm tries to incorporate novelties developed outside the firm. Zaltman et al. (1973) differentiated between two stages in this innovation process: initiation and implementation. The former includes activities that deal with the perception of the problem, the gathering of information and the development of an attitude that promotes and assesses innovation. The implementation stage is related to the decision making process regarding the adoption of innovation. In the initiation stage, competences possessed by individuals are essential. Group competences however, are crucial in the implementation stage as it is more systemic and involves the internalization of the innovation (Damanpour 1992). As user of innovation, the firm follows an inward innovation process.

The ambidextrous theory focuses on the process of adoption of innovation and proposes two stages in the process: the initiation stage and the implementation stage. The theory posits that high complexity (e.g. organic features) facilitate the initiation of innovations, while low complexity (e.g. mechanistic features) facilitates the implementation of innovations (Zaltman et al. 1973). Although very simplistic, this theory has not been assessed on a process perspective involving SMED projects.

Clearly, although process innovation is related to new elements, equipment or manufacturing methods that improve the production of a product and provide a better service (Damanpour 1991; Damanpour and Gopalakrishnan 2001; Dantas and Moreira 2011), firms play a different role as users or providers of innovation.

Process innovations are less tangible and more difficult to implement than product innovations (Ettlie and Reza 1992; Frost and Egri 1991). Ettlie and Reza (1992) and Damanpour (1996) state that product innovations are easier to imitate whereas process innovations are more organization specific, given that they cannot be copied without implementing changes in the organizational structure or management system.

The introduction of process innovations normally involves organizational innovation. For example, the introduction of lean technologies such as the Toyota Production System (TPS) and the Single Minute Exchange of Die (SMED) are examples of organization specific process innovation. The main objective of the TPS is to increase manufacturing productivity using techniques such as small-sized production lots and non-stock production with a strong focus on organizational change (Pisano and Hayes 1995; Godinho Filho and Fernandes 2004; Moreira and Pais 2011). Therefore, it is the firm's capability of deploying innovation that influences and triggers productivity. Organizational innovation has a distinctive characteristic: firms need to behave as providers of innovation by exploring new ways to improve their activities. However, they must also act as users of innovation since they have to internalize new solutions developed outside their boundaries.

The implementation of process innovation based techniques, which strongly rely on organizational changes, are very important among industrial competitiveness (Cusumano 1989; Womack et al. 1990; Womack and Jones 1994; Monden 1984; Liker 2004; Holweg 2007). Although, many of these techniques were introduced by Japanese firms (Lamming 1993; Nishiguchi 1994; Shingo 1985; Monden 1984), the implementation of Japanese process innovation-based techniques have had limited impact on the West (Nishiguchi 1994; Freire 1995; Pardi 2005).

The implementation of organizational and process innovation is in high demand for industrial firms for several reasons (Holweg 2007). Firstly, as competition is stiff, there has been a growing focus on product cost cuts. Secondly, as time-to-market is growing in importance, the reduction of setup times is a key component of competitiveness. Thirdly, as product demand is shrinking, there is a growing focus on the reduction of stocks and on the increase of flexibility. Accordingly, the better the efficiency of the firm in managing its production resources, the better ability the firm has to respond to the competitive challenges through improved operational performance. Consequently, the above mentioned initiation and implementation stages are key components of firms' competitiveness as they spark unique, difficult to match competitive advantages.

Individuals inside organizations can influence the processes of innovation with their personality, motivation, knowledge, work characteristics, and mood states (Anderson et al. 2004). In terms of groups, team structure, team climate, team processes, member characteristics, and leadership style are all considered determinants for organizational innovation (Anderson et al. 2004).

Damanpour (1991) settled thirteen organizational determinants of innovation: specialization; functional differentiation; professionalism; formalization; centralization; managerial attitude; managerial tenure; technical knowledge resources; administrative intensity; slack resources; external communications; internal communications; and vertical differentiation. The existence of two other determinants of innovation was also put forward: structural complexity, and organization size (Damanpour 1996). Pittaway et al. (2004) advanced five impediments of organizational innovation: inter-firm conflicts, lack of infrastructure, lack of scale, displacement, and external disruption. Klein and Knight (2005) focused on four different determinants affecting the implementation stage of organizational innovation, namely: positive climate; management support; learning orientation; and financial resources.

Clearly, organizational innovations are inherently complex. As SMED is based on a set of procedures that are used as a continuous improvement process, in this chapter the following definition of organizational innovation is going to be followed (Damanpour and Evans 1984): *organizational innovation is the use of new managerial and working concepts and practices*, which can be differentiated in structural and procedural changes.

23.3 Single-Minute Exchange of Die

Single-Minute Exchange of Die (SMED), also known as Quick Changeover of Tools, refers to the theory and techniques used to reduce setup times. It can be applied in any industrial unit and to any machine. It is defined as the minimum amount of time necessary to change the type of production activity and it takes into consideration the moment at which the last piece of a previous lot was produced *vis-à-vis* the first piece produced by the subsequent lot (Shingo 1985).

Traditionally, firms regarded setup times as one of the most expensive costs they had to face. Accordingly, many firms opted for minimizing the number of setups implemented. This resulted in very large production lots (Holweg 2007), contributing both to excessive inventory and low productivity (Liker 2004).

According to Shingo (1985), the development of the SMED methodology took place in Hiroshima, Japan during the 1950s. It was initially developed at the premises of Mazda Toyo Koyio and Mitsubishi Heavy Industries. Nevertheless, it was only in the 1970s, as part of the Toyota Production System, that the SMED methodology was widely acknowledged. Cusumano (1989), on the other hand, defends that the concept of Quick Changeover was already applied in the US during the fifties.

The SMED methodology was developed in order to reduce and simplify the setup time during changeover (Shingo 1985). SMED, which is a Japanese process-based innovation, makes it possible to respond to fluctuations in demand and results in lead time reductions. Furthermore, it eliminates wastefulness during changeover and diminishes lot sizes (Shingo 1985; Moreira and Pais 2011).

With the growing trend of product customization, mass production became obsolete. As a consequence, firms deploy strategies to compete, simultaneously, in terms of price, product quality, product differentiation, and delivery time. Accordingly, in order to improve production processes, it is necessary to analyze the value added by each activity and eliminate all those that do not add value to the product (Levinson 2002). Therefore, the SMED methodology becomes extraordinarily important.

When firms face demands with large diversity of products, it is imperative to be prepared to produce smaller lots without jeopardizing their global productivity (Blau 1994). Thus, firms must be capable of producing a large diversity of products in small quantities and, consequently, must have the competencies necessary for changeover activities. In order to compete, firms need to find ways to reduce setup times and eliminate wastefulness and non-added value activities. Furthermore, they must be able to convert idle setup time into regular production time. Therefore, a strong focus on process and organizational innovation is needed.

This type of problem can be successfully addressed following the SMED methodology (Shingo 1985). The main challenge is to implement a process-based innovation in which setup operations need to be standardized and properly documented. In this manner, production workers can follow all the procedures of a certain process, resulting in the reduction (optimization) of setup times.

Plenty of research on SMED have been widely described and presented with the exchange of Dies as the main focus (Monden 1984; Johansen and McGuire 1986; Sepheri 1987; Quinlan 1987; Noaker 1991; Gilmore and Smith 1996; McIntosh et al. 2000; Fogliatto and Fagundes 2003; Satolo and Calarge 2008). Van Goubergen and Van Landeghem (2002) analyzed how equipment design can improve existing setup times. In their study they analyzed more than 60 cases and concluded that up to 90 % of setup times could be improved. Neumann and Ribeiro (2004) analyzed how a supplier development program achieved a 50 % of improvement in the set up time of the firm. Sugai, McIntosh and Novaski (2007), addressing a single case study, concluded that the sequencing of production lots, the acceleration (during the setup) and deceleration (during post-setup) periods and the need to maintain rigorous set up times achieved are very important topics. Clearly, studies concerning SMED implementation and innovation process are in short supply.

As a consequence, and taking into account a set of seven projects from business-university partnerships in industrial firms, the main objectives of this chapter are: firstly, to cover the main results achieved from the implementation of the SMED methodology in the reduction of setup times in firms producing several types of products (dies; polyurethane and polyester foam; sanitary products; corrugated cardboards; rims for bicycles; and bolts and rivets); secondly, to provide examples of the SMED methodology outside the typical exchange of dies; thirdly, to address the importance of organizational innovation namely the ambidextrous theory, in achieving continuous intangible improvements.

23.4 The Operationalization of SMED

Traditionally, the minimization of the costs of idle machines during setup operations was to produce large lots, in order to obtain the lowest possible percentage of idle time per unit produced. As Toyota's inventory costs for their vehicles were extremely high, they decided to reduce setup times (Shingo 1985). Accordingly, if production changes could be done in less time, the ideal amount of production could be smaller decreasing the costs involved.

As the unitary costs are directly proportional to the setup time and to the production time, Shingo (1985) defends that firms need to have a clear strategy to reduce setup times, otherwise they can face the following disadvantages:

- The need for larger client orders, which is very negative and counter intuitive, as the SMED was developed to face the reduction of order sizes due to the growing customization;
- Longer lead times, which jeopardizes competitive responses to main competitors;
- Larger costs with inventory, pallets, forklifts, labor, among other things, which hinders business competitiveness;
- Larger quality problems, as we return to mass production techniques;
- Loss of money with inventory amortization, which hinders firm competitiveness;
- More labor linked to transport and inventory, which hinders firm competitiveness;
- More frequent refunds due to larger amounts of defects (probable).

According to Shingo (1985), one can extract direct and indirect benefits from the SMED application. The reduction of inventory, the increase of production flexibility and the rationalization of tools are among the indirect benefits. The direct benefits include the reduction of setup time, the reduction of time spent with fine tuning the machines, the reduction of errors during changeovers, the improvement of product quality and increased safety.

The principle behind the setup time reduction introduced by the SMED methodology is simple: the elimination of wastefulness related to the change of tools. To achieve this, Shingo (1985) applied a systematic approach in order to separate internal operations—namely, the Die exchange or the fitting of equipment, which must be performed with the machine in switched off mode—from external operations—namely, those performed with the machine in normal operation mode, as is the case of the preparation of tools. In their improvement process, firms normally go through the following four different phases (Shingo 1985):

- Phase A: the firm makes no distinction between internal and external setup operations and, consequently, machines remain idle for very long time periods. The main objective in implementing the SMED methodology is to study the shop floor conditions in great detail through a production analysis, interviews with workers and videotaping of setup operations.
- Phase B: the firm separates internal from external setup operations. Usually, this action saves 30 to 50 % of the time needed for the setup operation. Mastering this distinction is a key issue to achieving success in implementing SMED.
- Phase C: the firm converts the maximum internal setup operations to external ones. In this phase, it is important to re-examine all operations in order to assess if they were wrongly assumed as internal ones and, if needed, convert them to external ones.
- Phase D: Streamlining all aspects of the setup operation. This phase seeks the systematic improvement of each basic operation of internal and external setup, developing solutions to accomplish the different tasks in an easier, faster and safer way.

For firms to reach global success during the SMED implementation, Shingo (1985) describes, quite exhaustively, a set of procedures that must be followed:

- To analyze the actual procedure;
- To classify the several operations performed as internal or external ones;
- To convert internal operations into external ones;
- To develop brand new solutions in order to reduce both the time of internal operations and the time delays in external operations;
- Creating rigorous procedures in order to reduce flaws during the setup;
- Returning to the beginning of the process and repeating the whole procedure in order to continuously reduce the setup time.

This set of procedures requires a continuous analysis of the process in order to obtain good results. Whenever the method is applied, new and improved solutions must be obtained. Thus, a focus on process and organizational innovation is mandatory.

Several studies have implemented the SMED methodology. For example, Monden (1984) defends the simultaneous analysis of all internal and external operations and the standardization of all functions. On the other hand, Gilmore and Smith (1996) defend that Shingo's (1985) procedures can be applied even when not following his logical sequence. Moxan and Greatbanks (2001) analyzed the prerequisites for the implementation of the SMED methodology and found that the methodology might be very ineffective due to cultural, process-based management barriers. They defend the use of a preparatory/learning phase in order to reach a better implementation of SMED. Fogliatto and Fagundes (2003) identify four types of activities when implementing SMED: strategic, preparatory, operational and confirmatory activities. All these activities have different purposes and are part of a broader scope of teamwork, management involvement, training, visual management and internal communication of the results achieved. All these activities are necessary to fine-tune Japanese techniques to non-Japanese environments.

McIntosh et al. (1996) put forward an important contribution regarding changeover improvements. They experienced that even when firms conduct initiatives to improve setup time, and they successfully achieve them, the levels of performance of the initiatives kept sliding away to the same levels before the initiatives were taken. Although several difficulties were measured (motorization of the setup period, lack of production and quality measures, insufficient attention to setup time *vis-à-vis* to product quality on production rate, lack of improvement targets, lack of training and lack of goal orientation), the main problem is the lack of an organizational strategy in which a directive management style was used instead of a participative one.

Another important contribution is referred by McIntosh et al. (2001), who applied SMED to Total Productive Maintenance (TPM). They refer that the better the planning of the maintenance intervention, the lower the setup time of changeover activities. The achievements of their analysis are very clear. A good planning is essential to deploy organizational improvements, which will be reflected in the SMED results. Nevertheless, the message is very clear: without the proper training and human involvement it is difficult to achieve results.

Instead of focusing on SMED results, Gest et al. (1995) gathered information on the different specific techniques that might be used when implementing SMED. They conclude that although adjustments can be observed, the main contributory factor to setup problems firms face is the lack of clear instructions, namely due to the wide spectrum of machines they work with.

From a historical perspective Holweg's (2007) work is very important: SMED is just a part of a lean thinking philosophy, the Toyota Production Systems. Accordingly, mere transplants will certainly not achieve the results of an integrated perspective involving Kaizen, Six Sigma Value Stream Mapping and continuous improvement. This perspective is also supported by Hicks (2007) from a information management perspective and by Pardi (2005) from a socioeconomic perspective.

Satolo and Calarge (2008) analyzed the applicability of the SMED methodology and concluded that there are large differences among the firms that implemented it. More importantly, the differences were based on organizational barriers, resistance to change and difficulties in identifying opportunities for improvement. For Satolo and Calarge (2008), implementing the SMED methodology is doomed without proper staff preparation and training, and without the publication of results among those involved. Their contribution is very important in organizational terms, complementing Moxan and Greatbanks' (2001) and McIntosh et al.' (2001) studies about the difficulties in achieving results due lack of organizational strategy.

In order to circumvent the difficulties of the implementation of changeover activities, McIntosch et al. (2001) use a set of leveraging tools and a set of evolutionary steps for each of the four phases referred above:

- Phase A, which is the SMED project kick off: the analysis of the shop floor activities in order to differentiate internal from external operations.
- Phase B, which separates internal from external operations: the use of check-lists; the definition of functions for each worker; and the improvement of transportations tools.
- Phase C, which converts internal to external operations: the previous preparation of setup operations; the automation of operations; and the utilization of different tools.
- Phase D involves the improvement of all aspects of the setup operation: the improvement of tool transportation and warehousing; the elimination of settings, calibrations and adjustments; and the the automation of operations.

Finally, an important issue that deserves some attention is that most classical studies have addressed the implementation of SMED methodologies in die casting activities for the automotive industry. However, newer experiences have implemented SMED projects in other types of industries (Moreira and Garcez 2013; Satolo and Calarge 2008). One important characteristic of these studies is that the results vary quite broadly. Thus, as mentioned above, the main objective of this chapter is to provide examples of the implementation of the SMED methodology in SMEs outside the traditional applications in the exchange of dies, taking into account organizational innovation.

23.5 Research Methodology

In order to achieve the goals proposed, it was decided to use a case study approach (Yin 1989). The methodology used is based on the SMED characteristics addressed in the last section. During the implementation of the SMED methodology, we followed the phases proposed above as well as the techniques presented by Shingo (1985). However, some of them could not be directly applied as the machines have different characteristics from those of Die casts.

For simplicity reasons, the firms will be presented first and the results afterwards. The seven cases reported herein involved Business-University partnerships. Due to confidentiality concerns, the name of firms cannot be disclosed.

Firm A involved the study of SMED implementation in a medium-sized moldmaker that uses a wide range of Dies: from 80 Tons to a maximum of 1100 Tons. Firm B produces foam in polyurethane polyether and polyester for several markets. The firm had a sales volume close to 27 million euros and employed approximately 140 people, in 2010. Firm B transforms 60-m-blocks of foam into 5-mm-wide foam rolls. This operation takes place in a Looper, which is the case study covered here.

Firm C is a medium-sized enterprise from Aveiro producing plastic sanitary products. Firm C embarked on a 3-year project involving 5Ss and SMED methodologies following a continuous improvement approach. The analysis of the implementation of SMED was based on setup time reduction in 47 different plastic molds injection machines.

Firm D involves a case study in a SME from the north of Portugal producing corrugated plates and corrugated carton packaging. The analysis involved SMED and Overall Equipment Efficiency (OEE) implementation on a pilot machine producing corrugated plates with several thicknesses. Firm E is a metal mechanic firm from the Aveiro region. It produces a wide range of products, from bowls and dishwasher in stainless steel to aluminum rims and wheels. The case reported herein involves the production of rims for bicycles and the analysis was based on SMED and 5Ss techniques implemented for two types of products.

Firm F is a SME with 100 employees producing metallic bolts and rivets. Their main clients are in the auto industry. Firm F witnessed a productivity improvement through a just-in time-project and decided to be involved in a setup time reduction project involving SMED methodologies. This case involved the analysis of 118 records of setup time improvement projects for a specific type of equipment. Finally, case G involves a SME producing corrugated cardboard. Two types of SMED projects were analyzed: corrugated cardboard and high quality printed corrugated cardboard packaging. The analysis reports two different types of results involving behavioral responses to the SMED projects.

The analysis of the seven case studies followed a similar pattern and the approach was very implementation-oriented. This analysis was divided in the following steps:

1. Describing and analyzing the setup operations on the shop floor, tracking setup times and measuring all operative movements;
2. Separating internal from external operations;
3. Converting internal to external operations;
4. Streamlining all aspects of the setup operation in order to accomplish the different tasks in an easier, faster and safer way;
5. Assessing the impact of the methodology implemented;
6. Preparing the diffusion of the new SMED methodology to the other firms of the economic group.

For operational reasons, it was only possible to record on tape the whole set up process for A, B, D and E.

23.6 Results

The initial analysis is very crucial for obtaining a correct diagnosis as it marks the beginning of a new production system. The results obtained in this phase are also important for a subsequent assessment of the impact of the adopted solutions. Accordingly, the main objective of this phase was to gather information regarding the setups, namely: the sequence of shop floor operations; the timings of different tasks and operations; the organization of workers during the setup and the machine work rates; and the identification of critical points that reduce the effectiveness of the production system, as well as their causes.

The analysis of the production system took place during the setups and involved the following aspects: the analysis of the standard procedures, if any; the communication among workers; the difficulties felt by workers during setup operations; the settings, calibrations and adjustments during the setup; and the coordination among the various departments involved in the setup operations.

During the participation in the several setups, it was possible to identify several common problems. The following are among the most important ones:

- Poor organization, since the people involved were inadequately prepared for the setup and the necessary material for performing the operations was not ready;
- Lack of knowledge of the procedures for carrying out the complete setup in time;
- Lack of an established check-list of activities for carrying out the setup;
- The carrying out of external operations as if they were internal ones;
- Incorrect assignment of tasks during setup;

- Lack of planned procedure to deploy setup operations;
- Strong separation between shop floor activities and other activities that affected the setup time;
- Lack of knowledge of the consequences of poor setup operations.

After this initial stage/step, through the analysis of the data gathered and the video recordings, when allowed, the internal and external operations were separated and the set time of each of them was analyzed. Afterwards, the analysis of the several operations that can be transformed in external ones took place, i.e., performing procedures with the machine in operation as opposed to performing them with the machine in standby. Departing from the data gathered in the previous stage, a thorough analysis of the setup operations was accomplished in order to identify which operations could change their status.

Subsequently, in order to improve all internal and external operations that compose the changeover, we analyzed the records of the operations previously accomplished and identified how we could reduce all operations. In this way a new, post-SMED operation mode for the setup series was prepared as a result of the work developed by the entire team involved in the reduction of the setup time project. This new SMED operation mode was defined as the new standard.

The results achieved in each case studied were different. For example, Firm A managed to achieve an average of 39.1 % of setup time improvement. Using only an incremental innovation approach with no investments performed on equipment, they not only managed to use and master innovation, but also enhanced the solution to other sister firms. Moreover, they developed brainstorming sessions to implement changes in the plant layout, the management of quality and maintenance activities and in the design of their equipment. They appointed an improvement team to be responsible for the continuous improvement of the setup times and regular training is provided to all staff involved in SMED activities. Clearly, Firm A managed to implement both the initiation and the implementation stages, proving to be an ambidextrous organization.

The average setup time improvement of the Loopers in Firm B was 65 %. They managed to decrease from a setup time of 114 to 40 min. As a consequence of the improvement, the firm regarded SMED as a moving target: all SMED team members need to include in their portfolio of activities the identification of internal and external setup operations, the conversion of internal setup operations to external ones and the reassessment of all conventional procedures. As occurred in Firm A, Firm B managed to implement organizational innovations in such a way that ambidextry is present through the initiation and implementation phases.

The results obtained by Firm C are, by no means, the best of all seven cases analyzed: the average setup time improvement was 80 % and the number of changeovers increased more than 3 times. This achievement is the outcome of a 3-year project involving 47 plastic mold injection machines. Indeed, Firm C has managed to internalize knowledge generated during the initiation phase (perception of the problem, gathering information and promote change) and has managed to deploy this knowledge through routines and procedures to the rest of the organization. They have managed to improve the working environment, to reduce product unitary cost as well as to reduce the lot size (although data was not released). Clearly, Firm C has managed to deploy both the initiation and implementation stages that characterize ambidextrous organizations.

During the SMED project implemented in Firm D, it was possible to conclude that 31.42 % of production time stoppages were due to changeover operations. With the project, the setup time reduction was reduced in 22.3 %, and several modes of failure were detected. After analyzing them it was possible to intertwine several setup operations with other modes of failure. In the end, the three main modes of failure were responsible for 43.6 % of un-programmed stoppages.

Firm G, that also produces corrugated cardboard, was involved in two different projects: one involving the packaging and the other one involving merely corrugated cardboard. Two different types of results were found. One project team managed to reduce the setup time in 43 %, increase efficiency in 32 % and reduce the un-programmed stoppages from 15 to 11.1 %. At the end of the project, working conditions and a growing motivation were achieved and improvements rapidly assimilated. With the second project team the results were astonishingly fuzzy: the setup time increased only 8 %, efficiency decreased 2 % and the time of un-programmed stoppages increased. This second team showed high resistance in internalizing new operational methods. The different outcomes clearly show that even with the same training and procedures, teams might perform differently. In order to sort this incongruence out, Firm G decided to swap several team members and to create a task force team to deploy best practices to subsequent projects, without disregarding the leadership style to make projects successful.

The implementation of the SMED project in Firm F, was very straightforward as they simply analyzed 118 records of a type of equipment, and following the SMED procedures, they managed to improve setup time in 33 % without being involved in further investment, i.e., they used only a continuous innovation approach. On the other hand, Firm E managed to embark on a SMED project and used the 5Ss approach in the production of rims for bicycles. The average setup time improvement was 56 % and the lot size was reduced in 49 %.

23.7 Conclusion

One of the main conclusions is that the results of setup time improvement among the seven cases presented vary extensively. Apparently, some firms manage to muddle through the intricacies of the SMED methodology more successfully than others.

The deployment of the SMED methodology, though not very complex, involves the training and the active involvement of all SMED team members as well as the other production employees. Once the improvements are being achieved, team members feel more comfortable with the implementation.

Although, all firms implemented the same type of project, all of them did follow different paths. For example, Firms A, B, E and G decided to deploy the SMED methodology as part of an incremental innovation process with no investments. In the end, firms A and B needed to invest in minor tools. On the other hand, firms C and E complemented the SMED methodology with 5Ss methods. While some of them implemented focused projects (Firms B, D, E and G) others followed a wide perspective (Firms A, C and F). With the exception of Firm G, workers could internalize the knowledge and the intricacies of this organizational-based methodology and improve the results with other process innovation investments.

Another important conclusion is that the deployment of the SMED methodology should always have in consideration the ideas of those directly involved in the process. Often, these ideas are simple, effective and involve low costs. In SMED, as in other organizational innovation tools, those directly involved in the process must be empowered to find the best solutions to solve the problems. As can be observed in Firms G, even when workers have the knowledge and procedural tools to make organizational innovation happen, it might not be enough for firms due to natural resistance to change and to the not-invented-here syndrome.

As in continuous improvement methodologies, the SMED should be regarded as a moving target. Once an objective is achieved, new and more challenging objectives should be defined. Therefore, SMED teams should include in their portfolio of activities the identification of internal and external setup operations, the conversion of internal setup operations to external ones and the reassessment of all conventional procedures. In this way, it is possible to generate the improvement of setup times and the deployment of brainstorming sessions. Such brainstorming sessions aid in the identification of aspects not yet included in the analysis of SMED activities such as plant layout, total quality management, total maintenance management and equipment design changes.

Ambidextrous organizations are those that perform well at both the initiation and implementation stage. Although plenty of studies relate the former with product and the later with process innovation, organizational innovations as SMED need to have both of them to be successfully implemented. For example, the problem perception, the gathering of information and attitude formation might take place even before the beginning of the SMED project and involves the modification of the internal operations in external ones. The definition of new procedures and the involvement of workers in enduring new achievements are typical examples of how important the implementation stage is. In this regard only firm G failed in intertwining the initiation and the implementation stages. Moreover although all firms have initiation-implementation routines, there are clear differences among them as can be analyzed by the difference in the setup time improvements.

The implementation of the SMED methodology involving equipment and processes different from those originally used is still controversial. Several methodologies described by Shingo (1985) are not readily applicable for all types of equipment. Due to the large diversity of equipment and industries in which the methodology could be implemented, a ready-to-use set of guidelines or procedures seems to be quite generic. Nevertheless, contrary to what has been claimed by Sugai et al. (2007), this methodology can be implemented and deployed in any equipment as long as it is adapted to the team, to the machine and to the firm.

An important aspect that was not explicitly addressed in this chapter was how knowledge management among teams underpins performance. Thus, future works need to highlight the flexibility of SMED teams, the need to use a knowledge-based approach to properly disseminate the SMED methodology within the company, the consequences of SMED in the design of new machinery and, the economic importance of inventory reduction to the firm.

One of the limitations of this study that deserves a closer analysis is that the SMED teams were not addressed. In fact, most of the studies address the results achieved but not how the teams are organized and what types of teams are best suitable in SMED projects. Thus, future studies should address the type of organizational structure of SMED teams within the production/industrial organization structure, the role of SMED teams in the implementation of brand new solutions, and the knowledge absorption capacity of the SMED team.

Another important future challenge would be to address SMED projects within a broader focus of continuous improvement, such as, involving lean manufacturing techniques of visual stream mapping, pull systems, manufacturing cells, total productive maintenance, and 5S systems, among others.

References

Anderson, N., De Dreu, C., & Nijstad, B. (2004). The routinization of innovation research: A constructively critical review of the state-of-the-science. *Journal of Organizational Behavior, 25*(2), 147–173.

Blau, J. R. (1994). European carmakers turn lean to mean. *Machine Design, 66*(10), 26–32.

Cusumano, M. A. (1989). *The Japanese automobile industry: Technology and management at Nissan and Toyota*. Boston, Massachusetts: Harvard East Asia Monografs.

Damanpour, F. (1991). Organizational innovation: A meta-analysis of effects of determinants and moderators. *Academy of Management Journal, 34*(3), 555–590.

Damanpour, F. (1992). Organizational size and innovation. *Organization Studies, 13*(3), 375–402.

Damanpour, F. (1996). Organizational complexity and innovation: Developing and testing multiple contingency models. *Management Science, 42*(5), 693–716.

Damanpour, F., & Evan, W. (1984). Organizational innovation and performance: The problem of "organizational lag". *Administrative Science Quarterly, 29*(3), 392–409.

Damanpour, F., & Gopalakrishnan, S. (2001). The dynamics of the adoption of product and process innovations in organizations. *Journal of Management Studies, 38*(1), 45–65.

Dantas, J., & Moreira, A. C. (2011). *O Processo de Inovação*. Lisbon: Lidel.

Ettlie, J. E., & Reza, E. M. (1992). Organizational integration and process innovation. *Academy of Management Journal, 35*, 795–827.

Fogliatto, F. S., & Fagundes, P. R. (2003). Troca rápida de ferramentas: Proposta metodológica e estudo de caso. *Gestão e Produção, 10*(2), 163–181.

Freire, A. (1995). *Gestão Empresarial Japonesa. Lições para Portugal*. Lisbon: Verbo.

Frost, P. J., & Egri, C. P. (1991). The political process of innovation. In L. L. Cummings & B. M. Staw (Eds.), *Research in organizational behavior* (pp. 229–295). Greenwich, CT: JAI Press.
Gest, G., McIntosh, R. I., Mileham, A. R., & Owen, G. W. (1995). Review of fast tool change systems. *Computer Integrated Manufacturing Systems, 8*(3), 205–210.
Gilmore, M., & Smith, D. (1996). Setup reduction in pharmaceutical manufacturing: An action research study. *International Journal of Production Research, 16*(3), 4–17.
Godinho Filho, M., & Fernandes, F. C. (2004). Manufatura enxuta: Uma revisão que classifica e analisa os trabalhos apontando perspetivas futuras. *Gestão & Produção, 11*(1), 1–19.
Gopalakrishnan, S., & Damanpour, F. (1997). A review of innovation research in economics, sociology and technology management. *Omega, 25*(1), 15–28.
Hicks, B. J. (2007). Lean information management: Understanding and eliminating waste. *International Journal of Information Management, 27*, 233–249.
Holweg, M. (2007). The genealogy of lean production. *Journal of Operations Management, 25*, 420–437.
Johansen, P., & McGuire, K. J. (1986). A lesson in SMED with Shigeo Shingo. *Industrial Engineering, 18*, 26–33.
Klein, K. J., & Knight, A. P. (2005). Innovation implementation: Overcoming the challenge. *Current Directions in Psychological Science, 14*, 243–246.
Lamming, R. (1993). *Beyond partnership. Strategies for innovation and lean supply*. Cornwall: Prentice-Hall.
Levinson, W. A. (2002). *Henry Ford's lean vision: Enduring principles from the first Ford Motor Plant*. New York: Productivity Press.
Liker, J. K. (2004). *The Toyota way: 14 management principles from the world's greatest manufacturer*. New York: McGraw-Hill.
McIntosh, R., Culley, S., Gest, G., Mileham, T., & Owen, G. W. (1996). An assessment of the role of design in the improvement of changeover performance. *International Journal of Operations and Production Management, 16*(9), 5–22.
McIntosh, R., Culley, S., Mileham, T., & Owen, G. (2000). A critical evaluation of Shingo's "SMED" (Single Minute Exchange of Die) methodology. *International Journal of Production Research, 38*(11), 2377–2395.
McIntosh, R. I., Culley, S. J., Mileham, A. R., & Owen, G. W. (2001). Changeover improvement: A maintenance perspective. *International Journal of Production Economics, 73*(2), 153–163.
Moxan, C., & Greatbanks, R. (2001). Prerequisites for the implementation of SMED methodology. A study in the textile-processing environment. *International Journal of Quality & Reliability Management, 18*(4/5), 404–414.
Monden, Y. (1984). *Produção Sem Estoques: Uma Abordagem Prática ao Sistema de Produção da Toyota*. São Paulo: IMAM.
Moreira, A. C., & Garcez, P. (2013). Implementation of the single minute exchange of die (SMED) methodology in small to medium-sized enterprises: A Portuguese case study. *International Journal of Management, 30*(1), 66–87.
Moreira, A.C., & Pais, G. (2011). Single minute exchange of die. A case study implementation. *Journal of Technology Management and Innovation, 6*(1), 129–146.
Neumann, C., & Ribeiro, J. L. (2004). Desenvolvimento de fornecedores: Um estudo de caso utilizando a troca rápida de ferramentas. *Produção, 14*(1), 44–53.
Noaker, P. (1991). Pressed to reduce *setup*? *Manufacturing Engineering, 107*, 45–49.
Nishiguchi, T. (1994). *Strategic Industrial Sourcing. The Japanese Advantage*. Oxford: Oxford University Press.
Pardi, T. (2005). Crisis, paths dependency and social dynamics in the evolution of Toyota manufacturing UK. *Sociologie du Travail, 47*, 188–204.
Pittaway, L., Robertson, M., Munir, K., Denyer, D., & Neely, A. (2004). Networking and innovation: A systematic review of the evidence. *International Journal of Management Reviews, 5*(3–4), 137–168.

Pisano, G., & Hayes, R. (1995). *Manufacturing Renaissance*. Boston, Massachusetts: Harvard Business School Press.

Quinlan, J.P. (1987). Shigeo Shingo explains 'Single-minute Exchange of Die'. *Tooling and Production*, Feb., 67–71.

Satolo, G. E., & Calarge, F. A. (2008). Troca rápida de ferramentas: Estudo de casos em diferentes segmentos industriais. *Exacta, 6*, 283–296.

Shingo, S. (1985). *A revolution in manufacturing: The SMED system*. Cambridge, Massachusetts: Productivity Press.

Sepheri, P. (1987). Manufacturing revitalization at Harley Davidson motor company. *Industrial Engineering, 19*(8), 26–33.

Sugai, M., McIntosh, R., & Novaski, O. (2007). Metodologia Shigeo Shingo (SMED): Análise crítica e estudo de caso. *Gestão e Produção, 14*, 323–335.

Van Goubergen, D., & Van Landeghem, H. (2002). Rules for integrating fast changeover capabilities into new equipment design. *Robotic and Computer Integrated Manufacturing, 18*, 205–214.

Womack, J. P., & Jones, D. T. (1994). From lean production to lean enterprise. *Harvard Business Review, 72*(2), 93–103.

Womack, J. P., Jones, D. T., & Ross, D. (1990). *The machine that changed the world*. London: Macmillan.

Yin, R. (1989). *Case study research*. Beverly Hills: Sage.

Zaltman, G., Duncan, R., & Holbek, J. (1973). *Innovations and organizations*. New York: Wiley.

Chapter 24
Process Control Adjustment with Feedback Controller

Rey David Molina-Arredondo

Abstract Industrial Statisticians frequently face problems in their practice where adjustment of a manufacturing process is necessary (Del Castillo 2006). In this chapter, are presented some schemes of control settings from feedback controller, in the development we focus on drivers exponentially weighted moving average-simple (sEWMA) and doubles (dEWMA) for the single controllable factor (or input), single quality characteristic (or ouput) (SISO) and for multivariate double EWMA controller for squared multiple-input, multiple-output (MIMO) processes. One of the course in this chapter is that the parameters of the model are known and it is possible to have the characteristics of quality measures once it ends the run and before you start the next, in practice this is often not possible, some authors Khan et al. (2008) and Kang et al. (2011) use Virtual metrology as a method to augment existing metrology and has the potential to be used in control schemes for improved process control in terms of both accuracy and speed. The chapter starts with an introduction processes statistical adjustments, the effect of the auto correlation in statistical process control, an introduction to the engineering of process control to finish with EWMA control schemes.

Keywords Process optimization · Process control · Feedback controller

24.1 Introduction

Statistical Process Control (SPC) and Engineering Control Process (EPC) are two techniques widely used to control processes; the goal of these techniques is to control a process such that the average performance is as close as possible to a

R. D. Molina-Arredondo (✉)
Department of Industrial Engineering and Manufacturing,
Institute of Engineering and Technology—Autonomous University of Ciudad Juarez,
Av. Del Charro 450 Norte. Col. Partido Romero, Ciudad Juárez, 32310 Chihuahua, Mexico
e-mail: rey.molina@uacj.mx

J. L. García-Alcaraz et al. (eds.), *Lean Manufacturing in the Developing World*,
DOI: 10.1007/978-3-319-04951-9_24,

target value with a minimum variation. Some authors such as Del Castillo (2002) and Box and Luceno (2009) give a widespread introduction of these techniques. The main difference between these two techniques is that while the SPC monitors properties of the quality characteristic, and in case of an alarm signal, search for the causes of this variation to eliminate the root, the EPC makes adjustments in the control variable influencing the variable response and thus minimize the difference of this with the target value.

Within the EPC techniques, the feedback processing settings EWMA have gained popularity in the manufacture of semiconductors where production is batch-to-batch and have trends (drift). Although this control scheme can also be applied in some processes with batch control discrete parts manufacturing batch-to-batch where there are no elements of inertia, this type of control scheme is to be simple enough for the operator to implement and they are robust with respect to certain assumptions of the model Del Castillo (1999). Box and Luceño (2000) shows how these schemes of control can be used to keep the processes within 6 standard deviations. EWMA feedback settings belong to the field of adjustments **statistical** process (SPA), located at the intersection of Control Theory, Time Series Analysis, and Statistical Process Control (Del Castillo 2006).

Del Castillo (2006) defines the statistical process adjustment as the set of Statistical Techniques aimed at modelling, and hence, forecasting and controlling a dynamic process. Two distinctive characteristics of SPA are:

- That process the responses relate to quality characteristics of a product (or of a process producing it), and
- The implementation of the adjustments is not fully automatic since SPA corresponds to higher-level supervisory controller.

This chapter presents a brief description of the schemes of control by feedback EWMA, starting first for auto correlation and statistical process control, a brief introduction to the control engineering and finally we conclude with EWMA control schemes, some examples are presented.

24.2 Autocorrelation and Statistical Process Control

Statistical process control is based on the assumption that when the process is in control the average of the process remains constant and the variation is due only to the white noise (common causes of variation), when an observation is far from the average it is said that the process is out of control then seeks the cause that originated far out of control and is eliminated (Box and Luceno 2009). In a control chart type Shewart we must define what so far must be the observation to be considered a cause of variation attributable to it. Figure 24.1 shows an example of a process with a very distant point from the average, variation of quality feature is only due to the white noise to exception of point 10, here we must investigate the cause of this variation and eliminate it.

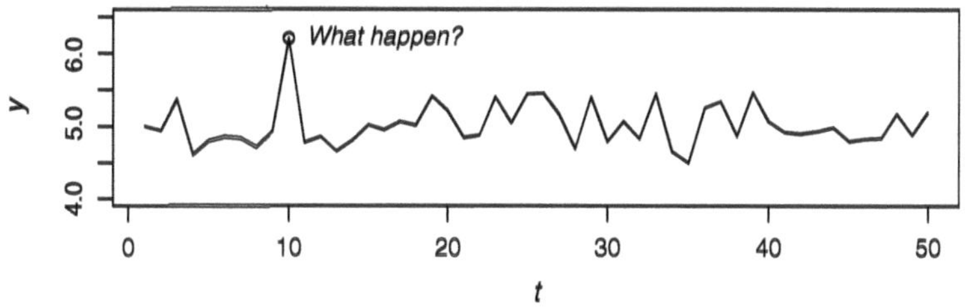

Fig. 24.1 Shewart process with one point out control

The model describing a process under control type Shewhart is model as Eq. 24.1.

$$y = \mu + \varepsilon \tag{24.1}$$

where y_t is the quality characteristic or response at time t, μ is the average of the process and ε_t is a sequence of random error (white noise) normally and independently distributed with mean zero and variance in σ^2, the same way you can say that $y \sim N(\mu, \sigma)$.

In some processes the effect of noise factors such as the wear of tool, environmental factors such as temperature, humidity, etc. can affect quality property and make this follow the same pattern of autocorrelation that the noise factor, even when the effect is a combination of more than one factor of noise (Molina et al. 2013).

When the response variable is autocorrelated, the dynamic model of the characteristic of quality can be represented as

$$y_t = \mu + N_t \tag{24.2}$$

where N_t is the noise not white (disturbance) of the process. Del Castillo (1999) presents three popular times series models for random perturbations found in some manufacturing processes, this models are the following:

- Deterministic trend (DT) model

$$N_t = \delta t + \varepsilon_t \tag{24.3}$$

- Random walk with drift (RWD) model

$$N_t = N_{t-1} + \delta + \varepsilon_t \tag{24.4}$$

- Intregated moving average IMA (1,1) model

$$N_t = N_{t-1} - \theta\varepsilon_{t-1} + \varepsilon_t \tag{24.5}$$

where ε_t denotes a sequence of normal and independently distributed white noise with $E[\varepsilon_t] = 0$ and $Var[\varepsilon_t] = \sigma^2$, the amount δ in DT and RWD models represents the drift expected per unit of time, which makes these model series with unidirectional tendency, in the IMA model, θ is the moving average parameters.

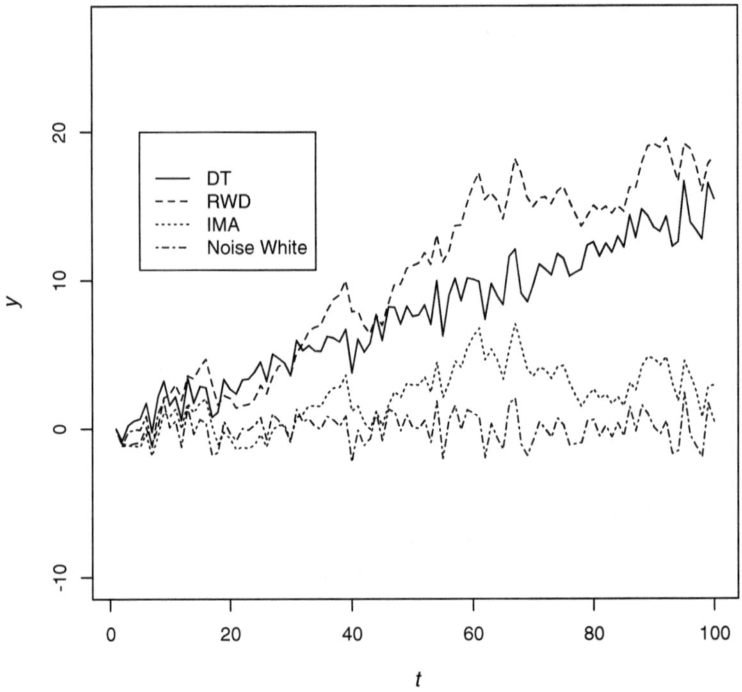

Fig. 24.2 Realization for DT, RWD and IMA disturbance

Figure 24.2 shows the simulation of a realizations with 100 points for each disturbance brought in Eqs. 24.3–24.5 for Dt, RWD and IMA models respectively, data were simulated with $\delta = 0.15$, $\theta = 0.2$ and the variance of the error $\sigma = 1$, in order to make comparison 3 realizations were simulated with the same random numbers, or white noise.

It is clear to note that the DT model adheres more to a line with slope δ the variations are due to the error, while the RWD model is more erratic but going adrift in the same address that DT model, instead the IMA model behaves erratically without any definite trend, the IMA can be downward or upward trend.

24.2.1 Effect of Correlation in an SPC Chart

As we was mentioned earlier one of the assumptions of the SPC is that the response variable is not autocorrelated, to give us an idea of the effect that has the autocorrelation in the performance of an SPC to consider the following example. Suppose that a process can be described as:

$$y_t = \mu + N_t \tag{24.6}$$

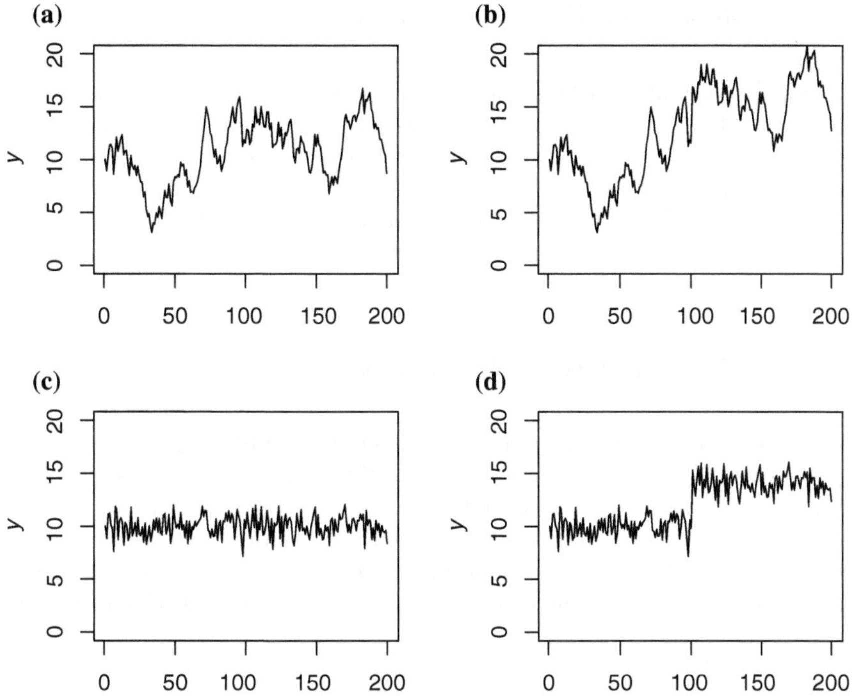

Fig. 24.3 Process with **a** Fixed average autocorrelation, **b** Autocorrelation, average change, **c** Without autocorrelation fixed average and **d** Autocorrelation and change in the average

with

$$N_t = \phi N_{t-1} + \varepsilon_t \tag{24.7}$$

The model of Eq. 24.7 is an autoregressive series model AR(1), the amount ϕ is the autoregressive parameter. It is clear that if $\phi = 0$ the process is a Shewhart process, in contrast if $\phi = 1$ the process is a random Walk. Now suppose that 200 observations are obtained, during the first 100 samples the process mean is maintained in 10 and after the process mean changes in 4σ with $\sigma = 1$. The Fig. 24.3 shows the process behavior when (a) with $\phi = 0$ and there is no change in the average, (b) with $\phi = 0$ and change in the average of 4σ, (c) with $\phi = 0.96$ and without change in the average and (d) with $\phi = 0.96$, and with the change in the average of 4σ. It is clear that when there is no correlation in the process a change of the average is easily detectable, while when there is autocorrelation change in the average is not so easily detected.

24.2.2 Engineering Processes Control

A controller is a rule, function or algorithm that describes as a control factor u_t in a process needs to be adjusted from observation to observation, on the application of quality control the controller are usually implemented manually by an operator once you have the data (Del Castillo 2002).

The controller aims to make adjustments a factor of control with the aim of influencing the response in such a way that this putting to the target value.

24.3 Feedback EWMA Adjustments

One of the schemes more used in process control is the EWMA controller, this is due to easy application, furthermore, the EWMA controller is robust to different disturbances of the process. The method consists of a forecast of the compensation for the run $t + 1$ necessary to reduce the difference between the response and its target value (τ), then, makes an adjustment to the control factor to offset the trend in each run, In the subsequent sections some variants of this controller are presents.

24.3.1 The Simple EWMA Controller

Assume that the input output of a manufacturing process can be modelled by:

$$y_t = \alpha + \beta u_{t-1} + N_t \tag{24.8}$$

where y_t denotes the response of the characteristic of quality in the run t, and N_t is the disturbance of the process. The Eq. 24.8 is a simple regression model where the parameters α and β are estimated offline. The simple EWMA controller is given by:

$$u_t = \frac{\tau - a_t}{b} \tag{24.9}$$

where

$$a_t = \lambda(Y_t - bu_{t-1}) + (1 - \lambda)a_{t-1}. \tag{24.10}$$

The amount b is an off line estimate of β, the Eq. 24.9 assumes that $b = \beta$ and a_t in the Eq. 24.10 is an EWMA to compensate the next run, in the Eq. 24.10 the current observation has a weight of λ and the forecast of the previous settlement a_{t-1} has an weighing of $(1 - \lambda)$, in other words, the amount $\alpha + N_t$ would be directly observable if we knew β perfectly.

Table 24.1 Results of the sEWMA controller for hipothetical case

t	y_t	a_t	u_t	∇u_t
0	–	32	0	–
1	33.04	32.41	−0.80	−0.80
2	31.33	32.95	−0.98	−0.18
3	31.78	33.66	−1.22	−0.24
4	31.02	34.06	−1.22	−0.14
5	30.49	34.26	−1.42	−0.06
6	30.72	34.55	−1.52	−0.10
7	30.57	34.78	−1.59	−0.08
8	29.96	34.77	−1.59	0.00
9	30.24	34.86	−1.62	−0.03
10	30.07	34.89	−1.63	−0.01
.	.	.	.	.
.	.	.	.	.
.	.	.	.	.

24.3.2 Example

Suppose that a manufacturing process a $\lambda = 0.4$ has been used to predict the compensation of a process with a $b = \beta = 2$ and $\alpha = 32$. The target value of the output of the process is 10. Table 24.1 shows the calculations made for these parameters of the hypothetical case, the initial value of a_1 is 32, it is assumed that this value was calculated offline, as every measure of the characteristic of quality is available the value a_t is updated with the Eq. 24.10. The setting of the factor of control for the next run is calculated accordance as:

$$u_t = \frac{30 - a_t}{3}.$$

Figure 24.4 shows the simulation of 100 runs for the hypothetical process described above; the top graph shows the process with the simple EWMA controller, while the bottom figure shows the performance without the controller (leaving the process to drift). It is clear that with the controller EWMA the response is closest to the target value of 30, while without the control strategy response goes adrift.

One of the main problems in the use of the EWMA controller is the choice of the weighting λ, this amount should be selected in such way that the mean square deviation MSD in Eq. 24.11 to be reduced, will be shown later as the weighting to reduce MSD value can be found when we know the time series modeling the perturbation.

$$MSD = E(y_t - \tau)^2 \tag{24.11}$$

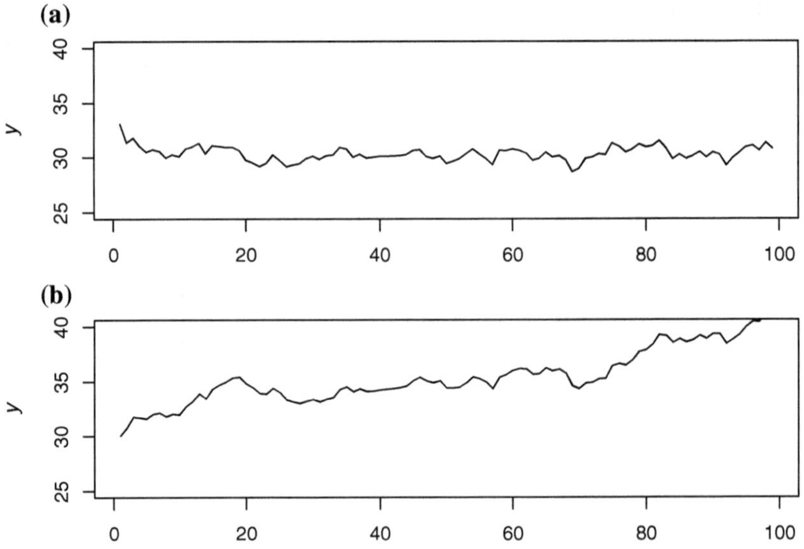

Fig. 24.4 Process with controller (**a**) and without controller (**b**)

Table 24.2 Cuadrado medio de la desviación calculado para diferentes ponderaciones

λ	0.25	0.30	0.35	0.40	0.45	0.50	0.55	0.60	0.65
MSD	0.80	0.67	0.59	0.53	0.48	0.44	0.41	0.39	0.37
λ	0.70	0.75	0.80	0.85	0.90	0.95	1.00		
MSD	0.35	0.34	0.33	0.32	0.31	0.30	0.30		

Table 24.2 shows the results of calculating the MSD for values of λ from 0.25 to 1.00, it is possible to notice that as it increases the value of the weighting the MSD decreases, which means that the estimate of a_t ponders more the first part of the Eq. 24.10 that the historical information of a_t.

When it comes to measuring the performance of a controller, it is best to use the asymptotic mean square deviation (AMSD) Eq. 24.2. The word *asymtotic* is used because this type of controller there are transient effects due to initialization of the EWMA equation than need to be accounted separately from steady-state behavior.

$$AMSD(y_t) = \lim_{t \to \infty} E(y_t - \tau)^2 \tag{24.12}$$

If the disturbance is RWD as of the Eq. 24.4, Del Castillo (1999) shows that the asymptotic square deviation is given by the Eq. 24.13

$$AMSD_{RWD} = \frac{\sigma^2}{\xi\lambda(1 - \xi\lambda)} - \frac{\sigma^2}{\xi^2\lambda^2} \tag{24.13}$$

where $\xi = \beta/b$ and the condition to obtain a output stable under the action of a controller is

$$|1 - \xi\lambda| < 1 \tag{24.14}$$

With the Eq. 24.13 it is possible to find the value of that minimizes the average asymptotic square deviation as

$$\lambda^* = \frac{4\delta^2 - \sigma^2 - \sigma\sqrt{8\delta^2 + \sigma^2}}{2(\delta^2 - \sigma^2)\xi} \tag{24.15}$$

In the above example the disturbance is RWD with $\sigma^2 = 0.2$ and assuming that $b = \beta$, from the Eq. 24.15 the sEWMA parameter λ^* is 0.97. The asymptotic mean square deviation obtained from Eq. 24.13 is AMSD = 0.22, very similar to the calculated in Table 24.2. When perturbations type are IMA or DT, equations to calculate the AMSD and weighting λ^* we can found it in (Del Castillo 2002). Some authors (Khan et al. 2008) show that the value of λ^* can be vary in the first runs to reduce MSD.

The variance of the adjustment to the control factor can be calculated as

$$Var(\nabla u_t) = \frac{\lambda}{b^2\xi(2 - \lambda\xi)}$$

Assuming that the cost of adjustment is important and that ρ is the relative cost of adjusting a unit of u. The choice of λ can be found to solve the following problem.

$$\operatorname*{Min}_{\lambda} J = \frac{AMSD(y_t)}{\sigma^2} + \rho\frac{\text{var}(\nabla u_t)}{\sigma^2}$$

Subject to

$$|1 - \xi\lambda| < 1$$

24.3.3 The Double EWMA Controller

When there is considerable offset that is not fully compensated with a simple EWMA is necessary to modify the controller to be removed. Stephanie and Jerry (1994) proposes the use of a double EWMA (dEWMA) for processes with considerable offset. The dEWMA driver is written as:

$$u_t = \frac{\tau - a_t - D_t}{b} \tag{24.16}$$

with

Table 24.3 Results of dEWMA controller

t	y_t	a_t	D_t	u_t
0	–	47	0	–
1	46.17	46.77	−0.06	0.82
2	50.57	46.91	−0.02	0.78
3	50.84	47.13	0.04	0.71
4	50.83	47.38	0.01	0.63
5	50.27	47.48	0.12	0.6
6	50.11	47.55	0.13	0.58

$$a_t = \lambda_1(y_t - bu_{t-1}) + (1 - \lambda_1)a_{t-1} \tag{24.17}$$

and

$$D_t = \lambda_2(y_t - bu_{t-1} - a_{t-1}) + (1 - \lambda_2)D_{t-1} \tag{24.18}$$

24.3.4 Example

Suppose that double-EWMA controller are used to control a process with weighing $\lambda_1 = 0.3$ and $\lambda_2 = 0.1$, the parameters of the model are $b = 5$, $\alpha = 47$ and the target value is $\tau = 50$. Table 24.3 shows the results obtained by using the dEWMA controller. The initial value of $a_t = 47$ is elected by assuming that it was off-line estimation of α, in Table 24.3 it is possible to observe the setting value diminishes with increasing periods t and the compensation D_t it increases with increasing time **also**.

In Fig. 24.5 you can appreciate as the line that represents the response using the Controller dEWMA this more attached to the target value that represents it to the sEWMA Controller, the difference is due to the adjustment D_t.

As before the weightings in the sEWMA λ_1 and λ_2 they must be chosen to reduce the AMSD. Del Castillo (2001) presents the conditions to have a stable output of the process when using a dEWMA Controller.

$$|1 - 0.5\xi(\lambda_1 + \lambda_2) + 0.5z| < 1$$
$$|1 - 0.5\xi(\lambda_1 + \lambda_2) - 0.5z| < 1$$

With

$$z = \sqrt{\xi\left[\xi(\lambda_1 + \lambda_2)^2 - 4\lambda_1\lambda_2\right]}$$

Thus the process will be asymptotically stable if and only if the previous conditions are met.

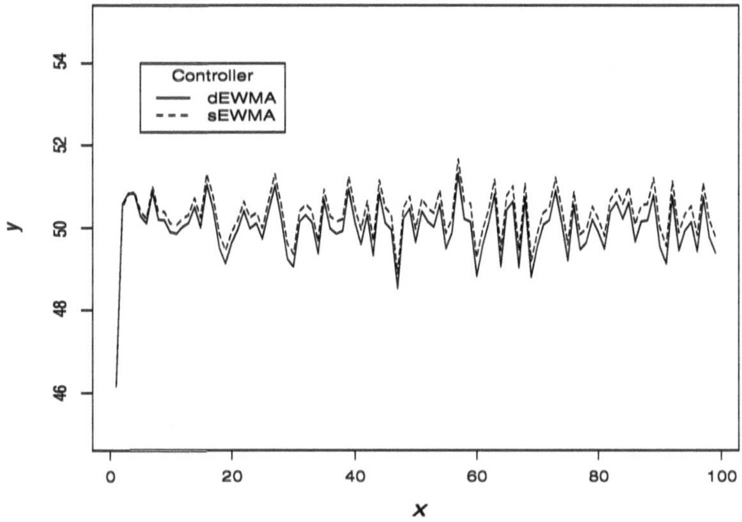

Fig. 24.5 Results for a sEWMA (*dotted line*) and a dEWMA (*solid line*)

24.4 The MIMO System

In many situations in statistical control there is more than one feature of quality and could also be more of one control factor that affects this feature of quality to this type of system is called multiple inputs, multiple outputs (MIMO), assuming the deterministic type perturbations trend (DT) the assumed model is like the one presented by Del Castillo and Rajagopal (2002):

$$Y_t = \alpha + \beta U_{t-1} + \delta_t + \varepsilon_t \tag{24.19}$$

where Y_t is a $p \times 1$ vector of responses measured in the run or moment of time t, U_t is a $m \times 1$ vector of levels of controllable factors adjusted at the end of the run $t - 1$, δ is a diagonal matrix which contains the average trend for run for each of the answers, and $\{\varepsilon\}_{t=1}^{\infty}$ is a sequence of multivariate white noise.

The MIMO controller dEWMA present by Del Castillo and Rajagopal (2002) and then spread to the dEWMA case by Rajagopal and Del Castillo (2003) is presented in Eq. 24.20

$$\mathbf{U_t}{=}(\mathbf{B}'\mathbf{B})^{-1}\mathbf{B}'(\mathbf{T} - \mathbf{A_t} - \mathbf{D_t}) \tag{24.20}$$

where **B** is the matrix of parameters estimator β, **T** is the vector of objective values for the response vector Y, and $\mathbf{A_t}$ and $\mathbf{D}_t$ are estimated on line using Eqs. 24.21 and 24.22 of the dEWMA at the end of each run.

$$A_t = \Lambda_1(\mathbf{Y_t}{-}\mathbf{BU_{t-1}}) + (1 - \Lambda_1)\mathbf{A}_{t-1} \tag{24.21}$$

Table 24.4 Results for a MIME dEWMA

t	y_t	A_t	D_t	U_t
0	–	$\begin{pmatrix}1600\\250\end{pmatrix}$	$\begin{pmatrix}0\\0\end{pmatrix}$	$\begin{pmatrix}3\\0.01\\0.22\\-1.05\end{pmatrix}$
1	$\begin{pmatrix}2013.84\\52.23\end{pmatrix}$	$\begin{pmatrix}1602.08\\242.83\end{pmatrix}$	$\begin{pmatrix}4.15\\-14.33\end{pmatrix}$	$\begin{pmatrix}2.68\\0.05\\0.28\\-0.88\end{pmatrix}$
2	$\begin{pmatrix}2012.56\\134.15\end{pmatrix}$	$\begin{pmatrix}1604.59\\242.83\end{pmatrix}$	$\begin{pmatrix}7.93\\-4.09\end{pmatrix}$	$\begin{pmatrix}2.85\\0.05\\0.23\\-0.99\end{pmatrix}$
3	$\begin{pmatrix}1936.33\\150.57\end{pmatrix}$	$\begin{pmatrix}1596.23\\252.78\end{pmatrix}$	$\begin{pmatrix}-11.16\\11.16\end{pmatrix}$	$\begin{pmatrix}3.23\\-0.03\\0.21\\-1.15\end{pmatrix}$
4	$\begin{pmatrix}2021.84\\56.48\end{pmatrix}$	$\begin{pmatrix}1597.83\\247.91\end{pmatrix}$	$\begin{pmatrix}-4.61\\-1.97\end{pmatrix}$	$\begin{pmatrix}2.96\\0.01\\0.25\\-1.17\end{pmatrix}$
5	$\begin{pmatrix}2100.40\\143.17\end{pmatrix}$	$\begin{pmatrix}1612.19\\254.09\end{pmatrix}$	$\begin{pmatrix}25.50\\10.98\end{pmatrix}$	$\begin{pmatrix}3.13\\-0.98\\0.11\\-1.16\end{pmatrix}$
6	$\begin{pmatrix}1971.16\\112.73\end{pmatrix}$	$\begin{pmatrix}1611.70\\257.65\end{pmatrix}$	$\begin{pmatrix}16.85\\14.80\end{pmatrix}$	$\begin{pmatrix}3.25\\-0.11\\0.10\\-1.22\end{pmatrix}$
7	$\begin{pmatrix}1911.88\\54.56\end{pmatrix}$	$\begin{pmatrix}1601.00\\253.06\end{pmatrix}$	$\begin{pmatrix}-9.587\\1.17\end{pmatrix}$	$\begin{pmatrix}3.07\\-0.01\\0.23\\-1.08\end{pmatrix}$

and

$$\mathbf{D}_t = \Lambda_2(\mathbf{Y_t} - \mathbf{BU}_{t-1} - \mathbf{A}_{t-1}) + (\mathbf{I} - \Lambda_2)\mathbf{D}_{t-1} \tag{24.22}$$

where Λ_1 and Λ_2 they are two matrices of weightings for the dEWMA controller. When the system is not square, i.e., the number of factors of control is different from the number of quality characteristics, the quantity $(\mathbf{B}'\mathbf{B})$ in 24.20 it is not invertible so that the controller cannot be calculated, Rajagopal and Del Castillo (2003) presents a solution to this problem using a ridge controller:

$$\mathbf{U}_t = (\mathbf{B}'\mathbf{B} - \mu\mathbf{I})^{-1}\mathbf{B}'(\mathbf{T} - \mathbf{A_t} - \mathbf{D_t}) \tag{24.23}$$

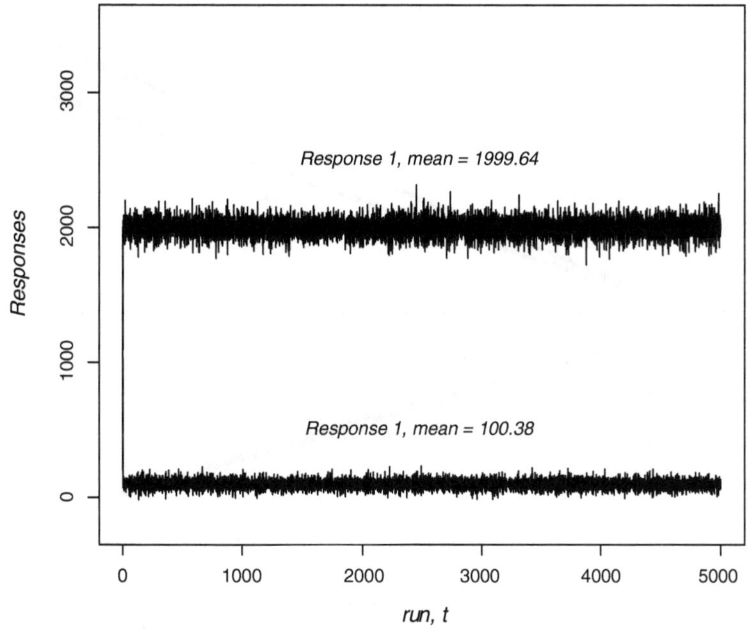

Fig. 24.6 Result of responses under the dEWMA Controller

The amount μI ago switchable to $(B'B - \mu I)$ as much as $\mu I \neq 0$, given that the choice of U_t is a problem of optimization, the amount μI is chosen such that the matrix is positive semi-defined.

24.4.1 Example

Rajagopal and Del Castillo (2003) presented an example of a semiconductor manufacturing process called Chemical—Mechanical Polishing (CMP). There are four controllable factors: plate speed (U_1), back (U_2) pressure, polishing down-force (U_3), and the profile of the conditioning system (U_4). The two responses of interest are removal rate (Y_1) and within-wafer non-uniformity (Y_2). The target values for the responses are 2000 and 100 respectively. Assuming the models for the answers are as follows

$$Y_1 = 1600 + 160U_1 - 40U_2 + 180U_3 + 25U_4 + N_{1,t} \tag{24.24}$$

$$Y_1 = 250 + 250U_1 + 100U_2 + 30U_3 + 35U_4 + N_{2,t} \tag{24.25}$$

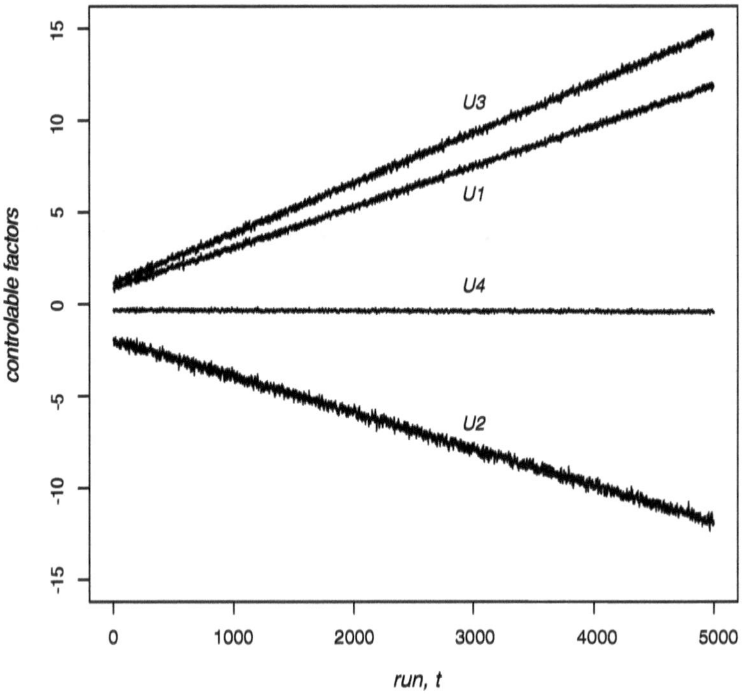

Fig. 24.7 Variation in control factors with the dEWMA controller

The parameter matrix is

$$\mathbf{B} = \begin{bmatrix} 150 & -40 & 180 & 25 \\ 250 & 100 & 30 & 35 \end{bmatrix}$$

The initial value of $\mathbf{A_t}$ was set at [1600, 250] based on Eqs. 24.24 and 24.25, and D_t was initialized at [0, 0]. The gain matrix, **B** was taken from Eqs. 24.24 and 24.25 too. Targets were set at 2000 for Y_1 and 100 for Y_2. The Lagrange multiplier μ was set at 0.001 in case of the ridge-solution controller. The process was simulated with EWMA weights $\Lambda_1 = (0.15)$ and $\Lambda_2 = (0.35)$. With the aim of showing the operation of the dEWMA controller for a MIMO system was simulated 5000 runs for each response using the same pseudo random numbers. Control factors were adjusted in each run using the dEWMA controller for MIMO. First 7 runs results are presented in Table 24.4, the initial value of the setting A_t is equal to the estimate of α, and D_0 values are zero, the control factors were adjusted to adhere to the target value according to the model of the Eqs. 24.24 and 24.25.

The results of simulations 5000 runs can be seen in Fig. 24.6, for the response 1, the average is 1999.64, very close to the average to the target value, while for the response 2 the average simulation is 100.38, also very close to the target value.

Assuming that the process is maintained for 5000 runs and that the control variables can be adjusted at any interval, the Fig. 24.7 shows the variation presented in the control variables, is possible to note that factor 3 and 1 increase to offset the trend, while the factor 4 varies very little and the factor 2 decreases to offset the trend.

References

Box, G. E., & Luceño, A. (2000). Quality quandaries: Six sigma, process drift, capability indices, and feedback adjustment. *Quality Engineering, 12*(3), 297–302. http://www.tandfonline.com/doi/abs/10.1080/08982110008962592#preview.

Box, G. E., & Luceno, A. (2009). *Statistical control by monitoring and feedback adjustment.* New York: Wiley.

Del Castillo, E. (1999). Long run and transient analysis of a double EWMA feedback controller. *IIE Transactions, 31*(12), 1157–1169.

Del Castillo, E. (2001). Some properties of EWMA feedback quality adjustment schemes for drifting disturbances. *Journal of Quality Technology, 33*, 153–166.

Del Castillo, E. (2002). *Statistical process adjustment for quality control.* New York: Wiley.

Del Castillo, E. (2006). Statistical process adjustment: a brief retrospective, current status, and some opportunities for further work. *Statistica Neerlandica, 60*(3), 309–326.

Del Castillo, E., & Rajagopal, R. (2002). A multivariate double EWMA process adjustment scheme for drifting processes. *IIE Transactions, 34*(12), 1055–1068. http://link.springer.com/article/10.1023%2FA%3A1019614312908.

Kang, P., Kim, D., Lee, H. J., Seungyong, D., & Cho, S. (2011). Virtual metrology for run-to-run control in semiconductor manufacturing. *Expert Systems with Applications, 38*(3), 2508–2522.

Khan, A., Moyne, J., & Tilbury, D. (2008). Virtual metrology and feedback control for semiconductor manufacturing processes using recursive partial least squares. *Journal of Process Control, 18*(10), 961–974.

Molina, R. D., Ríos, J., & Piña, M. (2013). Performance of the EWMA controller in the presence of noise factors. *Proceedings of the 2nd Annual World Conference of the Society for Industrial and Systems Engineering*, November 5–7, 2013.

Rajagopal, R., & Del Castillo, E. (2003). An analysis and MIMO extension of a double EWMA run-to-run controller for non-squared systems. *International Journal of Reliability, Quality and Safety Engineering, 10*(04), 417–428.

Stephanie, W. B., & Jerry, A. S. (1994). Supervisory run-to-run control of polysilicon gate etch using in situ ellipsometry. *IEEE Transactions on Semiconductor Manufacturing, 7*(2), 193–201.

Chapter 25
Techniques and Attributes Used in the Supply Chain Performance Measurement: Tendencies

Liliana Avelar-Sosa, Jorge Luis García-Alcaraz and Miguel Gastón Cedillo-Campos

Abstract The evaluation and performance measurement in supply chain (SC) is a key element in modern business competitiveness and it is an activity extremely important for improve its process, practices and common goals. The purpose of this chapter is to present a statistical analysis based in a literature review related to supply chain tendencies that includes 95 chapters published from January 2000 to June 2012. This research identifies the techniques and methodologies used in the supply chain performance evaluation process, and attributes most commonly evaluated as well as the industrial sector and regions where the solutions were applied, and finally the journals where they were published. The results show that multivariate analysis is the most important and frequent methodology for study supply chain performance, looking to find variables relations and association among them. Also, the finds indicate that most evaluated attributes in supply chain performance are the delivery, information flow and processes activities. The most important research groups that are investigating this area are located in Unites States of America, Taiwan and United Kingdom.

Keywords Supply chain management (SCM) · Supply chain performance · Attributes in supply chain

L. Avelar-Sosa (✉) · J. L. García-Alcaraz
Institute of Engineering and Technology, Autonomous University of Ciudad Juarez, Av. del Charro 450 Norte. Col. Partido Romero 32315, Ciudad Juárez, Chihuahua, Mexico
e-mail: liliana.avelar@uacj.mx

M. G. Cedillo-Campos
Department of Logistics and Supply Chain Management, Autonomous University of Nuevo Leon, Av. Universidad s/n. Ciudad Universitaria 6645, San Nicolás de los Garza, Nuevo León, Mexico

J. L. García-Alcaraz et al. (eds.), *Lean Manufacturing in the Developing World*,
DOI: 10.1007/978-3-319-04951-9_25,

25.1 Introduction

25.1.1 Concepts

The supply chain (SC) concept appeared in the 1960s when Forrester suggested that the success of companies depended on the interaction among then, and the success of one company dependents from the success in others; and that interactions are given as information flow, materials flow, orders and money flow (Forrester 1961). Forrester (1961) has stated that these flows in SC have to be analyzed from a systemic approach, and recently Cedillo and Sánchez (2008) has confirmed that. Forty-four year later, Surana et al. (2005) have defined SC as a dynamic process that involves a complex flow of information and materials that are achieved through multiple functional areas within and outside enterprises, and as can see, those definitions are very similar, including only the areas involved. That kind of information, materials and money flow must be controlled by the enterprises and then some companies have created a department specialized in the Supply chain management (SCM) that is responsible for raw material procurement, inventory control into the production process, finished distributions process and the use of information technologies that allow a better integration of marketing, finance, engineering and operations within the company and outside the company (among partners), having the main goal to create or increase the organization's competitiveness.

Figure 25.1 shows an overview for a traditional supply chain with suppliers, customers, manufacturers and retailers involved in every part of the production process. The arrows indicate the flow of material or information that flows through companies, but when the arrow is not continuous, that phenomena can be considered as an interruptions flow, representing an opportunity area to be improved (Hishamuddin et al. 2013; Schmitt and Singh 2012).

25.1.2 Reasons to Study Methodologies and Techniques Applied for Performance Measurement in SC

Due that contemporary enterprises have become dependent on a large amount of goods and services to complete their own value-adding activities, most companies have developed extensive networks of suppliers and customers (Schliephake et al. 2009; Kim et al. 2013); but, to the extent that there are more partners in the supply chain network, then the number of production process increases and its complexity, causing vulnerability to disruptions in materials and information flow. Thus, a company that breaks these flows means that interrupt sources for another, because both are part of the same supply chain network and that is a critical problem that has direct and indirect effect among partners (Revilla and Sáenz 2013). This complexity and increasing globalization have lead many companies worldwide to be members

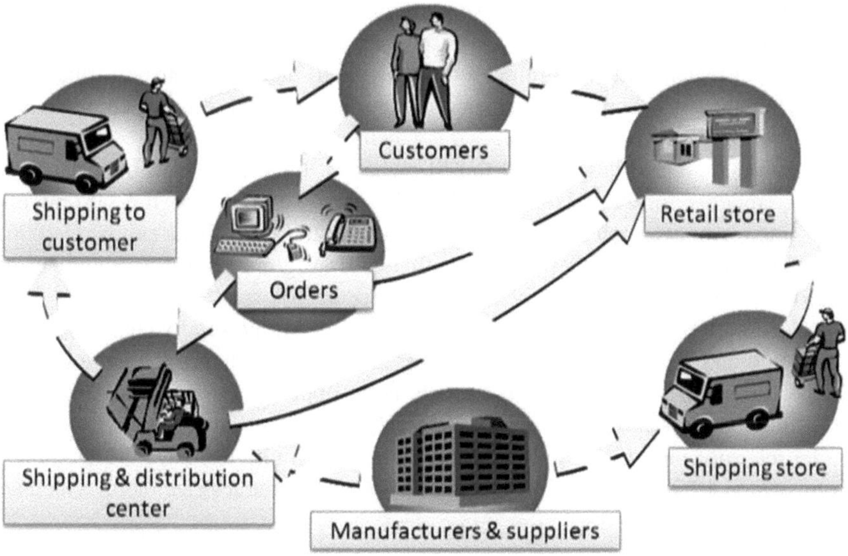

Fig. 25.1 General scheme of SC modified of (Santos 2010)

of any SC, which forces them to improve their internal structure, production processes, information systems and technology, in order to have more control and synchronization at any stage with other members (Świerczek 2013; Caniato et al. 2013).

So, the new business models must aim to provide effective and efficient predictable lead times among partners in SC (Díaz and Pérez 2002), avoiding any kind of disruptions flows. Then, this phenomenon is necessary to be studied, looking to have good enterprise performance and for achieve it, it's necessary to study the methodologies and techniques applied in SC for identify important procedures, activities and elements within the supply chain (Chen and Yan 2011; Vlachos 2014).

25.1.3 Reasons for Study the Performances in Supply Chain

According to last paragraphs, supply chain is a research area because it is a key element in the globalization of production operations and for it contribution to competitive advantage. Nowadays, companies do not compete in their production process; the hard competition is among their SC (Feng 2012). Importantly, the assessment of the competitiveness and performance of the supply chain must include suppliers and customers, due to their influence on the performance of the company (Yu et al. 2013; Droge et al. 2012). However, the performance should be measured not only within the company and its own SC; the performance must be extended to all companies within the supply chain to ensure sustainable and joined growth; but usually, the first thing companies do to measure their performance in supply chain, is to compare their performance indices with those of companies in the same industry (Elgazzar et al. 2012; Chen and Gong 2013).

Existing literature has studied the SC where it have noted the importance of support models for the design of a supply chain, considering the current and emerging elements, such as cost (Otto and Kotzab 2003; Guiffrida and Nagi 2006), collaboration among partners (Ramanathan 2014), globalization (Caniato et al. 2013) and corporate social responsibility (Cruz 2013).

All these previous models and approaches show that the supply chain should be monitored and should be generated indexes for it control; hence the evaluation of SC performance indices in all its aspects is an opportunity area to improve and keep the company in the desired competitive level.

To measure the SC performance there are used multiple indicators or dimensions, which are represented by a set of attributes that characterize them, which are discussed below.

25.1.4 Measured Attributes in Supply Chain

An attribute for SC performance measure is a set of metrics that are used to express a competitive strategy (SCC 10.0, 2010) and Green JR. and Inman (2005) had defined it as the SC ability for deliver products and services with good quality, on time, in precise amounts and minimizing costs. The control of that attributes must provide direct or indirect feedback information that help to have an adequate control of business processes (Abu-Suleiman et al. 2004; Chan 2003).

There two types of measurements to evaluate the overall performance of companies, financial and non-financial. The first considers indicators identified from economic and financial attributes, and second considered indicators regarding operational attributes. However, authors like Gunasekaran et al. (2001) notes that companies cannot reach equilibrium between those attributes classification because administrators and researchers have focused on measuring financial performance and others are focused in operative measurements only. Bhatnagar and Sohal (2005) had mentioned that literature has tended to emphasize on quantitative factors, but there is limited research with measures focused in operational competitiveness because some times are qualitative factors. Then, to diagnose a supply chain requires a performance analysis, utilizing assessment techniques include not only quantitative attributes, but also qualitative.

Jiménez and Hernández (2002) have mentioned attributes oriented to planning, material procurement, production and distribution as well as customer service. Where performance indicators are both qualitative and quantitative, to name a few: delivery time, cost of production, number of deliveries on time, transport cost and product quality, level of collaboration and integration, degree of effectiveness in programming systems and quality service.

Undoubtedly, having a monitoring system for attributes and SC performance indicators will allows companies to track the strategy that will help them:

- Transforming strategic planning into specific actions across the organization.
- Communicate the cause-effect that is present in the management process.
- Provide indicators consistent with the objectives and strategic initiatives.
- Improve the feedback level to correct the direction and intensity of resources allocated to each task, according to the gaps between goals and actual results observed.

25.1.5 The Problem and Research Objectives

There is enough previous research reports related to SC performance indicators, but have been directed to study attributes and/or variables separately using traditional methodologies and techniques, and are focused to inform a descriptive analysis (Boddy et al. 1998; Chen et al. 2011; Chong et al. 2009; Wu et al. 2006); however, those research has failed to point out which methods and techniques are the most used in the SC performance analysis. In addition, previous works are particular cases for the study of some attribute, and this work is grouping the studies of other authors and is aimed to identify which attributes are most important and studied in supply chain performance measurement. Besides the identification of methodologies and techniques used in supply chain evaluation, the industrial sectors and journals that publish supply chain studies.

In the literature review was considering works published from January 2000 to June 2012, this is due because previous studies cover the literature review regarding supply chain performance before 2000. The chapter is organized into four sections: after this introduction, appears the methodology used, which describes the development of this research, the third section shows the results, and finally the comments, discussion and conclusions are presented in the fourth section.

25.2 Methodology

This research was developed in three phases: the first phase was a literature review that involved the identification of items of interest in scientific database; the second was the creating a database for save information and the third was the information analysis.

25.2.1 *Literature Review*

The literature review was developed via journal databases review, where the search was performed by using the keywords "supply chain performance", "critical success factors", "supply chain" and "supply chain management". The articles were seeded according to objective research and based on the limitation of database that the authors had access, eliminating those chapters with low relevance.

25.2.1.1 Data Capture Using SPSS

A database was created in *Statistical Package for the Social Sciences* (SPSS®) because this software allows capturing the information and gets data analysis in an easy way.

The variables analyzed were the article's name, author's name, country to which belonged the first author, year of publication, journals' name and techniques used to measurement any performance attribute, the industrial sector in which the technique or group of techniques were applied, the university and department for the first author. Each item represented a variable investigated in the database designed to capture the information obtained.

25.2.2 *Information Analysis*

After all the data was input in SPSS®, the information was analyzed, bar charts were created for some of the most important variables, contingency tables were performed between variables and some interesting facts were obtained for important variables or elements.

25.3 Results

25.3.1 *General Information*

A total of 95 scientific articles were identified and selected from January 2000 to June 2012 and related to SC performance. As seen in Fig. 25.2, it is noted that a total of 21 articles were published until June 2012, while in 2011 there were a total of 18 in whole year. It's important to mention that for 2010 there were a total of 14 articles, and the amount decreases thereafter.

If the rate of scientific production in 2012 in relation to the evaluation of the performance of supply chain is maintained in this year, then surely this year will be

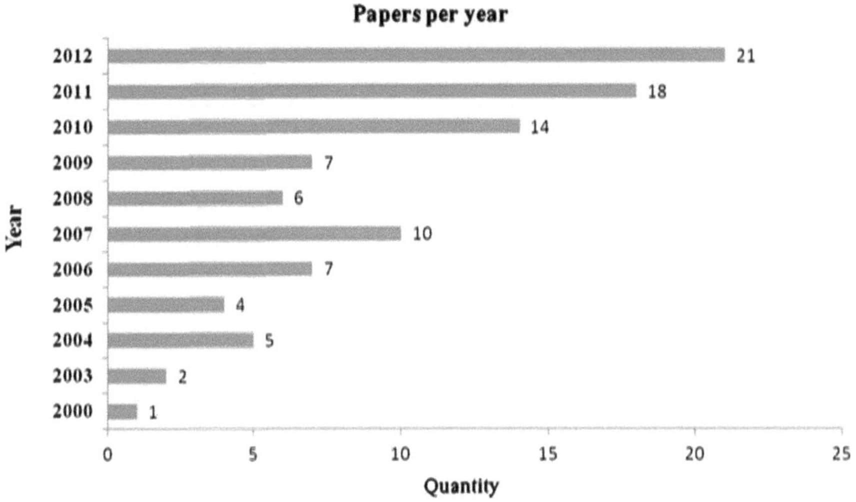

Fig. 25.2 Quantity of chapters per year for SC performance

more academic research in SC performance area than previous years, denoting a fast increase in scientific production rate due to a global interest in that research area.

25.3.2 The Methodologies Used for Supply Chain Performance Evaluation

This section is regarding the main methodologies used for performance evaluation in supply chain. Table 25.1 illustrates the main methodologies utilized per year. The multivariate analysis is the most widely used group of methodologies with 30 quotes in last 13 years. The second place was occupied by cases of study in different industrial sectors with 29 quotes and the third place is occupied by reviews with 17 quotes. There are others techniques applied for supply chain performance that have a low quotation, like quantitative analysis, multicriteria analysis and six sigma.

The Fig. 25.3 illustrates the techniques applied individually and how were used more than once by the researchers during their investigations to measure SC performance and note that the multivariate techniques has been ungrouped. The technique here is a procedure or set of rules, standards or protocols, which aim to achieve a particular result, whether in the field of science, education or any other activity. Is easy to see that *structural equations* is one of the most used multivariate technique used to study the supply chain performance of different problem according to our literature review, and this can be because using this technique is

Table 25.1 First methodologies of evaluation in supply chain

Methodology	Year of publication											Total
	2000	2003	2004	2005	2006	2007	2008	2009	2010	2011	2012	
Multivariate analysis	0	0	0	1	1	2	2	3	5	6	10	30
Case of study	1	2	3	0	0	4	1	1	4	6	7	29
Review	0	0	1	3	1	2	2	2	3	2	1	17
Quantitative analysis	0	0	0	0	1	1	1	0	0	3	2	8
Multicriteria analysis	0	0	1	0	2	0	0	1	0	0	0	4
Qualitative analysis	0	0	0	0	2	0	0	0	2	0	0	4
Others	0	0	0	0	0	1	0	0	0	0	0	1
Six sigma	0	0	0	0	0	0	0	0	0	1	1	2
Total	1	2	5	4	7	10	6	7	14	18	21	95

possible to find causal relations among latent variables, where the performance is a dependent latent variable that is measure by another items. The second group of techniques is the *empirical analysis*, related to cases of studies in different sectors and this technique is very important because usually are referred to comparisons among companies. Another big group of techniques is the descriptive analysis, where some authors are only describing some basic statistics from the SC, as central tendency measures and dispersion measures related to some attributes.

Simulation was another technique widely used and that is because the SC is a dynamic system and frequently the authors used it for find some dependence among independent variables and dependent variables. Here is important to say that dynamic systems was considerate has a simulation technique.

The most important multicriteria techniques that were used is *Analytic Hierarchy Process* (AHP) and *Neural Hierarchy Process* (NHP), with five and one quotes, respectively. Related the multivariate techniques, although *structural equations* were the most important, the *Factorial Exploratory Analysis* (AFE), *Linear Regression*, *Design of Experiments* and *Discriminant Analysis* were also used.

25.3.3 Authors Studying SC Performance

A lot authors have evaluated the SC performance; for example Ranganathan et al. (2011) have studied the impact of information networks, modeling four influential elements and the relationship between these. Zhang and Dhaliwal (2009) have analyzes the factors affecting the adoption of technologies in operations and administration of SC in China. Also, a lot doctoral theses have focused on the

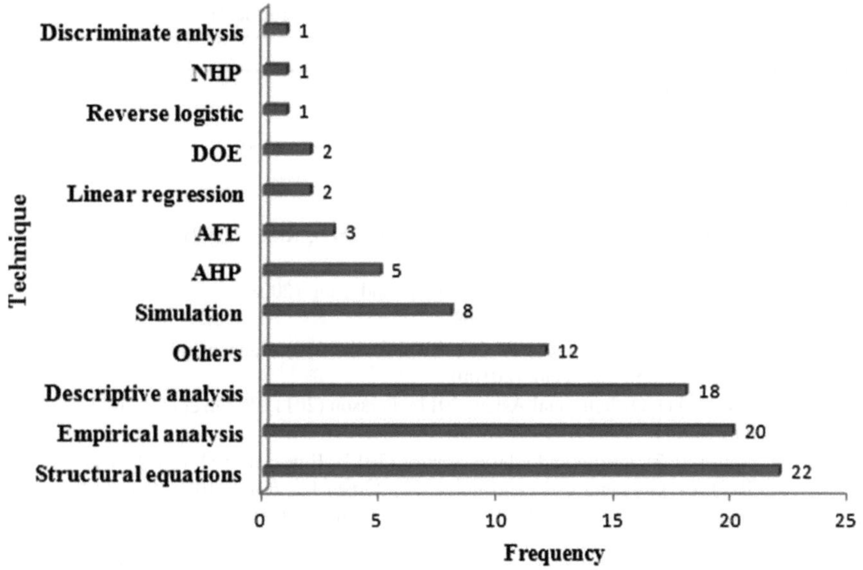

Fig. 25.3 Techniques used individually

quality of information displayed in SC; for example Zhou (2003) and Kroes (2007) makes emphasizes declaring that suppliers' management has a competitive advantage.

Other authors in recent dates have evaluating the performance in SC studying the collaboration (Ramanathan and Gunasekaran 2014), the environmental management (De Giovanni and Esposito 2012; Perotti et al. 2012), the dynamism in SC (Wiengarten et al. 2012), the responsibility, partnership and performance (Gallear et al. 2012) and other attributes. Also, there are previous studies covering literature review regarding supply chain, but sometimes the authors have been conducted their study to some attributes only, as the case of (Gunasekaran and Ngai 2004, 2009; Power 2005; Arun Kanda and Deshmukh 2008), whom studied only the coordination and collaboration.

The operations and processes in SC performance were studied for authors like Dorling et al. (2006), Meixell and Gargeya (2005), Papageorgiou (2009), Nath and Standing (2010), Tang and Musa (2011), and Janvier-James (2012); while Haytko and Kent (2007) and Jeong and Hong (2007) have reported their finds related to management commitment and confidence between member of SC; but Young and Esqueda (2005), Nath and Standing (2010), Jeong and Hong (2007) have reported the importance of security networks in relation to information exchanged between partners; and finally, Kannan and Tan (2010), Askarany et al. (2010), Arun Kanda and Deshmukh (2008), Sarimveis et al. (2008) have reported the importance of supplier in SC evaluating processes. Table 25.2 illustrates the main authors that studied supply chain performance and appears in year order.

Table 25.2 Year and authors that studied the supply chain performance

Year	Authors
2000	Falk and Hogström (2000)
2003	Zhou (2003)
2004	Jiménez (2004), Lockamy and McCormack (2004), Soin (2004)
2005	Meixell and Gargeya (2005)
2006	Dorling et al. (2006)
2007	Forslund and Jonsson (2007), Jeong and Hong (2007), Kroes (2007), Wong and Wong (2007)
2008	McCormack et al. (2008), Theeranuphattana and Tang (2008)
2009	Papageorgiou (2009), Ryu et al. (2009)
2010	Autry et al. (2010), Cedillo and Pérez (2010), Choi (2010), Kannan and Tan (2010), Lassar et al. (2010), Su and Yang (2010b)
2011	Lin et al. (2011), Özdemir and Aslan (2011), Persson (2011), Singh et al. (2011), Tang and Musa (2011)
2012	Cho et al. (2012), Cirtita and Glaser-Segura (2012), Elgazzar et al. (2012), Gallear et al. (2012), De Giovanni and Esposito (2012), Green JR. et al. (2012), Janvier-James (2012), Perotti et al. (2012), Arlbjørn and Lüthje (2012), Vilko and Hallikas (2012), Whitten et al. (2012)

It can be seen that during the year 2000–2006 there were fewer studies of performance. Compared to the next 6 years can be seen a large increase in interest by researchers in the area and who analyze performance from different points and approaches. This trend in increasing the performance analysis shows the importance that represents the supply chain for the current business competitiveness.

25.3.4 Journals that Publish Supply Chain Performance Matters

In Fig. 25.4 appears the journals that have most publications related the SC performance in the period from January 2000 to June 2012. The International Journal of Productions Economics has published 19 of 95 articles analyzed in this review, representing the 20 % of total. Others journals have a few publications, such as European Journal of Operational Research with 7, representing 7.37 %; Industrial Management and Data Systems with 5, representing 5.26 %; Supply Chain Management: an International Journal with 4, corresponding 4.21 %; and International Journal of Operations Production Management with 3 and representing the 3.16 %.

Others journals have only two publications, as the Journal of Manufacturing Technology Management, Industrial Marketing Management, Expert Systems with Applications, International Journal of Physical Distributions and Logistics Management, Journal of Operations Management, Production Planning and Control

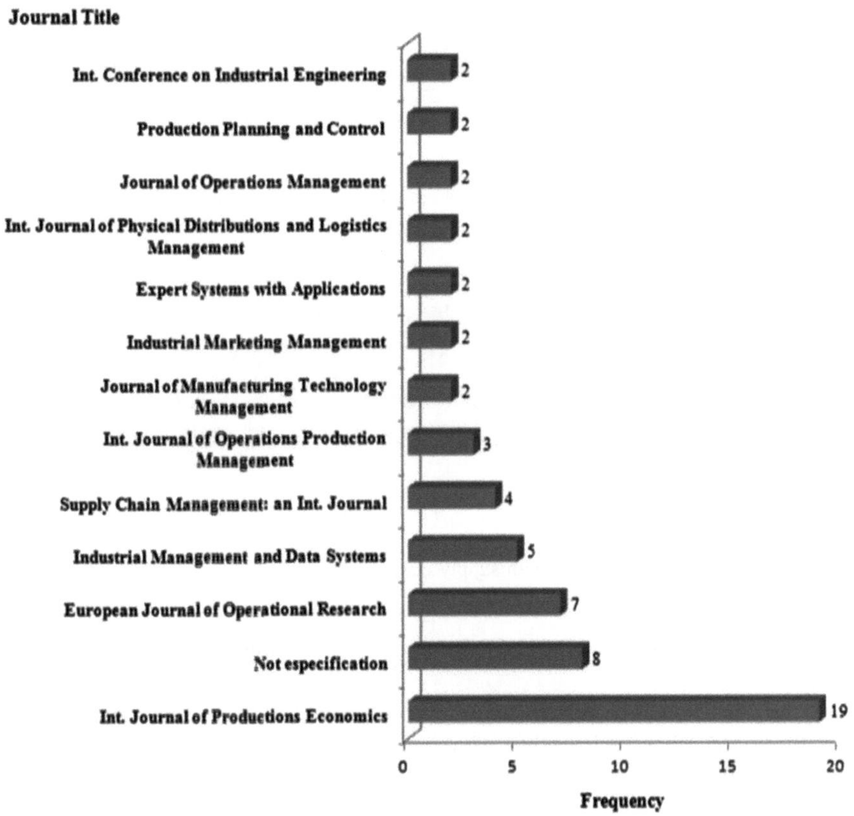

Fig. 25.4 Number chapters per journal

and finally, there is an annual International Conference on Industrial Engineering, representing 2.11 % respectively for everyone.

25.3.5 Industrial Sector of Application

Although all industrial sectors need to improve the supply chain process to be competitive, according to Feng (2012), who say that current competition is between supply chains and not between companies, this improvement is a real challenge because the global environment is changing and companies that have agility and flexibility to overcome then can remains on the market. Figure 25.5 illustrates the main sector in which the study of supply chain attributes and performance evaluation is applied. A total of 57 articles do not show information of industrial applications or are theoretical proposals (simulations); seven articles published are related to manufacturing industry and it is the most important

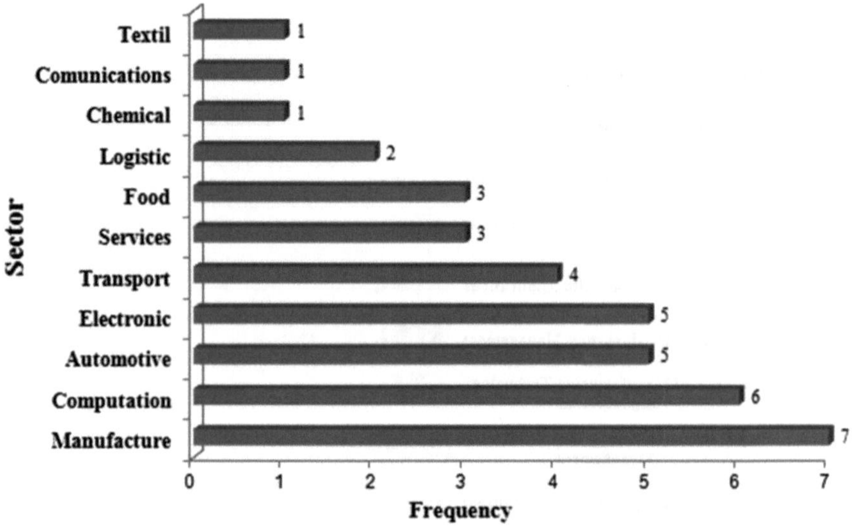

Fig. 25.5 Main industrial sectors with supply chain performance

industrial sector, followed by computation sector, automotive and electronic sector.

Finally the less important were transport sector, services, food and logistics; and there are other sectors that have only one reported case, like the textile, communications and chemical sectors.

25.3.6 Where is the Supply Chain Performance Being Investigated?

In this section is reported a summary of countries for the first author for 95 researched articles in order to identify the location of the research groups that are focused on supply chain performance research. Figure 25.6 shows these locations where there are at least two or more authors working on the topic.

It is clearly shown that United States of America and Taiwan are the countries with more published chapters related to supply chain performance methodologies and techniques, followed by China, Sweden, Mexico, Australia and Spain.

There are others countries that have also had worked on supply chain but having only one article published, like Greece, Denmark, Poland, Costa Rica, Brazil, Hong Kong, Romania, Finland and Germany.

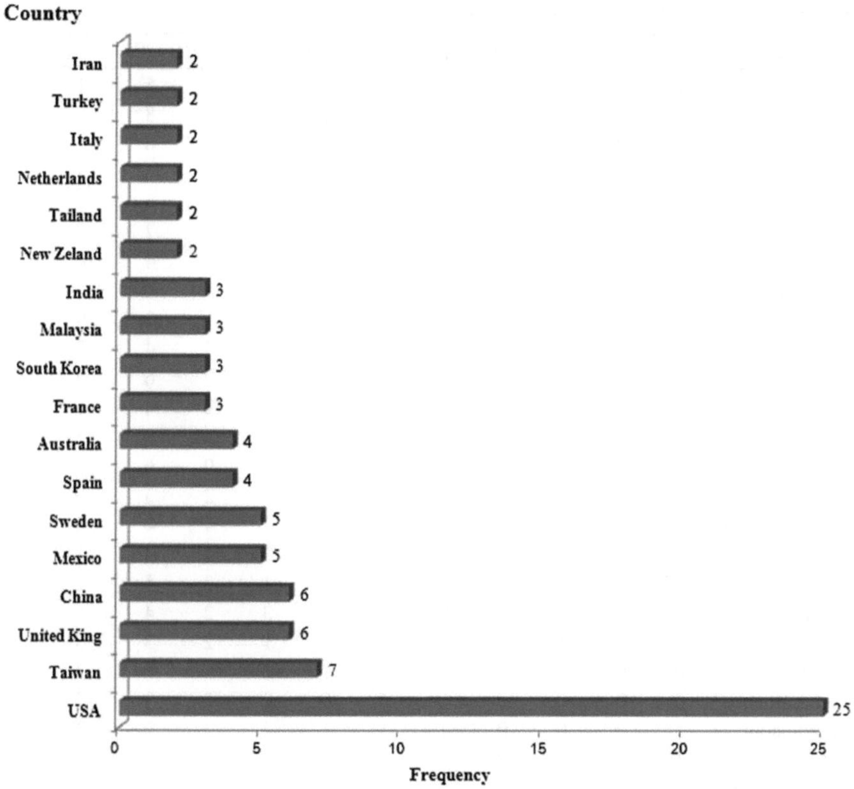

Fig. 25.6 Number of chapters published per country of first author

25.3.7 Attributes Used and Evaluated in Supply Chain Performance

The main attributes measured in supply chain performance were also investigated and Table 25.3 shows a list of those that are the most commonly used when evaluating, identifying a total of seventeen used for this type of investigation. Can be seen that the supply chain performance and delivery are the attributes that are most widely evaluated, 37 and 36 of 95 chapters have quoted it. The next attributes evaluated in SC are processes, operations, cost, demand, information and service.

Attributes like information technologies appear in 26 chapters; others like distribution, administration, planning, provision, communication and quality are present in 22, 21 and 20 academic chapters.

There are other attributes that are gaining importance nowadays, such as environmental impact; in this literature review has found that only seven chapters investigating it, however, those references appears in 2011 and June 2012. This means that the evaluation of this attribute within the supply chain is increasing.

Table 25.3 Main attributes in supply chain

Attributes	2000	2003	2004	2005	2006	2007	2008	2009	2010	2011	2012	Total
Performance SC	1	1	3	1	1	4	2	2	6	5	11	37
Delivery	1	2	3	1	1	5	4	0	3	8	8	36
Information	0	2	1	1	3	5	1	3	5	6	6	33
Processes	1	0	3	0	4	3	3	2	3	6	8	33
Operations	0	1	1	1	2	3	2	0	6	6	9	31
Service	1	0	3	0	1	6	0	1	4	7	7	30
Cost	1	0	2	1	2	3	1	1	4	4	10	29
Demand	0	0	1	1	3	3	3	2	5	6	3	27
Suppliers	0	1	2	2	1	3	2	2	1	3	9	26
IT	0	2	1	1	3	4	1	3	4	4	3	26
Flexibility	0	2	1	1	3	2	1	0	2	6	7	25
Provision	0	0	2	0	2	2	3	1	2	4	6	22
Planning	0	1	1	1	1	3	2	1	4	3	5	22
Distribution	0	0	3	3	1	1	2	2	0	6	3	21
Quality	0	0	1	1	1	2	1	0	2	3	9	20
Communication	1	0	1	1	2	2	0	3	4	4	2	20
Administration	0	0	2	2	2	2	1	2	4	4	1	20
Integration	0	1	1	2	2	2	0	1	3	5	2	19
Confidence	0	0	0	0	0	1	0	3	2	4	5	15
Inventory	0	0	1	0	0	1	4	1	2	3	3	15
Capacity	0	0	1	0	1	1	1	2	1	4	3	14
Commitment	0	0	0	0	0	2	0	2	3	3	3	13
Coordination	1	0	0	0	1	1	1	2	3	2	1	12
Networks	1	0	1	0	1	2	0	2	2	3	0	12
Collaboration	0	0	2	1	0	1	0	2	1	2	2	11
Productivity	0	0	1	0	0	0	1	1	1	4	3	11
Human Resources	0	0	0	1	1	1	0	1	2	2	1	9
Resources	0	0	0	1	1	2	0	0	3	2	0	9
Agility	0	0	1	0	1	1	0	0	0	2	2	7
Environment	0	0	0	0	0	0	0	0	0	2	5	7
Innovation	0	1	0	0	0	1	0	0	1	2	0	5
Competitiveness	0	0	0	0	0	0	0	0	1	1	3	5
Training	0	0	0	0	0	0	0	0	0	1	0	1

25.3.8 Attributes with More Than Twenty Citations

This subsection presents a summary of the main attributes that were investigated. Table 25.4 shown a list with the most commonly used attributes in SC performance measurement and authors with more than 20 citations. If the reader want to know the attributes that has less than 20 quotations, please contact to corresponding author.

Table 25.4 Summary of evaluating attributes in supply chain

Performance	Elgazzar et al. (2012), Autry et al. (2010), Cedillo and Pérez (2010), Choi (2010), Cirtita and Glaser-Segura (2012), Cho et al. (2012), Dorling et al. (2006), Falk and Hogström (2000), Forslund and Jonsson (2007), Gallear et al. (2012), De Giovanni and Esposito (2012), Green JR. et al. (2012), Janvier-James (2012), Jeong and Hong (2007), Jiménez (2004), Kannan and Tan (2010), Kroes (2007), Lassar et al. (2010), Lin et al. (2011), Lockamy and McCormack (2004), McCormack et al. (2008), Meixell and Gargeya (2005), Özdemir and Aslan (2011), Papageorgiou (2009), Perotti et al. (2012), Persson (2011), Ryu et al. (2009), Singh et al. (2011), Soin (2004), Arlbjørn and Lüthje (2012), Su and Yang (2010b), Tang and Musa (2011), Theeranuphattana and Tang (2008), Vilko and Hallikas (2012), Whitten et al. (2012), Wong and Wong (2007), Zhou (2003)
Delivery	Akkermans et al. (2003), Cedillo and Pérez (2010), Cho et al. (2012), Cirtita and Glaser-Segura (2012), Falk and Hogström (2000), Feng (2012), Green JR. et al. (2012), Gunasekaran et al. (2004), Huang et al. (2012), Hui and Nuo (2011), Ip et al. (2011), Jeong and Hong (2007), Jiménez (2004), Kannan and Tan (2010), Khajir and Shafaei (2011), Kroes (2007), Lin et al. (2011), Lockamy and McCormack (2004), McCormack et al. (2008), Nath and Standing (2010), Persson (2011), Quesada and Gazo (2007), Sánchez et al. (2007), Sánchez et al. (2008), Sarimveis et al. (2008), Singh et al. (2011), Arlbjørn and Lüthje (2012), Tang and Musa (2011), Theeranuphattana and Tang (2008), Vilko and Hallikas (2012), Whitten et al. (2012), Wiengarten et al. (2012), Wong and Wong (2007), Wu et al. (2006), Young and Esqueda (2005), Zhou (2003)
Information	Agarwal et al. (2006), Akkermans et al. (2003), Autry et al. (2010), Babak and Saeid (2012), Blome and Schoenherr (2011), Büyüközkan and Vardaloglu (2012), Cedillo et al. (2006), Chen et al. (2011), Choi (2010), Chong et al. (2009), Cannella et al. (2010), Fawcett et al. (2006), Feng (2012), Forslund and Jonsson (2007), Gunasekaran and Ngai (2009), Gunasekaran et al. (2004), Hui and Nuo (2011), Janvier-James (2012), Jeong and Hong (2007), Kannan and Tan (2010), Khajir and Shafaei (2011), Le Dain et al. (2010), Lu et al. (2006, 2007), Lee and Lee (2012), Nath and Standing (2010), Ranganathan et al. (2011), Ribas and Companys (2007), Ryu et al. (2009), Sánchez et al. (2007), Youn et al. (2012), Young and Esqueda (2005), Zhou (2003)
Processes	Agarwal et al. (2006), Akkermans et al. (2003), Babak and Saeid (2012), Blome and Schoenherr (2011), Chong et al. (2009), Cedillo et al. (2006), Cedillo and Pérez (2010), DaeSoo (2006), Elgazzar et al. (2012). Falk and Hogström (2000), Feng (2012), Green JR. et al. (2012), Huang et al. (2012), Ip et al. (2011), Janvier-James (2012), Jeong and Hong (2007), Jiménez (2004), Kisperska-Morón (2011), Lockamy and McCormack (2004), McCormack et al. (2008), Le Dain et al. (2010), Lu et al. (2006), Meixell and Gargeya (2005), Papageorgiou (2009), Persson (2011), Sánchez et al. (2007), Sarimveis et al. (2008), Su and Yang (2010b), Soin (2004), Tang and Musa (2011), Teller et al. (2012), Theeranuphattana and Tang (2008), Whitten et al. (2012), Wong and Wong (2007), Zhou (2003)
Operations	Autry et al. (2010), Babak and Saeid (2012), Cedillo et al. (2006), Choi (2010), Dorling et al. (2006), Fawcett et al. (2006), Feng (2012), Gallear et al. (2012), Green JR. et al. (2012), Hui and Nuo (2011), Janvier-James (2012), Jeong and Hong (2007), Khajir and Shafaei (2011), Le Dain et al. (2010), Meixell and Gargeya (2005), Nath and Standing (2010), Perotti et al. (2012), Persson (2011), Ribas and Companys (2007), Sánchez et al. (2007), Singh et al. (2011), Soin (2004), Su and Yang (2010b), Su and Yang (2010a), Tang and Musa (2011), Teller et al. (2012), Theeranuphattana and Tang (2008), Vilko and Hallikas (2012), Wiengarten et al. (2012), Zhou (2003)

(continued)

Table 25.4 (continued)

Service	Autry et al. (2010), Babak and Saeid (2012), Cambra-Fierro and Polo-Redondo (2011), Cedillo and Pérez (2010), Falk and Hogström (2000), Green et al. (2012), Gunasekaran et al. (2004), Huang et al. (2012), Ip et al. (2011), Janvier-James (2012), Jeong and Hong (2007), Jiménez (2004), Kannan and Tan (2010), Kroes (2007), Kumar et al. (2011), Lockamy and McCormack (2004), Lu et al. (2006, 2007), Lee and Lee (2012), Mendoza (2007), Merschmann and Thonemann (2011), Nath and Standing (2010), Özdemir and Aslan (2011), Ribas and Companys (2007), Ryu et al. (2009), Sánchez et al. (2007), Singh et al. (2011), Olugu et al. (2011), Whitten et al. (2012), Cho et al. (2012)
Cost	Elgazzar et al. (2012), Askarany et al. (2010), Babak and Saeid (2012), Cedillo and Pérez (2010), Cirtita and Glaser-Segura (2012), Glaser-Segura (2012), Cruz (2009), Falk and Hogström (2000), De Giovanni and Esposito (2012), Green JR. et al. (2012), Gunasekaran et al. (2004), Jeong and Hong (2007), Jiménez (2004), Kannan and Tan (2010), Kroes (2007), Lin et al. (2011), Lu et al. (2006), Meixell and Gargeya (2005), Nath and Standing (2010), Özdemir and Aslan (2011), Perotti et al. (2012), Singh et al. (2011), Arlbjørn and Lüthje (2012), Theeranuphattana and Tang (2008), Olugu et al. (2011), Whitten et al. (2012), Wiengarten et al. (2012), Wong and Wong (2007), Wu et al. (2006)
Demand	Agarwal et al. (2006), Büyüközkan and Vardaloglu (2012), Cedillo and Pérez (2010), Cedillo et al. (2006), Cannella et al. (2010), Cruz (2009), Fawcett et al. (2006), Feng (2012), Forslund and Jonsson (2007), Gunasekaran and Ngai (2009), Huang et al. (2012), Hui and Nuo (2011), Kannan and Tan (2010), Khajir and Shafaei (2011), Kumar et al. (2011), Lin et al. (2010). Merschmann and Thonemann (2011), Özdemir and Aslan (2011), Power (2005), Quesada and Gazo (2007), Sánchez et al. (2007), Sánchez et al. (2008), Sarimveis et al. (2008), Singh et al. (2011), Soin (2004), Thomassey (2010), Wu et al. (2006)
Flexibility	Elgazzar et al. (2012), Agarwal et al. (2006), Akkermans et al. (2003), Cedillo and Pérez (2010), Choi (2010), Cirtita and Glaser-Segura (2012), Cho et al. (2012), Ip et al. (2011), Janvier-James (2012), Khajir and Shafaei (2011), Kroes (2007), Kumar et al. (2011), Lu et al. (2006), Meixell and Gargeya (2005), Mendoza (2007), Özdemir and Aslan (2011), Singh et al. (2011), Soin (2004), Arlbjørn and Lüthje (2012), Swafford et al. (2006), Theeranuphattana and Tang (2008), Olugu et al. (2011), Whitten et al. (2012), Wiengarten et al. (2012), Zhou (2003)
IT	Akkermans et al. (2003), Arun Kanda and Deshmukh (2008), Autry et al. (2010), Babak and Saeid (2012), Blome and Schoenherr (2011), Büyüközkan and Vardaloglu (2012), Cedillo et al. (2006), Chen et al. (2011), Chong et al. (2009), Cannella et al. (2010), Cunha and Zwicker (2009), DaeSoo (2006), Gunasekaran and Ngai (2009), Gunasekaran and Ngai (2004), Janvier-James (2012), Jeong and Hong (2007), Kisperska-Morón (2011), Lu et al. (2007), Mendoza (2007), Nath and Standing (2010), Power (2005), Quesada and Gazo (2007), Ranganathan et al. (2011), Wu et al. (2006), Zhou (2003)
Suppliers	Blome and Schoenherr (2011), Cedillo et al. (2006), Cho et al. (2012), Forslund and Jonsson (2007), Gunasekaran et al. (2004), Haytko and Kent (2007), Huang et al. (2012), Kannan and Tan (2010), Khajir and Shafaei (2011), Le Dain et al. (2010), Lockamy and McCormack (2004), McCormack et al. (2008), Meixell and Gargeya (2005), Mendoza (2007), Merschmann and Thonemann (2011), Papageorgiou (2009), Persson (2011), Ryu et al. (2009), Sánchez et al. (2008), Arlbjørn and Lüthje (2012), Tang and Musa (2011), Olugu et al. (2011), Vilko and Hallikas (2012), Whitten et al. (2012), Zhou (2003)

(continued)

Table 25.4 (continued)

Planning	Babak and Saeid (2012), Cedillo and Pérez (2010), DaeSoo (2006), Feng (2012), Forslund and Jonsson (2007), Green JR. et al. (2012), Hui and Nuo (2011), Kisperska-Morón (2011), Lockamy and McCormack (2004), McCormack et al. (2008), Papageorgiou (2009), Persson (2011), Power (2005), Ramanathan and Gunasekaran (2014), Ribas and Companys (2007), Sánchez et al. (2007), Arlbjørn and Lüthje (2012), Su and Yang (2010b), Su and Yang (2010a), Theeranuphattana and Tang (2008), Zhou (2003)
Provision	Arun Kanda and Deshmukh (2008), Askarany et al. (2010), Blome and Schoenherr (2011), Cedillo et al. (2006), Feng (2012), Green JR. et al. (2012), Gunasekaran and Ngai (2009), Huang et al. (2012), Janvier-James (2012), Kannan and Tan (2010), Khajir and Shafaei (2011), Lockamy and McCormack (2004), McCormack et al. (2008), Mendoza (2007), Persson (2011), Sánchez et al. (2007), Sánchez et al. (2008), Soin (2004), Arlbjørn and Lüthje (2012), Swafford et al. (2006), Tang and Musa (2011), Vilko and Hallikas (2012).
Distribution	Arun Kanda and Deshmukh (2008), Blome and Schoenherr (2011), Feng (2012), Green et al. (2012), Gunasekaran and Ngai (2009), Hui and Nuo (2011), Jiménez (2004), Khajir and Shafaei (2011), Kisperska-Morón (2011), Lockamy and McCormack (2004), Lu et al. (2007), Meixell and Gargeya (2005), Papageorgiou (2009), Persson (2011), Power (2005), Sánchez et al. (2008), Soin (2004), Swafford et al. (2006), Tang and Musa (2011), Vilko and Hallikas (2012), Young and Esqueda (2005)
Communication	Autry et al. (2010), Babak and Saeid (2012), Blome and Schoenherr (2011), Büyüközkan and Vardaloglu (2012), Cambra-Fierro and Polo-Redondo (2011), Choi (2010), Chong et al. (2009), Cannella et al. (2010), Cunha and Zwicker (2009), DaeSoo (2006), Falk and Hogström (2000), Gunasekaran and Ngai (2004), Jeong and Hong (2007), Le Dain et al. (2010), Lu et al. (2006), Mendoza (2007), Power (2005), Ryu et al. (2009), Schotanus et al. (2010), Singh et al. (2011)
Quality	Elgazzar et al. (2012), Babak and Saeid (2012), Cedillo and Pérez (2010), Cho et al. (2012), Cirtita and Glaser-Segura (2012), Green JR. et al. (2012), Jeong and Hong (2007), Jiménez (2004), Kannan and Tan (2010), Kroes (2007), Lin et al. (2011), Lee and Lee (2012), Merschmann and Thonemann (2011), Perotti et al. (2012), Arlbjørn and Lüthje (2012), Theeranuphattana and Tang (2008), Olugu et al. (2011), Wiengarten et al. (2012), Wu et al. (2006)
Administration	Askarany et al. (2010), Blome and Schoenherr (2011), Choi (2010), DaeSoo (2006), Dorling et al. (2006), Fawcett et al. (2006), Gunasekaran and Ngai (2004), Haytko and Kent (2007), Laosirihongthog et al. (2011), Quesada and Gazo (2007), Ranganathan et al. (2011), Ryu et al. (2009), Soin (2004), Su and Yang (2010a), Tang and Musa (2011), Youn et al. (2012), Young and Esqueda (2005), Zhang and Dhaliwal (2009)

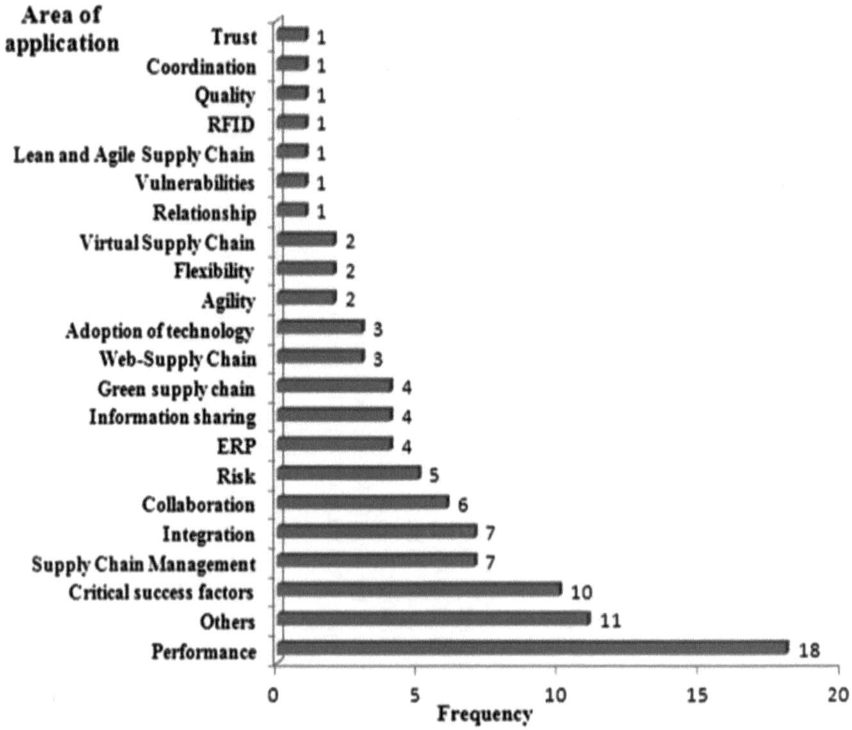

Fig. 25.7 Main areas of applications in supply chain

25.3.9 Areas of Application in Supply Chain

Figure 25.7 shows the main areas of application that were investigated in this period were, the performance with 18 quotes, the critical success factors in SC with 10 quotes, and in the case "other" with 11 was because authors do not specificity the area. Can be seen that the agility, flexibility, virtual-web supply chain, adoption of technology and ERP are areas are in growth for explore (See Fig. 25.7).

25.4 Conclusion

In this chapter we present a comparison statistical of supply chain techniques and attributes commonly used in evaluating performance. It was found that numerous researches were made and certain Journals investigate more about the subject than others, it was also notice that methodologies and combined techniques were utilized to evaluating the supply chain performance. The most used methodology is multivariate analysis and the most used technique approach is structural equation.

It was also observed that performance is the most popular criteria used for evaluating the supply chain, the analysis let us come to the conclusion that cost should be used along with quality and lead time in contemporary supply management as a robust method to follow. Something important that was found six attributes most commonly used are the processes, delivery, service, information, performance and demand, proving that companies in the present time are starting to consider elements of globalization as an important factor in the improvement of supply chain. It is noticed the introduction of new attributes that care about the environment, agility and innovation, although there is little investigation on the subject it is starting to take importance. One of the key objectives of this summary was to identified the regions of the globe that are developing research regarding supply chain, it was found that universities located in Unites States of America, Taiwan and United Kingdom are top producers of articles that handle the subject. To maintain this position economically in the industry they need to continuously improve and establish new procedures and methods one of these methods is selecting the right supplier and be part of any SCM to position their products in different markets. Applications of supply chain performance evaluating are used in a wide industrial sector how manufacture, computation, electronics, automotive, transport, food and services, due to the importance these have at present time.

The literature between January 2000 to June 2012 shows that there is a growth in the study of the supply chain performance using different methodologies and techniques and it is foreseen that the number of articles will keep increasing due to the importance that this topic has in the SCM performance.

References

Abu-Suleiman, A., Boardman, B. & Priest, J. (2004). A framework for an integrated Supply Chain Performance Management System. *Industrial Engineering research Conference*. University of Texas Arlington, Arlington, TX.

Agarwal, A., Shankar, R., Tiwari, M. K. (2006). Modeling the metrics of lean, agile and leagile supply chain: An ANP-based approach. *European Journal of Operational Research, 173*, 211–225.

Akkermans, H. A., Bogerd, P., Yücesan, E., & van Wassenhove, L. N. (2003). The impact of ERP on supply chain management: Exploratory findings from a European Delphi study. *European Journal of Operational Research, 146*, 284–301.

Arlbjørn, J. S., & Lüthje, T. (2012). Global operations and their interaction with supply chain performance. *Industrial Management & Data Systems, 112*(7), 1044–1064.

Arun Kanda, A., & Deshmukh, S. G. (2008). Supply chain coordination: Perspectives, empirical studies and research directions. *International Journal of Productions Economics, 115*, 316–335.

Askarany, D., Yazdifar, H., & Askary, S. (2010). Supply chain management, activity-based costing and organizational factors. *International Journal of Production Economics, 127*, 238–248.

Autry, C. W., Grawe, S. J., Daugherty, P. J., & Richey, R. G. (2010). The effects of technological turbulence and breath on supply chain technology acceptance and adoption. *Journal of Operations Management, 28*, 522–536.

Babak, J. N., & Saeid, I. (2012). Analyzing effective elements in agile supply chain. *Management Science Letters, 24*, 369–378.

Bhatnagar, R., & Sohal, A. S. (2005). Supply chain competitiveness: Measuring the impact of location factors, uncertainty and manufacturing practices. *Technovation, 25*(5), 443–456.

Blome, C., & Schoenherr, T. (2011). Supply chain risk management in financial crises: A multiple case-study approach. *International Journal of Productions Economics, 134*(1), 43–57.

Boddy, D., Cahill, C., Charles, M., Fraser-Kraus, H., & Macbeth, D. (1998). Success and failure in implementing supply chain partnering an empirical study. *European Journal of Purchasing and Supply Management*, *4*(1), 143–151.

Büyüközkan, G., & Vardaloglu, Z. (2012). Analyzing CPFR success factors using fuzzy cognitive maps in retail industry. *Expert Systems with Applications, 39*(12), 10438–10455.

Cambra-Fierro, J. J., & Polo-Redondo, Y. (2011). Post-satisfaction factors effecting the long-term orientation of supply relationships. *Journal of Business and Industrial Marketing, 26*(6), 395–406.

Cannella, S., Ciancimino, E., Framian, J. M., & Disney, S. M. (2010). The four arquetypes in supply chain. *Universia Business Review*, Second quarter, (26), 134–149, Universia Portal, Spain (In Spanish).

Caniato, F., Golini, R., & Kalchschmidt, M. (2013). The effect of global supply chain configuration on the relationship between supply chain improvement programs and performance. *International Journal of Production Economics, 143*(2), 285–293.

Cedillo, M. G., Sánchez, J., & Sánchez, C. (2006). The new relational schemas of inter-firms cooperation: The case of the Coahuila automobile cluster in Mexico. *International Journal Automotive Technology and Management, 6*(4), 406–418.

Cedillo, M. G., & Sánchez, C. (2008). *Dynamic analysis of industrial systems*. Mexico: In Spanish. Trillas Publishers.

Cedillo, M. G., & Pérez, A. (2010). Hybrid supply chains in emerging markets the case of the Mexican auto industry. *South African Journal of Industrial Engineering, 21*(1), 193–206.

Chan, F. T. S. (2003). Performance measurement in a Supply Chain. *International Journal Advanced Manufacturing of Technology, 21*, 534–548.

Chen, D. C., Rajkumar, T. M., & Tomochko, N. A. (2011). The antecedent factors on trust and commitment in supply chain relationships. *Computer Standards & Interfaces, 33*, 262–270.

Chen, T., & Gong, X. (2013). Performance evaluation of a supply chain network. *Procedia Computer Science, 17*, 1003–1009.

Chen, J. V., Yen, D. C., & Tomochko, N. A. (2011). The antecedent factors on trust and commitment in supply chain relationships. *Computers Standards and Interfaces, 33*, 262–270.

Chen, C., & Yan, H. (2011). Network DEA model for supply chain performance evaluation. *European Journal of Operational Research, 213*(1), 147–155.

Cho, D. W., Lee, Y. H., Ahn, S. H., & Hwang, M. K. (2012). A framework for measuring the performance of service supply chain management. *Computers and Industrial Engineering, 62*(3), 801–818.

Choi, S. (2010). *Key success factor of supply chain relationships: multiple case studies in China from buyer's and supplier's perspective*. Master Thesis, University of Gävle, Sweden.

Chong, A. Y., Ooi, K., & Sohal, A. (2009). The relationship between supply chain factors and adoption of e-collaboration tools: An empirical examination. *International Journal of Production Economics*, *122*(1), 150–160.

Cirtita, H., & Glaser-Segura, D. A. (2012). Measuring downstream supply chain performance. *Journal of Manufacturing Technology Management, 23*(3), 299–314.

Cruz, J. M. (2009). The impact of corporate social responsibility in supply chain management: Multicriteria decisión-making approach. *Decision Support Systems, 48*(1), 224–236.

Cruz, J. M. (2013). Modeling the relationship of globalized supply chains and corporate social responsibility. *Journal of Cleaner Production, 56*(October), 73–85.

Cunha, V., & Zwicker, R. (2009). The forerunners of relationship and performance in supply chain companies: Structuring and applying structural equations. *RAE: Revista de Administração de Empresas, 49*(2), 147–161, (in Portuguese).

DaeSoo, K. (2006). Process chain: A new paradigm of collaborative commerce and synchronized supply chain. *Kelly School of Business, 49*(5), 359–367.

De Giovanni, P., & Esposito Vinzi, V. (2012). Covariance versus component-based estimations of performance in green supply chain management. *International Journal of Production Economics, 135*(2), 907–916.

Díaz M. A., & Pérez, C. C. (2002). *Logistics practices in Venezuela an exploratory study.* Working paper, Enterprise Institute, Madrid.

Dorling, K., Scott, J., & Deakins, E. (2006). Determinants of successful vendor managed inventory relationships in oligopoly industries. *International Journal of Physical Distribution and Logistics Management, 36*(3), 176–191.

Droge, C., Vickery, S. K., & Jacobs, M. (2012). Does supply chain integration mediate the relationships between product/process strategy and service performance? An empirical study. *International Journal of Production Economics, 137*(2), 250–262.

Elgazzar, S. H., Tipi, N. S., Hubbard, N. J., & Leach, D. Z. (2012). Linking supply chain processes' performance to a company's financial strategic objectives. *European Journal of Operational Research, 223*(1), 276–289.

Falk, H., & Hogström, L. (2000). *Key success factors for a functioning supply chain in e-commerce B2B*. Master Thesis, Göteborg University, Sweden.

Fawcett, S. E., Magnan, G. M., & McCarter, M. W. (2006). Benefits, barriers, and bridges to effective supply chain management. *Supply Chain Management: an International Journal, 13*(1), 35–48.

Feng, Y. (2012). System dynamics modeling for supply chain information sharing. *Physics Procedia, 25*, 1463–1469.

Forrester, J. W. (1961). *Industrial Dynamics*. Portland (Or): Productivity Press.

Forslund, H., & Jonsson, P. (2007). The impact of forecast information quality on supply chain performance. *International Journal of Operations and Production Management, 27*(1), 90–107.

Gallear, D., Ghobadian, A., & Chen, W. (2012). Corporate responsibility, supply chain partnership and performance: An empirical examination. *International Journal of Production Economics, 140*(1), 83–91.

Green, JR., K. W., & Inman, R. A. (2005). Using a just in time selling strategy to strengthen supply chain linkages. *International Journal of production Research, 43*(16), 3437–3453.

Green JR., K. W., Whitten, D., & Inman, R. A. (2012). Aligning marketing strategies throughout the supply chain to enhance performance. *Industrial Marketing Management, 41*(6), 1008–1018.

Guiffrida, A. L., & Nagi, R. (2006). Cost characterizations of supply chain delivery performance. *International Journal of Production Economics, 102*(1), 22–36.

Gunasekaran, A., Patel, C., & Tirtiroglu, E. (2001). Performance measures and metrics in a supply chain environment. *International Journal of Operations and Production Management, 21*(1,2), 71–87.

Gunasekaran, A., Patel, C., & McGauhhey, R. E. (2004). A framework for supply chain performance measurement. *International Journal of Productions Economics, 87*(3), 333–347.

Gunasekaran, A., & Ngai, E. W. T. (2004). Virtual supply chain management. *Production Planning and Control, 15*(6), 584–595.

Gunasekaran, A., & Ngai, E. W. T. (2009). Modeling and analysis of built-to-order supply chains. *European Journal of Operational Research, 195*(2), 319–334.

Haytko, D. L., & Kent, J. L. (2007). Mexican maquiladoras: helping or hurting the US/Mexico cross-border supply chain? *The International Journal of Logistics Management, 18*(3), 347–363.

Huang, M., Yang, M., Zhang, Y., & Liu, B. (2012). System dynamics modeling-based study of contingent sourcing under supply discruptions. *Systems Engineering Procedia, 4*, 290–297.

Hishamuddin, H., Sarker, R. A., & Essam, D. (2013). Recovery model for a two-echelon serial supply chain with consideration of transportation disruption. *Computers and Industrial Engineering, 64*(2), 552–561.

Hui, L., & Nuo, L. (2011). System dynamics modeling and simulation of multi-stage supply chain: The value of information sharing. *Energy Procedia, 13*, 4861–4867.

Ip, W. H., Chan, S. L., & Lam, C. Y. (2011). Modeling supply chain performance and stability. *Industrial Management and Data Systems, 111*(8), 1332–1354.

Janvier-James, A. M. (2012). A new introduction to supply chains and supply chain management: Definitions and theories perspective. *International Business Research, 5*(1), 194–207.

Jeong, J. S., & Hong, P. (2007). Customer orientation and performance outcomes in supply chain management. *Journal of Enterprise Information Management, 20*(5), 578–594.

Jiménez, J. E. (2004). *Critical success factors in supply chain* (Technique Publication No. 237). Mexican Institute of Transport, Sanfandila, Querétaro, Mexico.

Jiménez J. E., & Hernández, S. (2002). *Conceptual framework of the supply chain: A new logistics focus* (Technique Publication No. 215). Mexican Institute of Transport, Sanfandila, Querétaro, Mexico.

Kannan, V. R., & Tan, K. C. (2010). Supply chain integration: cluster analysis of the impact of span of integration. *Supply Chain Management: an International Journal, 15*(3), 207–215.

Khaji, M., & Shafaei, R. (2011). A system dynamics approach for strategic partnering in supply networks. *International Journal of Computer Integrated Manufacturing, 24*(2), 106–125.

Kim, D., Cavusgil, S. T., & Cavusgil, E. (2013). Does IT alignment between supply chain partners enhance customer value creation? An empirical investigation. *Industrial Marketing Management, 42*(6), 880–889.

Kisperska-Morón, D. (2011). Virtual logistics as a support for the decomposition process of supply chain (conceptual reflections). *Scientific Journal of Logistics, 7*(5), 49–60.

Kroes, J. R. (2007). *Outsourcing of supply chain processes: Evaluating the impact of congruence between outsourcing drivers and competitive priorities on performance*. PhD. Thesis, Georgia Institute of Technology, United States of America.

Kumar, S., McCreary, M. L., & Nottestad, D. A. (2011). Quantifying supply chain trade-offs using six sigma, simulation, and designed experiments to develop a flexible distribution network. *Quality Engineering, 23*(2), 180–203.

Laosirihongthog, T., Punnakitikashem, P., & Adebanjo, D. (2011). Improving supply chain operations by adopting RFID technology: Evaluation and comparison of enabling factors. *Production Planning and Control, 1*, 1–20.

Lassar, W., Haar, J., Montalvo, R., & Hulser, L. (2010). Determinants of strategic risk management in emerging markets supply chain: Case of Mexico. *Journal of Economics, Finance and Administrative Science, 15*(28), 25–140.

Le Dain, M., Calvi, R., & Cheriti, S. (2010). Measuring supplier performance in collaborative design: Proposition of a framework. *R&D Management, 41*(1), 61–79.

Lee, M. S., & Lee, S. (2012). Success factors of open-source enterprise information systems development. *Industrial Management & Data Systems, 112*(7), 1065–1084.

Lin, C., Wing, S., Madu, C. N., Kuei, C., & Yu, P. (2005). A structural equation model of supply chain quality management and organizational performance. *International Journal of Productions Economics, 96*(3), 355–365.

Lin, R., Chen, R., & Nguyen, T. (2011). Green supply chain management performance in automobile manufacturing industry under uncertainty. *Procedia-Social and Behavioral Sciences, 25*, 233–245.

Lin, Y., Wang, Y., & Yu, C. (2010). Investigating the drivers of the innovation in cannel integration and supply chain performance: A strategy orientated perspective. *International Journal of Productions Economics, 127*(2), 320–332.

Lockamy, A, I. I. I., & McCormack, K. (2004). Linking SCOR planning practices to supply chain performance: An exploratory study. *International Journal of Operations and Production Management, 24*(12), 1192–1218.

Lu, X., Huang, L., & Heng, M. S. H. (2006). Critical success factors of inter-organizational information systems: A case study of Cisco and Xiao Tong in China. *Information and Management, 43*(3), 395–408.

Lu, C., Lai, K., & Chen, T. C. E. (2007). Application of structural equation modeling to evaluate the intention of shippers to use internet services in linear shipping. *European Journal of Operational Research, 180*(2), 845–867.

McCormack, K., Bronzo, M. L., & Valadares, M. P. (2008). Supply chain maturity and performance in Brazil. *Supply Chain Management: An International Journal, 13*(4), 272–282.

Meixell, M. J., & Gargeya, V. B. (2005). Global supply chain design: A literature review and critique. *Transportation Research Part E, 41*(6), 531–550.

Mendoza, E. (2007). *Uncertainty, integration and supply flexibility*. PhD. Thesis, Universitat Pompeu Fabra, Department of Economics and Business, Barcelona, Catalonia, Spain.

Merschmann, U., & Thonemann, U. W. (2011). Supply chain flexibility, uncertainty and form performance: An empirical analysis of German manufacturing firms. *International Journal of Productions Economics, 130*(1), 43–53.

Nath, T., & Standing, C. (2010). Drivers of information technology use in the supply chain. *Journal of Systems and Information Technology, 12*(1), 70–84.

Olugu, E. U., Wong, K. Y., & Shaharoun, A. M. (2011). Development of key performance measures for the automobile green supply chain. *Resources, conservation and Recycling, 55*(6), 567–579.

Otto, A., & Kotzab, H. (2003). Does supply chain management really pay? Six perspectives to measure the performance of managing a supply chain. *European Journal of Operational Research, 144*(2), 306–320.

Özdemir, A. I., & Aslan, E. (2011). Supply chain integration, competition capability and business performance: A study on Turkish SMEs. *Asian Journal of Business Management, 3*(4), 325–332.

Papageorgiou, L. G. (2009). Supply chain optimization for the process industries: Advances and opportunities. *Computers and Chemical Engineering, 33*(12), 1931–1938.

Perotti, S., Zorzini, M., Cagno, E., & Micheli, G. J. L. (2012). Green supply chain practices and company performance: the case of 3PLs in Italy. *International Journal of Physical Distribution and Logistics Management, 42*(7), 640–672.

Persson, F. (2011). SCOR template: A simulation based dynamic supply chain analysis tool. *International Journal of Production Economics, 131*(1), 288–294.

Power, D. (2005). Supply chain management integration and implementation: A literature review. *Supply Chain Management: an International Journal, 10*(4), 252–263.

Quesada, H., & Gazo, R. (2007). Methodology for determining key internal business processes based on critical success factors. *Business Process Management Journal, 13*(1), 5–20.

Ramanathan, U. (2014). Performance of supply chain collaboration—a simulation study. *Expert Systems with Applications, 41*(1), 210–220.

Ramanathan, U., & Gunasekaran, A. (2014). Supply chain collaboration: Impact of success in long-term partnerships. *International Journal of Production Economics, 147* (Part B), 252–259.

Ranganathan, C., Teo, T. S. H., & Dhaliwal, J. (2011). Web-enabled supply chain management: Key antecedents and performance impacts. *International Journal of Information Management, 31*(6), 533–545.

Revilla, E., & Sáenz, M. J. (2013). Supply chain disruption management: Global convergence versus national specificity. *Journal of Business Research*, (in press), Corrected Proof.

Ribas, I., & Companys, R. (2007). Estado del arte de la planificación colaborativa en la cadena de suministro: contexto determinista e incierto. *Intangible capital*, 3(003), ISSN: 1697-9818, Universitat Politecnica de Catalunya.

Ryu, I., So, S., & Koo, C. (2009). The role of partnership in supply chain performance. *Industrial Management and Data Systems, 109*(4), 496–514.

Sánchez, C., Cedillo, M. G., & Piña, M. R. (2007). Model for dynamic analysis of industrial clusters: the case of supply chain in the automotive clusters in the southeast region of Coahuila. *International Conference on Industrial Engineering. Proceedings of the 13th Annual.*

Sánchez, C., Cedillo, M. G., & Pérez, V. P. (2008). Sensitivity analysis of the impact of inventory and cycle time on performance of the automotive supply chain. *International Conference on Industrial Engineering. Proceedings of the 13th Annual.*

Santos, J. C. (2010). http://ciclog.blogspot.mx/2010/11/decisiones-el-flujo-escondido-en-la.html. Accessed 15 May 2012.

Sarimveis, H., Patrinos, P., Tarantilis, C. D., & Kiranoudis, C. T. (2008). Dynamic modeling and control of supply chain systems: a review. *Computers and Operations Research, 35*(11), 3530–3561.

SCC, (2010). SCOR model. http://www.SupplyChain.org. Accessed 20 June 2012.

Schliephake, K., Stevens, G., & Clay, S. (2009). Making resources work more efficiently—the importance of supply chain partnerships. *Journal of Cleaner Production, 17*(14), 1257–1263.

Schmitt, A., & Singh, M. (2012). A quantitative analysis of disruption risk in a multi-echelon supply chain. *International Journal of Production Economics, 139*(1), 22–32.

Schotanus, F., Telgen, J., & de Boer, L. (2010). Critical success factors for managing purchasing groups. *Journal of Purchasing and Supply Management, 16*(1), 51–60.

Singh, R., Singh, S. H., Metri, B. A., & Kaur, R. (2011). Organizational performance and retail challenges: A structural equation approach. *Scientific Research, 3*, 159–168.

Soin, S. S. (2004). *Critical success factors in supply chain management at high technology companies*. PhD. Thesis, University of Southern Queensland, Australia.

Su, Y., & Yang, C. (2010a). A structural equation model for analyzing the impact of ERP on SCM. *Expert Systems with Applications, 37*(1), 456–469.

Su, Y., & Yang, C. (2010b). Why are enterprise resource planning systems indispensable to supply chain management? *European Journal of Operational Research, 203*(1), 81–94.

Surana, A., Kumara, S., Greaves, M., & Raghavan, N. (2005). Supply chain networks: A complex adaptive system perspective. *International Journal of Production Research, 43*(20), 4235–4265.

Swafford, P. M., Ghosh, S., & Murthy, N. (2006). The antecedents of supply chain agility of a firm: scale development and model testing. *Journal of Operations Management, 24*(2), 170–188.

Świerczek, A. (2013). The impact of supply chain integration on the "snowball effect" in the transmission of disruptions: An empirical evaluation of the model. *International Journal of Production Economics*, (in press), Corrected Proof.

Tang, O., & Musa, N. (2011). Identifying risk and research advancements in supply chain risk management. *International Journal of Production Economics, 133*(1), 25–34.

Teller, C., Kotzab, H., & Grant, D. B. (2012). Improving the execution of supply chain management in organizations. *International Journal of Production Economics, 140*(2), 713–720.

Theeranuphattana, A., & Tang, J. C. S. (2008). A conceptual model of performance measurement for supply chains. *Journal of Manufacturing Technology Management, 19*(1), 125–148.

Thomassey, S. (2010). Sales forecasts in clothing industry: The key success factor of the supply chain management. *International Journal of Productions Economics, 128*(2), 470–483.

Vilko, J. P. P., & Hallikas, J. M. (2012). Risk assessment in multimodal supply chains. *International Journal of Production Economics, 140*(2), 586–595.

Vlachos, I. P. (2014). A hierarchical model of the impact of RFID practices on retail supply chain performance. *Expert Systems with Applications, 41*(1), 5–15.

Whitten, G. D., Green, K. W, Jr, & Zelbst, P. J. (2012). Triple-A supply chain performance. *International Journal of Operations and Production Management, 32*(1), 28–48.

Wiengarten, F., Pagell, M., & Fynes, B. (2012). Supply chain environmental investments in dynamic industries: Comparing investment and performance differences with static industries. *International Journal of Production Economics, 135*(2), 541–551.

Wong, W. P., & Wong, K. Y. (2007). Supply chain performance measurement system using DEA modeling. *Industrial Management and Data Systems, 107*(3), 361–381.

Wu, T., Blackhurst, J., & Chidambaram, V. (2006). A model for inbound supply risk analysis. *Computers in Industry, 57*(4), 350–365.

Youn, S., Yang, M. G., & Hong, P. (2012). Integrative leadership for effective supply chain implementation: An empirical study of Korean Firms. *International Journal of Production Economics, 139*(1), 237–246.

Young, R. R., & Esqueda, P. (2005). Supply chain vulnerability: Considerations of the case of latin America. *Revista Latinoamerican of Administration Journal, 34*, 63–77.

Yu, W., Jacobs, M., Salisbury, W., & Enns, H. (2013). The effects of supply chain integration on customer satisfaction and financial performance: An organizational learning perspective. *International Journal of Production Economics, 146*(1), 346–358.

Zhang, C., & Dhaliwal, J. (2009). An investigation of resource-based and institutional theoretic factors in technology adoption for operations and supply chain management. *International Journal of Productions Economics, 120*(1), 252–269.

Zhou, H., B.S., M.S., M.A. (2003). *The role of supply chain processes and information sharing in supply chain management.* PhD. Thesis. The Ohio State University, United States of America.

Chapter 26
Design of Experiments and Statistical Optimization in Manufacturing

María Blanca Becerra-Rodríguez, Jorge Domínguez-Domínguez, Roberto Zitzumbo-Guzmán and Jorge Luis García-Alcaraz

Abstract In this chapter is described the design of experiments methodology as tool widely used in the identification of sources of variation in manufacturing production processes, helping to improve quality systems. It also reports a case study, in which a production process is characterized and the main components to be integrated into a final product identified.

Keywords Design of experiments · Process control · Vulcanization

26.1 Introduction

It is common in different areas of study to consider problems represented by many features of interest based on a set of control factors. An experimental strategy is used to get the response value of these features. The type of design used involves selecting a set of factors of interest such that the combination of values matches the

M. B. Becerra-Rodríguez (✉)
Instituto Tecnológico de San Juan del Río, Av. Tecnológico # 2, Esq. Av. Paseo Central, San Juan del Río, 76800 Querétaro, Querétaro, México
e-mail: mblca@hotmail.com

J. Domínguez-Domínguez
Centro de Investigación en Matemáticas, Unidad Aguascalientes, F. Bartolomé de las Casas # 314, 20259 Aguascalientes, Aguascalientes, México

R. Zitzumbo-Guzmán
Centro de Innovación Aplicada en Tecnologías Competitivas, Omega # 201, Fracc. Industrial Delta, 37545 León, Guanajuato, México

J. L. García-Alcaraz
Department of Industrial Engineering and Manufacturing - Institute of Engineering and Technology, Autonomous University of Ciudad Juarez, Av. del Charro # 450 Norte, Col. Partido Romero, 32310 Ciudad Juárez, Chihuahua, México

J. L. García-Alcaraz et al. (eds.), *Lean Manufacturing in the Developing World*,
DOI: 10.1007/978-3-319-04951-9_26, © Springer International Publishing Switzerland 2014

best to all the features of a product. This process is known as a multi-response design optimization, since the characteristics of interest are defined for several responses.

Particularly, in industrial processes is important to have mathematical models to explain one or more characteristics, called responses, of the quality of a product. The purpose of the model is to study the relationship between k factors $X = (X_1, \ldots, X_k)$ (control variables) and r responses $y = (y_1, \ldots, y_r)$, for each of which a model is proposed. In this paper we consider linear or quadratic models. The general quadratic model for two factors and two responses is

$$\hat{y}_1 = \hat{\beta}_0 + \hat{\beta}_1 X_1 + \hat{\beta}_1 X_2 + \hat{\beta}_{12} X_1 X_2 + \hat{\beta}_{11} X_1^2 + \hat{\beta}_{22} X_2^2 \quad (26.1)$$

$$\hat{y}_2 = \hat{\alpha}_0 + \hat{\alpha}_1 X_1 + \hat{\alpha}_1 X_2 + \hat{\alpha}_{12} X_1 X_2 + \hat{\alpha}_{11} X_1^2 + \hat{\alpha}_{22} X_2^2 \quad (26.2)$$

The information generated by constructing these models is obtained through experimental design (Castaño and Domínguez 2010). Once obtained, the models are evaluated using statistical techniques of squared minimums or maximum likelihood. If the values $\hat{\beta}_{12}, \hat{\beta}_{11}, \hat{\beta}_{22}, \hat{\alpha}_{12}, \hat{\alpha}_{11}\ y\hat{\alpha}_{22}$ and $\hat{\alpha}_{22}$ are statistically significant equal to zero, then the models are linear. Using the models (26.1) and (26.2) different optimization models could be raised, for example:

$$\begin{aligned} &\text{Optimize} \quad y_1 \\ &\text{Subject to} \quad y_2 \leq l, \text{ where } l \text{ is a value of interest in study} \end{aligned} \quad (26.3)$$

X is in the experimental region.

This approach is an interesting application of mathematics to problems of linear programming and optimization (Ortiz et al. 2004), since the models come from engineering projects of interest or from actual cases of industrial manufacturing processes. Mathematical programming is used for decision making in various management, production, and research levels among others; in these cases the objective functions and constraints are used. In actual problems there are terms where the functions express issues such as gains and losses. The restrictions frame investment related issues. At present there are different lines of research in dynamic optimization, fuzzy theory, geneticist algorithms, among others, to study the optimization model (26.3). In summary, the objective is to define a set of factors X providing best mediating simultaneously to the r responses. Below is a brief presentation of the optimization problem for multiple-responses.

A common approach to solve problems of multi-response design is as follows: Initially the individual response variables are modeled to create an experimental design response surface. A transformation is applied to each response variable so that all the responses can be combined into a single function called objective function. From there the factors levels are varied so that can better meet the individual optimize to achieve a global optimum $x_0 = (x_{10}, \ldots, x_{k0})$. Here the optimum word is used as a reference to consider the most desirable or acceptable values of responses according to certain conditions. The process of multi-response optimization has application in many areas of knowledge and more frequently in engineering design problems (Ames et al. 1997). The multi-response optimization requires finding characteristics of the control variables to generate an optimal or near-optimal value, as to produce the best values for each of the responses that are being considered. The techniques of multi-response optimization can be studied by graphical and analytical methods.

This work aims to present the basic ideas to perform experiments in manufacturing, modeling responses in order to implement a process of multi-objective optimization using the method of desirability function to obtain a global optimum. The strategy of this presentation is to make the approach practical to situations of experimental design applied to manufacturing.

26.2 Design of Experiments and Optimization

26.2.1 Manufacturing

The experimental design has been applied to significantly improve the quality of processes and products and further, to make these products robust to extreme conditions.

The engineering and technology of manufacturing advance rapidly boosting the economy. Manufacturing is influenced by materials and processing parameters. Other factors considered in manufacturing include: design, product quality, and manufacturing cost.

The purpose of industrial engineering and manufacturing is industrial production systems of goods and services. That is, studying how to organize the physical and human resources for the transformation of raw materials into products and services. The production of goods and services converted through industrial processes.

Manufacturing is the process of converting raw materials into products. Includes (1) product design, (2) selection of raw materials, and (3) the sequence of processes through which the product is manufactured.

This word "manufacture" is derived from Latin *factus*, meaning handmade. The word manufacture first appeared in 1567, and the word manufacturing in 1863. Then manufacturing involves processing of products from raw materials by processes, machines and operations through a well-organized plan for each required activity.

Because a manufactured item has gone through a number of processes in which raw materials has become a useful product, it has value, defined as monetary value or market price.

Manufacturing, in the modern context, can be defined in two ways: Technological and economic.

Technologically is the application of chemical and physical processes that alter the geometry, properties, or appearance of a material to produce finished parts or products. The manufacturing processes involve a combination of machines, tools, power, and manual labor. Manufacturing is almost always performed as a sequence of operations.

This means that there must be an integration of processes in product design, material selection, with considerations on the methods of manufacturing, process planning, assembly, testing and quality assurance; in addition all manufacturing areas have the responsibility to meet the following objectives:

- A product must meet design requirements, specifications, and standards.
- A product should be performed in the most efficient way possible (financially) and with minimally complex methods.
- Quality should be included in each stage of product manufacturing, from design to assembly. This is extremely preferable to failures in quality testing after manufacture.
- Manufacturing activities must meet several requirements and trends: A product shall be manufactured using the most economically and environmentally friendly methods.
- In terms of appropriate, timely and cost-effective implementation in manufacturing organization new developments in materials, production methods, and computer integration should be evaluated constantly.
- A manufacturing organization should try to consistently achieve the highest levels of quality and productivity (optimum use of resources, materials, machines, energy, capital, labor and technology).

Some manufacturing processes are:

Injection molding: high heat is transferred in injection machines to the polymers due to frictional heating. The particles (pellets) or grains are fed to a heated barrel and the melt mass is forced through a camera with matrix by a hydraulic piston system with a rotating screw extruder. The products manufactured by injection molding are: Mugs, bowls, boxes, tool handles, knobs, electrical components, media, and toys.

Lathe machining: Is a cylinder operation, the work piece is clamped with a device. The long, slender parts must bear a fixed bezel and a mobile window placed in the bed; if this is not done, the piece will flex due to shear forces. These bezels are equipped with three adjustable rollers that hold the piece but letting it rotate freely. The cutting tool, attached to the tool holder is driven by the lead screw, and removes the material as it travels the bench. The lathe is used to machine small parts with precision to close tolerances. Lathes are used for working metal or plastic feeding an auger, controls are manipulated manually or automatically to improve efficiency and productivity.

Manufacturing chrome parts: The process begins with the arrival of the keys and fire handles as forging to the baths; the next step is the polishing of parts with special brushes and paste reviewing the parts to leave a mirror finish. The last pass is racking where the parts are places piece by piece in special racks for keys and fire handles passing through 36 tubs; some tubs are for washing, other tubs to paste nickel. Each tub with a different chemical to affix the chrome to the piece. Time and temperature must be controlled in the tubs because if time and temperature are exceeded the pieces with burn and the chromium will not adhere so the parts become defective with no option to be recovered.

The information is generated by an experimental scheme, and is presented by a matrix $D(n \times k)$, where n is the number of combinations (treatments), of the values of the k factors $(X_1, \ldots, X_k)$. The $X's$ are input variables to the process, and can be numeric (continuous or discrete), or not-numeric (ordinal or nominal). For example, for a specific process, if a factor is the temperature, then the variable X can take values between 30 and 120 °C. Usually in an experimental strategy only two or three values in that range are taken.

Several experimental schemes that can be used, such as factorial design, fractional factorial designs, design Box—Behnken, central composite design and optimal designs, among others (Box and Draper 1987). Suppose we have r responses for each of the n combinations. With the data generated by the experiment each response r can be modeled individually. In general these models are linear and quadratic, y are a function of the k factors. Thus for r responses there are r models. The j-th model, a polynomial of order$s$, for response Y_j is written as:

$$Y_j = Z^t(x)\delta_j + \varepsilon_j, \qquad j = 1, \ldots, r, \tag{26.4}$$

where $Z(x)$ is a matrix of order $(n \times q)$, which represents the n treatments, q elements consisting of a constant term in powers and products of powers (greater or equal to 1). In the case $s = 2$,

$$Z(x) = \left(1, X_1, \ldots X_k, X_1^2, \ldots, X_k^2, X_1X_2, X_1X_3, \ldots, X_{k-1}X_k\right)$$

and

$$\delta_j = \left(\beta_{j0}, \beta_{j1}, \ldots, \beta_{jk}, \beta_{j11}, \ldots, \beta_{jkk}, \beta_{j12}, \beta_{j13}, \ldots, \beta_{jk-1k}\right)^t.$$

In the case $s = 2$, the model (26.4) can be rewritten as:

$$Z(x)\delta_j = \beta_{j0} + X\beta_j + X^t\beta_j X, \qquad j = 1, \ldots, r, \tag{26.5}$$

where $X^t = (X_1, \ldots X_k)$ is a vector of k control variables, β_0 is constant, $\beta^t = (\beta_1, \ldots, \beta_k)$ a vector of parameters and $B = (\beta_{11}, \ldots, \beta_{1k}, \beta_{k1}, \ldots, \beta_{kk})$ a symmetric matrix of second-order parameters. In this situation $q = \frac{k(k+3)}{2} + 1$, corresponds to the number of terms: constant, linear and of second order; ε_j is a random variable. The theory that will be discussed in this presentation will be made under the assumption that ε_j follows a normal distribution with mean zero and variance σ_j^2, that is: $\varepsilon_j \sim \mathrm{N}\left(0, \sigma_j^2\right)$.

As a result of the experimental strategy, the y_j models are replaced by the best models generated by the minimum squares method, (Box and Draper 1987; Khuri and Cornell 1996) and the j-th simplified model is expressed by:

$$\hat{Y}_j = Z(x)\hat{\delta}_j. \tag{26.6}$$

The solution by the method of minimum squares is $\hat{\delta}_j = \left[(Z(x))'Z(x)\right]^{-1}$ $(Z(x))^t Y_j$, being $Z(x)$ a matrix of $(n \times q)$, where n is the number of treatments and, for the case where the model is of second order, that is $s = 2, q = \frac{k(k+3)}{2} + 1$ is the number of terms, constant, linear, quadratic, and second order interactions each Y_j is a vector of values predicted by the model in n treatments and with which the optimization procedure is followed.

In a process that refers to the manufacturing process of vulcanization. The proposed strategy is to select important factors in the process, considering scorch time, maximum tear strength, and cost of formulation as variable responses.

The experimental strategy is: The simultaneous optimization of three response variables in the vulcanization process.

1. Selection of factors, levels, and response variables.
2. Selection of experimental design and execution of experiment.
3. Statistical data analysis made through ANalysis Of VAriance (ANOVA), for each variable response
4. Obtaining a first order model that represents the behavior of the process between the values of the factors being studied.

$$Y = \beta_0 + \sum_{i=1}^{k} \beta_0 X_i + \varepsilon \tag{26.7}$$

5. Process optimization through the multi-response desirability function

$$d_i(X) \begin{cases} 0 & \text{si } \hat{Y}_i(X) < EI_i \quad \text{o} \quad \hat{Y}_i(X) > ES_i \\ \left[\frac{\hat{Y}_i(X) - Es_i}{T_i - ES_i}\right]^t & \text{si } T_i < \hat{Y}_i(X) \le ES_i \\ \left[\frac{\hat{Y}_i(X) - EI_i}{T_i - EI_i}\right]^s & \text{si } EI_i \le \hat{Y}_i(X) \le T_i \end{cases} \tag{26.8}$$

$$D = \sqrt{[n]} d_1 \; x \; d_2 \; x \ldots x d_n$$

26.2.2 Fractional Factorial Design

The experimental design chosen for the important variables in the vulcanization process was a fractional factorial design $2_{III}^{11-7} = 16$ Castaño (Castaño and Domínguez 2010). The quantitative factors, levels and response variables are presented in Table 26.1.

Table 26.1 Experimental factors, levels and measurement units

Factors	Code	Levels		Measurement units
		(−1)	(1)	
Carbon black	NH	40	80	PCR[a]
Paraffinic oil	AP	15	40	PCR
Stearic acid	AE	0.5	2	PCR
Zinc oxide	ZnO	3	5	PCR
EPDM-MA	EM	2	7	PCR
MBTS	MB	0.7	2	PCR
TMTD	TM	0.1	0.6	PCR
Fiber	FI	10	40	PCR
Sulphur	AZ	1	2.5	PCR
Vulcanization time	TV	7	10.5	min[b]
Mixgin temperature	TMEZ	105	120	°C[c]

[a] Resin parts per one hundred, [b] Minutes, and [c] Degrees Celsius

26.3 Solution

The experiments were made in a random order, and the results are shown in Table 26.2.

26.3.1 Analysis of Response Variables ts1, EMD, and Cost

Table 26.3 shows the ANOVAs of the contributing factors (Castaño and Domínguez 2010) to explain the response variables.

From the table above for each ANOVA we discussed the factors that play an important role in each response variable as shown in Table 26.4.

26.3.2 Multi-response Optimization

With the desirability function obtained from Eq. (26.8), the feasible point where the response variables have the optimal value was found. Table 26.5 shows the summary of the obtained function desirability.

Table 26.5 shows 9 of the de initial 11 factors because two factors were not significant. The columns present the factors and the lines three response variables. For the response variable (ts1) significant factors are: NH, AP, MB, MT and AZ which have an effect to change level −1 to level 1. In optimizing the response variable (EMD) we found that significant factors are: NH, AP, AZ and ZnO. And as the response variable (Cost) we obtained as major factors the significant effects: NH, AP, EM, FI y TV. However, the TV factor has a significant effect in terms of

Table 26.2 Results of experimental design 2_{III}^{11-7} and response variables

Exp	Coded variables											Quality characteristics		
	NH	AP	AE	ZnO	EM	MB	TM	FI	AZ	TV	TMEZ	ts1[a]	EMD[b]	Cost[c]
1	−1	−1	−1	−1	−1	−1	−1	−1	1	1	1	2.54	41.50	15.95
2	1	−1	−1	−1	1	−1	1	1	−1	−1	−1	2.27	46.26	14.01
3	−1	1	−1	−1	1	1	−1	1	−1	−1	1	5.29	35.90	13.95
4	1	1	−1	−1	−1	1	1	−1	1	1	−1	2.4	34.71	15.00
5	−1	−1	1	−1	1	1	1	−1	−1	1	−1	3.56	40.57	17.27
6	1	−1	1	−1	−1	1	−1	1	1	−1	1	3.33	43.18	13.05
7	−1	1	1	−1	−1	−1	1	1	1	−1	−1	2.38	33.31	12.86
8	1	1	1	−1	1	−1	−1	−1	−1	1	1	3.44	35.27	15.89
9	−1	−1	−1	1	−1	1	1	1	−1	1	1	3.13	40.01	14.65
10	1	−1	−1	1	1	1	−1	−1	1	−1	−1	2.92	40.45	15.14
11	−1	1	−1	1	1	−1	1	−1	1	−1	1	2.69	28.81	15.11
12	1	1	−1	1	−1	−1	−1	1	−1	1	−1	3.92	31.84	14.01
13	−1	−1	1	1	1	−1	−1	1	1	1	−1	1.85	33.27	15.64
14	1	−1	1	1	−1	−1	1	−1	−1	−1	1	1.27	47.19	14.07
15	−1	1	1	1	−1	1	−1	−1	−1	−1	−1	5.23	31.37	14.01
16	1	1	1	1	1	1	1	1	1	1	1	2.56	31.36	14.90

ts1[a] Scorch time = measurement unit minutes
EMD[b] Tear maximum strength = measurement unit KN/m
Cost[c] Cost of formulation = measurement unit Mexican pesos

Table 26.3 Results of ANOVAS

Factors	gl	ts1 P-Value-P	EMD Value-P	Cost Value-P
NH	1	0.0390[a]	0.0148[a]	0.0024[a]
AP	1	0.0066[a]	0.0001[a]	0.0009[a]
ZnO	1	0.3806	0.0128[a]	0.5265
EM	1	0.8337	0.1873	0.0000[a]
MB	1	0.0035[a]	0.9906	0.5444
TM	1	0.0031[a]	0.2571	0.7429
FI	1	0.7084	0.5535	0.0000[a]
AZ	1	0.0051[a]	0.0277[a]	0.7644
TV	1	0.2968	0.0549	0.0000[a]
Total error	6			
Total (Corr.)	15			
R^2		93.65 %	95.50 %	99.13 %
R^2 adjusted		84.14 %	88.77 %	97.84 %

[a] significant

cost. The value d of the graph is equal to the degree of desirability of the response variables, that is, the d value close to 1 means that the response is desirable. Regarding the response variables they have an acceptable level of desirability. Regarding the overall desirability D, it refers to having a suitable value which

Table 26.4 Discussion of ANOVA of ts1, EMD, and Cost

ANOVA	Significant factors	Discussion	Regression equations derived from the Eq. (26.7)
ts1	NH, AP, MB, TM and AZ	NH load for altering the physical and chemical properties, imparts certain process properties, and reduces development costs. AP is an oil that improves the processability and fluidity of the compounds. MB primary accelerator, used for its efficiency and gives a good process reliability. TM secondary accelerator which promotes the time flow. AZ vulcanizing agent holding the vulcanization, because with no crosslinking (crosslinking of polymer chains), it would be impossible to obtain the improvements involved in the vulcanization	$ts1 = 3.04 - 0.28NH + 0.44AP + 0.50\,MB - 0.51TM - 0.46AZ$ (26.10)
EMD	NH, AP, ZnO and AZ	NH conducive to partial immobilization of rubber chains increasing tear resistance. AP oil was used to provide moderate flexibility to the product. ZnO is a trigger to optimize accelerators, increasing cure rates, helping to achieve better properties of the final product. AZ vulcanizing agent used to carry out the vulcanization reaction	$EMD = 37.18 + 1.59NH - 4.36AP - 1.65ZnO - 1.36AZ$ (26.11)
Cost	NH, AP, EM, FI and TV	NH is used to reduce the formulation cost. AP was used to decrease formulation costs. EM compatibilizing agent is used to make improve the adhesion of the natural fiber with rubber. FI is a natural fortifier to help improve product properties, used to reduce formulation costs. Besides having advantages such as low cost, low density, renewable resource and is abundant in Mexico. TV optimum cure time, achieving the highest number of crossovers and the faster crosslinking	$Cost = 14.71 - 0.21NH - 0.25AP + 0.51EM - 0.58FI + 0.69TV$ (26.12)

Table 26.5 Results of the desirability function for (ts1) (EMD) and (Cost)

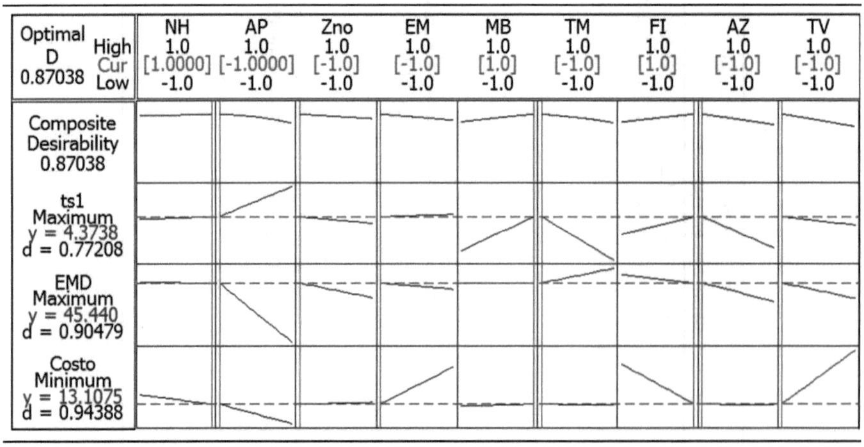

Table 26.6 Characteristics of control factor X

X\x		$\alpha = -\sqrt{2}$	-1	0	1	$\alpha = \sqrt{2}$
X_1	Calcium chloride	2.6	8.0	21.0	34.0	39.4
X_2	Texturing	2.5	6.5	16.2	25.9	29.9

shows that the vulcanization process achieves optimal global response. The values shown in brackets [ ] in the lines of factors correspond to levels that meet the response variables vulcanization process.

In order to illustrate the general approach to the problem and to show the optimization process using different methods, will present a classic example that has frequently been used in the literature.

An example at laboratory level before moving to the manufacturing process is shown. It is about the making of cheese and the aim is to know the combined effects of cysteine (rennet): X_1 and calcium chloride: X_2 in the texturing, as well as the features of hot dialyzed water in a protein concentration of serum on a gel. In this experimental procedure a central composite design was applied, where each $X_i (i = 1, 2)$ factor has five values, as shown in Table 26.6. Texture characteristics are measured by the hardness Y_1, cohesiveness (consistency) Y_2, elasticity Y_3, and compressible water Y_4. This study was developed by (Schmidt et al. 1979) and the expert in such processes aimed at the objective of simultaneous maximum for the four variables. The design used in this study was the composite core. In order to provide greater clarity to this scheme, the main characteristics of this type of design are presented and are used in this exhibit. For more information about this design please check Reference Myers and Montgomery (2002), and Khuri and Cornell (1996).

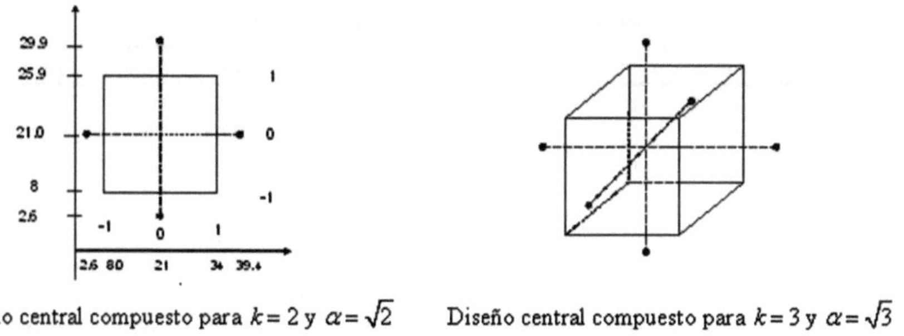

Fig. 26.1 Central composite designs for two and three factors, respectively. **a** for $k = 2$ and $\alpha = \sqrt{2}$, **b** The central composite design for $k = 3$ and $\alpha = \sqrt{3}$

26.3.3 Central Composite Design

1. A factorial design $2k$, where the levels of the factors (values) are coded values, usually -1 and 1, as will be seen later.
2. Two axial points on the axes of each design factor to distance α from the center of the design, in total of $2k$ points.
3. A number n_0 of points in the center of the design ($n_0 \geq 1$). The total number of experimental tests performed in the central composite design is: $2^k + 2k + n_0$.

Figure 26.1 illustrates the arrangement of these points for the cases $k = 2$ and $k = 3$. In the chart on the left in Fig. 26.1, the table corresponds to points 2^2, points $-\alpha$ and α (called axial points) are the $2k = 2(2)$ points, two axial points for each factor and n_0 is the number of treatments in the center.

Similarly, in the chart on the right in Fig. 26.1, we have the description to three factors. The value of α corresponds to the rotatability properties of the design or the orthogonality, here will consider only the first one, in such a case $\alpha = \sqrt{4}(2^k)$, the value of n_0 is also chosen based on these two properties. Particularly for the rotatability $n_0 = 4$ if $k = 2$ and $n_0 = 6$ if $k = 3$, the original details of this design were introduced (Box and Wilson 1951).

For the data of Example 1, in Table 26.6 the actual values of the two factors are shown. Since the factors are expressed in different units, the original values are transformed by the equation

$$x_i = \frac{X_i - (\max(X_i) + (\min(X_i))/2}{0.5 * [\max(X_i)] - \min(X_i)]}, i = 1, \ldots, k. \tag{26.9}$$

These x_i are known as coded values and are shown in the first row of Table 26.6.

Table 26.7 Experiment schematic and responses in each treatment

Treatment	Cte.	x_1	x_2	x_1x_2	x_1^2	x_2^2	Y_1	Y_2	Y_3	Y_4
1	1	−1	−1	1	1	1	2.48	0.55	1.95	0.22
2	1	1	−1	−1	1	1	0.91	0.52	1.37	0.67
3	1	−1	1	−1	1	1	0.71	0.67	1.74	0.57
4	1	1	1	1	1	1	0.41	0.36	1.20	0.69
5	1	$-\sqrt{2}$	0	0	2	0	2.28	0.59	1.75	0.33
6	1	$\sqrt{2}$	0	0	2	0	0.35	0.31	1.13	0.67
7	1	0	$-\sqrt{2}$	0	0	2	2.14	0.54	1.68	0.42
8	1	0	$\sqrt{2}$	0	0	2	0.78	0.51	1.51	0.57
9	1	0	0	0	0	0	1.50	0.66	1.80	0.44
10	1	0	0	0	0	0	1.66	0.66	1.79	0.50
11	1	0	0	0	0	0	1.48	0.66	1.79	0.50
12	1	0	0	0	0	0	1.41	0.66	1.77	0.43
13	1	0	0	0	0	0	1.58	0.66	1.73	0.47

Table 26.8 Regression coefficients for each one of the four models and CM_{error}

Coefficients	Response			
	$\hat{Y}_1$	$\hat{Y}_2$	$\hat{Y}_3$	$\hat{Y}_4$
$\hat{\beta}_0$	1.526[a]	0.66[a]	1.78[a]	0.47[a]
$\hat{\beta}_1$	−0.575[a]	−0.092[a]	−0.25[a]	0.13[a]
$\hat{\beta}_2$	−0.524[a]	−0.010	−0.078[a]	0.073[a]
$\hat{\beta}_{12}$	0.318[b]	−0.070[a]	0.01	−0.082[a]
$\hat{\beta}_{11}$	−0.171	−0.096[a]	−0.16[a]	0.026
$\hat{\beta}_{22}$	−0.098[c]	−0.058[a]	−0.08[a]	0.024
CM_{error}	0.040	$0.5\text{x}10^{-3}$	0.003	0.002
R^2	0.952	0.987	0.977	0.949

[a] highly significant, [b] significant, [c] marginally significant

Table 26.7 describes in the columns corresponding to x_1 and x_2 the central composite design for two factors and in this case $n_0 = 5$, and in the last four columns the values of the four responses are shown for each of the treatments.

The $Z(x)$ matrix for this example corresponds to $n = 13$ rows corresponding to treatments and to columns 2 to 7, that is $q = 6$, ($q = 2(2 + 3)/2 + 1$). The results of minimum squares for each of the four patterns is obtained by $\hat{\delta}_j = \left[(Z(x))^t Z(x)\right]^{-1}(Z(x))^t Y_j$. Table 26.8 shows the regression coefficients for the four models, equivalent to graphing the Eq. (26.6), plus statistical complementary information is shown such as the mean quadratic error CM_{error} and the coefficient of determination R^2.

The first is an estimator of σ^2 and the second indicates what percentage of the total variability is explained by the model. The corresponding expressions are:

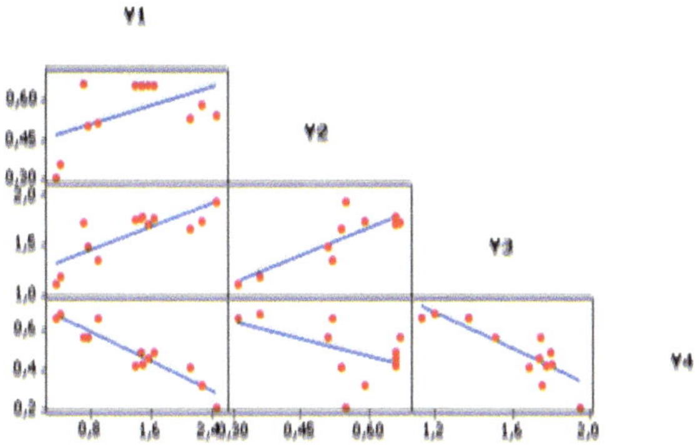

Fig. 26.2 Matrix describing the relationship between the four pairs of response variables

$$CM_{error}(j) = \frac{(Y_j - \hat{Y}_j)^t (Y_j - \hat{Y}_j)}{N - q} \tag{26.10}$$

and

$$R^2 = 1 - \frac{(Y_j - \hat{Y}_j)^t (Y_j - \hat{Y}_j)}{(Y_j - \bar{Y}_j)^t (Y_j - \bar{Y}_j)} \tag{26.11}$$

26.3.4 Solution

Graphical description of the models presented in this section to assess whether there is any relationship between them. The models adjusted by minimum squares and their individual optimums are shown. In Fig. 26.2 the relationship between the four variables of response is shown.

It can be noted in Fig. 26.2 that there is a consistent inverse relationship between the response variables Y_1 and Y_4 and between Y_3 and Y_4 and this trend is milder between Y_2 and Y_4; in that sense when you want to maximize the responses Y_1, Y_2, and Y_3 then Y_4 will decrease further in the case of Y_1 and Y_3. In the context of the problem, it can be said that the response Y_4 (the compressible water) is inversely related to the responses Y_1 (hardness) Y_2 (cohesiveness) and Y_3 (elasticity). In this situation you have to evaluate the importance of each response in the actual process to check what if it is more convenient to sacrifice response Y_4 or find a balance between the four responses. This scenario will be considered in the process of simultaneous optimization. In phase of maximizing responses Y_1, Y_2 and Y_3 there is no major complication because there is a direct relationship between them, however, it is important to consider the efficiency of optimization methods to the correlation of the response variables. The optimization approaches consider the importance of the functions with weight w_j, which will be discussed later.

Table 26.9 Statistical summary for response Y_1

Term	Coefficient	ES	t	p
Constante	1.526	0.089	17.071	0.000[a]
X_1	−0.575	0.071	−8.136	0.000[a]
X_2	−0.524	0.071	−7.417	0.000[a]
X_{12}	0.318	0.100	3.177	0.016[b]
x_1^2	−0.171	0.076	−2.250	0.059[c]
x_2^2	−0.098	0.076	−1.293	0.237

[a] highly significant, [b] significant, [c] marginally significant

The principle of minimum squares is applied to each of the responses, to fix ideas in Table 26.9 we show a summary of the report and statistical analysis for the response variables Y_1. In column 2 show the value of the coefficient that generates minimum squares $\hat{\beta} = (X^tX)^{-1}X^tY$ and the matrix of variance–covariance of the $\hat{\beta}$ vector is given by the expression $Var\left(\hat{\beta}\right) = (X^tX)^{-1}\hat{\sigma}^2$; where, in the main matrix $C = (X^tX)^{-1}\hat{\sigma}^2$, the input ii is $c_{ii}\hat{\sigma}^2$ and corresponds to the variance of $\hat{\beta}_i$, the i-th element of $\hat{\beta}$. The standard error of the regression coefficient of $\hat{\beta}_i$, is $ES\left(\hat{\beta}_i\right) = \sqrt{Var\left(\hat{\beta}_i\right)} = \sqrt{(X^tX)^{-1}\hat{\sigma}^2}$, where $\hat{\sigma}^2 = CM_{error}$. In column 3 of Table 26.9 are the results, and the $CM_{error} = 0.040$, which is obtained from the expression (26.10). In column 4, t corresponds to a value of the t distribution of Student $N - q$ degrees of freedom and it is obtained from the ratio between the regression coefficient and the standard error $\frac{\hat{\beta}}{ES\left(\hat{\beta}\right)}$, the probability left to the right in this distribution is shown in column 5 and to assess the significance this value is compared with a reference value, conveniently set, $\alpha = 0.05$. If $p < \alpha$, it is said that the corresponding factor is significant, as noted in Table 26.9. (Note: All calculations are rounded to thousandths).

A similar analysis is done for the other 3 responses, Table 26.8 shows the regression coefficients for the four models with an indication if were significant. Note that although the statistical analysis of the model has more details, here this statistical summary is sufficient for the implementation of the multi-response optimization.

The model for response 1 is:

$$\hat{Y}_1 = 1.53 - 0.57x_1 - 0.52x_2 + 0.32x_1x_2 - 0.17x_1^2 - 0.10x_2^2, \qquad (26.12)$$

Similarly the three other models are written. The graph on the right in Fig. 26.3 shows the response surface model for response Y_1, to the left are the response regions and the lines between regions are level curves. At first glance, it can be seen that the maximum response is approximately the point $x_0 = (-1.41, -1.41)$, this point coincides with the optimum in the optimization process $\ddot{Y}(x_O) = 3.178$

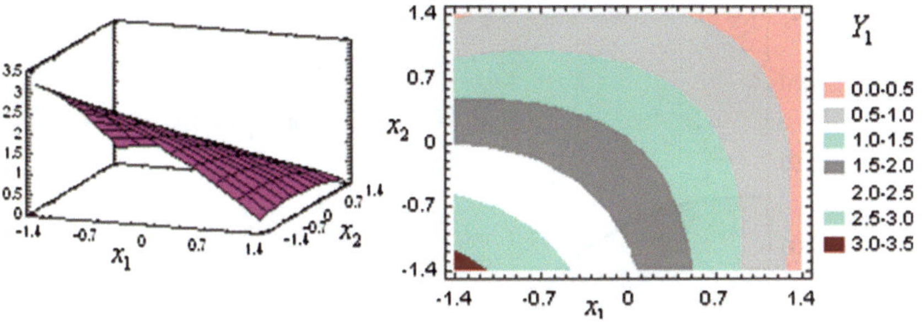

Fig. 26.3 Response surface and level lines for the model $\hat{Y}_1$

Table 26.10 Optimal solution for each response

Optimums	$\hat{Y}_1(x_{jo})$	$\hat{Y}_2(x_{jo})$	$\hat{Y}_3(x_{jo})$	$\hat{Y}_4(x_{jo})$
$x_{10} = (-1.41, -1.41)$	3.178	0.36	1.78	0.11
$x_{20} = (-0.58, 0.26)$	1.60	0.685	1.84	0.44
$x_{30} = (-0.82, -0.54)$	2.29	0.63	1.899	0.31
$x_{40} = (1.41, -1.41)$	0.29	0.37	1.04	0.815

(Table 26.10). The optimum is obtained by deriving of the model (26.16); in general for the second order model x_0 is

$$x_0 = -0.5\,\hat{\beta}^t\,\hat{\beta}^{-1} \tag{26.13}$$

where

$$\hat{\beta}^t = (-0.575, -0.524)\ y\ B = \begin{pmatrix} -1.17 & 0.16 \\ 0.16 & -0.10 \end{pmatrix}$$

A summary of the optimal response for each response is given in Table 26.10.

In the main diagonal of columns 2 through 5 in Table 26.10 are the maximum, the objective of the problem, for each of the four responses. Out of that diagonal three responses evaluated at the maximum of the other, and can be noticed that while one peaks the other are far from their peak. Particularly in the last row of Table 26.10, while Y_4 reaches its maximum the other three responses are far from their peak. A "watch!" in row 3 a value in x_0 is obtained which is suitable for this process, since in this case the four responses show a global optimum, as indicated below. In the global optimization process we will see if it is possible to get a better value for the four common responses. The next step is to find the global optimum that best satisfies at a maximum for all the responses.

First the optimization methods and approaches will be discussed, the first one corresponds to what is understood by linear programming and the second one

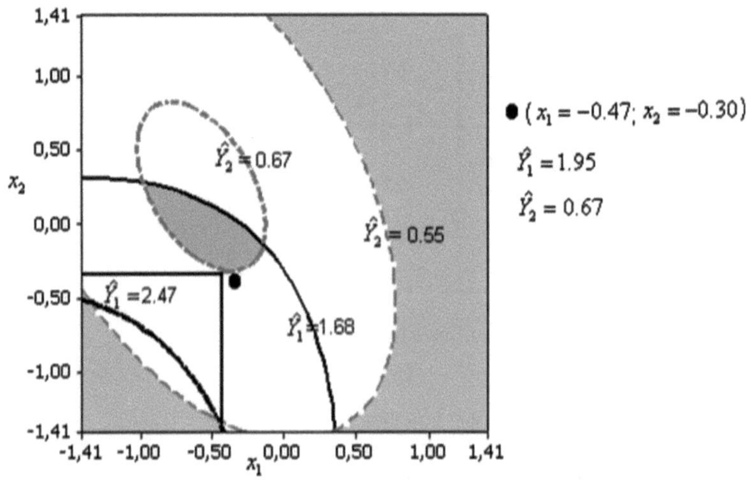

Fig. 26.4 Overlay of the models $\hat{Y}_1$ and $\hat{Y}_2$, description of a specific optimal solution for these two responses

considers the construction of a simple response function representing the multi-response.

When dealing with more than one response variable, regions or level curves can be overlaid, as shown in Fig. 26.3. To fix ideas, the first two responses are used, as described in Fig. 26.4. The shaded region indicates a common feasible solution for these two responses, having $\hat{Y}_1 \geq 1.6$ and $\hat{Y}_2 \geq 0.67$. In this situation the cohesiveness response is $\hat{Y}_2$, practically peaks, and the hardness response $\hat{Y}_1$ may still have better values. The point in Fig. 26.4 is $(x_1, x_2) = (-0.47, -0.30)$ was obtained by moving the level curves of the response Y_1. At this point the level curves corresponding to the two responses Y_1 and Y_2intercept, the values are $\hat{Y}_1$=1.95 and $\hat{Y}_2$=1.95 respectively. Regarding the problem, these values of hardness and cohesiveness are achieved when the X_1 factor has a level calcium chloride equivalent to 14.89 units and the X_2 factor equivalent to a level of texturization of 13.29 units, these values are obtained in reference to Table 26.6. It is observed from the chart that if Y_1 increases the other variable decreases.

In Fig. 26.5, the four response variables overlap, point • is a common value for the four responses that approaches to potentially being a global optimum, $(x_1, x_2) = (0.33, -1.40)$. Applying the expression (26.10), the actual optimum value 25.29 for calcium chloride and 2.62 texturization is obtained. As shown in the chart several regions of possible optimal solutions are generated.

It is common in the area of food biotechnology that experiments are carried with several response variables, for example (Tseo et al. 1983) evaluated several washing treatments for improving the quality of minced salmon meat. Then will result in the manufacture of this food.

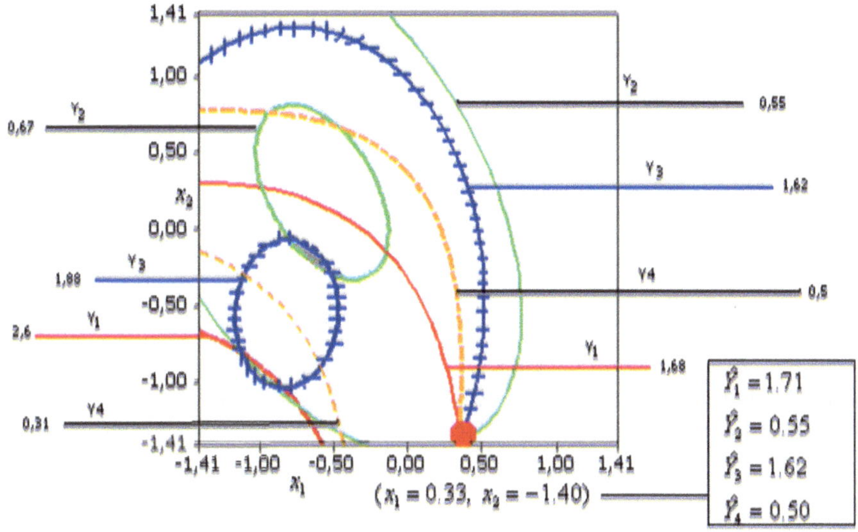

Fig. 26.5 A common optimal solution to overlay the four response variables

26.3.5 Approach to Multi-response Optimization

Below we show the expressions for the optimization by desirability function. This is an efficient analytical optimization procedure; its principle is to transform the j-th response variable described in model (26.6) to a set of values, called desirable value u with values between 0 and 1. This value grows as the best value of the corresponding response variable is required:

$$u_j\left(\hat{Y}_j(x)\right) = \begin{cases} 0 & si\ \hat{Y}_j(x) < Y_j^{min}\ o\ \hat{Y}_j(x) > Y_j^{max}, \\ 1 - \frac{M_j - \hat{Y}_j(x)}{M_j - Y_j^{min}} & si\ Y_j^{min} \le \hat{Y}_j(x) \le M_j, \\ 1 - \frac{\hat{Y}_j(x) - M_j}{Y_j^{min} - M_j} & si\ M_j < \hat{Y}_j(x) < Y_j^{max}, \end{cases} \tag{26.14}$$

where M is a target value set according to the interest of the researcher and Y_j^{min}, Y_j^{max} are two bounds of the j-th response. These must be set at the beginning. The first chart in Fig. 26.6 shows this situation. There are several criteria for determining these bounds, for example, the specification limits of a product, regulations or standards of a company, or just subjectively. If you need to determine the bounds based on a physical range of responses, it is reasonable to consider the minimum and maximum of the estimated individual responses, that is

$$Y_j^{min} = \min_{x \in R}[\hat{Y}_j(x)], \max_{x \in R}[\hat{Y}_j(x)]$$

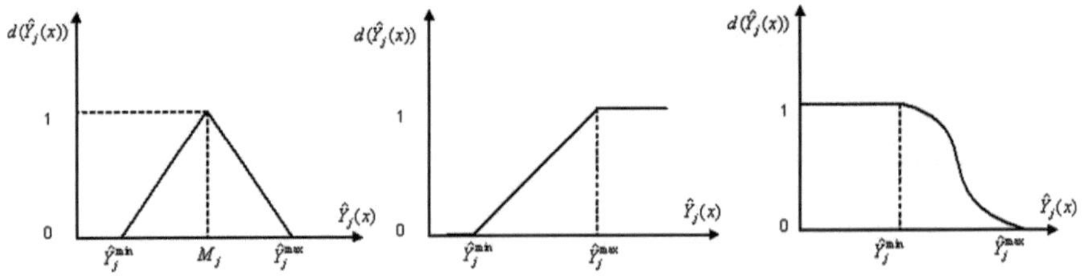

Fig. 26.6 Three representations to achieve the degree of satisfaction with an response, the first two linear and nonlinear second

The $u_j(\hat{Y}_j(x))$ function depends on the conditions of the process and one may also want to minimize their response, in the first case $M_j = Y_j^{min}$ is replaced in the third function in the expression (26.16):

$$u_j(\hat{Y}_j(x)) = \begin{cases} 1 & si\ \hat{Y}_j(x) < Y_j^{min}, \\ 1 - \frac{\hat{Y}_j(x) - Y_j^{max}}{Y_j^{max} - Y_j^{min}} & si\ Y_j^{min} \leq \hat{Y}_j(x) \leq Y_j^{max}, \\ 0 & si\ \hat{Y}_j(x) > Y_j^{max}, \end{cases} \tag{26.15}$$

If you want to maximize, $M_j = Y_j^{max}$, it is on the second function in (26.16). This situation is shown in the middle graph in Fig. 26.6, and the expression is:

$$u_j(\hat{Y}_j(x)) = \begin{cases} 0 & si\ \hat{Y}_j(x) < Y_j^{min}, \\ 1 - \frac{Y_j^{max} - \hat{Y}_j(x)}{Y_j^{max} - Y_j^{min}} & si\ Y_j^{min} \leq \hat{Y}_j(x) \leq Y_j^{max}, \\ 1 & si\ \hat{Y}_j(x) > Y_j^{max}, \end{cases} \tag{26.16}$$

The third graph in Fig. 26.6 shows the non-linear case of function $u_j(\hat{Y}_j(x))$.

26.3.6 Desirability Function

The desirability function (DE) was proposed by Harrington (1965) and its classical expression is obtained from the expression (26.16) that is, $d_j = d(\hat{Y}_j(x)) = u_j(\hat{Y}_j(x))$. In this case it is assumed that the degree of satisfaction of an experimenter, with respect to the *j*-th variable is maximized when $\ddot{Y}_j(x)$ equals its value M_j and decreases as $\hat{Y}_j(x)$ moves away from M_j. If Y_j^{min} and Y_j^{max} represent respectively the minimum and maximum bounds, then the solution point x is not accepted for $\ddot{Y}_j(x) < Y_j^{min}$ or $\ddot{Y}_j(x) > Y_j^{max}$. So the degree of acceptance regarding the response is modeled as a monotonically decreasing function of 1 in $\ddot{Y}_j(x) = M_j$ to 0 in $\hat{Y}_j(x) < Y_j^{min}$ or $\hat{Y}_j(x) > Y_j^{max}$. The overall desirability is obtained by the geometric mean:

$$Maximize\ \lambda, \tag{26.17}$$

Subject to $d(\hat{Y}_j(x)) \geq \lambda, j = 1, 2, \ldots, r,$

$$X \in RR : \text{experimental region}$$

The main purpose of this formulation is to find a point x_0 that maximizes the minimum degree of satisfaction λ with respect to all responses within the experimental region.

References

Ames, A. E., Mattucci, M., Stephen, M., Szonyi, G., & Hawkins, D. M. (1997). Quality loss functions for optimization across multiple response surfaces. *Journal of Quality Technology, 29*, 339–346.

Box, G. E. P., & Draper, N. R. (1987). *Empirical model building and response surfaces*. New York: Wiley.

Box, G. E. P., & Wilson, K. B. (1951). On the experimental attainment of optimum conditions (with discussion). *Journal of the Royal Statistical Society, B13*, 195–241.

Castaño T. E., Domínguez, D. J. (2010). Design of experiments: Strategy and analysis in science and technology. UAQro., CIMAT, México (In Spanish).

Harrington, E. (1965). The desirability function. *Industrial Qualty Control, 21*, 494–498.

Khuri, A. I., Cornell, J A. (1996). *Response surface, designs and analysis*. New York: Marcel Dekker, Inc.

Myers, R., & Montgomery, D. C. (2002). *Response surface methodology: Process and product optimization using designed experiments*. New York: Wiley Series in Probability and Statistics.

Ortiz, F., Simpson, J., & Pignatiello, J. (2004). A genetic algorithm approach to multiple-response optimization. *Journal of Quality Technology, 36*(4), 432–450.

Schmidt, R. H., Illingworth, B. L., Deng, J. D., & Cornell, J. A. (1979). Multiple regression and response surface analysis of the effects of calcium chloride and cysteine on heat-induced whey protein gelation. *Journal of Agriculture and Food Chemistry, 27*, 529–532.

Tseo, C. L., Deng, J. C., Cornell, J. A., Khuri, A. I., & Schmidt, R. H. (1983). Effect of washing treatment on quality of minced mullet flesh. *Journal of Food Science, 48*, 163–167.

Chapter 27
Dynamic Analysis of Inventory Policies for Improving Manufacturing Scheduling

Cuauhtémoc Sánchez-Ramírez, Miguel Rendón-Sagardi, Guillermo Cortes-Robles, Dulce Mota-López and Giner Alor-Hernández

Abstract Many researchers cite the automotive industry to study the application of Lean Manufacturing in reducing waste and improving productivity. However, in practice, the use of Lean Manufacturing techniques has spread into other industrial and service sectors, such as health and food, because of the benefits that this practice can achieve. Furthermore, different studies demonstrate that Lean Manufacturing combined with others techniques, such as simulation, produces benefits that impact on the key performance indicators of a company. Thus, in this study we analyze the combination of a simulation approach as System Dynamics on Lean Manufacturing practice in order to improve procurement policies and reduce the inventory in a livestock feed company.

Keywords Lean manufacturing · System dynamics · Inventory policies

27.1 Introduction

Lean Manufacturing (LM) is one of the most widely accepted practices in the automotive industry. The reason of LM application lies on its capacity for improving competitiveness without reducing the quality standards. Moreover, LM increases the options in vehicles assembling, which is an important competitive advantage for any automotive company (Holweg 2007).

Womack et al. (1990) described the concept of Lean Manufacturing, but the source of this practice is the Toyota Production System developed in Japan by Taiichi Ohno and Shigeo Shingo (Spear and Bowen 1999). This production philosophy of Lean Manufacturing is based on waste reduction in all operations

C. Sánchez-Ramírez (✉) · M. Rendón-Sagardi · G. Cortes-Robles · D. Mota-López · G. Alor-Hernández
Division of Research and Postgraduate Studies, Instituto Tecnológico de Orizaba, Av. Oriente 9, No. 852, Orizaba, Veracruz, Mexico
e-mail: csanchez@itorizaba.edu.mx

J. L. García-Alcaraz et al. (eds.), *Lean Manufacturing in the Developing World*,
DOI: 10.1007/978-3-319-04951-9_27,

through a series of tools and techniques (e.g., cellular manufacturing, synchronous manufacturing, Poka-Yoke, Standardized working, and work place organization, among others) (Bicheno 2000). All these tools and techniques have a positive impact on several industrial activities or factors; however, one of them is of special importance: inventories (e.g., work in process, raw materials inventories, and finished goods inventories). LM aims to eliminate global inventories as strategy to reduce the manufacturing cost (Watanabe and Hiraki 1997).

However, the study of Lean Manufacturing has used other analysis approaches such as the simulation technique (Robinson et al. 2012; Azlina et al. 2012), in order to complement the lean practice and improve the manufacturing process in any company. Thus, in this study is described a model which combines the concept of Lean Manufacturing and System Dynamics approach (Forrester 1961). The objective of this integration is to improve the procurement policies of the raw material inventory, since these policies seem to be empirically applied in a company and this produces higher inventories levels, which also generates an inappropriate use of facilities.

The chapter is organized as follows: Sect. 27.2 describes the state of the art from different perspectives (i.e., Lean Manufacturing and System Dynamics). Section 27.3 briefly describes the methodology and the implications of the proposed model, but also defines the context to develop a System Dynamics model simulation. Section 27.4 summarizes results and discussions. Finally, Sect. 27.5 describes the conclusion and future research.

27.2 State of Art

John Krafcik was the first person to define the term Lean Manufacturing (LM) in 1988, when he was a researcher at the MIT International Motor Vehicle Program (IMVP) (Jurado and Moyano 2011). However, according to Voss (1995), the origin of LM is the Toyota Production System (TPS). LM could be used in different processes within a company (see Fig. 27.1) but first, the company should define a systematic lean implementation and evaluation as is proposed in (Amin and Karim 2012).

LM has been used in different industrial and service sectors. On the one hand, Robison et al. (2012) used LM concepts and discrete event simulation (DES) to analyze the improvement of health services. This work demonstrated that the combination of LM and DES could improve the decision making process in this sector. On the other hand, Riezebos et al. (2009) combined Information Technology (IT) and LM to improve three activities in a company: the use of IT in production logistics, production systems, and computer-aided advanced maintenance plant.

Moreover, according to Dombrowskia et al. (2012) the most important aspect for a successful implementation of LM must be a change in the knowledge of people; otherwise, changes in the company will not be sustained. Meanwhile, Krogstie and

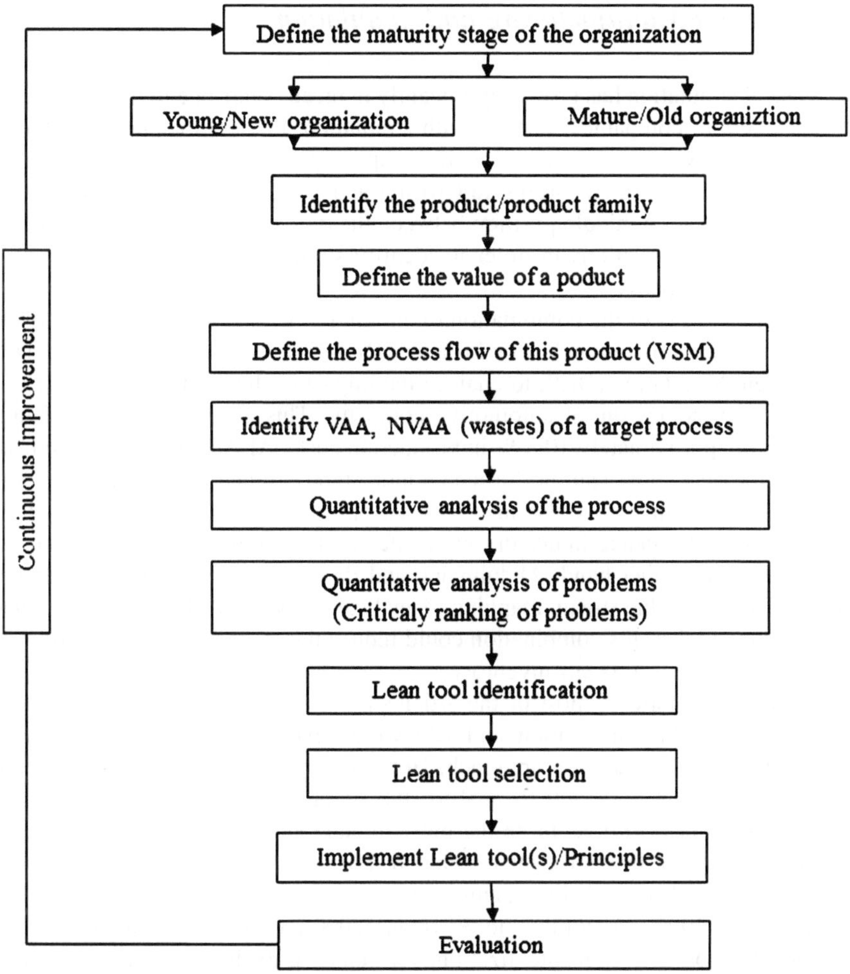

Fig. 27.1 A systematic lean implementation and evaluation process (Amin and Karim 2012)

Martinsen (Krogstie and Martinsen 2013) used Six Sigma and Lean Manufacturing to reduce variation and improve performance in manufacturing processes; however, authors found that there is an opportunity for combining both approaches, since they have not been widely explored yet. Also, Chen et al. (2013) discussed the use of Radio Frequency Identification (RFID) and Lean Manufacturing to improve efficiency and effectiveness in the Supply Chain Management. Results showed that the use of the RFID can reduce the cost of production significantly and, consequently, increase return on investment (ROI). Results also demonstrated that this combination could be effective and feasible. A key performance indicator in the case of Lean Manufacturing is an expected reduction of inventories (Eroglu and Hofer 2011; Demeter and Matyusz 2011; Hofer et al. 2012).

27.2.1 Lean Manufacturing and Simulation

The use of simulation has grown significantly in recent years due to the advantages that it offers to companies. In fact, with simulation, these companies are able to visualize, analyze, and improve their complex processes of production (Sandanayake et al. 2008). According to GröBler and Schieritz (2005), the simulation approach provides a middle position between the pure mathematical modeling and the empirical observations in order to identify strategies that can improve some areas of the company.

The advantages of the combination of simulation with LM have been analyzed in several pieces of research. For instance, Diaz-Elsayed et al. (2013) used Discrete Event Simulation (DES) to analyze the impact of the implementation of LM and green strategies on an automotive company. This experiment resulted in a decrease of approximately 10.8 % in production costs. Al-Aomar (2011) also used DES to measure three lean performance indicators: productivity, cycle time, and work in process inventory. He also used an optimization model to identify how these three performance indicators were affected by variability in the production processes. On another hand, Abdulmalek and Rajgopal (2007) combined simulation (DES) with value stream mapping to analyze the benefits of LM. The obtained results showed that this combination could reduce the cycle time of production, as well as the work in process inventory.

As it can be noticed, most of the articles that make reference to some advantages of integrating simulation and LM were developed using Discrete Event Simulation (DES) as a main approach. However, other authors have used a continuous simulation approach to analyze the production processes. In this case, System Dynamics (SD) could be the most suitable approach.

MIT researcher Jay Forrester developed the SD approach, which is useful to understand the characteristic dynamics of complex systems through a simulation model (Ford 1999). According to Sterman (2000), four stages can be used to develop SD models: conceptualization, formulation, evaluation, and implementation. In the conceptualization stage, the system under study is defined through a Causal Loop Diagram (CLD) in order to represent the relationships between the variables that create the feedback loops. A feedback loop can be of two types: while a balancing feedback loop seeks balance in a system, a reinforcing one generates growth and amplifies deviations (Sterman 2000). The second stage, formulation, includes: (1) the use of techniques to define the parameters of the variables influencing the system under study, and (2) the system modeling in specialized software, where the CLD is converted into a Flow and Stock Diagram (FSD), which is driven by a set of differential equations. In the evaluation stage the verification and the validation of the model is carried out. Finally, in the implementation stage, the model can generate results and help support the decision making process.

The following section provides a brief comparison between both approaches: Continuous Simulation and Discrete Simulation.

Table 27.1 Comparison between continuous and discrete simulation approaches

Features	SD	DES
Objective of model	Strategic–Tactical	Tactical–Operational
Elements of construction	Stocks and flows	Queues and activities
Time of simulation	Continuous	Discrete
Feedback	Explicit	Implicit
Perspective	Holistic	Analytical
Data input	Quantitative–Qualitative	Quantitative
Validation	White box approach	Black box approach
Relation among variables	Linear–Nonlinear	Linear

27.2.2 Continuous and Discrete Simulation Approaches

Although different simulation methodologies and meta-heuristics have been applied to the analysis of LM, in this section we analyzed the differences between System Dynamics (SD), which is a continuous simulation approach, and Discrete Event Simulation (DES).

Several comparisons between both approaches were proposed by Sweeter (1999), Scholl (2001), Morecroft and Stewart (2005), and Tako and Robinson (2009), among others. A synthesis of these works is described in Table 27.1.

27.3 Integrating System Dynamics on Lean Manufacturing

Since one of the advantages, from SD approach, is to identify the interactions between the system's variables. This approach considers several features that can benefit the Lean Manufacturing (LM) practice. For instance, the dynamic hypothesis can help prove different policies in order to support decision-making process, and eventually reduce processing time, inventories, or human resources, which are performance indicators in LM practice.

SD has been used in the analysis and improvement of supply chains and logistic processes; however, only a few studies have considered SD in lean production systems. For instance, Zhang et al. (2013) developed a SD model to analyze the sustainability of a production process, and Elmoselhy (2013) analyzed the way in which a lean-agile production system can be implemented in the automotive sector.

The main objective in this case study is to use the SD approach to reach a fundamental goal in LM: which is to evaluate inventory policies as strategy for inventory and cost reduction.

The production system under study belongs to a livestock feed company. The inventory current policies in this company were proposed and implemented empirically, producing several problems such as disruptions or overstocking of raw material inventory, and even in finished goods inventory.

27.3.1 Case of Study: Inventories and Scheduling in a Livestock Feed Plant

This study analyzes the situation of a livestock feed plant located in the state of Veracruz, Mexico. The plant produces 41 different animal feeds to raise and fatten poultry, pig, and cattle.

Each product has a particular nutritional formula involving the use of a variety of ingredients (components of the formulas). However, the whole manufacturing method uses batches, and each product follows the same sequence of unit operations: (1) Formulation, (2) Mixing, (3) Pelletizing, and (4) Bagging. The general production process is outlined in Fig. 27.2.

The company uses a sales monthly forecast, in order to create a production program at the end of each month. The forecast and the corresponding nutritional formulation generate the necessities for each raw material. All this information is useful to estimate the daily consumption rate, which is calculated by dividing the average use of raw material by the number of weekdays. This information and each product's priority are combined in the master production schedule. However, the prioritization is a subjective process conducted frequently by the company's executives.

27.3.2 Describing the Problem

Inventories rule the production schedule, and the company uses a safety stock of 5 days of production for each component in any given formula. The company wishes to know whether the current inventory policy is suitable for all raw materials. Therefore, SD approach is used as a tool to analyze if the raw material inventory can meet the production schedules. To reach this objective, five activities are necessary: (1) evaluate the efficacy of assigned priorities, (2) analyze in real time the movement of all their inventories (raw material, finished goods, and work in process), (3) determine the number of complete orders and orders to complete, (4) identify non-productive time by machine, and finally (5) propose alternative production schedules.

27.3.3 Causal Loop Diagram

The Causal Loop Diagram (CLD) is used to schematize the relationships among the variables of the system under study. Thus, the CLD can guide the model construction in the simulation software, and it is also useful to verify the model. The developed CLD is presented in Fig. 27.3.

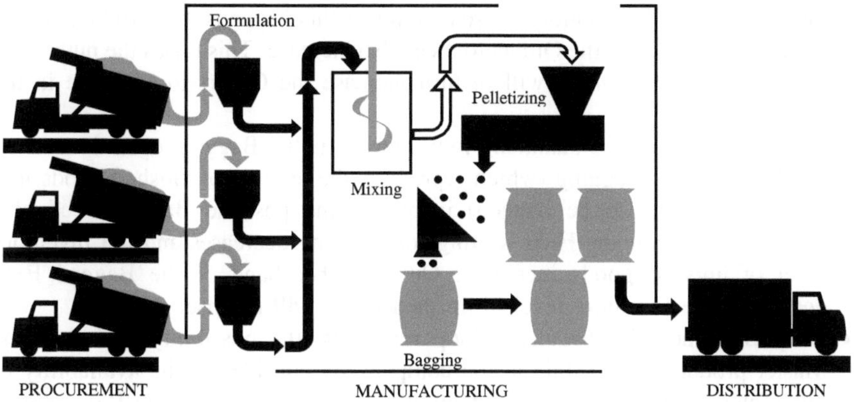

Fig. 27.2 Unit operations in livestock feed manufacturing

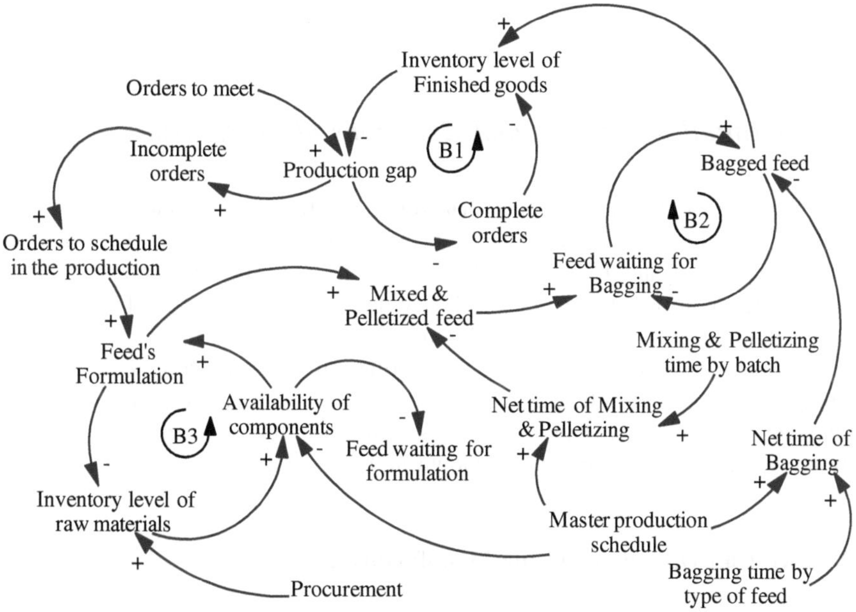

Fig. 27.3 Causal loop diagram of the dynamics of livestock feed manufacturing

One of the most important characteristics of CLD are feedback loops. The following points describe the behaviors of each of these loops:

- Loop B1: This balancing loop represents the need of the company to manufacture certain feeds when the finished goods inventory cannot meet an order. Thus, if the Production Gap (the difference between what is demanded and what can be supplied) decreases, the number of Complete orders increases. Thereby, the finished goods inventory decreases, but if this happens, the Production gap increases, as much as the Orders to meet also increase. This raises the number of incomplete orders, which will in turn increase the Orders to schedule in the production.

- Loop B2: This second balancing loop represents the Bagging operation. If the Feed waiting for Bagging (which represents the amount of finished goods in a container waiting to be bagged) increases, the possible Bagged Feed also increases; however the Feed waiting for Bagging decreases and the Inventory level of finished goods increases. On the other hand, if the Bagged Feed decreases, the Inventory level of finished goods will not rise.
- Loop B3: The third balancing loop represents the possibility to follow the Master production schedule in the company by considering the availability of raw materials. If the Master Production Schedule has an efficient Procurement process, the Inventory level of raw materials will rise. This will produce higher Availability of components for the manufacturing process, and will also make possible the Feed's Formulation. Similarly, when the Feed's Formulation increases, the Inventory level of raw materials decreases.

27.3.4 Initial Conditions and Equations of the System

The company produces 41 products. However, a Pareto analysis showed that 18 of these 41 products reach the 80 % of the company's income. Therefore, the information associated to these products was collected. The information included: the products nutritional formula, batch size, processing times, size of the bags for sale, and initial inventory of finished goods. The bill of materials was obtained by decomposing nutritional formulas. The order quantity (procurement policies), storage capacity, and initial inventory for each raw material were obtained in the company's warehouse. The list of raw materials, initial inventories, and the monthly consumption average are shown in Table 27.2.

27.3.4.1 Unit Operations in the Manufacturing

The unit operations to manufacture the livestock feed in the company under study are:

Formulation: In order to manufacture a given product, the specific components of the formula are taken from raw material inventories. In this study, the time spent on this operation is added to the time for mixing and pelletizing.

Mixing and pelletizing: in these operations, the components are integrated and molded to take the shape of a pellet. The operation time per batch was obtained using historical data. As it was expected, there is a different processing time for every product. Thus, even for the same product, the mixing and pelletizing time differs, which is why a probability distribution based on the relative frequency is

Table 27.2 Initial raw materials inventory and average consumption

ID	Ingredient	Initial inventory (Kg)	Average monthly usage (Kg)
1.	Corn	824,622	1,869,750
2.	Gluten	18,849	16,860
3.	Soy base	973,612	277,490
4.	DDG's	72,087	135,490
5.	Canola base	33,729	458,330
6.	Meat meal	61,300	39,970
7.	Chicken meal	10,000	132,110
8.	Calcium	55,613	116,410
9.	Kaolin	24,810	63,440
10.	Corn bran	45,170	145,490
11.	Rice bran	34,299	402,910
12.	Wheat bran	51,802	400,070
13.	Barley bran	27,398	89,990
14.	Soy bran	8,000	75,070
15.	Liquid methionine	8,363	1,390
16.	Liquid choline chloride	13,376	1,230
17.	Liquid L-Lysine	26,766	9,260
18.	Molasses	64,654	193,000
19.	Propionic acid	4,483	3,610
20.	Micro-components	10,000	94,050

used. For instance, Table 27.3 presents the time distribution to mix and pelletize the product 7 (intended for fattening pigs).

Bagging: in this operation, a machine fills the bags with the finished feed. In order to determine the operation time, historical data of all selected products was analyzed to define one or a set of probability functions. As an example, Table 27.4 shows the bagging time for pig feed.

27.3.4.2 Simulation Model

In response to market changes; the company's consumption projections, sales forecasts, and product formulations are made on a monthly basis. Therefore, the horizon of simulation was set at a month.

In the simulation model, a daily production program is the main tool to manage the manufacturing. Also, any given production plan depends on the raw material availability. For this reason, the model was divided in two sectors to observe: (1) the projected flow of raw material inventory, and (2) the actual movement according to the daily consumption. Two equations rule these sectors (Eqs. 27.1 and 27.2):

$$\frac{d}{dt}(\mathrm{PRMI_i}) = \mathrm{RMI_i}|_{t=0} + \mathrm{OQ_i} - \frac{\mathrm{AMU_i}}{\mathrm{WD_i}} \tag{27.1}$$

Table 27.3 Formulation, mixing and pelletizing time for product 7 (ID company: 204)

Class	Time (hours/batch) Lower bound	Frequency Upper bound	Probability	Cumulative probability	
1.	0.19	0.23	9	0.23	0.23
2.	0.24	0.28	18	0.45	0.68
3.	0.29	0.33	8	0.2	0.88
4.	0.34	0.38	3	0.08	0.95
5.	0.39	0.43	2	0.05	1
Total	40	1.0			

Table 27.4 Bagging time for product 7 (ID company: 204)

Class	Time (hours/batch) Lower bound	Frequency Upper bound	Probability	Cumulative probability	
1.	1.62	1.7	6	0.17	0.17
2.	1.71	1.79	10	0.29	0.46
3.	1.8	1.88	10	0.29	0.74
4.	1.89	1.97	7	0.2	0.94
5.	1.98	2.06	1	0.03	0.97
6.	2.07	2.15	1	0.03	1
Total	35	1.0			

$$\frac{d}{dt}(ARMI_i) = RMI_i|_{t=0} + OQ_i - F_i - UOP_i \tag{27.2}$$

In (27.1), $PRMI_i$ represents the projected raw material inventory (expected inventory), AMU_i is the monthly consumption average of raw materials, and WD_i quantifies working days during a month. In (27.2), $ARMI_i$ represents the actual raw material inventory, Fi is the daily use (formulation feed) to manufacture each product under study; and finally, UOP_i stands for the daily use of raw materials in other feed manufacturing. In both equations, $RMI_i|_{t=0}$ represents the inventory level at the beginning of the simulation, and OQ_i stands for the planned order quantity. All this is possible for the component i, where i = {1, 2, …, 20}.

As example, Fig. 27.4 shows the Flow and Stock Diagram (FSD) to represent the beginning of the Formulation according to the projected schedule, but considering the restriction of inventory.

It is important to highlight that this model is linked to a spreadsheet containing information about the production program priorities and production scheduling. The information of April 2013 was used for the purpose of this study; however, Table 27.5 shows a fragment of the production plan and the priorities for April 2nd due to the length of the information report.

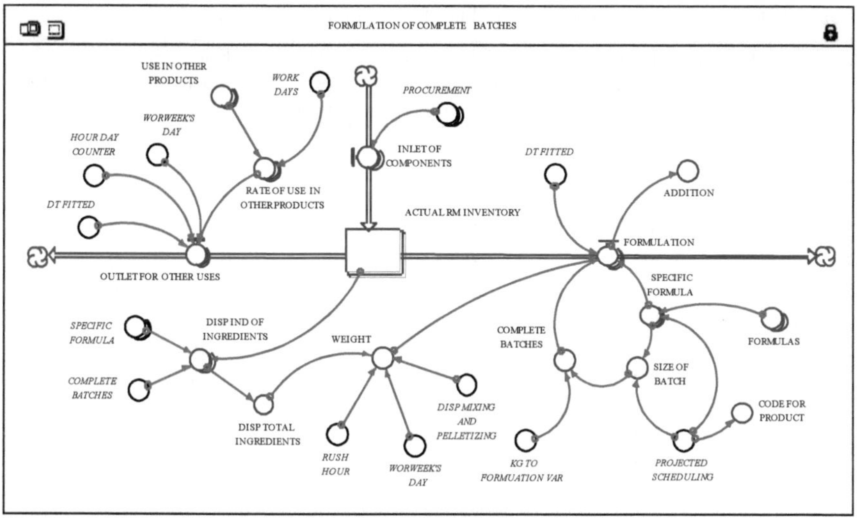

Fig. 27.4 Flow and stock diagram for formulation

Table 27.5 Production plan and priority for April 2nd

Product	ID company	Name of product (feed)	Priority	Kilograms to manufacture
5	182	Poultry feed (raising)	2	20,000
14	256	Fattening feed for cattle	3	3,000
18	292	Poultry feed (final step)	1	20,000

27.4 Results

Table 27.6 introduces the production schedule of April 2nd considering that any working day starts at 7:00 a.m. Products are sequenced according to the priorities, and they remain a certain amount of time in each unit operation until the end of the manufacturing process. The described production schedule in Table 27.6 is schematized in Fig. 27.5 using a timeline.

In the timeline (Fig. 27.5), the product ID 292 begins its manufacturing process at 7:00 a.m. due to its priority, and it stays in the process of mixing and pelletizing during 1 h and 45 min. At the end of this process, the product moves to the bagging operation at 8:45 a.m. and remains there for 1 h. At 9:45 a.m., once the bagging operation is concluded, the product ID 292 is considered as a finished good, and consequently, it is available to meet the customer's orders.

The product ID 182 is scheduled for the mixing and pelletizing process 15 min after the product ID 292 has left the process. These 15 min of pause represent the setup time or time needed to prepare the machine for the manufacturing of a different product. Therefore, from 9:00 to 10:45 a.m., the product ID 182 is mixed and pelletized. The following operation (bagging) is executed from 10:45 to 11:45 a.m. At 11:45 a.m. the product is ready to meet the customer's orders.

Table 27.6 Production sequencing according the scheduling for April 2nd

Day	Month	Hour	ID product to mix and pelletize	Kilograms to mix and pelletize	ID product to bag	Kilograms to bag	Kilograms of bagged feed
2	4	7	292	2,857.14	0	0	0
2	4	7.25	292	2,857.14	0	0	0
2	4	7.5	292	2,857.14	0	0	0
2	4	7.75	292	2,857.14	0	0	0
2	4	8	292	2,857.14	0	0	0
2	4	8.25	292	2,857.14	0	0	0
2	4	8.5	292	2,857.16	0	0	0
2	4	8.75	0	0	292	5,000	0
2	4	9	182	2,857.14	292	5,000	0
2	4	9.25	182	2,857.14	292	5,000	0
2	4	9.5	182	2,857.14	292	5,000	20,000
2	4	9.75	182	2,857.14	0	0	0
2	4	10	182	2,857.14	0	0	0
2	4	10.25	182	2,857.14	0	0	0
2	4	10.5	182	2,857.16	0	0	0
2	4	10.75	0	0	182	5,000	0
2	4	11	256	1,500.00	182	5,000	0
2	4	11.25	256	1,500.00	182	5,000	0
2	4	11.5	0	0	182	5,000	20,000
2	4	11.75	0	0	256	1,500	0
2	4	12	0	0	256	1,500	3,000

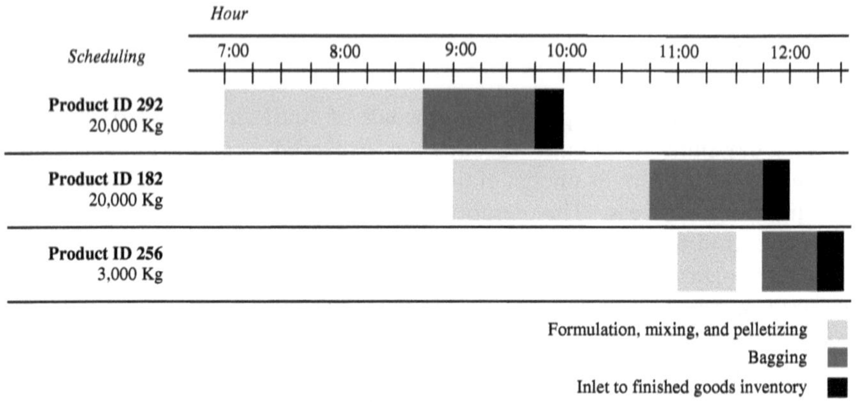

Fig. 27.5 Timeline for sequencing and scheduling

Finally, the product ID 256, which has the lowest priority, begins the mixing and pelletizing operations at 11:00 a.m. (the setup time is already considered). The product remains 30 min in these operations, since this product is requested in less quantity than the previous ones. The next operation (bagging) cannot be carried

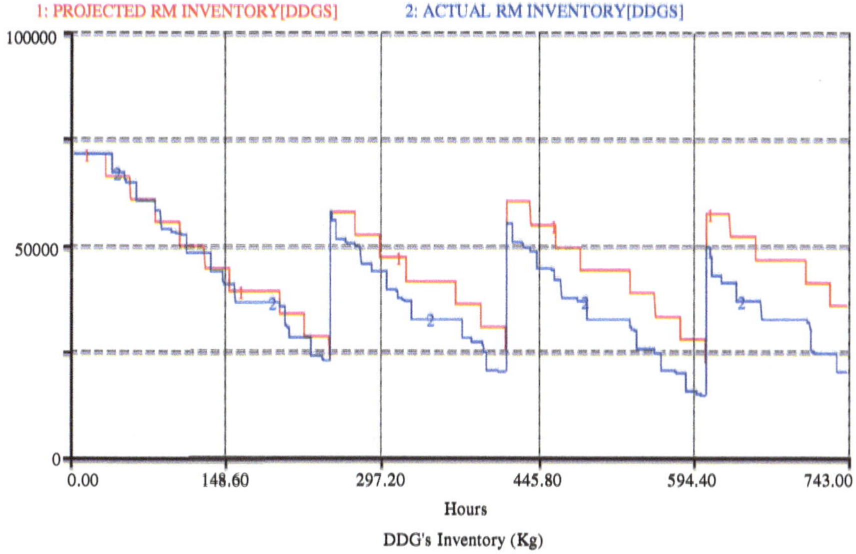

Fig. 27.6 Actual and predicted inventory of dried distillers

out, because the product ID 182 has not finished this process and is still making use of the facilities. Therefore, the product ID 256 needs to wait until 11:45 a.m. to continue with the bagging operation, and it will be considered as a finished good after 12:15 p.m.

The provided information in this section is highly important, since it facilitates the testing of different production schedules scenarios and permits evaluating the expected takt time for the production of any given product. The scheduling information is also useful to conceive optional production scenarios (and eventually to achieve a Lean Manufacturing).

Therefore, the model is able to test different scenarios and consider the level of inventories, the orders to meet, and the assigned priorities under one particular condition: when there is a lack of one or more raw materials, or when there are several restrictions to manufacture a certain product.

An example of additional information that can be obtained from the simulation model is presented Fig. 27.6, in which the current (actual) and the predicted (expected) consumption of Dried Distillers Grains (DDG's) is compared. Thus, earlier April shows that both the forecasted and the current consumptions are similar; however, at the end of the month, the actual consumption happens to be higher than it was expected; hence, the actual inventory level is progressively decreasing below the forecasted amount. These situations require a detailed analysis, since an unexpected decrease in the inventory level would cause a delay in the manufacturing.

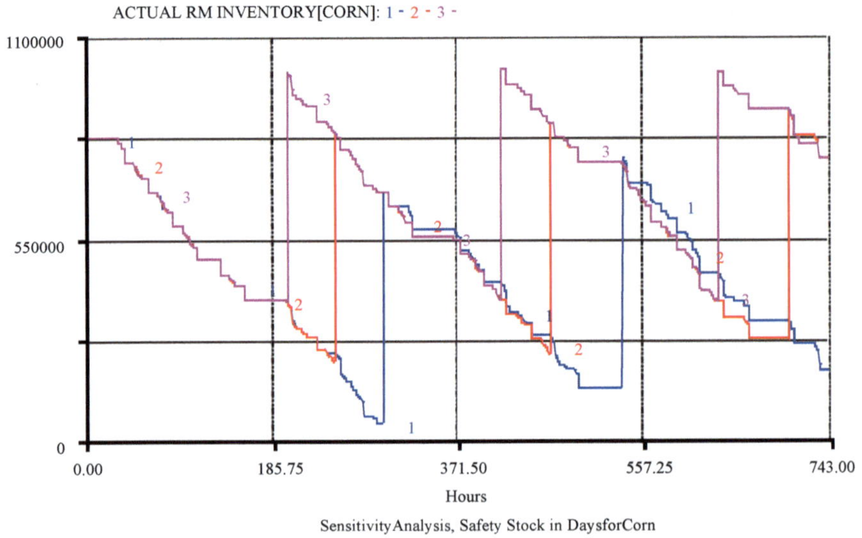

Fig. 27.7 Sensitivity analysis for days of safety stock, component: corn

27.4.1 Sensitivity Analysis of Inventory Policies

The company follows a 5 days safety stock policy for all raw materials. This restriction was defined based on the company's experience and, thus it is mainly a subjective criterion, which needs to be evaluated to observe its impact over global performance. The purpose of this section is to determine the effectiveness of this policy with the aim to change the number of days as safety stock for raw materials. This evaluation considers the associated cost and the risk of inventory shortage.

As example, Figs. 27.7 and 27.8 explore the inventory response of two raw materials: corn and DDG's. Line 3 represents the inventory level considering 5 days of safety stock, while line 2 considers 3 days of stock, and finally, line 1 stands for only 1 day.

Figure 27.7 illustrates that even with a safety stock of 1 day, the corn inventory does not generate any supply problems, but in exchange, it could decrease cost. However, for the case of the DDG's (Fig. 27.8), a safety stock of 1 day generates shortage, and this condition increases the risk in the manufacturing scheduling.

Table 27.7 illustrates the maximum and minimum inventory levels according to a safety stock for 7 of the 18 components that may be changed without incurring shortages in manufacturing. However, executives need to consider a particular situation: a lower inventory level reduces the associated production cost, but also reduces the company's capacity to respond to an increasing demand for a product. Consequently, the production system becomes more sensitive to uncertainties in procurement and market variations.

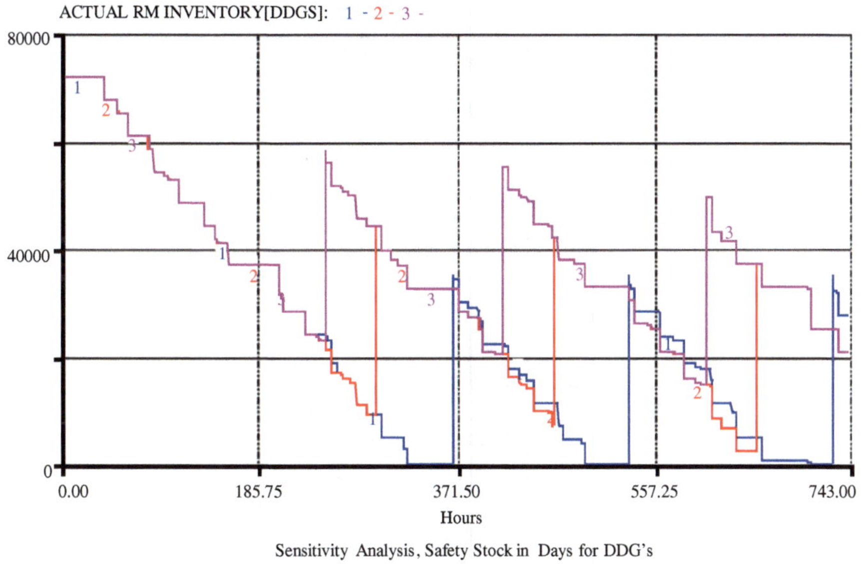

Fig. 27.8 Sensitivity analysis for days of safety stock, component: DDG's

27.4.2 Graphical User Interface

An interface was created to avoid modifying every single parameters in the Flow and Stock Diagrams (simulation model); i.e., to facilitate the testing of any given procurement and manufacturing scenario.

Figure 27.9 depicts the main window from the interface. The first step is to select and modify the variables of interest, which may include: (1) Times and dates of manufacture, (2) Nutritional formula to use, (3) Raw material inventory at the beginning of the month, or (4) Finished goods inventory at the beginning of the month. Then, when the NEXT button is clicked on, the window shown in Fig. 27.10 is displayed, in which the user can select a specific raw material or finished goods to analyze their movement through out the month of simulation.

In the options window depicted in Fig. 27.10: (1) when a button from the RAW MATERIAL INVENTORY area is clicked, a graph is displayed (as it is presented in Fig. 27.6); (2) when TIME TABLE is clicked, a table such as Table 27.6 (but including the total days in the month) is displayed; and finally, (3) when a button from the FINISHED GOODS INVENTORY area is clicked, a graph showing the movement of inventory according to the schedule of customer orders is displayed (Fig. 27.11 shows an example for the product ID 202).

Table 27.7 Analysis of inventory policies and reduction of inventory (waste)

Safety stock	5 days (current policy)	3 days	Average reduction of waste (%)	1 day	Average reduction of waste (%)			
Components	Maximum inventory (Kg)	Minimum inventory (Kg)	Maximum inventory (Kg)	Minimum inventory (Kg)		Maximum inventory (Kg)	Minimum inventory (Kg)	
Corn	1,011,225	374,963	898,735	211,676	14.65	769,484	46,501	33.30
Gluten	39,657	4,657	39,300	4,300	0.90	18,849	3,573	29.11
Kaolin	31,273	9,739	26,861	7,132	11.91	21,691	1,219	34.96
Corn bran	68,353	29,993	57,177	16,489	19.98	48,217	599	45.21
Barley bran	31,085	14,578	23,372	7,762	26.99	18,281	3,508	50.87
Soy bran	41,631	8,000	36,253	7,440	7.23	29,396	205	27.95
Molasses	83,970	32,665	66,779	19,725	19.93	52,835	1,474	48.06

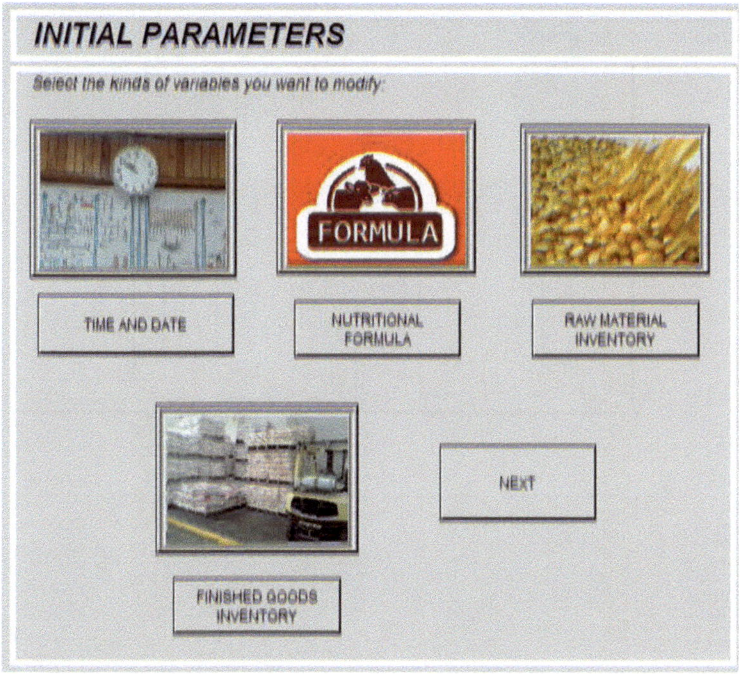

Fig. 27.9 Main window from graphical user interface

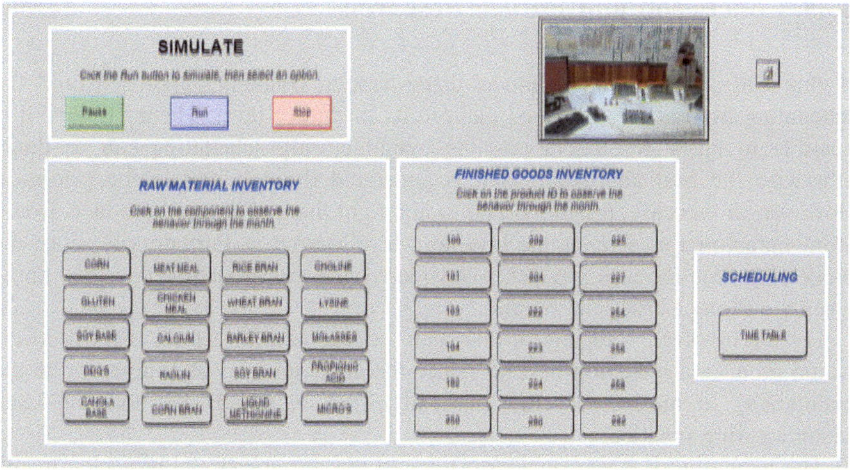

Fig. 27.10 Options window from graphical user interface

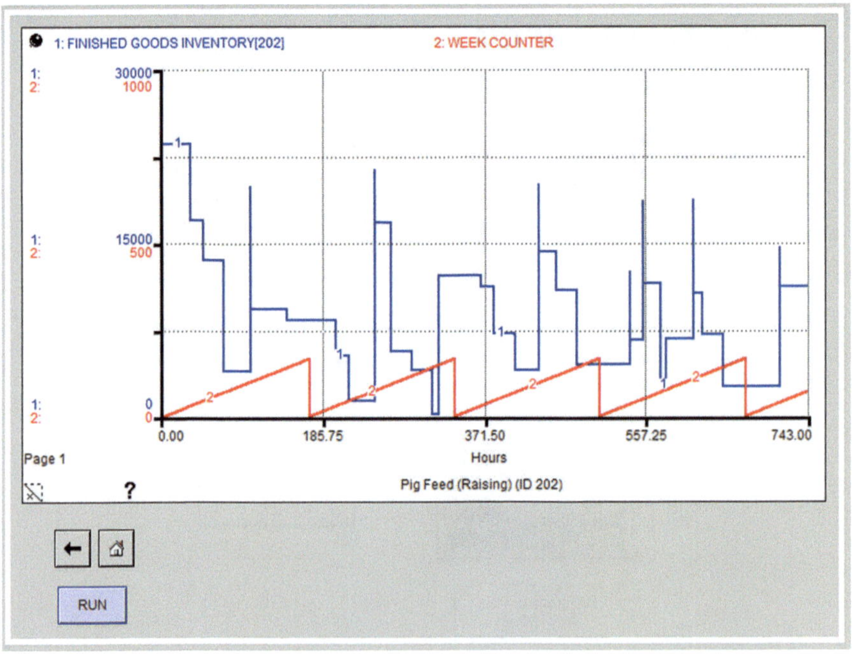

Fig. 27.11 Inventory's movement for product ID 202 in April 2012

27.5 Conclusion and Future Research

In this study, the simulation model using System Dynamics approach links the scheduling from a livestock feed plant to its inventory level. Thanks to this, it is possible to know whether a possible manufacturing scheduling can be done; otherwise the best alternative plan is generated through reassigning priorities. However, in this case, it is important to highlight that the delayed products would be manufactured as soon as the inventory level permit it. Furthermore, since the processing time for each product is considered, the model lets the company know when a customer order can be met.

For the involved company in this study, a reduction in certain security levels would produce a reduction in costs (since the company may stop investing in the maintaining of unnecessary raw materials), and eventually promote the Lean Manufacturing practice.

The reduction of waste (inventory) would not put at risk the projected manufacturing scheduling. However, future work may involve an economic analysis to prove that the reduction of safety stock can reduce costs without affecting scheduling.

Acknowledgments This work was supported by the General Council of Superior Technological Education of Mexico (DGEST). Additionally, this work was sponsored by the National Council of Science and Technology (CONACYT) and the Public Education Secretary (SEP) through PROMEP.

References

Al-Aomar, R. (2011). Handling multi-lean measures with simulation and simulated annealing. *Journal of the Franklin Institute, 348*, 1506–1522.

Abdulmalek, F. A., & Rajgopal, J. (2007). Analyzing the benefits of lean manufacturing and value stream mapping via simulation: A process sector case study. *International Journal of Production Economics, 107*, 223–236.

Amin, M. A., & Karim, M. A. (2012). A systematic approach to evaluate the process improvement in lean manufacturing organizations. In G. Seliger (Ed.), *Sustainable manufacturing* (pp. 65–70). Berlin: Springer.

Azlina, N., Salleh, M., Kasolang, S., & Jaffar, A. (2012). Simulation of integrated total quality management (TQM) with lean manufacturing (LM) practices in forming process using Delmia Quest. *Procedia Engineering, 41*, 1702–1707.

Bicheno, J. (2000). *The lean toolbox* (2nd ed.). Buckingham: PICSIE Books.

Chen, J. C., Cheng, C. H., & Huang, P. B. (2013). Supply chain management with lean production and RFID application: A case study. *Expert Systems with Applications, 40*, 3389–3397.

Demeter, K., & Matyusz, Z. (2011). The impact of lean practices on inventory turnover. *International Journal of Production Economics, 133*, 154–163.

Dombrowskia, U., Mielkea, T., & Engela, C. (2012). Knowledge management in lean production systems. *Procedia CIRP, 3*, 436–441.

Diaz-Elsayed, N., Jondral, A., Greinacher, S., Dornfeld, D., & Lanza, G. (2013). Assessment of lean and green strategies by simulation of manufacturing systems in discrete production environments. *CIRP Annals-Manufacturing Technology, 62*, 475–478.

Elmoselhy, S. (2013). Hybrid lean–agile manufacturing system technical facet, in automotive sector. *Journal of Manufacturing Systems, 32*(4), 598–619.

Eroglu, C., & Hofer, C. (2011). Lean, leaner, too lean? The inventory-performance link revisited. *Journal of Operations Management, 29*, 356–369.

Forrester, J. (1961). *Industrial dynamics*. Portland: Productivity Press.

Ford, D. (1999). A behavioral approach to feedback loop dominance analysis system. *Dynamics Review, 15*, 3–36.

GröBler, A., & Schieritz, N. (2005). Of stocks, flows, agents and rules—strategic simulation in supply chain research. In H. Kotzab et al. (Eds.), *Research Methodologies in Supply Chain Management* (pp. 445–460). Heidelberg: Physica-Verlag.

Hofer, C., Eroglu, C., & Hofer, A. R. (2012). The effect of lean production on financial performance: The mediating role of inventory leanness. *International Journal of Production Economics, 138*, 242–253.

Holweg, M. (2007). The genealogy of lean production. *Journal of Operations Management, 25*, 420–437.

Jurado, P. J., & Moyano, J. (2011). Lean production y gestión de la cadena de suministro en la industria aeronáutica. *Investigaciones Europeas de Dirección y Economía de la Empresa, 17*(1), 137–157.

Krogstie, L., & Martinsen, K. (2013). Beyond lean and six sigma; cross-collaborative improvement of tolerances and process variations—A case study. *Procedia CIRP, 7*, 610–615.

Morecroft, J., & Stewart, R. (2005). Explaining puzzling dynamics: A comparison the use of system dynamics and discrete event simulation. *Proceedings of System Dynamics Society*.

Robinson, S., Radnor, Z. J., Burgess, N., & Worthington, C. (2012). SimLean: Utilizing simulation in the implementation of lean in healthcare. *European Journal of Operational Research, 219*, 188–197.

Riezebos, J., Klingenberg, W., & Hicks, C. (2009). Lean production and information technology: Connection or contradiction? *Computers in Industry, 60*, 237–247.

Spear, S., & Bowen, K. H. (1999). Decoding the DNA of the Toyota production system. *Harvard Business Review, 77*(5), 97–106.

Sandanayake, Y. G., Oduoza, C. F., & Proverbs, D. G. (2008). A systematic modelling and simulation approach for JIT performance optimization. *Robotics and Computer-Integrated Manufacturing*, *24*, 735–743.

Sterman, J. (2000). *Business Dynamics: Systems Thinking and Modeling for a Complex World*. Boston: Irwin McGraw-Hill.

Sweeter, A. (1999). A comparison of system dynamics (SD) and discrete event simulation (DES). *Proceedings of System Dynamics Society*.

Scholl, H. (2001). Looking across the fence: comparing finding from SD modeling efforts with those other modeling techniques. *International Conference of the System Dynamics Society, Atlanta, GA: System Dynamics Society*.

Tako, A., & Robinson, S. (2009). Comparing model development in discrete event simulation and system dynamics. *Proceedings of the 2009 Winter Simulation Conference*.

Voss, C. A. (1995). Alternative paradigms for manufacturing strategy. *International Journal of Operations and Production Management, 15*(4), 5–16.

Womack, J., Jones, D., & Roos, D. (1990). *The machine that changed the world*. New York: Rawson Associates.

Watanabe, N., & Hiraki, S. (1997). A modeling approach to a JIT-based ordering system. *Annals of Operations Research, 69*, 379–403.

Zhang, H., Calvo-Amodio, J., & Haapala, K. R. (2013). A conceptual model for assisting sustainable manufacturing through system dynamics. *Journal of Manufacturing Systems, 32*(4), 543–549.

About the Editors

Jorge Luis García-Alcaraz

Jorge Luis García-Alcaraz received his BSc and MS degree in industrial engineering from the Technological Institute of Colima in Mexico in 1993. He obtained a Ph.D. in Industrial Engineering from the Technological Institute of Ciudad Juarez in Mexico. His doctoral thesis combined multicriteria decision techniques for Advanced Manufacturing Technologies selection. He is currently a full time professor in Autonomous University of Ciudad Juarez in Juarez, Mexico in de Department of Industrial and Manufacturing Engineering. His research area is the production systems optimization using parametric and nonparametric statistical methodologies, as regression techniques, structural equation moles, discriminant analysis, factor analysis, among others. He collaborate with the Mexican Association of Operation Research, the Mexican Academy of Industrial Engineering, EURO Working Group Multicriteria Decision Aiding (EWG-MCDA), among others.

Aidé Aracely Maldonado-Macías

Dra. Maldonado-Macías received her M.S and Ph.D. degrees in Industrial Engineering by the Technological Institute of Juarez City in Mexico. She is currently a full-time professor-investigator at the Autonomous University of Juarez City. Her academic career as a university professor covered a span of more than 12 years in Juarez City institutions. She is a certified professional in Ergonomics by the College of Ergonomists in Mexico. Her research main interest are the ergonomic evaluation and ergonomic design, anthropometric studies and work-related stress phenomena. Also, structural equations models, multi-attribute and multi-criteria decision-making studies with fuzzy logic and axiomatic design applications. She has made several researches about ergonomic evaluation models for the selection of advanced manufacturing technology and manufacturing systems, ergonomic evaluation of workplaces and burnout syndrome evaluation among the automotive industry management personnel.

She received the national first place award for her doctoral dissertation in Mexico and has received multiple recognitions for her academic excellence. She has been participating actively in undergraduate committees and panels in several institutions. She also collaborates as a reviewer for the International Journal of

DOI: 10.1007/978-3-319-04951-9, © Springer International Publishing Switzerland 2014

Advanced Manufacturing Technology, Ergonomics, Applied Ergonomics and International Journal of Uncertainty Fuzziness and Knowledge Based Systems. She has participated in multiple national and international Conferences. She represented Mexico in the world IEA (International Ergonomics Association) committee held in Recife Brazil in the year 2012, within the framework of the World Congress of Ergonomics and actively collaborates with the Ergonomists Society and College of Mexico.

Guillermo Cortes-Robles
Guillermo Cortes-Robles received his BSc degree in electronic engineering from the Instituto Tecnologico de Orizaba (ITO) in 1995. He obtained his master in science of industrial engineering from the same institution and a Ph.D. in Industrial Systems from the Institut National Polytechnique de Toulouse in France. His doctoral thesis combined the TRIZ theory with the Case-Based Reasoning solving process. In 2003, this was the first model proposing this kind of hybridization. The model was supported by an implementation oriented to validate the synergy TRIZ-CBR. Later, this model will be the foundation for a research topic focused on impelling innovation through knowledge capitalization. He is today professor of engineering science at the ITO's postgraduate school, specifically in the engineering management master and the science engineering Ph.D. His research interests are application of TRIZ theory combined with several approaches: knowledge management, decision support systems, lean manufacturing, and creativity techniques, among other approaches with the aim to accelerate the innovation process. He is today the general secretary of the Mexican TRIZ association (AMETRIZ) and responsible of the www.innovasolver.com initiative.

Back cover copy: This book presents some definitions and concepts applied in Latin America on lean manufacturing (LM), the LM tools most widely used and human and cultural aspects that most matter in this field. The book contains a total of 14 tools used and reported by authors from different countries in Latin America, with definition, timeline with related research, benefits that have been reported in literature, and case studies implemented in Latin American companies. Finally, the book presents a list of software available to facilitate the tools' implementation, monitoring and improvement.

Information text (to occur on Springer website, amazon etc.): This book presents some definitions and concepts applied in Latin America on lean manufacturing (LM), the LM tools most widely used and human and cultural aspects that most matter in this field. The book contains a total of 14 tools used and reported by authors from different countries in Latin America, with definition, timeline with related research, benefits that have been reported in literature, and case studies implemented in Latin American companies. Finally, the book presents a list of software available to facilitate the tools' implementation, monitoring and improvement.